Nanomineralogy

Nanomineralogy

Special Issue Editors

Yiwen Ju
Quan Wan
Michael F. Hochella, Jr.

MDPI • Basel • Beijing • Wuhan • Barcelona • Belgrade • Manchester • Tokyo • Cluj • Tianjin

Special Issue Editors

Yiwen Ju
University of Chinese Academy of Sciences
China

Quan Wan
Institute of Geochemistry,
Chinese Academy of Sciences
China

Michael F. Hochella, Jr.
Virginia Polytechnic Institute and State University
USA

Editorial Office
MDPI
St. Alban-Anlage 66
4052 Basel, Switzerland

This is a reprint of articles from the Special Issue published online in the open access journal *Minerals* (ISSN 2075-163X) (available at: https://www.mdpi.com/journal/minerals/special_issues/Nanomineralogy).

For citation purposes, cite each article independently as indicated on the article page online and as indicated below:

LastName, A.A.; LastName, B.B.; LastName, C.C. Article Title. *Journal Name* **Year**, *Article Number*, Page Range.

ISBN 978-3-03936-599-9 (Hbk)
ISBN 978-3-03936-600-2 (PDF)

Cover image courtesy of Yiwen Ju and Hongjian Zhu.

Contents

About the Special Issue Editors

Yiwen Ju was born in Anhui, China, in 1963. He is a professor of Geology and a doctoral supervisor at the University of the Chinese Academy of Sciences. He obtained his B.S. from Anhui University of Science and Technology in 1985. From 1985 to 2000, he served as a senior engineer for Huaibei Mining (Group) Co. Ltd., China. He received his M.S. and Ph.D. from the China University of Mining and Technology in 1994 and 2003, respectively. He became a postdoctoral researcher in 2003 and a full professor in 2010 at the University of the Chinese Academy of Sciences and worked as a senior visiting scholar at Virginia Tech, USA, in 2010–2011. His main research interests are structural geology and energy basin geology, unconventional energy geology, coal and gas geology, and nanogeology. Currently, he serves as the director of the Nanogeology Specialized Committee of the Geological Society of China and as the vice president of the Chinese Sub-society for Soft Rock Engineering & Deep Disaster Control. He serves as an associate editor of the Journal of Nanoscience and Nanotechnology, a guest editor for a Special Issue of Minerals, and an editorial board member of the China Geology, Journal of China Coal Society and Journal of Natural Gas Geoscience, among other publications. He has played the role of principal leader for more than 20 national scientific research projects in China, including Key Programs for the National Natural Science Foundation of China, National Major Science and Technology Projects of China, National Basic Research Programs of China (973 Program), and a Strategic Priority Research Program of the Chinese Academy of Sciences. He has authored more than 200 publications, including more than 90 SCI-indexed scientific papers and 3 books, and has won seven scientific and technological awards at the provincial and national levels. In addition to maintaining academic communication and cooperation with many international universities and research agencies in the United States, Australia, France, Japan, South Africa, and Nepal, he has organized and chaired more than 20 conferences, workshops, and symposia and has given more than 20 talks at scientific conferences. He was the proposer and executive chairman of the Chinese Xiangshan Science Conferences on "Deep Coal Mine Gas Disasters and Coalbed Methane Development" and "Frontier Science Problems of Nanogeology and Accumulation of Metallogenic" in 2012 and 2013, respectively. He served as session chairman of the shale gas branch of the AAPG Annual Meeting, the unconventional oil and gas branch of the Euro–Asia Economic Forum, and the coal and carbon chemistry symposium in China and Japan in 2013; as an organizing committee member and session chairman of the International Geological Congress in Nepal in 2015 and 2016; as the chairman of the International Symposium on Nanogeoscience in 2016 and 2018; and as a member of the technical committee of the International Chinese Shale Gas Conference in 2019.

Quan Wan is a professor at the Institute of Geochemistry, Chinese Academy of Sciences (CAS). He was born in Sichuan, China, in 1971. He received his B.S. in Chemistry in 1994 and his M.S. in Polymer Science and Engineering in 1997 from Peking University, Beijing, China. In 2002, he received his Ph.D. in Materials Science and Engineering from the State University of New York at Stony Brook, USA. From 2005 to 2010, he worked in the Center for Bioengineering and Biomaterials at Temple University, Philadelphia, USA, as a postdoctoral researcher and then as an associate scientist. In 2010, he joined the Institute of Geochemistry, CAS, as a professor through the CAS Hundred Talent Program. He has broad scientific interests in the fields of nanogeoscience and mineralogy and currently serves as vice director of the Specialized Committee on Nanogeology of the Geological

Society of China and the Specialized Committee on Environmental Mineralogy of the Chinese Society of Mineralogy, Petrology and Geochemistry. Ongoing research topics in his group include sorption behavior and occurrence state of gold (nanoparticles) on pyrite/arsenian pyrite/clay minerals, nanoconfinement effects of silica and mineral micro/mesopores, and pore characteristics and gas enrichment in shale gas reservoirs. By combining laboratory simulation and field sample characterization, he and his collaborators have enhanced mechanistic understandings of the origin, special properties, and interfacial reactivity of nanoparticulate and nanoporous materials, as well as the associated implications for natural resources (such as Carlin-type gold ore and shale gas reserves) and the environment. In 2018, he joined his colleagues in organizing the first international conference dedicated to nanogeoscience, for which he served as local host and secretary general.

Michael F. Hochella, Jr. was born in Yokohama, Japan, in 1953 to American parents. He is currently a University Distinguished Professor (Emeritus) at Virginia Tech in Blacksburg, Virginia, USA, and also a Laboratory Fellow at the Pacific Northwest National Laboratory in Richland, Washington, USA. He is a specialist in nanoscience and nanotechnology related to geological, biological, and environmental science at local, regional, and global levels. He is the Founder and Former Director of NanoEarth, the National Center for Earth and Environmental Nanotechnology, funded by the U.S. National Science Foundation and located at Virginia Tech. Hochella received his B.S. and M.S. degrees from Virginia Tech in 1975 and 1977, respectively, and his Ph.D. from Stanford University in 1981. He has been a professor, first at Stanford and then at Virginia Tech, for a total of 31 years. He has been a Fulbright Scholar and has won a Humboldt Award, the Brindley Lecture Award, the Dana Medal, the Virginia Scientist of the Year Award, and the Virginia Outstanding Faculty Award. He has been elected Fellow in nine international scientific societies and laboratories, including the American Association of the Advancement of Science (AAAS) and the American Geophysical Union (AGU). He has held the Presidencies of both the Geochemical Society and the Mineralogical Society of America. He has also won the Distinguished Service Medal of the Geochemical Society. He has served on high-level advisory boards to the National Science Foundation and the Department of Energy in the United States. He has a total of 203 professional peer-reviewed publications, including five papers in the journal Science, and has been cited over 16,500 times in the professional literature, a tally that is currently increasing at a rate of 1400 new citations per year (Google Scholar). He has given hundreds of invited lectures in 17 countries (Europe, Far East, Middle East, and Africa). Professor Hochella has taught a total of 11 different courses at Stanford and Virginia Tech, from introductory freshman classes to advanced classes for Ph.D. students. He has developed brand new courses and also new curricula at the undergraduate and graduate levels. He has also closely advised and mentored over 100 students, each for multiple years, and moved them towards the specific careers of their choice. Twenty-four of his former graduate students and postdocs have been professors at universities in five countries, while many others hold prominent positions in national laboratories, industry, and scientific publishing.

Preface to "Nanomineralogy"

Over several hundred years of development, Earth science has made great advances at the largest of scales (e.g., tectonics) and the smallest of scales (e.g., molecular geochemistry). In the 21st century, the latest technological revolutions (e.g. molecular biology and precise GPS measurement) continue to add important components of how we understand Earth. Even more recently, these latest advancements also have included nanogeoscience, which has clearly become another international frontier now ushering in revolutionary developments. If Earth science is to continue to see breakthroughs at ultrasmall scales, then geoscience problems and phenomena must continue to be studied at the nanoscale in order to reveal and better understand the mechanisms and processes at the larger scales of Earth. Nanomineralogy, an important aspect of nanogeoscience, focuses on the composition, structure, and physical and chemical properties of nanoscale minerals and their interrelations with other Earth critical components. Nanomineralogy is characterized by high-resolution transmission electron microscopy (HRTEM), scanning probe microscopy (SPM), and other high-resolution microscopies, several of them synchrotron-based. In combination with an array of spectroscopic and gas adsorption methods, nanomineralogy reveals the mesoscopic structure, morphology, interface relationships, and formation mechanism of minerals at the nanoscale.

Nanomineralogy allows the study of a series of interfacial interactions of nanominerals, mineral nanoparticles, and nanopore structures, including the growth, dissolution, transformation, interaction, and evolution of fluids, organisms, and minerals. In addition, exploration and application of the physicochemical properties of nanomineral resources can help broaden mineralogy's scope for development and prospects for practical applications.

In 2018, the International Symposium on Nanogeoscience was held in Guiyang, China. Scholars from around the globe gathered to discuss recent progress and development trends in various aspects of nanogeoscience, including nanomineralogy. To give a sampling of the latest progress in nanomineralogy and related fields, we offer this Special Issue, whose 20 papers describe a full range of recent nanomineralogic achievements relating to everything from nanominerals and geochemistry, mineral nanostructures, and nanomineral deformation, to nanopores in oil and gas reservoirs, nanomineral deposits, and nanomineral material. More specifically, this issue includes the following:

(1) Nanominerals and geochemistry: The morphology and structural phase transition of various natural nanominerals are described, as are the experimental synthesis of nanomineral crystals; the adsorption, percolation, dissolution, metasomatism of nanominerals; and the environmental conditions of nanomineral composition transformation.

(2) Structural and seismic geology: The mechanism of the movement of nanominerals is added to the ductile shear zone theory. The seismogenic fault zone is revealed to be a viscoelastic friction belt of nanoparticles, with the macroscopic system's instability possibly caused by the release of energy from microscopic particles.

(3) Unconventional energy geology: The structure and composition of nanominerals in shale or tight reservoirs and the change of nanopores and their fractal dimensions are described, and structural characteristics such as the size, shape, and connectivity of nanopores in unconventional oil and gas reservoirs, along with their evolutionary laws and control factors, are preliminarily clarified.

(4) Ore deposits and mineral materials: Genesis of Carlin-type gold deposits is analyzed at the nanoscale; the significance of using metallic nanoparticles to search for concealed deposits is

discussed; and the preparation, characterization, and physicochemical properties of nanomineral materials are explored.

Today, nanomineralogy faces a new strategic opportunity as well as a revolutionary challenge. We look forward to future developments of nanomineralogy made with the aid of cutting-edge research methods, based on the experience and achievements made possible by nanoscience and geoscience through the study of the application and interaction of mineralogy research techniques and nanotechnological methods in the Earth's lithosphere, atmosphere, hydrosphere, and biosphere. We also anticipate detailed interpretations of geological evolution and information about Earth's development on the nanoscale, as well as further clarification of the basic connotations and main research directions of nanomineralogy. These include the following topics and more: nanostructure and surface effects of minerals and genesis of rock, evolutionary mechanisms of nanoparticles and nanopores in natural geological bodies, characteristics and formation mechanisms of nanodeformation, nanogeochemical processes and their behaviors, nanometer effects of hydrocarbon accumulation and their dynamic mechanisms, nanometallogenesis and associated patterns, nanominerals and new composite materials, and nanomineralogy and disaster/environmental problems. We intend to make full use of the advantages offered by multidisciplinary study, cooperating extensively with global geoscientists while promoting research into nanomineralogy and its associated resources, such as through environment- and disaster-related science and technology projects. In doing so, we aim to spread the global word about the latest research achievements in nanomineralogy while identifying key scientific issues in the field and enriching and developing the theory and methods of nanogeoscience. In this way, we will be able to provide an important theoretical basis for use of minerals and materials, exploration and development of energy resources and mineral resources, prevention of disasters, and protection of the environment.

We thus present this special nanomineralogy-focused issue of Minerals with the aim of encouraging our colleagues to familiarize themselves with current developments, trends, and directions in nanomineralogy, enabling an understanding of the potential of the field as a whole. We look forward to developing further scientific research and cooperation in nanomineralogy, hoping thereby to attract and guide young scholars to participate in this field. We thank the Editor-in-Chief of Minerals for his encouragement and support, and we are also grateful to the other editors for their help with English-language revision. We thank Ziyuan Ouyang, Youwei Du, Jinxing Dai, Jishun Ren, Guowei Zhang, Yaolin Shi, Zhijun Jin, Zhenqi Song, Caineng Zou, Yanxin Wang, et al., the academicians of the Chinese Academy of Sciences, for their guidance. We are also particularly grateful to the scientific research institutions, universities, and state-owned enterprises in China that extended their support and assistance. Finally, we thank the authors and reviewers for their efforts and contributions. Without their help and support, it would have been impossible to produce this invaluable Special Issue.

Yiwen Ju, Quan Wan, Michael F. Hochella, Jr.
Special Issue Editors

Editorial

Editorial for Special Issue "Nanomineralogy"

Yiwen Ju [1,*], Quan Wan [2] and Michael F. Hochella, Jr. [3]

[1] Key Laboratory of Computational Geodynamics, College of Earth and Planetary Sciences, University of Chinese Academy of Sciences, Beijing 100049, China
[2] State Key Laboratory of Ore Deposit Geochemistry, Institute of Geochemistry, Chinese Academy of Sciences, Guiyang 550081, China; wanquan@vip.gyig.ac.cn
[3] Department of Geosciences, Virginia Polytechnic Institute and State University, Blacksburg, VA 24061, USA; hochella@vt.edu
* Correspondence: juyw03@163.com

Received: 28 May 2020; Accepted: 3 June 2020; Published: 5 June 2020

Nanoscience and nanotechnology study the properties of materials within the range 0.1~100 nm. Materials at the nanoscale have special physicochemical properties owing to their surface and interface, small size, quantum tunneling, and quantum size effects. Nanotechnology has crossed over into many areas of Earth science, driving important advances in mineralogy, petrology, geochemistry, structural geology, energy geology, mineral deposits, geologic hazards, and environmental geology [1–7]. Studying problems and phenomena at the nanometer scale, which reveals mechanisms as well as processes, is the inevitable future of developments in Earth science. Nanogeosciences are the global frontier for the contemporary cross-development of geosciences and nanotechnology, and the rise of nanoscience is set to bring about a revolutionary leap in the development of Earth science in the 21st century, driving breakthroughs in Earth science on the ultra-micro scale [1–3].

Nanomineralogy, an important aspect of nanogeoscience, is characterized by high-resolution transmission electron microscopy (HRTEM); scanning probe microscopy (SPM); and other high-resolution microscopies, several of them synchrotron-based, in combination with an array of spectroscopic and gas adsorption methods. It focuses on the composition, structure, and physical and chemical properties of nanoscale minerals and their interrelations with other Earth critical components [2–4], involving everything from nanominerals and geochemistry, mineral nanostructures, and nanomineral deformation to nanopores in oil and gas reservoirs, nanomineral deposits, and nanomineral materials. The exploration and application of the physical and chemical properties of nanomineral resources and environment broaden mineralogy's scope for the development of practical applications as well as the prospects of such applications.

Nanominerals have both resource and environmental attributes [8–11] and offer an extensive record of nanoscale substances and geological processes. With recent developments in nanoscience, research into nanomineral particles, nanomineral solids, nanomineral structures, and resources and the environment have also deepened [1–13]. In many cases, nanominerals or minerals with nanostructures appear in aggregate form. These include mineral particles with a grain size at the nanoscale, minerals of one-dimensional nanostructure (such as halloysite of nanotube structure), and layered minerals of two-dimensional nanostructure (such as clay minerals). Studies of actual minerals by HRTEM, STM, and AFM have demonstrated the existence of nanoparticles and nanostructures, with such nanophenomena more common in crystal surfaces and interfaces. The special nanostructure and composition of a crystal's surface determine its surface properties, which are significant in discussing the physical and chemical environment for the formation and genesis of minerals. Between two mineral phase transitions, not only do nanoscale particles exist, but also nanoparticle aggregates appear under particular physical and chemical conditions—including clay minerals, zeolites, colloidal minerals, volcanic glass, meteorite glass, fusion crust, and tectonite—and fall primarily into four categories: (1) clay minerals, (2) quasicrystal nanostructures, (3) nanostructures in colloids, and (4) nanostructures in crystalline rocks. Minerals with grain size larger than 1 µm can provide information about minerals' late growth stage, whereas those with grain size between 0.1 and 100 nm can provide records about their initial crystallization. The polymorph, polytype, multi-body, and micro-intergrowth of minerals at the nanoscale are the actual minimum bodies capable of reflecting geological information at present. Accordingly, this kind of information must be integrated to more fully reflect the physical and chemical environment in which minerals were formed. At present, the active exploration of techniques and methods for developing and using nanoscale minerals, together with the quest for breakthroughs in both theory and practice, is heightening mineralogists' ability to understand and transform nature [2,4].

Some of the papers [14–20] in this issue deal with nanominerals and geochemistry. They describe the morphology and structural phase transition of various natural nanominerals, as well as the experimental synthesis of nanomineral crystals; the adsorption, percolation, dissolution, and metasomatism of nanominerals; and the environmental transformation conditions of nanomineral composition. Hong and coauthors [14] have reported two structural phase transitions and metallization for nanocrystalline rutile using a diamond anvil cell at ~7.2, ~12.3, and ~14.5 GPa. During compression, a structural phase transition from rutile to baddeleyite occurs, with the high electrical conductivity value offering a crucial clue regarding metallization, as confirmed by temperature-dependent electrical conductivity measurements. On decompression, a structural phase transformation from baddeleyite to columbite occurs. HRTEM observations regarding the starting and recovered samples indicate that phase transformations from rutile to baddeleyite to columbite are irreversible under high pressure. Bjorn von der Heyden and coauthors [15] investigated analytical approaches to characterizing natural Fe nanoparticles before detailing a dedicated synchrotron-based X-ray spectromicroscopy investigation into the speciation of suspended Fe nanoparticles collected from fluvial, marine, and lacustrine surface waters. They identified ferrous, ferric, and magnetite classes of Fe nanoparticles (10–100 nm), all of which exhibited a high degree of heterogeneity in the local bonding environment around the Fe center. The results provide an important baseline for natural nanoparticle speciation in pristine aquatic systems, highlighting the degree of interparticle variability. Nie and coauthors [16] employ such an experimental approach to study the facile hydrothermal synthesis of nanocubic pyrite crystals using greigite (Fe_3S_4) and thiourea (NH_2CSNFH_2) as precursors. Nanocubic pyrite (FeS_2) crystals with exposed (100) crystal faces and sizes of 100–200 nm were successfully synthesized via a facile hydrothermal method using greigite as the iron precursor and thiourea as the sulfur source. These results demonstrate the critical influence of sulfur source on pyrite morphology. These findings also provide new insights into the formation environments and pathways of nanocubic pyrite under hydrothermal conditions. Gu and coauthors [17] explore the variability in rare earth elements (REEs) and trace metal concentrations in dolostone samples of the Cryogenian Nantuo Formation of the Neoproterozoic Era, taken from a drill core in South China. Petrological evidence suggests that the dolostone in the Nantuo Formation was formed in nearshore waters. All the examined dolostone samples featured a significant enrichment of manganese and middle rare earth elements (MREEs), suggesting that the dolostone samples were deposited from suboxic to iron-enriched and anoxic waters. The MREE-enriched dolostone were transported by colloids and nanoparticles in meltwaters. Liu and coauthors [18] report the results of the geochemical alteration and mineralogy of coals under the influence of fault motion. Five coal samples were carefully collected from a reverse fault zone in Qi'nan colliery, China. Systematic detection methods were employed to analyze the different chemical and physical characteristics of the fault-related coal samples. The frictional heat and strong ductile deformation generated by the fault motion led to the dissociation of phenol and carboxyl groups in coal molecules, which sharply decreased the concentrations of elements Co and Mo bound to these functional groups. The disruption and delamination of laminar clay minerals by strong compression–shear stress significantly increased the adsorption sites for related elements, especially heavy rare earth elements (HREE) and MREE. Nanoscale clay minerals resulting from stress-induced scaly exfoliation also enhanced the retention capability of REE. Zhang and coauthors [19] present a model for the metal-bearing nanoparticles observed in soils and fault gouges over the Shenjiayao gold deposit and their significance. Their findings indicate that ore-forming elements in soils can come only from deep-seated ore bodies and that they occur in nanoparticles, and an obvious relationship between metal nanoparticles in fault gouges and soils, with the metallic nanoparticles in fault gouges representing a transitional phase along the whole vertical migration process. Pyrite is the most common authigenic mineral preserved in many ancient sedimentary rocks. Liu and coauthors [20] demonstrate pyrite morphology to be an indicator of paleoredox conditions and shale gas content in the Longmaxi and Wufeng shales in the Middle Yangtze area, South China. Based on the formation mechanism, pyrites in the Longmaxi and Wufeng shales can be divided into syngenetic pyrites, early diagenetic pyrites, and late diagenetic pyrites. The size of the pyrite framboid microcrystals is used to distinguish syngenetic pyrite and diagenetic pyrite, owing to their different formation mechanisms. The sedimentary environment of the Longmaxi and Wufeng shales is mainly a euxinic environment with good reduction conditions, but some parts of the Longmaxi Formation reflect an oxic–dysoxic environment. The process of thermochemical sulfate reduction (TSR), accompanied by the formation of diagenetic pyrite, results in oxidation of organic matter (OM) and depletion of the H bond of OM, which influences the amounts of alkane gas produced by organic matter during thermal evolution. Consequently, the size of pyrite framboids and microcrystals could be widely used to rapidly evaluate paleoredox conditions and the gas content in shales.

Verberne and coauthors [21] deal with structural geology, including the mechanism of nanomineral movement in ductile shear zone theory. Principal slip zones (PSZs) are narrow (<10-cm) bands of localized shear deformation that occur in the cores of upper-crustal fault zones, where they accommodate the bulk of fault displacement. Natural and experimentally formed PSZs consistently show the presence of nanocrystallites in the <100-nm size range. The physical properties of nanocrystalline materials may profoundly affect fault rheology and the structural characteristics of nanocrystalline (NC) PSZs observed in natural faults and in experiments. Numerous reports in the literature show that such zones form in a wide range of faulted rock types, under a wide range of conditions pertaining to seismic and aseismic upper-crustal fault slip, as well as that they frequently show an internal crystallographic-preferred orientation (CPO) and partial amorphization and form glossy or "mirrorlike" slip surfaces. It is suggested that the widespread occurrence of NC PSZs in upper-crustal faults is of general significance. Specifically, the generally high rates of (diffusion) creep in nanocrystalline fault rock may play a key role in controlling the depth limits of the seismogenic zone.

Several papers [22–27] in this special issue focus on energy geology, addressing the structure and composition of coal and shale nanominerals as well as changes in nanopores and their fractal dimensions. They also describe structural characteristics such as the size, shape, and connectivity of nanopores in unconventional oil and gas reservoirs, along with their evolutionary laws and control factors. Chen and coauthors [22] focus on an experimental methodological issue involving total porosity measured for shale gas reservoir samples from the Lower Silurian Longmaxi formation in southeast Chongqing, China. Measuring the total porosity in shale gas reservoir samples remains a challenge because of their fine-grained texture, low porosity, ultra-low permeability, and high content of organic matter (OM) and clay mineral. The composition content porosimetry method, which is a new method for evaluating the porosity of shale samples, was used in the study to measure the total porosity of shale gas reservoir samples, based on bulk and grain density values. The findings indicated that the composition content porosimetry porosity generally increases with increasing OM and clay content and decreases with increasing quartz and feldspar content. Huang and coauthors [23] evaluated the pore structure characteristics of Lower Silurian Longmaxi shale gas reservoirs in Well LD1 of the Laifeng–Xianfeng Block, Upper Yangtze region in South China. N_2 adsorption and helium ion microscope (HIM) were used to investigate various pore features, including pore volume, pore surface area, and pore size distribution. The calculated results show a good hydrocarbon storage capacity and development potential for the shale samples. Meanwhile, the reservoir space and migration pathways may be affected by the small pore size. The authors found that large pores usually have large values of fractal dimensions and identified a strong positive correlation between the fractal dimensions and pore volume, as well as pore surface area. Fractal dimensions can be taken as a visual indicator of the degree of development of a pore structure in shale. Zhu and coauthors [24] studied tectonic and thermal controls on the nano–micro structural characteristics of a Cambrian organic-rich shale. Naturally deformed samples from the main source rocks are selected from the Lower Cambrian Lujiaping Formation in the Dabashan Thrust–Fold Belt to investigate nanometer to micrometer-sized structures. Their findings indicated that fracture-related pores predominate over mineral-hosted pores, as well as that these two pore types account for 90% of total pore space. Overall, Lujiaping deformed over-mature samples have abundant nano to micro-sized inorganic pores. High-resolution SEM images provide direct evidence of the formation of nano- and microsized structures such as OM–clay aggregates and silica nanograins. The characteristics of nanopore structure in shale play a crucial role in methane adsorption and determination of the occurrence and migration of shale gas. Using an integrated approach of X-ray diffraction (XRD), N_2 adsorption, and field emission scanning electron microscopy (FE-SEM), Yu and coauthors [25] systematically focused on eight drilling samples of marine Taiyuan shale from well ZK1 in southern North China to study the characteristics and heterogeneity of their nanopore structure. Their findings indicated that different sedimentary environments may control the precipitation of clay and quartz between transitional shale and marine shale, leading to different organic matter (OM)–clay relationships and different correlations between total organic carbon (TOC) and mineral content. Taiyuan shale of higher heterogeneity is highly fractal, and its fractal dimensions are principally related to micropores. Li and coauthors [26] present a thorough investigation of the impacts of matrix compositions on the nanopore structure and fractal characteristics of lacustrine shales from the Changling Fault Depression (CFD), Southern Songliao Basin. The Lower Cretaceous Shahezi shales are targets for lacustrine shale gas exploration. The Shahezi shales were investigated to explore the impacts of rock composition, including organic matter and minerals, on pore structure and fractal characteristics. Seven lithofacies can be identified on a mineralogy-based shale classification scheme. Inorganic mineral–hosted pores are the most abundant pore type, whereas relatively few organic matter (OM) pores are observed in FE-SEM images of the Shahezi shales. The primary controlling factors for pore structure in Shahezi shales are clay minerals rather than OM. Clay content is the most significant factor controlling the fractal dimensions of the lacustrine Shahezi shale. Pore connectivity of lacustrine shales has been inadequately documented in previous papers. Li and coauthors [27] examine findings relating to lacustrine shales from the lower Cretaceous Shahezi Formation in the Changling Fault Depression (CFD) based on investigations using field emission scanning electron microscopy (FE-SEM), mercury intrusion capillary pressure (MICP), low-pressure gas (CO_2 and N_2) sorption (LPGA), and spontaneous fluid imbibition (SFI) experiments. Their findings indicate that pores observed from FE-SEM images are primarily interparticle (interP) pores in clay minerals and organic matter (OM) pores. Pore connectivity proceeds in the calcareous shale > argillaceous shale > siliceous shale. The connected pores of Shahezi shales are affected mainly by the abundance and coexistence of OM pores and clay carbonate mineral host pores.

The other articles [28–32] in this special issue are devoted to ore deposits and mineral materials. The genesis of Carlin-type gold deposits is analyzed at the nano scale; the significance of using metallic nanoparticles to search for concealed deposits is discussed; and the preparation, characterization, and physicochemical properties of nanomineral materials are explored. Tan and coauthors [28] present two hydrothermal events at the Shuiyindong Carlin-type gold deposit in southwestern China from Sm–Nd dating of fluorite and calcite. The Shuiyindong Gold Mine hosts one of the largest and highest-grade strata-bound Carlin-type gold deposits discovered to date in Southwestern China. The ore is hosted mainly in Upper Permian bioclastic limestone near the axis of an anticline. The gold is hosted mainly in arsenian pyrite and arsenopyrite, primarily as crystal lattice gold, submicroscopic particles, and nanoparticles. Fluorite commonly occurs at the vicinity of an unconformity between the Middle–Upper Permian formations, proposed as the structural conduit that fed the ore fluids. Calcite commonly fills fractures at the periphery of decarbonated rocks, which contain high-grade orebodies. Their study was intended to verify the occurrence of two distinct hydrothermal events at the Shuiyindong, based on

Sm–Nd isotope dating of the fluorite and calcite, which contains considerable concentrations of REE, and exhibit variable Sm/Nd ratios, facilitating the direct dating of associated hydrothermal events. Two groups of Sm–Nd isochron ages suggest two episodes of hydrothermal events in Shuiyindong. The age of the calcite likely represents the late stage of the gold mineralization period. Initial Nd isotopic compositions indicate that the Nd in the fluorite and calcite was likely derived from mixtures of basaltic volcanic tuff and bioclastic limestone of the Permian formations. Lu and coauthors [29] contribute to a better understanding of the green preparation and performance and mechanism of nanoporous pyrrhotite by thermal treatment of pyrite as an effective Hg(II) adsorbent. The removal of Hg(II) from aqueous solutions using pyrrhotite derived from the thermal activation of natural pyrite was explored by batch experiments. The adsorption isotherms demonstrated that the sorption of Hg(II) by modified pyrite (MPy) can be fitted well by the Langmuir model. The removal of Hg(II) by MPy was attributed mainly to a chemical reaction that resulted in cinnabar formation as well as electrostatic attraction between the negative charges in MPy and positive charges of Hg(II). The results of this work suggest that the thermal activation of natural pyrite is highly important for effective use of ore resources to remove Hg(II). Ca-bentonite was used as the feedstock material for the synthesis of hydroxysodalite because of its high Al, Si content, good chemical reactivity, and natural abundance. Liu and coauthors [30] report the results of one-step synthesis of hydroxysodalite using natural bentonite at moderate temperatures. The Na/Si molar ratio and reaction temperature both played important roles in controlling the degree of crystallinity of the synthetic hydroxysodalite. Although clays are widely used as sorbents for heavy metals due to their high specific surface areas, low cost, and ubiquity in most soil and sediment environments, they still face inherent limitations owing to the low loading capacity of heavy metals. Yao and coauthors [31] studied enhanced potential toxic metal removal using a novel hierarchical SiO_2–$Mg(OH)_2$ nanocomposite derived from sepiolite. The structural characteristics of the resulting modified samples were characterized by a wide range of techniques, including field emission scanning electron microscopy (FESEM), transmission electron microscopy (TEM), energy dispersive X-ray spectroscopy (EDX), X-ray diffraction (XRD), and nitrogen physisorption analysis. The results show that a hierarchical nanocomposite constructed by loading $Mg(OH)_2$ nanosheets onto amorphous SiO_2 nanotubes can be successfully prepared, with the nanocomposite having a high surface area ($377.3 \ m^2/g$) and pore volume ($0.96 \ cm^3/g$). Batch removal experiments indicate that the nanocomposite exhibits high removal efficiency toward Gd(III), Pb(II), and Cd(II). Zhou and coauthors [32] address the characterization and amoxicillin adsorption activity of mesopore $CaCO_3$ microparticles prepared using rape flower pollen. $CaCO_3$ adsorbent was characterized using X-ray diffraction (XRD) and scanning electronic microscopy (SEM). The equilibrium adsorption data on amoxicillin were explained using the Langmuir, Freundlich, and Temkin adsorption isotherm models, whereas equilibrium adsorption of as-prepared $CaCO_3$ was better depicted using the Langmuir adsorption model, indicating the favorable adsorption of amoxicillin.

The understanding and cognition of the Earth directly affect human beings' ability to survive and develop. The exploration and development of energy and mineral deposits, disaster and environmental prediction and management, and the like are all urgent problems in Earth science and relate to its resources, disasters, and environment. Using the nanoscale to re-understand and analyze these problems will open new perspectives and directions for researchers, whose object in conducting nanomineralogy is related chiefly to minerals and their pores in different spheres of the Earth system.

As a geoscientific frontier, nanomineralogy faces novel opportunities and challenges. Inevitably, some results of prior research, and the knowledge based on them, will be overturned. Nanomineralogy is still in its infancy, however, and although certain achievements have already been made, understanding is still lacking concerning the formation mechanism of mineral elements; the migration process, structural form, and disaster-related and environmental effects of nanoparticles; and the mineralization of nanopores. Taking full advantage of the advantages of multidisciplinary cross-development; promoting international cooperation among scientists; and systematically researching nanomineralogy, including by seeking concentrated solutions to key scientific problems, will enrich and spur the development of nanomineralogic theory and methods. Such an approach will provide an important theoretical basis for use of new mineral- and carbon-based materials, the exploration and exploitation of energy and mineral resources, environmental protection, and disaster prediction.

Acknowledgments: The authors would like to express their deep appreciation for having been invited to serve as guest editors for the Nanomineralogy special issue of *Minerals*. Special thanks also go to the article authors and their sponsoring organizations. The authors thank the Editors-in-Chief, assistant editors, editors, and reviewers for their informative comments and constructive suggestions, which helped the contributing authors improve the quality of their manuscripts.

Conflicts of Interest: The authors declare no conflicts of interest.

References

1. Hochella, M.F., Jr. Nanoscience and technology: The next revolution in the Earth sciences. *Earth Planet. Sci. Lett.* **2002**, *203*, 593–605. [CrossRef]
2. Ju, Y.W.; Sun, Y.; Wan, Q.; Lu, S.F.; He, H.P.; Wu, J.G.; Wang, G.C.; Huang, C. Nanogeology: A Revolutionary Challenge in Geosciences. *Bull. Mineral. Petrol. Geochem.* **2016**, *35*, 1–20. (In Chinese with English abstract).
3. Ju, Y.W.; Huang, C.; Sun, Y.; Wan, Q.; Lu, X.C.; Lu, S.F.; He, H.P.; Wang, X.Q.; Zou, C.N.; Wu, J.G.; et al. Nanogeosciences: Research History, Current Status, and Development Trends. *J. Nanosci. Nanotechnol.* **2017**, *17*, 5930–5965. [CrossRef]
4. Hochella, M.F., Jr.; Lower, S.K.; Maurice, P.A.; Penn, R.L.; Sahai, N.; Sparks, D.L.; Twining, B.S. Nanominerals, mineral nanoparticles, and Earth systems. *Science* **2008**, *319*, 1631–1635. [CrossRef] [PubMed]
5. Hochella, M.F., Jr. Nanogeoscience: From origins to cutting-edge applications. *Elements* **2008**, *4*, 373–379. [CrossRef]
6. Ju, Y.W.; Huang, C.; Sun, Y.; Zou, C.N.; He, H.P.; Wan, Q.; Wang, X.Q.; Lu, X.C.; Lu, S.F.; Wu, J.G.; et al. Nanogeology in China: A review. *China Geol.* **2018**, *1*, 286–303. [CrossRef]
7. Hochella, M.F., Jr.; Mogk, D.W.; Ranville, J.; Allen, I.C.; Luther, G.W.; Marr, L.C.; McGrail, B.P.; Murayama, M.; Qafoku, N.P.; Rosso, K.M.; et al. Natural, incidental, and engineered nanomaterials and their impacts on the Earth system. *Science* **2019**, *363*, 1414–1423. [CrossRef]
8. Banfield, J.F.; Zhang, H.Z. Nanoparticles in the environment. *Rev. Mineral. Geochem.* **2001**, *44*, 1–58. [CrossRef]
9. Buseck, P.R.; Adachi, K. Nanoparticles in the Atmosphere. *Elements* **2008**, *44*, 389–394. [CrossRef]
10. Hassellöv, M.; von der Kammer, F. Iron Oxides as Geochemical Nanovectors for Metal Transportin Soil-River Systems. *Elements* **2007**, *4*, 401–406. [CrossRef]
11. Theng, B.K.G.; Yuan, G.D. Nanoparticles in the Soil Environment. *Elements* **2007**, *4*, 395–399. [CrossRef]
12. Verberne, B.A.; Plümper, O.; De Winter, D.A.M.; Spiers, C.J. Superplastic nanofibrous slip zones control seismogenic fault friction. *Science* **2014**, *346*, 1342–1344. [CrossRef] [PubMed]
13. Viti, C. Exploring Fault Rocks at the Nanoscale. *J. Struct. Geol.* **2011**, *33*, 1715–1727. [CrossRef]
14. Hong, M.L.; Dai, L.D.; Li, H.P.; Hu, H.Y.; Liu, K.X.; Yang, L.F.; Pu, C. Structural Phase Transition and Metallization of Nanocrystalline Rutile Investigated by High-Pressure Raman Spectroscopy and Electrical Conductivity. *Minerals* **2019**, *9*, 441. [CrossRef]
15. Von der Heyden, B.; Roychoudhury, A.; Myneni, S. Iron-Rich Nanoparticles in Natural Aquatic Environments. *Minerals* **2019**, *9*, 287. [CrossRef]
16. Nie, X.; Luo, S.X.; Yang, M.Z.; Zeng, P.; Qin, Z.H.; Yu, W.B.; Wan, Q. Facile Hydrothermal Synthesis of Nanocubic Pyrite Crystals Using Greigite Fe3S4 and Thiourea. *Minerals* **2019**, *9*, 273. [CrossRef]
17. Gu, S.Y.; Fu, Y.; Long, J.X. Predominantly Ferruginous Conditions in South China during the Marinoan Glaciation: Insight from REE Geochemistry of the Syn-glacial Dolostone from the Nantuo Formation in Guizhou Province, China. *Minerals* **2019**, *9*, 348. [CrossRef]
18. Liu, H.W.; Jiang, B. Geochemical Alteration and Mineralogy of Coals under the Influence of Fault Motion: A Case Study of Qi'nan Colliery, China. *Minerals* **2019**, *9*, 389. [CrossRef]
19. Zhang, B.M.; Han, Z.X.; Wang, X.Q.; Liu, H.L.; Wu, H.; Feng, H. Metal-Bearing Nanoparticles Observed in Soils and Fault Gouges over the Shenjiayao Gold Deposit and Their Significance. *Minerals* **2019**, *9*, 414. [CrossRef]

20. Liu, Z.Y.; Chen, D.X.; Zhang, J.C.; Lü, X.X.; Wang, Z.Y.; Liao, W.H.; Shi, X.B.; Tang, J.; Xie, G.J. Pyrite Morphology as an Indicator of Paleoredox Conditions and Shale Gas Content of the Longmaxi and Wufeng Shales in the Middle Yangtze Area, South China. *Minerals* **2019**, *9*, 428. [CrossRef]

21. Verberne, B.A.; Plümper, O.; Spiers, C.J. Nanocrystalline Principal Slip Zones and Their Role in Controlling Crustal Fault Rheology. *Minerals* **2019**, *9*, 328. [CrossRef]

22. Chen, F.W.; Lu, S.F.; Ding, X.; Zhao, H.Q.; Ju, Y.W. Total Porosity Measured for Shale Gas Reservoir Samples: A Case from the Lower Silurian Longmaxi Formation in Southeast Chongqing, China. *Minerals* **2019**, *9*, 5. [CrossRef]

23. Huang, C.; Ju, Y.W.; Zhu, H.J.; Qi, Y.; Yu, K.; Sun, Y.; Ju, L.T. Nano-Scale Pore Structure and Fractal Dimension of Longmaxi Shale in the Upper Yangtze Region, South China: A Case Study of the Laifeng–Xianfeng Block Using HIM and N2 Adsorption. *Minerals* **2019**, *9*, 356. [CrossRef]

24. Zhu, H.J.; Ju, Y.W.; Huang, C.; Qi, Y.; Ju, L.T.; Yu, K.; Li, W.W.; Su, X.; Feng, H.Y.; Qiao, P. Tectonic and Thermal Controls on the Nano-Micro Structural Characteristic in a Cambrian Organic-Rich Shale. *Minerals* **2019**, *9*, 354. [CrossRef]

25. Yu, K.; Ju, Y.W.; Qi, Y.; Qiao, P.; Huang, C.; Zhu, H.J.; Feng, H.Y. Fractal Characteristics and Heterogeneity of the Nanopore Structure of Marine Shale in Southern North China. *Minerals* **2019**, *9*, 242. [CrossRef]

26. Li, Z.; Liang, Z.K.; Jiang, Z.X.; Gao, F.L.; Zhang, Y.H.; Yu, H.L.; Xiao, L.; Yang, Y.D. The Impacts of Matrix Compositions on Nanopore Structure and Fractal Characteristics of Lacustrine Shales from the Changling Fault Depression, Songliao Basin, China. *Minerals* **2019**, *9*, 127. [CrossRef]

27. Li, Z.; Liang, Z.K.; Jiang, Z.X.; Yu, H.L.; Yang, Y.D.; Xiao, L. Pore Connectivity Characterization of Lacustrine Shales in Changling Fault Depression, Songliao Basin, China: Insights into the Effects of Mineral Compositions on Connected Pores. *Minerals* **2019**, *9*, 198. [CrossRef]

28. Tan, Q.P.; Xia, Y.; Xie, Z.J.; Wang, Z.P.; Wei, D.T.; Zhao, Y.M.; Yan, J.; Li, S.T. Two Hydrothermal Events at the Shuiyindong Carlin-Type Gold Deposit in Southwestern China: Insight from Sm-Nd Dating of Fluorite and Calcite. *Minerals* **2019**, *9*, 230. [CrossRef]

29. Lu, P.; Chen, T.H.; Liu, H.B.; Li, P.; Peng, S.C.; Yang, Y. Green Preparation of Nanoporous Pyrrhotite by Thermal Treatment of Pyrite as an Effective Hg(II) Adsorbent: Performance and Mechanism. *Minerals* **2019**, *9*, 74. [CrossRef]

30. Liu, B.; Sun, H.J.; Peng, T.J.; He, Q. One-Step Synthesis of Hydroxysodalite Using Natural Bentonite at Moderate Temperatures. *Minerals* **2018**, *8*, 521. [CrossRef]

31. Yao, Q.Z.; Yu, S.H.; Zhao, T.L.; Qian, F.J.; Li, H.; Zhou, G.T.; Fu, S.Q. Enhanced Potential Toxic Metal Removal Using a Novel Hierarchical SiO_2 –$Mg(OH)_2$ Nanocomposite Derived from Sepiolite. *Minerals* **2019**, *9*, 298. [CrossRef]

32. Zhou, L.S.; Peng, T.J.; Sun, H.J.; Guo, X.G.; Fu, D. The Characterization and Amoxicillin Adsorption Activity of Mesopore $CaCO_3$ Microparticles Prepared Using Rape Flower Pollen. *Minerals* **2019**, *9*, 254. [CrossRef]

Article

Structural Phase Transition and Metallization of Nanocrystalline Rutile Investigated by High-Pressure Raman Spectroscopy and Electrical Conductivity

Meiling Hong [1,2], Lidong Dai [1,*], Heping Li [1], Haiying Hu [1], Kaixiang Liu [1,2], Linfei Yang [1,2] and Chang Pu [1,2]

[1] Key Laboratory of High-Temperature and High-Pressure Study of the Earth's Interior, Institute of Geochemistry, Chinese Academy of Sciences, Guiyang 550081, China
[2] University of Chinese Academy of Sciences, Beijing 100049, China
* Correspondence: dailidong@vip.gyig.ac.cn; Tel.: +86-851-8555-9715

Received: 19 May 2019; Accepted: 1 July 2019; Published: 18 July 2019

Abstract: We investigate the structural, vibrational, and electrical transport properties of nanocrystalline rutile and its high-pressure polymorphs by Raman spectroscopy, and *AC* complex impedance spectroscopy in conjunction with the high-resolution transmission electron microscopy (HRTEM) up to ~25.0 GPa using the diamond anvil cell (DAC). Experimental results indicate that the structural phase transition and metallization for nanocrystalline rutile occurred with increasing pressure up to ~12.3 and ~14.5 GPa, respectively. The structural phase transition of sample at ~12.3 GPa is confirmed as a baddeleyite phase, which is verified by six new Raman characteristic peaks. The metallization of the baddeleyite phase is manifested by the temperature-dependent electrical conductivity measurements at ~14.5 GPa. However, upon decompression, the structural phase transition from the metallic baddeleyite to columbite phases at ~7.2 GPa is characterized by the inflexion point of the pressure coefficient and the pressure-dependent electrical conductivity. The recovered columbite phase is always retained to the atmospheric condition, which belongs to an irreversible phase transformation.

Keywords: nanocrystalline rutile; phase transition; metallization; high pressure; diamond anvil cell

---

1. Introduction

As a typical transition-metal oxide, titanium dioxide (TiO_2) has received extensive attention in recently several decades due to its widespread applications in the field of photocatalysis, dye-sensitized solar cells (DSCs), transparent conducting oxide (TCO) films, etc. [1–3]. In ambient conditions, it is well known that TiO_2 crystallizes in three representative polymorphs: rutile, anatase, and brookite. In light of its unique physicochemical characterizations with relatively high brightness, large refractive index (n = 2.75), chemical inertness, and large dielectric constant, rutile has been widely applied, such as in white pigment, opacifiers, and thin film capacitors [4,5].

A large quantity of high-pressure experimental and theoretical investigations has been employed to explore the phase stabilities and structural transitions for rutile by the synchrotron X-ray diffraction, Raman spectroscopy, and first-principles theoretical calculations. Previous results have already confirmed that there existed many high-pressure polymorphs for rutile, e.g., the columbite phase (α-PbO_2, space group *Pbcn*) and the baddeleyite phase (MI, $P2_1/c$). However, till now, the high-pressure structural phase transition sequence and the pressure point of rutile to the baddeleyite phase transition has remained controversial. Some researchers think that rutile transformed directly to the baddeleyite phase without undergoing the intermediate phase of the columbite [6–12]. Furthermore, there exist considerable disputes regarding the pressure point of rutile and the baddeleyite phase transition.

Machon et al. [6] have investigated the Raman spectroscopy of rutile nanorods with a diameter of around 6–8 nm using a diamond anvil cell and revealed the phase transition of rutile and baddeleyite at a pressure of ~16.0 GPa. However, when the pressure was released, the new columbite phase appeared at ~0.2 GPa and remained stable in atmospheric conditions, whereas a similar study reported that synchrotron X-ray diffraction results on the rutile-to-baddeleyite phase transition for nanocrystalline rutile with an average grain size of 30 nm occurred at ~8.7 GPa by virtue of a diamond anvil cell [12]. Previous high-pressure Raman spectroscopy experiments in the diamond anvil cell have already confirmed that one available intermediate phase of columbite existed during the process of the phase transformation between rutile and baddeleyite at ~10.4 GPa with an initial grain diameter of 20–30 um, and further, the baddeleyite phase appeared at ~20.0 GPa [13].

As usual, the pressure-induced structural phase transition, metallization, and amorphization are accompanied by the variation of electrical transport characteristics for some engineering materials [14–18]. To the best of our knowledge, only one high-pressure electrical resistivity experiment on the synthetic rutile with various Ni-doped concentrations was reported under a limited pressure range by using a Bridgman opposed anvil setup [19]. They observed that the electrical resistivity of sample decreased drastically under the conditions of 4.0 GPa and 500 °C, and then became constant at the pressure range of 4.0–8.0 GPa, which indicated the occurrence of the semiconductor-to-metal phase transition in synthetic rutile. As for the nanocrystalline rutile, no relative high-pressure electrical transport properties have been reported so far. Therefore, a systematic study on the electrical transport characteristic for the nanocrystalline rutile is crucial under high pressure.

In the present work, we report two structural phase transitions and metallization for nanocrystalline rutile at pressures of up to ~25.0 GPa using the diamond anvil cell in conjunction with Raman spectroscopy, AC complex impedance spectroscopy, and high-resolution transmission electron microscopy. Furthermore, two correspondent structural phase transitions and metallization for nanocrystalline rutile under high pressure are discussed in detail.

2. Experimental Procedure

Natural rutile with a gem-class single crystal was gathered from Xinyi city, Jiangsu province, China. The single crystal was crushed and ground into the fine particles in an agate mortar. X-ray diffraction (XRD) analysis of the starting sample was collected by an X'Pert Pro X-ray powder diffractometer (Phillips Company, Amsterdam, The Netherlands, the Cu Kα radiation with working voltage 45 kV and applied current 40 mA, respectively). Selected X-ray diffraction pattern was used to determine the lattice parameters of the starting sample by a Rietveld refinement as implemented in MDI Jade 6.5 software. Figure 1 shows the X-ray diffraction pattern of the starting sample; the observed XRD peaks are in good accordance with the tetragonal rutile in ambient conditions (space group: $P4_2/mnm$, JCPDS no. 88-1175). Some lattice parameters of rutile were calculated to be $a = b = 4.5933$ Å, $c = 2.9592$ Å, $\alpha = \beta = \gamma = 90°$, and $V = 62.43$ Å^3, which is close to the values in the International Centre for Diffraction Data (ICDD). The average particle size of the starting sample was calculated to be 72 nm by virtue of the Scherrer's equation, which is in good agreement with the result from the TEM observation (Figure S1).

High-pressure Raman spectroscopy measurements were performed using a DAC with an anvil culet of 300 μm. The ruby single crystal with its grain size of ~5 μm was applied to calibrate the pressure based on the shift of R1 photoluminescence line. To produce a hydrostatic environment, Helium was used as the pressure medium. Raman spectra were carried out using a Raman spectrometer (Invia, Renishaw, Wharton Anderch, UK) equipped with a confocal microscope (TCS SP8, Leica, Wetzlar, Germany) and a CCD camera (Olympus, Shinjuku, Tokyo, Japan). Spectra were taken in the backscattering geometry using an Argon ion laser (Spectra physics: 514.5 nm and power <1 mW) in the frequency shift range of 100–1000 cm^{-1} with a spectral resolution of 1.0 cm^{-1}. Each spectrum was collected for 450 s. To avoid pressure oscillation, the equilibrium time of 15 min was kept at each designated pressure point. The positions of Raman peaks were determined by fitting a Lorenz-type function using PeakFit software. The particle size and microstructure observations for the starting and recovered samples were investigated by the high-resolution transmission electron microscopy (HRTEM, Tecnai G2 F20 S-TWIN TMP, FEI, Hillsboro, OR, USA).

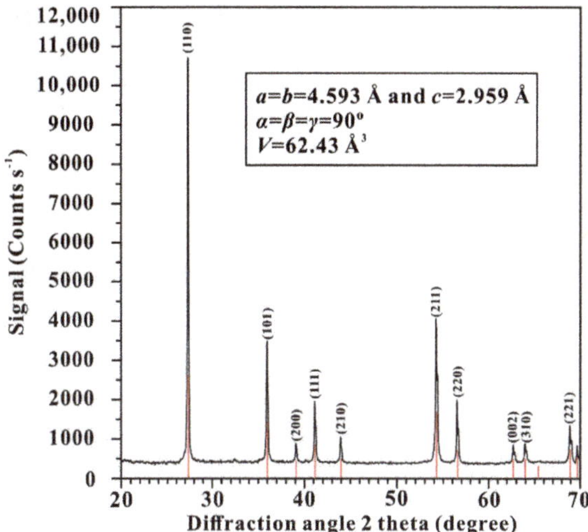

Figure 1. The X-ray diffraction (XRD) pattern of nanocrystalline rutile under ambient conditions.

High-pressure electrical conductivity experiments were conducted by a DAC with the anvil culet of 300 µm diameter. A T-301 stainless steel gasket was pre-indented into a thickness of ~40 µm, and a 180 µm center hole was drilled by a laser. Then, a mixture of boron nitride powder and epoxy resin was compressed into the hole, and another one 100 µm center hole was drilled as the insulating sample chamber. The AC complex impedance spectroscopy was measured using a Solartron-1260 impedance/gain phase analyzer in the frequency range of 10^{-1}–10^7 Hz. The plate electrode was integrated into both diamond anvils. A low temperature was obtained by liquid nitrogen and an experimental temperature was measured by a k-type thermocouple with an estimated accuracy of 5 K. Detailed descriptions of the high-pressure experimental procedures and measurement methods can be found elsewhere [14–18].

3. Results and Discussion

High-pressure Raman spectroscopy was performed to investigate the structural property of nanocrystalline rutile at room temperature up to ~25.0 GPa. In Figure 2a, four typical Raman vibration modes for nanocrystalline rutile are observed in ambient conditions, which can be assigned as 143 cm^{-1} (B_{1g}), 242 cm^{-1} (multi-phonon), 441 cm^{-1} (E_g), and 609 cm^{-1} (A_{1g}). The peaks at 143 cm^{-1} (B_{1g}) and 609 cm^{-1} (A_{1g}) are related to the O-Ti-O bond bending and Ti-O bond stretching modes, while the 441 cm^{-1} (E_g) peak is due to the oxygen atom liberation along the c-axis orientation [20]. An anomalously strong and broad peak at 242 cm^{-1} is a multi-phonon peak caused by the second-order Raman scattering experiment in rutile structure. All of these observed Raman characteristic peaks are in good agreement with previous studies in ambient conditions [21,22]. At the pressure range of 0–12.3 GPa, all of the Raman peaks for rutile phase shifted toward higher frequencies with increasing pressure, except for the B_{1g} soft mode. The red shift of the B_{1g} soft mode, which is characterized by the negative pressure-dependent Raman peak, can provide a clue to the instability of rutile structure under high pressure. Our observed phenomenon of red shift in rutile phase also existed in some similar rutile-structured compounds, such as SnO$_2$ and GeO$_2$ [23]. At ~12.3 GPa, six acquired new peaks at around 229, 278, 323, 445, 674, and 721 cm^{-1} were identified as the baddeleyite phase [24–26], which demonstrated the occurrence of phase transition from rutile to baddeleyite phases. When the pressure was continuously enhanced up to 13.8 GPa, the Raman peak at 494 cm^{-1} was split into two new separate peaks at 511 and 529 cm^{-1}, respectively. The splitting phenomenon in the baddeleyite phase was possibly related to the nanometer size effects [24]. Furthermore, the Raman peaks of baddeleyite phase shifted toward higher frequencies, which indicated the structure of baddeleyite phase remained stable up to the highest pressure of ~25.0 GPa.

The evolution of the Raman shift for nanocrystalline rutile under pressure (pressure coefficient, dv/dP) is plotted in Figure 2b. Two discrete pressure ranges can be identified by the variation of the slope of pressure coefficient: the pressure ranges from ambient to 12.3 GPa, and from 12.3 to 25.0 GPa, respectively. A discontinuous change in the pressure coefficient at ~12.3 GPa indicates the structural phase transition from rutile to baddeleyite phases. The fundamental structural units in rutile and columbite phases are of the TiO$_6$ octahedrons with totally different link modes. As for the high-pressure baddeleyite phase, each tetravalent titanium cation (Ti^{4+}) is coordinated with seven divalent oxygen anions (O^{2-}) and forms the distorted fluorite structure [27]. A discontinuous change in the pressure coefficient at ~12.3 GPa arises from the distortion and breakdown of TiO$_6$ octahedron during the process of phase transition. Thus, the occurrence of phase transition from rutile to baddeleyite phases is possibly related to the variation of coordination number in the tetravalent titanium cation (Ti^{4+}) [28].

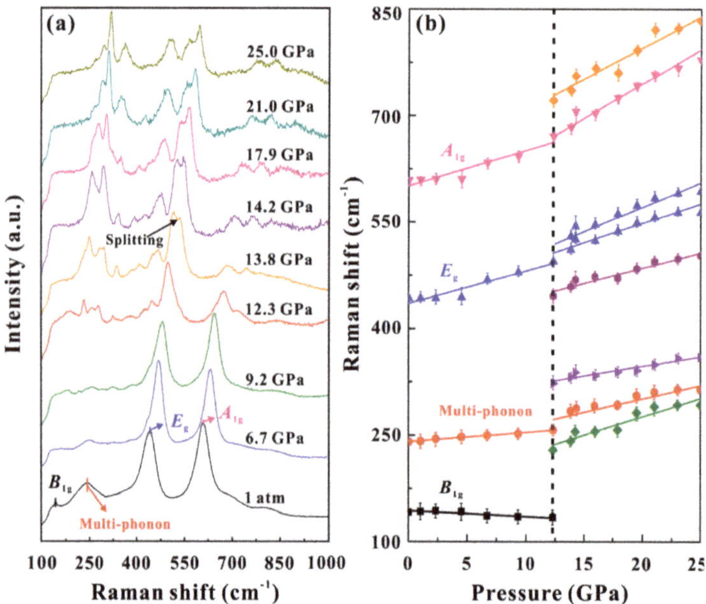

Figure 2. (a) Raman spectra of nanocrystalline rutile at representative pressures during compression process. (b) The evolution of the Raman shift with increasing pressure (dv/dP) at atmospheric temperature.

Upon decompression, the Raman peaks of baddeleyite phase continuously shifted toward lower frequencies in the pressure range of 25.0–7.2 GPa, as presented in Figure 3a. When the pressure was decreased to ~7.2 GPa, new Raman peaks appeared at the positions of 134, 230, 277, 320, 378, 442, 512, 645, and 706 cm^{-1}. All of these representative Raman peaks are the characteristic of the columbite phase [10,26,29,30], which suggests the occurrence of phase transition from baddeleyite to columbite phases at ~7.2 GPa. As the pressure was continuously reduced, all of these Raman intensities for the columbite phase became obviously stronger. Therefore, the phase transformations from rutile to baddeleyite to columbite phases were irreversible. The corresponding pressure-dependent Raman shift of nanocrystalline rutile during decompression is detailedly illustrated in Figure 3b. The available inflexion point of the pressure coefficient at ~7.2 GPa displays the structural phase transition from baddeleyite to columbite phases, which is possibly related to the variation of coordination number in the tetravalent titanium cation (Ti^{4+}) [28].

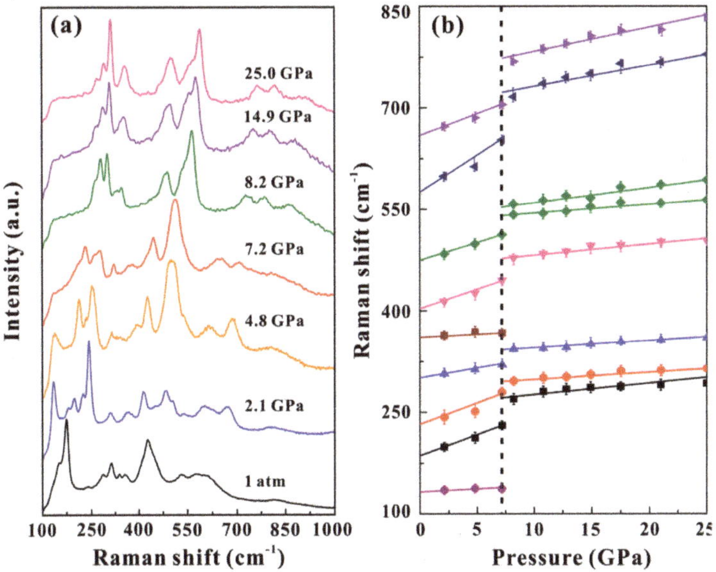

Figure 3. (a) Raman spectra of nanocrystalline rutile at selected pressures in the process of decompression. (b) The pressure dependence of the Raman shift ($d\upsilon/dP$) at room temperature.

The representative Nyquist diagrams of the impedance spectroscopy for nanocrystalline rutile at atmospheric temperature during compression are displayed in Figure 4a–c. At the pressure range of 1.6–12.3 GPa, the impedance spectra exhibit a semicircle within the high-frequency range and a low-frequency oblique line. Each impedance semicircular arc was fitted by the equivalent circuit consisting of a parallel resistor (R) and constant-phase element (CPE). Further increasing the pressure, the grain boundary effect of the sample became weaker gradually. When the pressure was higher than 14.5 GPa, the impedance arc only appeared in the fourth quadrant, and it could be fitted only by the simple resistor (R). The representative Nyquist diagrams of the impedance spectroscopy for baddeleyite and columbite phases upon compression are presented in Figure S2. Only one impedance semicircular arc of grain interior or one pure resistance was obtained among the phases of baddeleyite and columbite. The electrical conductivities of the samples can be calculated as follows:

$$\sigma = L/SR \tag{1}$$

where L is the distance between the two electrodes (cm), S is the cross-sectional area of the electrode (cm^2), R is the resistance of sample (Ω), and σ is the electrical conductivity of sample (S/cm). Figure 4d shows the pressure-dependent electrical conductivity of the grain interior and boundary for the nanocrystalline rutile in the process of compression and decompression at atmospheric temperature. During compression, the electrical conductivity of grain interior increases with increasing pressure, and three linear regions were obtained on the base of various slopes. At the pressure ranges of 1.6–12.3 GPa and 14.5–25.0 GPa, the grain interior electrical conductivity enhances slowly with increasing pressure at the rates of 0.032 and 0.041 S $cm^{-1} \cdot GPa^{-1}$, respectively. However, the grain interior electrical conductivity increases drastically by about four orders of magnitude at 12.3–14.5 GPa. The grain boundary electrical conductivity shows the opposite trend at the pressure range of 1.6–12.3 GPa, and then disappears above 12.3 GPa. The available discontinuities of electrical conductivity for both the grain interior and boundary at ~12.3 GPa are observed, which hint the occurrence of phase transition from rutile to baddeleyite phases. Above 14.5 GPa, the sample electrical conductivity within the range of 6–11 S cm^{-1} may be indicative of metallization. The electrical conductivity magnitude remains constant at ~12 S cm^{-1} at 25.0–7.2 GPa and then decreases within the range of 5–11 S cm^{-1} below ~7.2 GPa. The available inflexion point at ~7.2 GPa is consistent with our above-mentioned Raman scattering results, which can be ascribed to the occurrence of transformation from baddeleyite to columbite phases. In a similar study, Olsen et al. [31] observed the structural phase transition from rutile to baddeleyite at a higher pressure range of 20–30 GPa with an average grain size of 10 nm. This was possibly related to the different grain size, which may have resulted in a discrepancy of the pressure point of phase transition and the width of the phase coexistence regime reported by Olsen et al. and us.

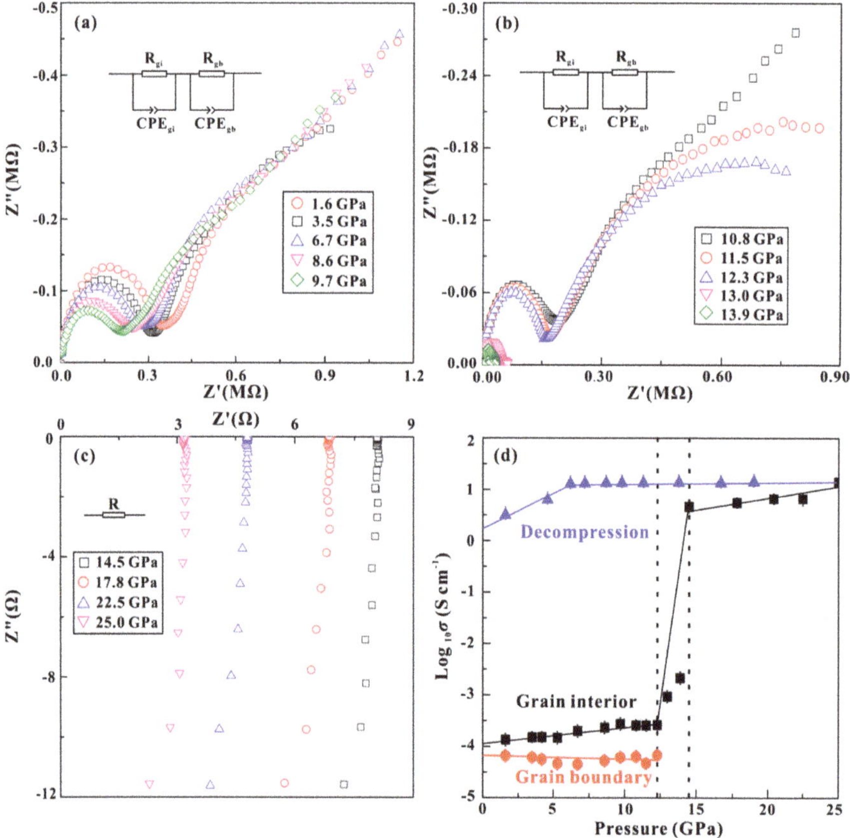

Figure 4. (**a–c**) Nyquist diagrams of the impedance spectra for nanocrystalline rutile at different pressures. R_{gi} and R_{gb} represent the resistance of the grain interior and boundary, respectively. CPE_{gi} and CPE_{gb} are the constant phase element of the grain interior and boundary, respectively. (**d**) The grain interior and boundary electrical conductivity of nanocrystalline rutile during compression and decompression process at atmospheric temperature.

To check the high-pressure metallization of nanocrystalline rutile, we performed temperature-dependent electrical conductivity measurements up to 25.0 GPa at 120–240 K. As usual, the electrical conductivity of sample increased with increasing temperature for semiconductor, whereas the metal exhibited a negative relation between the temperature and electrical conductivity [15–18]. The temperature-dependent electrical conductivity measurements of nanocrystalline rutile at selected pressures are plotted in Figure 5. Below 13.2 GPa, the electrical conductivity of sample increases with increasing temperature, displaying a typical characterization of semiconductor. A negative relation between electrical conductivity and temperature above 14.8 GPa indicates the occurrence of metallization. At 0.3 GPa, the recovered columbite phase also shows a typical metallic behavior. As usual, there are two dominant causes for the occurrence of metallization phenomenon in semiconducting materials: the closure of bandgap and the drastic increase of defect concentration under high pressure. In order to effectively distinguish the metallization mechanism in our present rutile sample, first-principles theoretical calculations were implemented to predict the electronic and structural evolutions of rutile and baddeleyite phases under high pressure in the Supplementary Information (Figures S3 and S4). This made it clear that the bandgap energy of rutile phase fells within the range of 1.99 eV to 1.96 eV when the pressure increased from 0 GPa to 12.0 GPa. As for the baddeleyite phase, the bandgap energy fells within the range of 2.21 eV to 2.17 eV when the pressure increased from 14.0 GPa to 25.0 GPa. Therefore, it is impossible that the occurrence of pressure-induced metallization for rutile is related to the closure of bandgap. An absolutely new experiment was performed to observe the variation of color for nanocrystalline rutile under high pressure using a diamond anvil cell, as shown in Figure S5. We found that there was no observable color change at the pressure range of 0–10.0 GPa. However, when the pressure increased to 15.0 GPa, one obvious variation of color in sample from almost white to black (dark) transition was observed. As a matter of fact, the colors from the shallow to deep variation in rutile stand for the enhancement of oxygen vacancies concentration in TiO_2 particles [32]. Therefore, the obvious color variation of nanocrystalline rutile is strongly related to the enhancement of the defect concentration under pressure, which results in the occurrence of metallization. Therefore, the metallization of nanocrystalline rutile is attributed to the enhancement of defect concentration rather than the closure of bandgap.

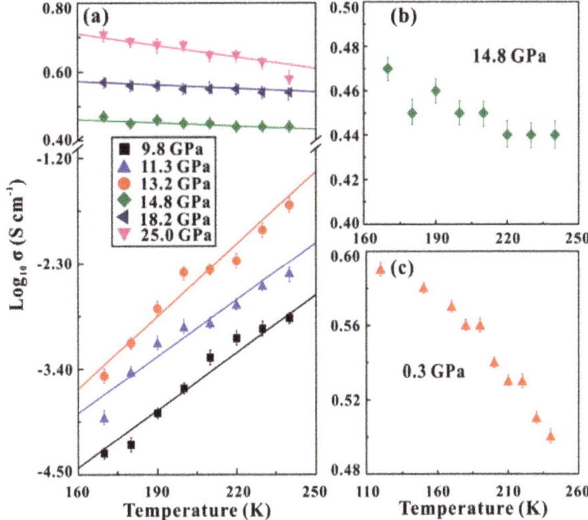

Figure 5. (a) The pressure dependence of electrical conductivity for nanocrystalline rutile as a function of temperature, the highest pressure achieved in the experiment is 25.0 GPa. The equilibrium time of 20 min was kept at each designated pressure point. (b) The metallic state of baddeleyite phase at ~14.8 GPa during compression process. (c) The metallic property of columbite phase after quenched down to 0.3 GPa. A relatively longer equilibrium time of 120 min was applied at almost atmospheric pressure in order to decrease the experimental uncertainty of electrical conductivity for sample.

In order to further investigate the reversibility of the structural phase transition for nanocrystalline rutile, HRTEM observation was performed for both of the starting and recovered samples. In initial HRTEM image in Figure 6a, the interplanar distance value is ~0.32 nm, which corresponds to the (110) plane of rutile phase. At the same time, the initially selected area electron diffraction (SAED) pattern (Figure 6c) consists of a series of rings with bright discrete diffraction spots, which can be identified as rutile phase. In Figure 6b of the recovered HRTEM image, the interplanar distance values are ~0.27 and ~0.35 nm, assigned to the (020) and (110) planes of the columbite phase, respectively. Meantime, the corresponding SAED of the recovered sample exhibits a few clear spots, which were confirmed as a columbite phase [33]. Thus, the nanocrystalline rutile eventually transformed and maintained the columbite phase under ambient conditions. In conclusion, all of these obtained results on nanocrystalline rutile from the Raman spectroscopy experiments and HRTEM observations revealed the irreversibility of the structural transformation under pressure.

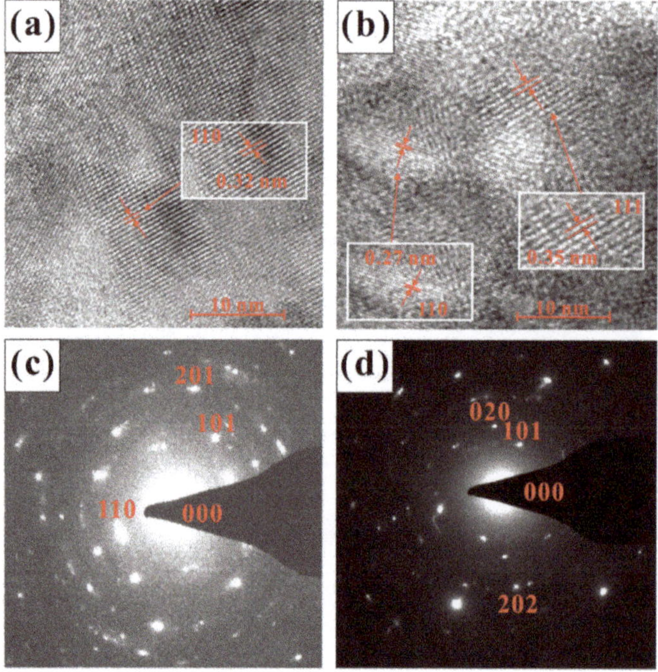

Figure 6. (**a,b**): high-resolution transmission electron microscopy (HRTEM) images of the starting and recovered samples, respectively. (**c,d**): the corresponding selected area electron diffraction (SAED) patterns.

4. Conclusions

We have reported two structural phase transitions and metallization for nanocrystalline rutile using the diamond anvil cell at around 7.2, 12.3, and 14.5 GPa, respectively. During compression, the structural phase transition from rutile to baddeleyite phases at ~12.3 GPa was disclosed by the appearance of new characteristic peaks in Raman spectroscopy, the inflexion point of the pressure coefficient, and pressure-dependent electrical conductivity. As the pressure was continuously increased up to ~14.5 GPa, the high electrical conductivity value provided a crucial clue regarding metallization, which was confirmed by the temperature-dependent electrical conductivity measurements. Upon decompression, the pressure-dependent Raman peaks and electrical conductivity for the columbite phase indicated the occurrence of structural phase transformation from baddeleyite to columbite phases at ~7.2 GPa. The HRTEM observations on the starting and recovered samples demonstrated that the phase transformations from rutile to baddeleyite to columbite phases were irreversible under high pressure.

Supplementary Materials: The following are available online at http://www.mdpi.com/2075-163X/9/7/441/s1, Figure S1: (a) and (b) are the TEM images of the starting sample. (c) and (d) the corresponding histograms of the particle size distribution. It is one of the potentially effective and good methods that the TEM observation can be used to determine the particle size distribution state in our starting sample. As shown in Figure S1 (a) and (b), the starting rutile particles with almost homogenous distribution state. We estimated roughly that there existed at least 20 and 8 particles in Figure S1 (a) and (b), respectively. Figure S1 (c) and (d) represent the corresponding histograms of the particle size distribution for the starting sample, most of the particle size are within the range of 70–80 nm. The average particle size of the starting sample was estimated to be 78 nm, which is in good consistent with the result from XRD; Figure S2: (a) The Nyquist diagram of the impedance spectra for baddeleyite phase at the pressure range of 19.0–8.7 GPa during decompression. (b) The Nyquist diagram of the impedance spectra for columbite phase at the pressure range of 7.2–1.6 GPa during decompression, the equivalent circuit of R stands for the resistance. Figure S3: (a) and (b) Calculated band structure for rutile phase at the pressures of 0 GPa and 10.0 GPa. The bandgap energy for rutile phase are 1.99 eV and 1.96 eV at the pressures of 0 GPa and 10.0 GPa, respectively. (d) and (e) The corresponding total density and projected density at the pressures of 0 GPa and 10.0 GPa for rutile phase. (c) Calculated band structure for baddeleyite phase at 25.0 GPa. The bandgap energy for baddeleyite phase is 2.17 eV at 25.0 GPa. (f) The corresponding total density and projected density at 25.0 GPa for baddeleyite phase. Figure S4: The calculated bandgap energy of rutile phase at the pressure range of 0–12.0 GPa and the baddeleyite phase within the pressure range of 14.0–25.0 GPa. Figure S5: (a) The optical microscope image of the starting material for nanocrystalline rutile. (b) and (c) The optical microscope images of the nanocrystalline rutile at the pressure points of 10.0 GPa and 15.0 GPa using the diamond anvil cell, respectively.

Author Contributions: L.D. designed the project. M.H. and L.D. wrote the initial draft of the work and the final paper). M.H., L.D., H.L., H.H., K.L., L.Y., and C.P. interpreted the results. L.D. corrected and recognized the final paper. M.H. and K.L. performed and interpreted the high-P experiments and the HRTEM images. All authors discussed the results and commented on the manuscript.

Funding: This research was financially supported by the strategic priority Research Program (B) of the Chinese Academy of Sciences (18010401), Key Research Program of Frontier Sciences of CAS (QYZDB-SSW-DQC009), Hundred Talents Program of CAS, NSF of China (41774099 and 41772042), Youth Innovation Promotion Association of CAS (2019390), Special Fund of the West Light Foundation of CAS, and Postdoctoral Science Foundation of China (2018M643532).

Acknowledgments: We thank the editor for kindly handling our paper, as well as two anonymous reviewers for their constructive and enlightened advice in the revising process.

Conflicts of Interest: The authors declare no conflict of interest.

References

1. Linsebigler, A.L.; Lu, G.; Yates, J.T. Photocatalysis on TiO_2 Surfaces: Principles, Mechanisms, and Selected Results. *Chem. Rev.* **1995**, *95*, 735–758. [CrossRef]

2. Son, H.-J.; Prasittichai, C.; Mondloch, J.E.; Luo, L.; Wu, J.; Kim, D.W.; Farha, O.K.; Hupp, J.T. Dye Stabilization and Enhanced Photoelectrode Wettability in Water-Based Dye-Sensitized Solar Cells through Post-assembly Atomic Layer Deposition of TiO_2. *J. Am. Chem. Soc.* **2013**, *135*, 11529–11532. [CrossRef]

3. Chen, D.-M.; Xu, G.; Miao, L.; Nakao, S.; Jin, P. Sputter deposition and computational study of M-TiO_2 (M=Nb, Ta) transparent conducting oxide films. *Surf. Coat. Technol.* **2011**, *206*, 1020–1023. [CrossRef]

4. Tian, C. Calcination intensity on rutile white pigment production via short sulfate process. *Dye. Pigment.* **2016**, *133*, 60–64. [CrossRef]

5. Middlemas, S.; Fang, Z.Z.; Fan, P. A new method for production of titanium dioxide pigment. *Hydrometallurgy* **2013**, *131*, 107–113. [CrossRef]

6. Machon, D.; Bail, N.; Hermet, P.; Cornier, T.; Daniele, S.; Vignoli, S. Pressure-induced phase transitions in TiO_2 rutile nanorods. *J. Phys. Chem. C* **2019**, *123*, 1948–1953. [CrossRef]

7. Gerward, L.; Olsen, J.S. Post-Rutile High-Pressure Phases in TiO_2. *J. Appl. Crystallogr.* **1997**, *30*, 259–264. [CrossRef]

8. Al-Khatatbeh, Y.; Lee, K.K.M.; Kiefer, B. High-pressure behavior of TiO_2 as determined by experiment and theory. *Phys. Rev. B* **2009**, *79*, 134114. [CrossRef]

9. Sato, H.; Endo, S.; Sugiyama, M.; Kikegawa, T.; Shimomura, O.; Kusaba, K. Baddeleyite-Type High-Pressure Phase of TiO_2. *Science* **1991**, *251*, 786–788. [CrossRef]

10. Li, Q.; Liu, R.; Liu, B.; Wang, L.; Wang, K.; Li, D.; Zou, B.; Cui, T.; Liu, J.; Chen, Z.; et al. Stability and phase transition of nanoporous rutile TiO_2 under high pressure. *RSC Adv.* **2012**, *2*, 9052–9057. [CrossRef]

11. Mei, Z.-G.; Wang, Y.; Shang, S.-L.; Liu, Z.-K. First-principles study of the mechanical properties and phase stability of TiO_2. *Comput. Mater. Sci.* **2014**, *83*, 114–119. [CrossRef]

12. Saxena, S.K.; Pischedda, V.; Liermann, H.P.; Zha, C.S.; Wang, Z. X-ray diffraction study on pressure-induced phase transformations in nanocrystalline anatase/rutile (TiO_2). *J. Phys. Condens. Matter* **2001**, *13*, 8317–8832.

13. Arashi, H. Raman spectroscopic study of the pressure-induced phase transition in TiO_2. *J. Phys. Chem. Solids* **1992**, *53*, 355–359. [CrossRef]

14. Hu, H.; Liu, K.; Dai, L.; Wu, L.; Li, H.; Zhuang, Y. Evidence of the pressure-induced conductivity switching of yttrium-doped $SrTiO_3$. *J. Phys. Condens. Matter* **2016**, *28*, 475501.

15. Yang, L.; Dai, L.; Li, H.; Hu, H.; Liu, K.; Pu, C.; Hong, M.; Liu, P. Pressure-induced metallization in MoSe2 under different pressure conditions. *RSC Adv.* **2019**, *9*, 5794–5803. [CrossRef]

16. Zhuang, Y.; Dai, L.; Wu, L.; Li, H.; Hu, H.; Liu, K.; Yang, L.; Pu, C. Pressure-induced permanent metallization with reversible structural transition in molybdenum disulfide. *Appl. Phys. Lett.* **2017**, *110*, 122103. [CrossRef]

17. Dai, L.; Liu, K.; Li, H.; Wu, L.; Hu, H.; Zhuang, Y.; Yang, L.; Pu, C.; Liu, P. Pressure-induced irreversible metallization accompanying the phase transitions in Sb_2S_3. *Phys. Rev. B* **2018**, *97*, 024103. [CrossRef]

18. Dai, L.; Zhuang, Y.; Li, H.; Wu, L.; Hu, H.; Liu, K.; Yang, L.; Pu, C. Pressure-induced irreversible amorphization and metallization with a structural phase transition in arsenic telluride. *J. Mater. Chem. C* **2017**, *5*, 12157–12162. [CrossRef]

19. Karthik, K.; Jaya, N.V. High pressure electrical resistivity studies on Ni-doped TiO_2 nanoparticles. *J. Alloy. Compd.* **2011**, *509*, 5173–5176. [CrossRef]

20. Parker, J.C.; Siegel, R.W. Raman microprobe study of nanophase TiO_2 and oxidation-induced spectral changes. *J. Mater. Res.* **1990**, *5*, 1246–1252. [CrossRef]

21. Li, J.; Li, F.; Li, C.; Yang, G.; Xu, Z.; Zhang, S. Evidences of grain boundary capacitance effect on the colossal dielectric permittivity in (Nb + In) co-doped TiO_2 ceramics. *Sci. Rep.* **2015**, *5*, 8295. [CrossRef]

22. Liu, G.; Fan, H.; Xu, J.; Zhao, Y. Colossal permittivity and impedance analysis of niobium and aluminum co-doped TiO_2 ceramics. *RSC Adv.* **2016**, *6*, 48708–48714. [CrossRef]

23. Mammone, J.; Nicol, M.; Sharma, S. Raman spectra of TiO_2-II, TiO_2-III, SnO_2, and GeO_2 at high pressure. *J. Phys. Chem. Solids* **1981**, *42*, 379–384. [CrossRef]

24. Swamy, V.; Dubrovinsky, L.S.; Dubrovinskaia, N.A.; Langenhorst, F.; Simionovici, A.S.; Drakopoulos, M.; Dmitriev, V.; Weber, H.-P. Size effects on the structure and phase transition behavior of baddeleyite TiO_2. *Solid State Commun.* **2005**, *134*, 541–546. [CrossRef]

25. Lagarec, K.; Desgreniers, S. Raman study of single crystal anatase TiO_2 up to 70 GPa. *Solid State Commun.* **1995**, *94*, 519–524. [CrossRef]

26. Sun, Q.; Zheng, C.; Huston, L.Q.; Frankcombe, T.J.; Chen, H.; Zhou, C.; Fu, Z.; Withers, R.L.; Norén, L.; Bradby, J.E.; et al. Bimetallic Ions Codoped Nanocrystals: Doping Mechanism, Defect Formation, and Associated Structural Transition. *J. Phys. Chem. Lett.* **2017**, *8*, 3249–3255. [CrossRef]

27. Muscat, J.; Swamy, V.; Harrison, N.M. First-principles calculations of the phase stability of TiO_2. *Phys. Rev. B* **2002**, *65*, 224112. [CrossRef]

28. Li, Q.; Cheng, B.; Yang, X.; Liu, R.; Liu, B.; Liu, J.; Chen, Z.; Zou, B.; Cui, T.; Liu, B. Morphology-Tuned Phase Transitions of Anatase TiO_2 Nanowires under High Pressure. *J. Phys. Chem. C* **2013**, *117*, 8516–8521. [CrossRef]

29. Zhang, Y.; Wang, Q.; Wu, X.; Ma, Y. An immutable array of TiO_2 nanotubes to pressures over 30 GPa. *Nanotechnology* **2017**, *28*, 145705. [CrossRef]

30. Dong, Z.; Song, Y. Size- and morphology-dependent structural transformations in anatase TiO_2 nanowires under high pressures. *Can. J. Chem.* **2015**, *93*, 165–172. [CrossRef]

31. Olsen, J.S.; Gerward, L.; Jiang, J.Z. High-Pressure Behavior of Nano Titanium Dioxide. *High Press. Res.* **2002**, *22*, 385–389. [CrossRef]

32. Kikuchi, T.; Yoshida, M.; Matsuura, S.; Natsui, S.; Tsuji, E.; Habazaki, H.; Suzuki, R.O. Rapid reduction of titanium dioxide nano-particles by reduction with a calcium reductant. *J. Phys. Chem. Solids* **2014**, *75*, 1041–1048. [CrossRef]

33. Hwang, S.; Shen, P.; Chu, H.; Yui, T. Nanometer-size α-PbO$_2$-type TiO$_2$ in garnet: A thermobarometer for ultrahigh-pressure metamorphism. *Science* **2000**, *288*, 321–324. [CrossRef]

Article

Iron-Rich Nanoparticles in Natural Aquatic Environments

Bjorn von der Heyden [1],*, Alakendra Roychoudhury [1] and Satish Myneni [2]

[1] Department of Earth Sciences, Stellenbosch University, Stellenbosch 7600, South Africa; roy@sun.ac.za
[2] Department of Geosciences, Princeton University, Princeton, NJ 08544, USA; smyneni@princeton.edu
* Correspondence: bvon@sun.ac.za; Tel.: +27-21-808-3122

Received: 8 April 2019; Accepted: 27 April 2019; Published: 11 May 2019

Abstract: Naturally-occurring iron nanoparticles constitute a quantitatively-important and biogeochemically-active component of the broader Earth ecosystem. Yet detailed insights into their chemical speciation is sparse compared to the body of work conducted on engineered Fe nanoparticles. The present contribution briefly reviews the analytical approaches that can be used to characterize natural Fe nanoparticles, before detailing a dedicated synchrotron-based X-ray spectro-microscopic investigation into the speciation of suspended Fe nanoparticles collected from fluvial, marine, and lacustrine surface waters. Ferrous, ferric and magnetite classes of Fe nanoparticles (10–100 nm) were identified, and all three classes exhibited a high degree of heterogeneity in the local bonding environment around the Fe center. The heterogeneity is attributed to the possible presence of nanoparticle aggregates, and to the low degrees of crystallinity and ubiquitous presence of impurities (Al and organic moieties) in natural samples. This heterogeneity further precludes a spectroscopic distinction between the Fe nanoparticles and the larger sized Fe-rich particles that were evaluated. The presented results provide an important baseline for natural nanoparticle speciation in pristine aquatic systems, highlight the degree of inter-particle variability, which should be parameterized in future accurate biogeochemical models, and may inform predictions of the fate of released engineered Fe nanoparticles as they evolve and transform in natural systems.

Keywords: iron; nanoparticle; Fe; nanomaterial; colloid; L-edge XANES; speciation

---

1. Introduction

Iron (Fe) is the most abundant transition metal in the Earth's crust, with important and biogeochemically-active repositories existing in all four of the Earth's major subsystems (viz. lithosphere, hydrosphere, atmosphere and biosphere). In these natural environments iron particles exist over a broad range of sizes, however with the recent advent of the so-called "nanotechnology revolution", it is the smallest size classes (i.e., nanoparticles defined as having one dimension <100 nm; e.g., [1]) that are receiving special scientific and industrial attention. Two key reasons for this gain in prominence include: (1) On account of their high surface area to size ratio and because of quantum effects, nanoparticles commonly exhibit differing chemical behavior to their bulk analogues; and (2) technological advancement and human activity is rampantly increasing the concentrations of incidental and engineered nanoparticles released into the natural environment (e.g., References [2–4]). Iron nanoparticles are of particular interest since their surface chemistry renders them as important transporting agents for nutrients and pollutants in natural systems [5,6], their reactivity has been exploited for environmental and hazardous waste site remediation (notably engineered Fe^0 nanoparticles [7]), and they may be actively cycled in biogeochemical systems where Fe is an essential nutrient for driving primary productivity (which in turn influences atmospheric CO_2 levels) [8,9].

Despite their ubiquity and importance, natural Fe nanoparticles have received relatively limited scientific attention relative to the large body of work that focusses on the synthesis, behavior, stability, toxicity, and fate of engineered Fe nanoparticles (e.g., References [7,10,11]). A recent review presented by Sharma and coworkers [12] highlights that, relative to engineered nanoparticles, environmental inorganic nanoparticles may differ in both their biogeochemical behavior and their respective ecotoxicities. For this to be validated however, a greater body of work focused specifically on identifying the speciation of natural Fe nanoparticles is required. Examples of previous work include Neubauer et al. [13] who use Transmission Electron Microscopy (TEM) to investigate soil Fe nanomineralogy and its association with ambient heavy metals; and Carbone et al. [14] who use Electron Paramagnetic Resonance (EPR) spectroscopy to investigate incidental Fe nanoparticles formed in an Acid Mine Drainage environment. The present contribution seeks to augment the existing body of work by (1) briefly reviewing the prominent analytical techniques capable of evaluating natural Fe nanoparticles; and (2) applying a newly-developed synchrotron-based soft X-ray spectromicroscopy approach towards evaluating natural Fe nanoparticles and nanoparticle aggregates collected from a selection of fluvial, marine, and lacustrine aquatic environments. The approach exploits the chemical information contained within the Fe $L_{2,3}$-edge X-ray Absorption Near-Edge Structure (XANES) spectra (pertaining to the local bonding environment around the Fe metal center [15]) and is augmented with additional information collected at the Aluminum K-edge and at the Carbon K-edge. It is foreseen that the new insights derived from this novel study approach will add meaningfully to our understanding of Fe nanoparticle speciation and biogeochemical behavior in natural systems. This, in turn, may better inform the development of future generations of biogeochemical models (e.g., Reference [16]) and may assist in predictions related to the fate of engineered Fe^0 nanoparticles, which readily experience surface oxidation in the natural environment.

2. Methods for Evaluating Fe Nanomineralogy

The current section diverges from a traditional "Methods" section in that it includes a brief but pertinent review of a range of analytical methodologies that can be and have been used to study natural Fe nanominerals and nanoparticles. This review is augmented with a detailed description of the Fe L-edge XANES research protocol that we have developed specifically for evaluating the chemistry, valence speciation and local coordination in natural Fe nanoparticles. For a more thorough assessment of the science underpinning this analytical technique (i.e., X-ray interactions with Fe metal centers), the reader is referred to Reference [15].

2.1. Review of Current Methods of Studying Fe Nanominerals

Iron nanoparticles have a complex composition and structure in nature, thus a diverse suit of methods is required to analyze them. Due to their effects on nanoparticle reactivity, the main analytical parameters of interest include size, shape, size distribution, aggregation, surface area, crystal structure, surface chemistry, chemical composition, speciation, and coordination. The methods that are commonly employed therefore are primarily for microscopy and imaging, for physico-chemical characterization or for a combination of both. A number of instrumental techniques (XRD, BET, NMR, SEM-EDS, AFM, SEM, TEM, FTIR, and 3-D Tomography; Table 1) are now readily available in common laboratories that together can characterize a large set of properties of interest rather quickly and economically.

Table 1. Overview of techniques used to study Fe nanoparticles and nanominerals.

Technique	Acronym	Signal Analyzed	Property/Information Retrieved	Relevant References
Micro-focused X-ray diffraction	μXRD	Diffraction pattern of X-rays	Crystal structure, Grain size, Lattice parameters	[17–19]
Brunauer-Emmett-Teller Surface area analysis	BET	Gas sorption	Surface area, Pore analysis	[20]
Nuclear Magnetic Resonance Spectroscopy	NMR	Electromagnetic signal	Purity, Molecular structure	[21]
Fourier Transform Infrared Spectroscopy	FTIR	Absorpiton of electromagnetic radiations	Surface composition, Ligand binding	[22]
Scanning Electron Microscopy - Energy dispersive spectrometry	SEM-EDS	Elemental X-ray flourescence	Size, Shape, Composition	[23]
Atomic Force Microscopy	AFM	Force spectroscopy, imaging	Atomic structure, Crystal latice defects, Youngs modulus, Surface potential	[24,25]
Transmission Electron Microscopy	TEM	Electron beam flourescence	Size, Shape, Composition, Aggregation, Size distribution, Bonding environment	[26–28]
3-D Tomography	μCT	X-ray Irradiation	Size, Shape, Aggregation, Porosity	-
Extended X-ray Absorption Fine Structure	EXAFS	X-ray absorption spectrum	X-ray absorption coefficent, Chemical state of species	[29,30]
X-ray Absorption Near Edge Structure	XANES	X-ray absorption spectrum, excitation of an inner shell electron	Vacant orbitals, Oxidation state, Chemical bonding environment	[31,32]
Scanning Transmission X-ray Microscopy	STXM	Elemental X-ray absroption and spectrum	Elemental concentration and speciation map, valency, coordination number	[33,34]
X-ray Photoelectron Spectroscopy	XPS	Kinetic energy of ejected photoelectron	Elemental concentration and speciation on particle surfaces	[31,35]

While hard to access, in recent years there is an increased use of synchrotron-based XRD and spectroscopy techniques to study Fe-nanoparticles. Some advantages of synchrotron techniques include a need for only small amounts of sample material, no need for sample homogenization or excessive preparation, the option to selectively target specific and sub-micrometer scale areas of interest, and in some cases, analyses can be carried out on wet samples at room temperature and pressure. These X-ray based techniques (e.g., EXAFS, XANES, STXM, and XPS; Table 1) encompass a wide range of approaches that measure the X-rays or particles emitted during electronic transition caused by Fe-nanoparticles exposed to high intensity energy beams. The techniques provide element specific data and offer information on chemical state of iron species, interatomic distances, electronic structure, oxidation states, and ligand bonding. For example, micro-focused synchrotron XRD has been used to understand the formation condition or role of surface reactions among iron minerals found in biofilms or Fe-rich nodules [18,19]. More recently and using TEM coupled with L-edge X-ray Absorption Spectroscopy (XAS), Hirst et al. [32] evaluated nanoparticles from a boreal river system and found that the Fe(III)-rich nanoparticles comprised predominantly ferrihydrite aggregates, with minor goethite and hematite also observed. A detailed discussion and application of analytical techniques to study Fe-nanoparticles is beyond the scope of this article and readers are referred to recent published reviews [36,37].

2.2. Fe L-Edge X-ray Absorption Spectroscopy as a Powerful Tool for Investigating Fe Nanomineralogy in Natural Systems

Synchrotron X-ray spectromicroscopy at the Fe $L_{2,3}$-absorption edge is well suited towards imaging and evaluating the chemical and mineralogical speciation of natural Fe particles. The technique is capable of providing spatial resolutions as small as 10 nm, and spectral resolutions on the order of 0.1 eV and smaller. Importantly, the Fe L-edge (~700 eV) falls within the soft X-ray range and thus provides complementary information to the Fe K-edge (~7100 eV) but without the significant beam damage artifacts that may be caused to natural samples by the more energetic X-rays. Soft X-ray spectroscopy is also favored over hard X-rays because of better resolution of fine structure, higher absorption cross section, and lower intrinsic lifetime broadening (e.g., Reference [38]). As the Fe L-edge probes chemical information related to the local coordination around the metal center, its application is not limited to crystalline mineral phases and it is thus our preferred analytical approach when evaluating natural nanoparticles, which are commonly amorphous or poorly crystalline (e.g., Reference [34]).

Data presented in this submission were collected using Scanning Transmission X-ray Microscopy (STXM) at the Molecular Environmental Sciences (MES) beamline 11.0.2 at the Advanced Light Source, Lawrence Berkeley National Laboratory, CA, USA [39]. This soft X-ray beamline is especially designed to ensure that natural particles can be investigated in their pristine geochemical state under ambient physicochemical conditions (i.e., without need for excessive sample drying which induces shrinkage and loss of water from chemically-active hydration shells). X-ray absorption near-edge structure (XANES) spectra were collected in transmission mode in a 1 atm. He environment at room temperature. XANES spectra were collected primarily at the Fe L_3-edge (703–715 eV), with supporting data related to elemental associations and substitutions collected at the C K-edge (283–300 eV) and the Al K-edge (1563–1572 eV). The spectral and spatial resolutions were 0.2 eV and 12 nm, respectively, and were achieved using a 17 nm zone plate, a 1200 L/mm grating and 25 mm exit slitsas key parameters in the experimental end-station. The Fe nanoparticles evaluated in this study were sampled from various Southern African marine and fluvial water masses, and the details of sample collection and preparation are described elsewhere [33,34,40].

The Fe $L_{2,3}$ X-ray absorption edge contains a wealth of information related to the mineral structure and chemistry of Fe containing minerals. A host of earlier workers have used Fe L-edge spectroscopy to identify the valence state of iron in mineral structures [41–43], to identify the structural coordination of Fe in nano-particulate-ferrihydrite [44], and to probe Fe speciation in samples collected from natural biogeochemical systems [45–48]. A recent systematic study into the Fe L-edge spectral response to Fe mineralogy has highlighted how the split peaks of the Fe L_3-edge (Figure 1) can be used to identify Fe mineral speciation [15]. Briefly, this peak splitting arises largely because of the effects of ligand field splitting which results in the Fe 3d orbitals separating into t_{2g} and e_g subsets occurring at different energy positions [15,38,49]. The peak splitting can be parameterized according to the energy difference between the two peaks (ΔeV) and according to the ratio of the respective peaks' intensities (Figure 1B). These two parameters vary in response to Fe mineralogy (i.e., factors inherent to mineral structure such as Fe valence state, coordination number, strength of the ligand field, levels of distortion to the local Fe polyhedron, nature of ligand bonding and the composition of the resultant molecular orbitals [15]), and a two parameter ΔeV versus intensity ratio plot can thus be used to identify Fe mineral speciation (Figure 2: black squares indicating standard Fe-oxide and Fe-oxyhydroxide spectral signatures). The two parameter plot has previously been used to investigate sub-micrometer sized environmental particles from glacial [50], lacustrine [40], marine [33], and fluvial systems [34], and the present study seeks to extend on this body of research by utilizing the spectroscopic tool specifically to evaluate environmental particles existing in the nano size-domain (i.e., having at least one size dimension <100 nm).

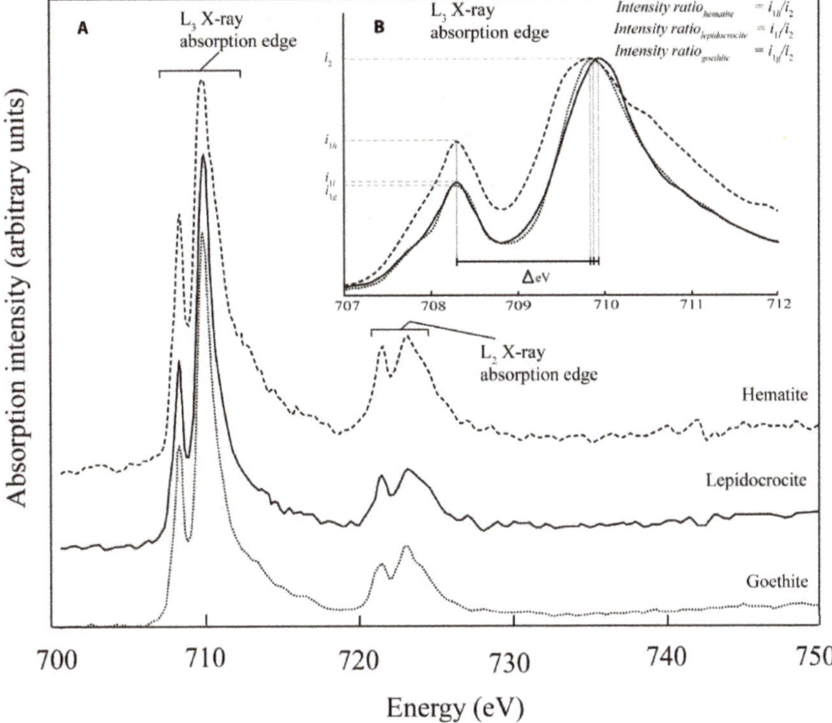

Figure 1. (**A**) The Fe $L_{2,3}$ X-ray Absorption Near-Edge Structure (XANES) spectrum for three standard iron oxide minerals (goethite (α-FeOOH); lepidocrocite (γ-FeOOH); and hematite (α-Fe$_2$O$_3$)) clearly illustrating the peak splitting that occurs in both the L_2-edge and the L_3-edge on account of ligand field splittingeffects. (**B**) The fine structure of the L_3-edge region of the spectrum (i.e., the degree of L_3-edge peak splitting (ΔeV) and the intensity ratio (*i*) between the split peaks) reflects the differences in chemistry and mineralogy of the three minerals. These differences can be quantified and represent a useful tool for Fe nanoparticle characterisation (e.g., Figure 2; Reference [15]).

3. Results

The complete dataset includes four fluvial nanoparticles, ten coastal marine nanoparticles, and 122 open ocean nanoparticles. Of these particles, the majority comprises Fe in its ferric form although three ferrous nanoparticles were also identified. Carbon K-edge XANES confirmed an association between these ferrous nanoparticles and organic carbon [40], suggesting that organic functional groupssuch as carboxamidemay play a role in stabilizing reduced forms of Fe in oxic water masses. Additionally, at one of the open ocean marine sampling sites (located in sub-Antarctica frontal zone of the Southern Ocean) the nanoparticle mineralogy predominantly matched that of magnetite (Figure 3A). Throughout the samples evaluated, the nanoparticle morphology was generally rounded, although several examples (e.g., Figure 3B,D—particle corresponding to spectrum F) were elongate and classified as nanoparticles because of a single dimension being in the <100 nm size range.

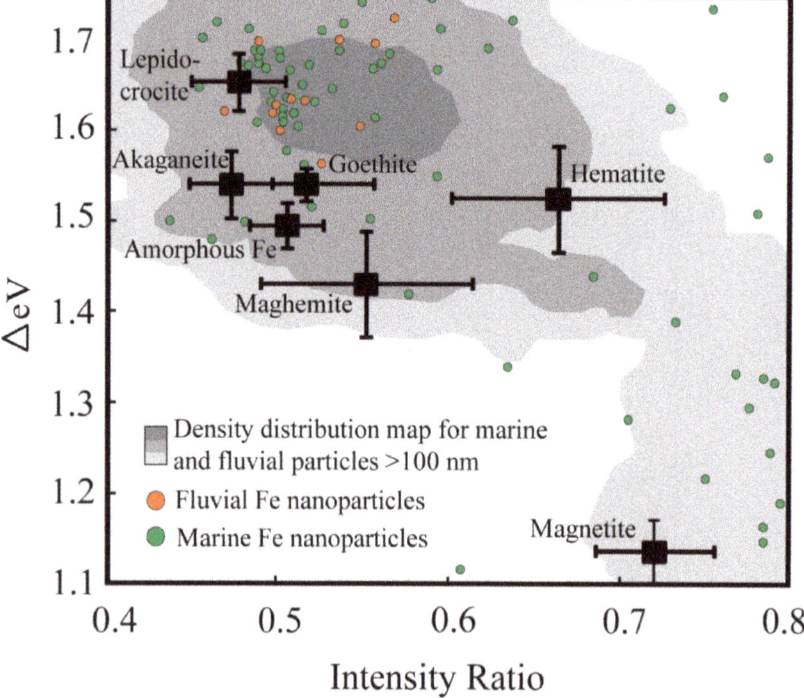

Figure 2. Plot of the spectral parameters ΔeV versus intensity ratio clearly illustrating the discrete fields occupied by synthetic Fe mineral standards (black squares and associated bars representing standard deviation (modified after Reference [15])). Colored point data represent the spectral signatures for individual Fe nanoparticles from fluvial (orange) and marine (green) systems, whereas the shaded region represents the density distribution of data points for sub-micrometer (larger than 100 nm) and micrometer sized particles collected from marine and fluvial environments (generated using ArcGIS point density function).

Figure 2 provides an overview of the spectral signatures of the nanoparticles identified in this study (green- and orange point data), and compares these data to larger μm-scale natural Fe particles (grey shading) and standard Fe oxides and Fe oxy-hydroxides [15]. Most of the data points fall outside of the fields for the standard Fe mineral phases, alluding to the fact that most of these evaluated natural nanoparticles are aggregates, are amorphous in nature or contain ionic substituents that render their chemistry impure relative to the standards. The data points are relatively scattered on the ΔeV versus intensity ratio diagram, although there exists some clustering near to the position of the lepidocrocite standard. This clustering agrees well with the density distribution of larger (>100 nm) natural Fe-rich particles indicating a degree of similarity in mineral chemistry between the two datasets. This clustering occurs at high ΔeV values (between 1.6 and 1.7) which suggests that the natural Fe nanoparticles predominantly comprise Fe(III) in octahedral coordination, experience a relatively high degree of ligand field splitting, and may be characterized by distorted (larger volume) local Fe coordination polyhedral [15]. Similarly, the clustering of data favours lower intensity ratio values (between 0.45–0.55) indicative that a significant concentration of hydroxyl ligands coordinate to the Fe metal center (i.e., compare FeOOH standards at low intensity ratio values to Fe_2O_3 standards at intensity ratios> 0.55). A secondary clustering (i.e., Δ1.1–1.3; intensity ratio >0.7; Figure 3E) occurs near to the standard magnetite data point. Again, the agreement between the standard data point and the natural particles is not perfect, likely reflecting structural and chemical impurities. The level of disagreement may also reflect a possible particle size effect in the nanoparticle size regime, since it is noted that a proportion of the >100 nm natural Fe particles (grey shading) do show good overlap with the standard magnetite data points.

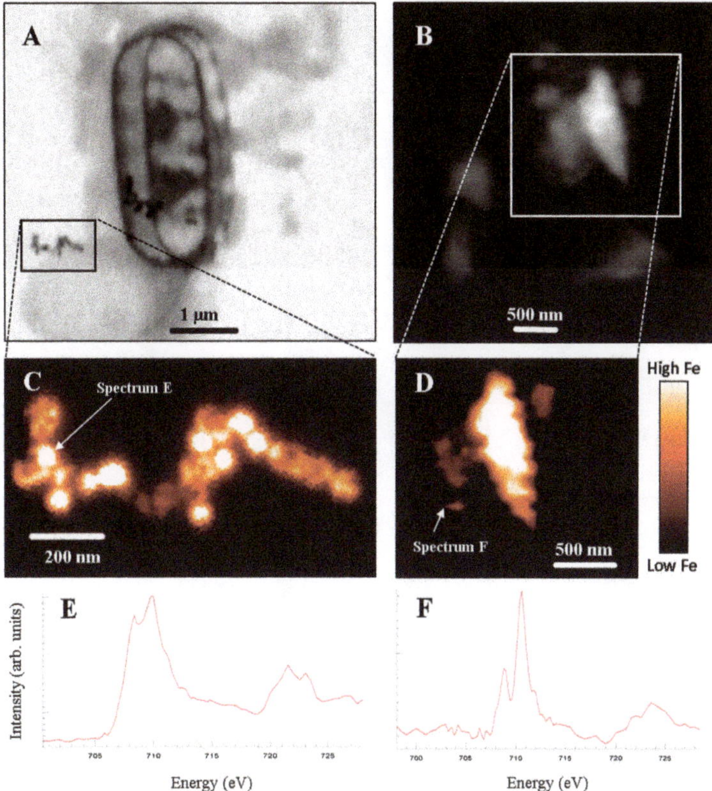

Figure 3. (**A,B**) X-ray absorption maps of marine nanoparticles collected at 709.8 eV using Scanning Transmission X-ray Microscopy (STXM)). (**C,D**) Fe distribution maps generated by subtracting the X-ray absorption map collected below the Fe L-edge (i.e., 703 eV) from the X-ray absorption map collected at the Fe L-edge (e.g., Figure 3A,B). Colour bar shows relative Fe concentration. (**E,F**) Fe $L_{2,3}$-edge spectra for identified Fe-rich nanoparticles. Spectrum E reflects magnetite chemistry, whereas Spectrum F is indicative of Fe(III)—likely in the form of an amorphous Fe oxy-hydroxide phase.

Similarities in cation size and charge result in Al-for-Fe substitution reactions being common phenomena in Fe oxides and Fe oxy-hydroxides. For example, natural soil Fe oxide minerals have previously been investigated and were found to contain levels of Al substitution reaching up to 36% [51]. Using Al subtraction maps generated by subtracting an X-ray absorption map collected below the Al K-edge (e.g., 1565 eV) from the X-ray absorption map collected at the Al K-edge (1572 eV), we were able to measure and compare the Al/Fe ratio in a subset of natural aquatic Fe nanoparticles to the Al/Fe ratio measured in natural aquatic particles >100 nm in size (Figure 4). The Al/Fe ratio in the nanoparticles ranges between 0.07 and 0.18; whereas the larger particles show a more extensive range spanning between 0.03 and 0.45. A student t-test conducted on the two datasets however, revealed that there is no significant difference ($\alpha = 0.05$) between the Al/Fe ratios measured in the two size fractions. The Al K-edge XANES spectra were also used to evaluate the ligand coordination around the Al metal center in the natural nanoparticles and larger particles [52]. The side panels in Figure 4 summarize these findings (further details in Reference [34]) and show that Al associated with Fe-rich nanoparticles is high hydroxylated (i.e., coordinated to between four (coordination environment similar to Al in 1:1 clays) and six ($Al(OH)_3$-like coordination) hydroxyl ligands). In comparison, the larger sized Fe-rich particles show a much more diverse array in the local coordination environment of the contained Al, with the number of hydroxyl ligands ranging between none (i.e., Al is coordinated only to oxygen ligands in a local bonding environment characteristic of Al in primary silicates) and six.

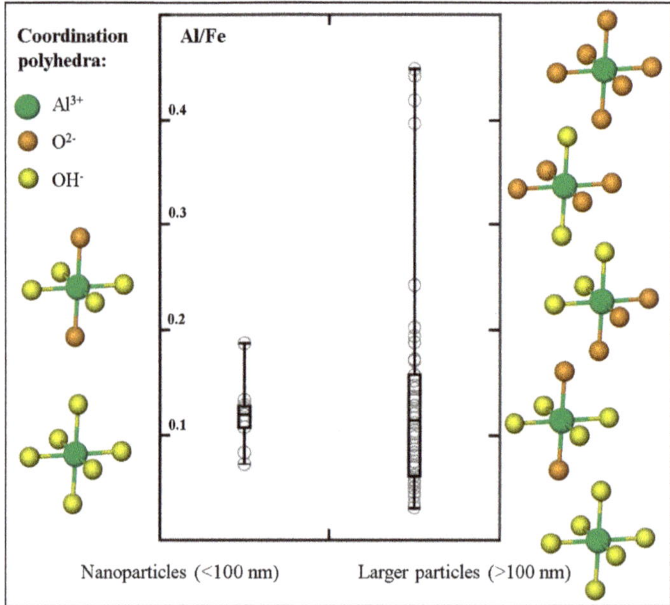

Figure 4. Box-and-whisker diagram comparing the Al/Fe ratio measured in nanoparticles versus the range measured in larger sub-micrometer (>100 nm) and micrometer sized particles (data extracted from [34]). Side panels show the diversity in the number of hydroxyl ligands coordinated to the Al metal centers in the two size fractions (from interpretations of Al K-edge XANES [34,52]; ligand positions are not absolute). These side panels highlight that hydroxyl ligands predominate over oxygen ligands in the Fe-rich nanoparticles, whereas larger-sized natural Fe particles are characterized by a prevalence of both oxygen and hydroxyl ligand bonding.

4. Discussion

The nanoparticles evaluated in this study were collected from relatively pristine aquatic systems and are thus expected to reflect the natural class of nanoparticles, although the presence of engineered and incidental nanoparticles cannot be completely ruled out. In contrast to the volume of work that has recently been conducted on engineered nanoparticles, natural inorganic nanoparticles have received a surprisingly lesser amount of scientific attention, despite some studies showing that the fate and toxicity of these natural nanoparticles differ from those of engineered ones (Reference [12], and references therein). Using a specialized synchrotron-based X-ray spectro-microscopic tool, this study thus seeks to provide new and additional insights into the characterization of biogeochemically-active natural Fe-rich nanoparticles.

Because of its combined imaging and local coordination environment probe capabilities, soft X-ray Fe L-edge XANES and STXM techniques have gained prominence as tools for studying natural sub-micrometer particles and nanoparticle in their pristine state [34,50]. Furthermore, and as correlative study approaches continue to evolve, it is foreseen that the detailed chemical information contained in the Fe L-edge spectrum will be used to complement other study methodologies (Table 1) to more fully explore the emerging field of natural nanoparticle research. In contrast to engineered nanoparticles, and in contrast to experimental results produced in controlled laboratory environments; natural nanoparticles are produced by a diverse array of formation mechanisms (including bottom-up (aggregation and growth of nanomaterial precursors); top-down (e.g., weathering reactions), and biologically-mediated reactions) and are influenced by variable environmental conditions (e.g., pH, Eh, temperature, salinity, solar radiation and the presence of organic matter) [1,2,12]. Given this vast range in influencing parameters, it is thus not surprising that our dataset shows a high degree of scatter on the ΔeV versus intensity ratio diagram (Figure 2). This scattering reflects divergence in the local Fe coordination environment away from what is expected for pure crystalline Fe mineral standards (Figure 2, black squares), and is caused for example, by variations in the ligand type and geometry and by structural distortions to the local Fe coordination polyhedra [15]. Given the measured presence of Al in the Fe-rich nanoparticles (Figure 4), we suggest that one of the reasons for the scatter is the ubiquitous presence of cation substitutions within the nanoparticle structure, including but not limited to Al and Si. Two other possible reasons for the scatter include: (1) The likelihood that natural particles are poorly-crystalline or amorphous, and (2) the possibility that what we have measured as discrete 10–100 nm sized particles actually represent aggregates of smaller (<10 nm) particles with variable chemistry and bound together by organic coatings (e.g., humic acids [53]).

Our research broadly identifies three groups of Fe nanoparticle mineralogy in the natural aquatic systems evaluated. A small number of Fe(II)-rich nanoparticles were identified despite all samples being collected from oxygen replete waters where Fe(II) is the thermodynamically less-favored form of Fe. Based on carbon K-edge XANES analyses, we have previously suggested that the Fe(II) presence may be a result of active reduction of Fe by microbial communities [40], or may be due to the stabilizing effect of Fe(II)-specific organic functional groups [48]. Recent work focused on Fe nanoparticulate speciation in glacial sediments invokes the role of amorphous silica in Fe(II) preservation [50], highlighting the need for further work to more firmly elucidate the mechanisms that stabilize ferrous Fe in natural nanoparticles and to constrain the interactions between these phases and the biosphere.

The second broad group identified in this study-in which Fe is present in its oxidized (Fe(III)) valence state-represents the dominant Fe nanoparticle speciation identified in natural aquatic environments. No distinctive difference was observed between the chemical speciation of the nanoparticles and that of the larger sized particles (Figure 2), likely reflecting that in heterogenous natural systems nanosize effects (e.g., quantum effects) have a secondary influence on XANES spectral signatures relative to the more obvious effects of inter-particle variability caused by differences in environmental parameters. Because of their strong surface complexing capacity, their high surface area to charge ratio and their resistance to sedimentation, ferric oxide nanoparticles have been highlighted as important contaminant and nutrient transporting agents in fluvial systems [1,5,6,54]. Furthermore, within the marine domain where important biologically-mediated air-sea CO_2 exchange takes place, Fe particles (including these Fe(III) nanoparticles) collected in the nominally-defined "dissolved" size fraction (<0.2 or <0.45 µm filter pore sizes) are suggested as being important repositories of potentially bioavailable Fe [8,9].The Fe L-edge spectral signatures (i.e., intensity ratio < 0.55) suggest that these phases are predominantly hydroxyl-bearing moieties, which is corroborated by the prevalence of highly-hydroxylated Al metal centers associated with these Fe nanoparticles (Figure 4). This Al is further noteworthy in that its presence (especially as a cation substituent) is expected to modify the particles' surface chemistry by decreasing the net concentration of reactive Fe sites through replacement by less-reactive Al [55,56].This impact on surface reactivity in turn affects the nutrient- and pollutant transporting capacity of the natural Fe nanoparticles, especially when compared to model experimental systems in which observations are based on the behavior of pure mineral standards. Aluminum substitution is further known to modify the solubility and dissolution kinetics of model Fe oxides (e.g., Reference [57], and references therein). The importance and accessibility of the marine Fe nanoparticulate pool as a repository of bioavailable Fe for driving primary productivity (and thus for influencing global climate feedback loops) must therefore be interpreted in a context of the observed and associated Al presence.

Finally, the presence of magnetite nanoparticles in marine surface waters of the Southern Ocean (Figures 2 and 3) has previously been related to Patagonian atmospheric Fe inputs [33], but may also derive from magnetotactic bacteria [58]. The possible presence of biogenic Fe nanoparticles may be extended to the Fe(III) phases discussed above, since certain genera of Fe-oxidising bacteria are known to produce ferrihydrite and Fe(III) oxyhydroxides [12].Magnetic Fe nanoparticles have recently received increased attention since they are known to cause complex and potentially harmful interactions with living matter (e.g., through the production of excess reactive oxygen species [59]). However, it is important to highlight that the ecotoxicology of natural magnetite nanoparticles and the other forms of natural Fe nanoparticles identified in this study is not well constrained. Indeed, by virtue of their small size, they have a much higher likelihood of crossing cellular barriers (e.g., Reference [60]). However, ecotoxicological effects of the natural nanoparticles are expected to be restricted (relative to engineered nanoparticles) due to the potential presence of organic surface coatings [12]—such as those identified from our carbon K-edge XANES spectro-microscopy.

5. Conclusions

The discussion above highlights the diversity in nanoparticle mineral chemistry and speciation in natural aquatic environments. Given that the production of engineered Fe nanoparticles is expected to increase significantly in forthcoming years (2010 estimates at up to 42,000 t per year [61]), such a study focused on relatively pristine aquatic systems forms an important baseline from which to interpret future environmental change. The most common type of engineered Fe nanoparticle is the Fe^0 class of nanoparticles, which are commonly used in environmental remediation efforts [7]. These engineered nanoparticles are however expected to experience significant levels of surface oxidation and surface reactions when released into natural aquatic environments [62]. The insights provided here related to the associations of natural Fe nanoparticles with Al and with organic functional groups, and may thus serve as a good proxy for the ultimate fate and speciation of engineered nanoparticles that evolve in heterogenous natural systems. Further work should continue to focus on the speciation, biogeochemical behavior and ecotoxicological impacts of both natural and engineered Fe nanoparticles existing in Earth's various subsystems (e.g., hydrosphere, biosphere, etc.).

Author Contributions: Conceptualization, S.M., A.R., B.v.d.H.; methodology, S.M., B.v.d.H.; formal analysis, investigation and data curation, B.v.d.H.; writing—original draft preparation, B.v.d.H., A.R.; writing—review and editing, S.M.; validation, supervision, project administration and funding acquisition, A.R., S.M.

Funding: This research is supported by grants from NRF, South Africa, the Stellenbosch University VR(R) fund, the National Science Foundation (chemical sciences), the U.S. Department of Energy (BES and SBR), and the Princeton-in-Africa program.

Acknowledgments: The authors would like to thank T. Tyliszczak (beamline scientist) for assistance with data collection at the Molecular Environmental Sciences beamline 11.0.2. at Lawrence Berkeley National Laboratory (Berkeley, CA, USA). Colleagues who assisted with sample collection from lacustrine and marine sampling sites are also duly acknowledged. Finally, the authors wish to thank the two anonymous reviewers and the editors for the comments which served to strengthen the manuscript.

References

1.   Lead, J.R.; Wilkinson, K.J. Aquatic colloids and nanoparticles: Current knowledge and future trends. *Environ. Chem.* **2006**, *3*, 159–171. [CrossRef]

2.   Hochella, M.F.J.; Mogk, D.W.; Ranville, J.; Allen, I.C.; Luther, G.W.; Marr, L.C.; Mcgrail, B.P.; Murayama, M.; Qafoku, N.P.; Rosso, K.M.; et al. Natural, incidental, and engineered nanomaterials and their impacts on the Earth system. *Science* **2019**, *363*, eaau8299. [CrossRef]

3. Nowack, B.; Bucheli, T.D. Occurrence, behavior and effects of nanoparticles in the environment. *Environ. Pollut.* **2007**, *150*, 5–22. [CrossRef] [PubMed]

4. Gottschalk, F.; Nowack, B. The release of engineered nanomaterials to the environment. *J. Environ. Monit.* **2011**, *13*, 1145–1155. [CrossRef]

5. Hassellöv, M.; von der Kammer, F. Iron oxides as geochemical nanovectors for metal transport in soil-river systems. *Elements* **2008**, *4*, 401–406. [CrossRef]

6. Hochella, M.F.; Moore, J.N.; Putnis, C.V.; Putnis, A.; Kasama, T.; Eberl, D.D. Direct observation of heavy metal-mineral association from the Clark Fork River Superfund Complex: Implications for metal transport and bioavailability. *Geochim. Cosmochim. Acta* **2005**, *69*, 1651–1663. [CrossRef]

7. Yan, W.L.; Lien, H.-L.; Koel, B.E.; Zhang, W.-X. Iron nanoparticles for environmental clean-up: Recent developments and future outlook. *Environ. Sci. Process. Impacts* **2013**, *15*, 63–77. [CrossRef] [PubMed]

8. Tagliabue, A.; Bowie, A.R.; Boyd, P.W.; Buck, K.N.; Johnson, K.S.; Saito, M.A. The integral role of iron in ocean biogeochemistry. *Nature* **2017**, *543*, 51. [CrossRef]

9. Boyd, P.W.; Ellwood, M.J. The biogeochemical cycle of iron in the ocean. *Nat. Geosci.* **2010**, *3*, 675–682. [CrossRef]

10. Li, L.; Fan, M.; Brown, R.C.; Van Leeuwen, J.H.; Wang, W.; Song, Y.; Zhang, P. Synthesis, Properties, and Environmental Applications of Nanoscale Iron-Based Materials: A Review Applications of Nanoscale Iron-Based Materials: A Review. *Crit. Rev. Environ. Sci. Technol.* **2006**, *36*, 405–431. [CrossRef]

11. Raychoudhury, T.; Scheytt, T. Potential of Zerovalent iron nanoparticles for remediation of environmental organic contaminants in water: A review. *Water Sci. Technol.* **2013**, *68*, 1425–1439. [CrossRef] [PubMed]

12. Sharma, V.K.; Filip, J.; Zboril, R.; Varma, R.S. Natural inorganic nanoparticles—Formation, fate, and toxicity in the environment. *Chem. Soc. Rev.* **2015**, *44*, 8410–8423. [CrossRef]

13. Neubauer, E.; Schenkeveld, W.D.; Plathe, K.L.; Rentenberger, C.; Von Der Kammer, F.; Kraemer, S.M.; Hofmann, T. The influence of pH on iron speciation in podzol extracts: Iron complexes with natural organic matter, and iron mineral nanoparticles. *Sci. Total Environ.* **2013**, *461*, 108–116. [CrossRef]

14. Carbone, C.; Di Benedetto, F.; Marescotti, P.; Sangregorio, C. Natural Fe-oxide and -oxyhydroxide nanoparticles: An EPR and SQUID investigation. *Mineral. Petrol.* **2005**, *85*, 19–32. [CrossRef]

15. Von der Heyden, B.P.; Roychoudhury, A.N.; Tyliszczak, T.; Myneni, S.C.B. Investigating nanoscale mineral compositions: Iron L3-edge spectroscopic evaluation of iron oxide and oxy-hydroxide coordination. *Am. Mineral.* **2017**, *102*, 674–685. [CrossRef]

16. Tagliabue, A.; Aumont, O.; DeAth, R.; Dunne, J.P.; Dutkiewicz, S.; Galbraith, E.; Misumi, K.; Moore, J.K.; Ridgwell, A.; Sherman, E.; et al. How well do global ocean biogeochemistry models simulate dissolved iron distributions? *Global Biogeochem. Cycles* **2016**, *32*, 149–174. [CrossRef]

17. Upadhyay, S.; Parekh, K.; Pandey, B. Influence of crystallite size on the magnetic properties of Fe_3O_4 nanoparticles. *J. Alloys Compd.* **2016**, *678*, 478–485. [CrossRef]

18. Marcus, M.A.; Manceau, A.; Kersten, M. Mn, Fe, Zn and As speciation in a fast-growing ferromanganese marine nodule. *Geochim. Cosmochim. Acta* **2004**, *68*, 3125–3136. [CrossRef]

19. Toner, B.M.; Santelli, C.M.; Marcus, M.A.; Wirth, R.; Chan, C.S.; McCollom, T.; Bach, W.; Edwards, K.J. Biogenic iron oxyhydroxide formation at mid-ocean ridge hydrothermal vents: Juan de Fuca Ridge. *Geochim. Cosmochim. Acta* **2009**, *73*, 388–403. [CrossRef]

20. Křížek, M.; Pechoušek, J.; Tuček, J.; Šafářová, K.; Medřík, I.; Machala, L. Iron oxide nanoparticle powders with high surface area. *AIP Conf. Proc.* **2012**, *1489*, 88–94.

21. Kenouche, S.; Larionova, J.; Bezzi, N.; Guari, Y.; Bertin, N.; Zanca, M.; Lartigue, L.; Cieslak, M.; Godin, C.; Morrot, G.; et al. NMR investigation of functionalized magnetic nanoparticles Fe_3O_4 as T1–T2 contrast agents. *Powder Technol.* **2014**, *255*, 60–65. [CrossRef]

22. Sabale, S.; Jadhav, V.; Khot, V.; Zhu, X.; Xin, M.; Chen, H. Superparamagnetic MFe_2O_4 (M= Ni, Co, Zn, Mn) nanoparticles: Synthesis, characterization, induction heating and cell viability studies for cancer hyperthermia applications. *J. Mater. Sci. Mater. Med.* **2015**, *26*, 127. [CrossRef] [PubMed]

23. Ribeiro, J.; Flores, D.; Ward, C.R.; Silva, L.F. Identification of nanominerals and nanoparticles in burning coal waste piles from Portugal. *Sci. Total Environ.* **2010**, *408*, 6032–6041. [CrossRef] [PubMed]

24. Pyrgiotakis, G.; Blattmann, C.O.; Pratsinis, S.; Demokritou, P. Nanoparticle–nanoparticle interactions in biological media by atomic force microscopy. *Langmuir* **2013**, *29*, 11385–11395. [CrossRef] [PubMed]

25. Raj, X.J.; Nishimura, T. Investigation of the Surface Potential on Iron Nanoparticles During the Corrosion by Atomic Force Microscopy (AFM) and Kelvin Probe Force Microscopy (KFM). *Int. J. Electrochem. Sci.* **2014**, *9*, 2090–2100.

26. Poulton, S.W.; Raiswell, R. Chemical and physical characteristics of iron oxides in riverine and glacial meltwater sediments. *Chem. Geol.* **2005**, *218*, 203–221. [CrossRef]

27. Allard, T.; Menguy, N.; Salomon, J.; Calligaro, T.; Weber, T.; Calas, G.; Benedetti, M.F. Revealing forms of iron in river-borne material from major tropical rivers of the Amazon Basin (Brazil). *Geochim. Cosmochim. Acta* **2004**, *68*, 3079–3094. [CrossRef]

28. Allard, T.; Weber, T.; Bellot, C.; Damblans, C.; Bardy, M.; Bueno, G.; Nascimento, N.R.; Fritsch, E.; Benedetti, M.F. Tracing source and evolution of suspended particles in the Rio Negro Basin (Brazil) using chemical species of iron. *Chem. Geol.* **2011**, *280*, 79–88. [CrossRef]

29. Liu, T.; Guo, L.; Tao, Y.; Hu, T.D.; Xie, Y.N.; Zhang, J. Bondlength alternation of nanoparticles Fe_2O_3 coated with organic surfactants probed by EXAFS. *Nanostruct. Mater.* **1999**, *11*, 1329–1334. [CrossRef]

30. Sundman, A.; Karlsson, T.; Laudon, H.; Persson, P. XAS study of iron speciation in soils and waters from a boreal catchment. *Chem. Geol.* **2014**, *364*, 93–102. [CrossRef]

31. Sun, Y.P.; Li, X.Q.; Cao, J.; Zhang, W.X.; Wang, H.P. Characterization of zero-valent iron nanoparticles. *Adv. Colloid Interface Sci.* **2006**, *120*, 47–56. [CrossRef]

32. Hirst, C.; Andersson, P.S.; Shaw, S.; Burke, I.T.; Kutscher, L.; Murphy, M.J.; Maximov, T.; Pokrovsky, O.S.; Mörth, C.M.; Porcelli, D. Characterisation of Fe-bearing particles and colloids in the Lena River basin, NE Russia. *Geochemica Cosmochim. Acta* **2017**, *213*, 553–573. [CrossRef]

33. von der Heyden, B.P.; Roychoudhury, A.N.; Mtshali, T.N.; Tyliszczak, T.; Myneni, S.C.B. Chemically and geographically distinct solid-phase iron pools in the Southern Ocean. *Science* **2012**, *338*, 1199–1201. [CrossRef]

34. Von Der Heyden, B.P.; Frith, M.G.; Bernasek, S.; Tylizszak, T.; Roychoudhury, A.N.; Myneni, S.C.B. Geochemistry of Al and Fe in freshwater and coastal water colloids from the west coast of Southern Africa. *Geochim. Cosmochim. Acta* **2018**, *241*, 56–68. [CrossRef]

35. Sheng, G.; Yang, P.; Tang, Y.; Hu, Q.; Li, H.; Ren, X.; Hu, B.; Wang, X.; Huang, Y. New insights into the primary roles of diatomite in the enhanced sequestration of UO_2^{2+} by zerovalent iron nanoparticles: An advanced approach utilizing XPS and EXAFS. *Appl. Catal. B Environ.* **2016**, *193*, 189–197. [CrossRef]

36. Mourdikoudis, S.; Pallares, R.M.; Thanh, N.T. Characterization techniques for nanoparticles: Comparison and complementarity upon studying nanoparticle properties. *Nanoscale* **2018**, *10*, 12871–12934. [CrossRef] [PubMed]

37. Shrivastava, M.; Srivastav, A.; Gandhi, S.; Rao, S.; Roychoudhury, A.; Kumar, A.; Singhal, R.K.; Kumar, S.; Singh, S.D. Monitoring of engineered nanoparticles in soil-plant system: A review. *Environ. Nanotechnol. Monit. Manag.* **2019**, *11*, 100218. [CrossRef]

38. De Groot, F.M.F.; Glatzel, P.; Bergmann, U.; Van Aken, P.A.; Barrea, R.A.; Klemme, S.; Haevecker, M.; Knop-Gericke, A.; Heijboer, W.M.; Weckhuysen, B.M. 1s2p Resonant Inelastic X-ray Scattering of Iron Oxides. *J. Phys. Chem. B* **2005**, *109*, 20751–20762. [CrossRef] [PubMed]

39. Bluhm, H.; Andersson, K.; Araki, T.; Benzerara, K.; Brown, G.E.; Dynes, J.J.; Ghosal, S.; Gilles, M.K.; Hansen, H.C.; Hemminger, J.C.; et al. Soft X-ray microscopy and spectroscopy at the molecular environmental science beamline at the Advanced Light Source. *J. Electron. Spectros. Relat. Phenom.* **2006**, *150*, 86–104. [CrossRef]

40. Von Der Heyden, B.P.; Hauser, E.J.; Mishra, B.; Martinez, G.A.; Bowie, A.R.; Tyliszczak, T.; Mtshali, T.N.; Roychoudhury, A.N.; Myneni, S.C.B. Ubiquitous Presence of Fe(II) in Aquatic Colloids and Its Association with Organic Carbon. *Environ. Sci. Technol. Lett.* **2014**, *1*, 387–392. [CrossRef]

41. Cressey, G.; Henderson, C.M.B.; van der Laan, G. Use of L-edge X-ray absorption spectroscopy to characterize multiple valence states of 3d transition metals; a new probe for mineralogical and geochemical research. *Phys. Chem. Miner.* **1993**, *20*, 111–119. [CrossRef]

42. van Aken, P.A.; Liebscher, B. Quantification of ferrous/ferric ratios in minerals: New evaluation schemes of Fe L_{23} electron energy-loss near-edge spectra. *Phys. Chem. Miner.* **2002**, *28*, 188–200. [CrossRef]

43. Calvert, C.C.; Brown, A.; Brydson, R. Determination of the local chemistry of iron in inorganic and organic materials. *J. Electron. Spectros. Relat. Phenom.* **2005**, *143*, 173–187. [CrossRef]

44. Peak, D.; Regier, T. Direct observation of tetrahedrally coordinated Fe(III) in ferrihydrite. *Environ. Sci. Technol.* **2012**, *46*, 3163–3168. [CrossRef] [PubMed]

45. Miot, J.; Benzerara, K.; Morin, G.; Kappler, A.; Bernard, S.; Obst, M.; Férard, C.; Skouri-Panet, F.; Guigner, J.M.; Posth, N.; et al. Iron biomineralization by anaerobic neutrophilic iron-oxidizing bacteria. *Geochim. Cosmochim. Acta* **2009**, *73*, 696–711. [CrossRef]

46. Chan, C.S.; De Stasio, G.; Welch, S.A.; Girasole, M.; Frazer, B.H.; Nesterova, M.V.; Fakra, S.; Banfield, J.F. Microbial polysaccharides template assembly of nanocrystal fibers. *Science* **2004**, *303*, 1656–1658. [CrossRef]

47. Chen, C.; Sparks, D.L. Multi-elemental scanning transmission X-ray microscopy-near edge X-ray absorption fine structure spectroscopy assessment of organo-mineral associations in soils from reduced environments. *Environ. Chem.* **2015**, *12*, 64–73. [CrossRef]

48. Toner, B.M.; Fakra, S.C.; Manganini, S.J.; Santelli, C.M.; Marcus, M.A.; Moffett, J.W.; Rouxel, O.; German, C.R.; Edwards, K.J. Preservation of iron(II) by carbon-rich matrices in a hydrothermal plume. *Nat. Geosci.* **2009**, *2*, 197–201. [CrossRef]

49. Todd, E.C.; Sherman, D.M.; Purton, J.A. Surface oxidation of chalcopyrite (CuFeS$_2$) under ambient atmospheric and aqueous (pH 2-10) conditions: Cu, Fe L- and O K-edge X-ray spectroscopy. *Geochim. Cosmochim. Acta* **2003**, *67*, 2137–2146. [CrossRef]

50. Hawkings, J.R.; Benning, L.G.; Raiswell, R.; Kaulich, B.; Araki, T.; Abyaneh, M.; Stockdale, A.; Koch-müller, M.; Wadham, J.L.; Tranter, M. Biolabile ferrous iron bearing nanoparticles in glacial sediments. *Earth Planet. Sci. Lett.* **2018**, *493*, 92–101. [CrossRef]

51. Schwertmann, U.; Carlson, L. Aluminum Influence on Iron Oxides: XVII. Unit-Cell Parameters and Aluminum Substitution of Natural Goethites. *Soil Sci. Soc. Am. J.* **1994**, *58*, 256–261. [CrossRef]

52. Myneni, S.; Hay, M.; Mishra, B. Applications of scanning transmission X-ray microscopy in studying clays and their chemical interactions. In *Special Volume, Advanced Applications of Synchrotron Radiation in Clay Science*; Clay Minerals Society: Chantilly, VA, USA, 2013; pp. 231–258.

53. Baalousha, M.; Manciulea, A.; Cumberland, S.; Kendall, K.; Lead, J.R. Aggregation and surface properties of iron oxide nanoparticles: Influence of pH and natural organic matter. *Environ. Toxicol. Chem.* **2008**, *27*, 1875–1882. [CrossRef] [PubMed]

54. Wigginton, N.S.; Haus, K.L.; Hochella, M.F. Aquatic environmental nanoparticles. *J. Environ. Monit.* **2007**, *9*, 1306–1316. [CrossRef] [PubMed]

55. Potter, H.A.B.; Yong, R.N. Influence of iron/aluminium ratio on the retention of lead and copper by amorphous iron–aluminium oxides. *Appl. Clay Sci.* **1999**, *14*, 1–26. [CrossRef]

56. Jentzsch, T.L.; Penn, R.L. Effects of aluminum doping on the reactivity of ferrihydrite nanoparticles. *J. Phys. Chem. B* **2006**, *110*, 11746–11750. [CrossRef]

57. Cornell, R.M.; Schwertmann, U. *The Iron Oxides: Structure, Properties, Reactions, Occurrences and Uses*, 2nd ed.; Wiley-VCH: Weinheim, Germany, 2003; ISBN 3527302743.

58. Lefevre, C.T.; Bazylinski, D.A. Ecology, diversity, and evolution of magnetotactic bacteria. *Microbiol. Mol. Biol. Rev.* **2013**, *77*, 497–526. [CrossRef] [PubMed]

59. Liu, G.; Gao, J.; Ai, H.; Chen, X. Applications and Potential Toxicity of Magnetic Iron Oxide Nanoparticles. *Small* **2013**, *9*, 1533–1545. [CrossRef]

60. Phogat, N.; Khan, S.A.; Shankar, S.; Ansary, A.A.; Uddin, I. Fate of inorganic nanoparticles in agriculture. *Adv. Mater. Lett.* **2016**, *7*, 3–12. [CrossRef]

61. Keller, A.A.; Lazareva, A. Predicted Releases of Engineered Nanomaterials: From Global to Regional to Local. *Environ. Sci. Technol. Lett.* **2013**, *1*, 65–70. [CrossRef]

62. Liu, A.; Liu, J.; Han, J.; Zhang, W. Evolution of nanoscale zero-valent iron (nZVI) in water: Microscopic and spectroscopic evidence on the formation of nano- and micro-structured iron oxides. *J. Hazard. Mater.* **2017**, *322*, 129–135. [CrossRef] [PubMed]

Article

Facile Hydrothermal Synthesis of Nanocubic Pyrite Crystals Using Greigite Fe$_3$S$_4$ and Thiourea as Precursors

Xin Nie [1], Suxing Luo [1,2,3], Meizhi Yang [1,2], Ping Zeng [1,2], Zonghua Qin [1], Wenbin Yu [1] and Quan Wan [1,4,*]

[1] State Key Laboratory of Ore Deposit Geochemistry, Institute of Geochemistry, Chinese Academy of Sciences, Guiyang 550081, China; niexin2004@163.com (X.N.); luosuxing123123@163.com (S.L.); yangmeizhi@mail.gyig.ac.cn (M.Y.); zengpeng@mail.gyig.ac.cn (P.Z.); qinzonghua@vip.gyig.ac.cn (Z.Q.); yuwenbin@mail.gyig.ac.cn (W.Y.)
[2] University of Chinese Academy of Sciences, Beijing 100049, China
[3] Zunyi Normal College, Zunyi 563002, China
[4] CAS Center for Excellence in Comparative Planetology, Hefei 230026, China
*   Correspondence: wanquan@vip.gyig.ac.cn; Tel.: +86-0851-85891928

Received: 2 April 2019; Accepted: 29 April 2019; Published: 1 May 2019

Abstract: Nanocubic pyrite (FeS$_2$) crystals with exposed (100) crystal faces and sizes of 100–200 nm were successfully synthesized via a facile hydrothermal method using greigite (Fe$_3$S$_4$) as the iron precursor and thiourea (NH$_2$CSNH$_2$) as the sulfur source. When the concentration of thiourea was 40 mmol/L, both pyrite and hematite were observed in the as-prepared sample, indicating incomplete conversion of greigite into pyrite. With an increased thiourea concentration to 80 mmol/L, pyrite was found to be the only crystalline phase in the synthesized samples. All greigite could be transformed to pyrite within 24 h via the hydrothermal method, while further prolonging the hydrothermal time had insignificant effect on the crystal phase composition, crystallinity, and morphologies of the prepared nanocubic pyrite crystals. In contrast, when a mixture of Na$_2$S and S powder was used to replace the thiourea as the sulfur source, tetragonal, orthorhombic, cubic, and irregular pyrite crystal particles with sizes of 100 nm–1 μm were found to co-exist in the prepared samples. These results demonstrate the critical influence of sulfur source on pyrite morphology. Furthermore, our hydrothermal process, using a combination of greigite and thiourea, is proved to be effective in preparing nanocubic pyrite crystals. Our findings can also provide new insight into the formation environments and pathways of nanocubic pyrite under hydrothermal conditions.

Keywords: nanocubic pyrite; hydrothermal synthesis; greigite; thiourea

---

1. Introduction

Iron sulfides, particularly pyrite (FeS$_2$), are ubiquitous in various hydrothermal ore deposits as well as Earth surface environments, and their scientific merits have been demonstrated in many fundamental studies. For example, pyrite may provide essential information for better understanding the origin and evolution of early life on the Earth's surface environment and the global biogeochemical cycling of sulfur and iron [1,2]. Because pyrite is preferentially formed in anoxic conditions, and its morphology and chemical composition highly depends on the formation conditions, pyrite can also be used as a key geochemical indicator of contemporary environmental conditions in hydrothermal systems or Earth's surface system [1,3]. In addition, previous studies have documented that pyrite plays a crucial role in the transport, fate, reactivity, and the associated ecological toxicity of various trace elements of economic or environmental importance, including the noble metal Au and toxic heavy metals [1,4–6].

On the application aspect, owing to its abundance, low cost, low toxicity, and high chemical reactivity, pyrite has been recognized as a promising material for effectively eliminating environmental contaminants in Earth's near-surface environment under anoxic and oxic conditions, including toxic heavy metals and metalloids, radionuclides, and organic pollutants (e.g., chlorinated organic pollutants, polycyclic aromatic hydrocarbons, organic dyes, and others) [7–15]. Moreover, thanks to its high optical absorption coefficient, unique electrical and semiconducting properties, and suitable band gap (0.95 eV), pyrite (especially micro-nanopyrite) has received extensive attention for its potential applications in electrocatalytic hydrogen evolution reactions (HERs), catalytic hydrogenation, high capacity lithium ion batteries, photovoltaics, photocatalysts, photoelectrochemical solar cells, and so on [16–19]. It should be noted the chemical composition (or purity), size, morphology, exposed surface facet, and microstructure can significantly affect the surface physiochemical properties of pyrite, and consequently impact its application performance [3]. Since naturally occurring pyrite inevitably contains significant quantities of impurities and crystal defects, undesirable variations in the physical and chemical properties are often presented in the application of natural pyrite. Therefore, synthesis of pure-phase pyrite with controllable morphology and specific facets is of great significance for their application and, thus, has attracted considerable research interest in recent years.

Over the past few decades, micro-nanopyrite crystals with various geometrical morphologies, microstructures (e.g., nanoparticles, nanowires, nanocubes, nano-octahedrons, and micro-spheres), and different sizes have been successfully synthesized using different synthetic methods, including hydrothermal methods [20–25], solvothermal synthesis [26–30], heating-up [31,32], the low-temperature aqueous method [33], sulphidation [34,35], chemical vapor deposition [36], hot-injection [31,37–39], electrochemical deposition [40], and sonosynthesis [41]. Among these fabrication methods, hydrothermal synthesis, usually conducted in an autoclave containing all precursors and a certain amount of water under high temperature and pressure, can provide excellent control over the size and morphology of pyrite and is relatively easy to implement [42]. Various Fe salts (including Fe(II) and Fe(III)), iron oxide (e.g., Fe_2O_3, Fe_3O_4), FeS, FeS_m, $(Fe(S_2CNEt_2)_3)$, $[(C_2H_5O)_2P(S)S]_3Fe$, Na_2S or H_2S, $Na_2S_2O_3$, S, and thiourea can be used as precursors to fabricate micro-nanopyrite [42]. The type of precursor, hydrothermal temperature, pH, and surfactant may significantly affect the size and morphology of prepared pyrite. In a polysulfide pathway to synthesize pyrite, the initially formed amorphous FeS can be converted to a metastable intermediate greigite (Fe_3S_4), which is then transformed to pyrite via further sulfidation [19]. However, most previous studies focused on the transformation of greigite to pyrite at low temperatures (<100 °C); little research has been conducted at hydrothermal temperatures. Thus, there is direct experimental evidence to support the hypothesis that greigite transforms to pyrite under hydrothermal conditions [1,43]. Therefore, we thought that greigite should be a critical intermediate or precursor to synthesize micro-nanopyrite, and it may help to provide a direct evidence to reveal the formation mechanism of pyrite under hydrothermal conditions. Although a lot of hydrothermal progresses has been developed in controllable synthesis of micro-nanopyrite crystals, simultaneously use of greigite as the iron precursor and thiourea (NH_2CSNH_2) as the sulfur source has not been previously attempted, and the relevant reaction mechanisms also need to be further investigated.

In order to enhance our understanding of the formation mechanism of pyrite under hydrothermal conditions and provide new insight into synthesis of pyrite crystals, nanocubic pyrite crystals with exposed (100) crystal faces were successfully synthesized in this work via a facile hydrothermal method with greigite as the iron precursor and thiourea as the sulfur source. Crystal phase compositions and morphologies were characterized by X-ray diffraction (XRD) and field emission scanning electron microscopy (FESEM), respectively. The influences of reaction conditions, such as hydrothermal time, precursor concentration, as well as type of sulfur source on composition and morphology of nanocubic pyrite crystals, were also systematically investigated.

2. Materials and Methods

2.1. Chemicals and Materials

Sodium sulfide ($Na_2S \cdot 9H_2O$), ferrous sulfate ($FeSO_4 \cdot 7H_2O$), thiourea (NH_2CSNH_2), sulfur (S) powder, hydrochloric acid (HCl) (36~38 wt%), and anhydrous ethanol (C_2H_5OH) were purchased from Sinopharm Chemical Reagent Co., Ltd. (Shanghai, China). All chemicals were of analytical or guaranteed reagent grade and were used without further purification. Deionized water with a resistivity of 18.2 MΩ·cm was obtained from a Millipore synergy UV system (Millipore Corporation, Molsheim, Alsace, France), and deoxygenated deionized water was used in all experiments.

2.2. Synthesis of the Greigite Precursor

Greigite precursor was prepared by a refluxing method. In a typical process, 100 mL of 0.1 mol/L $FeSO_4 \cdot 7H_2O$ solution was added to 100 mL of 0.1 mol/L $Na_2S \cdot 9H_2O$ boiling solution in a three-neck flask under vigorous stirring and degassing with Ar. A black precipitate appeared immediately. Subsequently, the solution was refluxed at 100 °C for 3 h and then naturally cooled to room temperature. The resulting black precipitation was centrifuged for 5 min (4000 rpm) and then washed with deoxygenated deionized water three times and anhydrous ethanol three times. Finally, the product was dried at 40 °C for 24 h in a vacuum oven (DZF-6050, Shanghai Shenxian Thermostatic Equipment, Shanghai, China).

2.3. Synthesis of Pyrite Nanocubic Crystals

Pyrite samples were prepared using a hydrothermal approach. In a typical process, 0.88 g Fe_3S_4 precursor was added to 60 mL of thiourea solution with different concentrations under vigorous stirring at room temperature. The mixture was then transferred to a 100 mL Teflon-line autoclave, and then hydrothermally treated at 200 °C in an oven for different times. After the hydrothermal reaction, the resulting black product was collected by centrifugation. Then it was thoroughly washed with 30 mL of 1 mol/L HCl, 1 mol/L of Na_2S boiling solution, and deoxygenated deionized water three times, respectively, and anhydrous ethanol ten times. Finally, the obtained black product was dried at 40 °C for 6 h in a vacuum oven and then stored in a glove box with an anaerobic environment. The crystal phase composition and crystallinity were determined by X-ray diffraction (XRD, Empyrean, PANalytical B.V, Almelo, The Netherlands) operating with Cu-Kα radiation. The surface morphologies of the resulting samples were characterized using scanning electron microscopy (SEM, Scios, FEI Company, Hillsboro, OR, USA) with an acceleration voltage of 30.0 kV. To investigate the effect of sulfur source on the structure and morphology of pyrite, thiourea was replaced by a boiling mixture of Na_2S and S powder.

3. Results and Discussion

3.1. Characterization of the Greigite Fe_3S_4 Precursor

The crystal phase composition and crystallinity of the as-prepared greigite product were characterized by X-ray diffraction (XRD), and the corresponding XRD pattern is shown in Figure 1. All diffraction peaks were consistent with cubic greigite (Fe_3S_4, JCPDS no. 16-0713) with $Fd\bar{3}m$ space group. The dominant characteristic diffraction peaks of the XRD pattern at $2\theta = 15.48°$, $25.43°$, $29.96°$, $36.34°$, $47.81°$, and $52.36°$ were attributed to the (111), (220), (311), (400), (511), and (440) planes of greigite (Fe_3S_4), respectively. Greigite was found to be the only crystalline phase in the sample, and the intermediate (e.g., FeS) was undetectable by XRD. The characteristic diffraction peaks appeared weak and broad, implying low purity and crystallinity of greigite precursor. Similar with its oxide analogue magnetite Fe_3O_4, greigite is an inverse spinel with a general formula AB_2X_4, where A is nominally a, Fe^{2+}, B is a, Fe^{3+} and X is a S^{2-}. Greigite shows typical ferromagnetic behavior because of the presence of unpaired electrons (data not shown here) [1]. The unit cell of greigite is face-centered cubic. It consists of 32 close packed atoms of sulfur and 24 atoms of iron with a Fe^{2+}/Fe^{3+} ratio of 1:2, where Fe^{2+} atoms occur in tetrahedral sites, and mixed Fe^{2+} and Fe^{3+} occur in the octahedral sites coordinated with S^{2-} [44]. Greigite could be formed directly through the rapid autoxidation reaction of pre-existing mackinawite (FeS_m, formed after mixing Fe(II) and S(-II) solutions) in anoxic H_2O at temperatures somewhat above 70 °C [1].

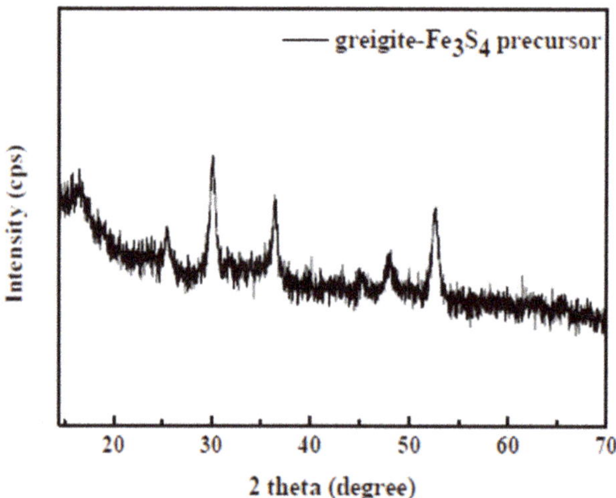

Figure 1. X-ray diffraction (XRD) pattern of the obtained greigite precursor.

3.2. Characterization of Pyrite Nanocubes Prepared with Thiourea as the Sulfur Source

3.2.1. Effect of Hydrothermal Time

Figure 2a shows the XRD patterns for the samples prepared at different hydrothermal times in 80 mmol/L thiourea solution. It can be seen that the XRD patterns of all samples exhibited similar characteristics. The diffraction peaks at 2θ = 28.51°, 33.08°, 37.11°, 40.78°, 47.41°, 56.28°, 59.02°, 61.69°, 64.28°, 76.60°, and 78.96° were well attributed to the (111), (200), (210), (211), (220), (311), (222), (023), (321), (331), and (420) planes of cubic pyrite (FeS_2) (JCPDS card no. 42-1340) with a space group of $Pa\overline{3}$, respectively [45–48]. Pyrite was found to be the only crystalline phase in all prepared samples at different hydrothermal times. No other impure phases (e.g., greigite (Fe_3S_4), pyrrhotite ($Fe_{(1-x)}S$), marcasite (FeS_2), or other impurities) appeared, implying high phase-purity of pyrite in these samples. The characteristic diffraction peaks were sharp and narrow, confirming good crystallization of these samples. Additionally, the diffraction peak intensities and the peak widths of pyrite were found to be almost identical, with increases of hydrothermal times from 24 to 168 h, indicating that all the greigite could be thoroughly transformed to pyrite within 24 h via hydrothermal method with thiourea as the precursor. Further prolonging the hydrothermal time had no significant effect on the crystal phase composition and crystallization of the product.

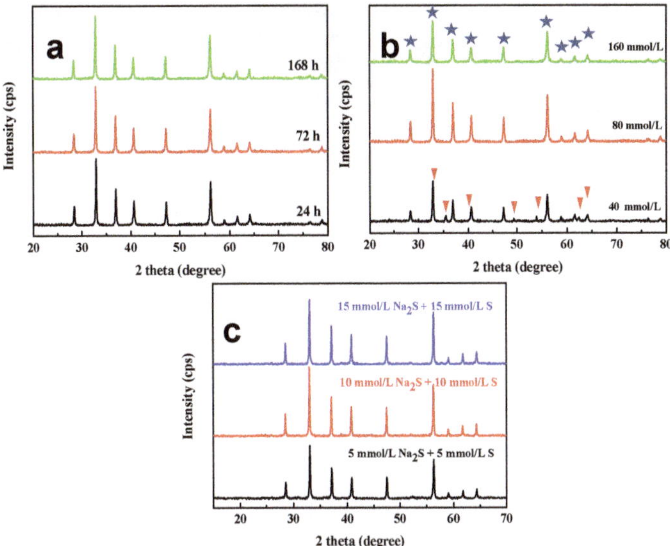

Figure 2. (**a**) XRD patterns of pyrite samples prepared at different hydrothermal times in 80 mmol/L thiourea solution; (**b**) XRD patterns of samples prepared at 24 h with different concentrations of thiourea, ★: pyrite, ▼: hematite; and (**c**) XRD patterns of the as-synthesized pyrite at 24 h with different concentrations of Na$_2$S and S powder as the sulfur source.

To investigate the effect of hydrothermal time on the morphologies and microstructures of the samples, the as-synthesized samples were characterized by SEM. As shown in Figure 3, pyrite in all samples obtained with a hydrothermal time from 24 to 168 h exhibited individual nanocubic shapes with smooth surfaces, implying that screw-dislocation or two-dimensional nucleation growth might be the dominant growth mechanisms of pyrite crystal [43]. The edge length of the nanocubes was approximately 100–200 nm. With hydrothermal time increased from 24 to 168 h, the morphologies and microstructures of nanocubic pyrite in all samples showed no noticeable changes. The average edge length of the nanocubes was statistically analyzed to be about 115, 112, and 120 nm for the samples prepared at 24, 72, and 168 h, respectively, indicating that the crystal growth process could be completed within 24 h. The EDX result (Figure 4) further confirmed that the pyrite was composed of Fe and S elements, with a molar ratio of S/Fe of approximately two, supporting that the prepared product was pure pyrite (FeS$_2$). Moreover, the dominant facet on nanocubic crystals pyrite was the (100) facet, indicating that the cubic structure was the most stable structure than others under our experimental conditions. In fact, the major surface crystallographic planes of pyrite included (100), (111), (210), and (110), and their surface energies were 1.06, 1.40, 1.50, and 1.68 J/m^2, respectively, suggesting that the surface energy of the (100) was the lowest. Thus, the (100) crystal face was considered to be the most stable [2,42]. Accordingly, crystal growth along the (110), (210), and (111) directions was expected to be more favorable than the (100) direction, leading to the formation of a cubic structure with exposed (100) crystal face, which was consistent with observations of the most common naturally occurring and synthetic pyrites [49].

Figure 3. Scanning electron microscopy (SEM) images of the as-synthesized pyrite nanocrystals with 80 mmol/L thiourea as the sulfur source at 24 (**a**,**b**), 72 (**c**,**d**), and 168 h (**e**,**f**), respectively.

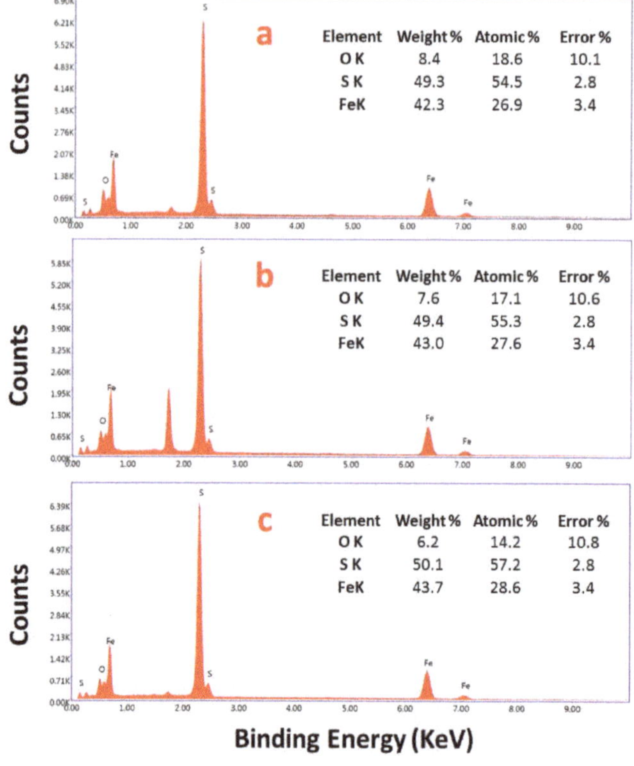

Figure 4. Energy-dispersive X-ray spectrum (EDS) of the as-synthesized pyrite nanocrystals with 80 mmol/L thiourea as the sulfur source at 24 (**a**), 72 (**b**), and 168 h (**c**), respectively.

3.2.2. Effect of the Thiourea Concentration

To further investigate the effect of thiourea on the pyrite prepared from greigite, hydrothermal experiments were conducted with different concentrations of thiourea at 24 h. Figure 2b shows the XRD patterns of the samples prepared with different thiourea concentrations. It was clearly seen that the phase compositional characteristics of the prepared product seemed to depend essentially on the thiourea concentration. When the thiourea concentration was 40 mmol/L, besides the clearly observed characteristic diffraction peaks of pyrite, several other diffraction peaks could be seen, which were attributed to a small amount of hematite (Fe_2O_3) formed under this condition. Specifically, the diffraction peaks agreed well with hexagonal scalenohedral hematite (JCPDS no. 33-0664). The peaks at $2\theta = 33.15°$, $35.61°$, $40.85°$, $49.48°$, $54.09°$, $62.45°$, and $63.99°$ can be indexed to the (104), (110), (113), (024), (116), (214), and (300) planes of hematite, respectively. This phenomenon suggests that a low concentration of thiourea was not sufficient to completely convert the greigite into pyrite. Increasing the concentration of thiourea resulted in an increase of pyrite content, simultaneously accompanied by a decrease in the amount of hematite in the as-synthesized samples. When the concentration of thiourea increased to 80 mmol/L, the XRD diffraction peaks of hematite vanished completely, and pyrite was found to be the only crystalline phase in the samples, suggesting that all greigite could be transformed into pyrite with the addition of 80 mmol/L thiourea. The XRD peak intensities of pyrite became remarkably stronger, and the peak widths became narrower, indicating an increase in the crystal size and crystallinity of pyrite. However, with a further increase in thiourea concentration to 160 mmol/L, the XRD peak intensities of pyrite decreased significantly, suggesting that high concentration of thiourea might inhibit the growth of pyrite crystals.

Figure 5 shows the effect of the thiourea concentration on the morphologies of the as-prepared product at 24 h. The morphologies and sizes of the obtained samples were also strongly dependent on the thiourea concentration used in the reaction. When the concentration of thiourea was 40 mmol/L (Figure 5a,b), the SEM images revealed that the resulting product consisted of nanocubic pyrite particles (with an average particle size of ~100 nm) with exposed (100) crystal faces and a fraction of hematite nanocrystals with polyhedral bipyramid (marked with yellow dotted circle) with the particle size of 100–200 nm. This indicated that a low concentration of thiourea resulted in conversion of a fraction of greigite into hematite Fe_2O_3 because of the shortage of the sulfur source. The EDX result (Figure 6a) also validated that a significantly higher content of O existed in this sample, owing to the presence of hematite Fe_2O_3. When the concentration of thiourea was increased to 80 mmol/L (Figure 5c,d), it was clearly seen that only nanocubic pyrite could be observed in the SEM images, and the content of O in the sample remarkably decreased (Figure 6b). The average particle size of nanocubic pyrite was found to increase significantly to ~115 nm, indicating an increase of thiourea content was conducive to the growth of nanocubic pyrite crystal, which was consistent with the XRD results. However, with the further increase of the thiourea concentration to 160 mmol/L (Figure 5e,f), no obvious morphological changes could be observed, except for a slight decrease of the particle size of nanocubic pyrite (with the average particle size of ~110 nm), and the individual nanocubic pyrite appeared more uniform. This may be attributed an overly high concentration of thiourea in the solution that could inhibit the growth of pyrite crystals, implying that the morphologies and microstructures of the as-synthesized product could be controlled by the concentration of thiourea used as the sulfur source.

Figure 5. SEM images of the as-synthesized pyrite nanocrystals at 24 h with different concentrations of thiourea as the sulfur source: (**a**,**b**) 40 mmol/L, (**c**,**d**) 80 mmol/L, and (**e**,**f**) 160 mmol/L.

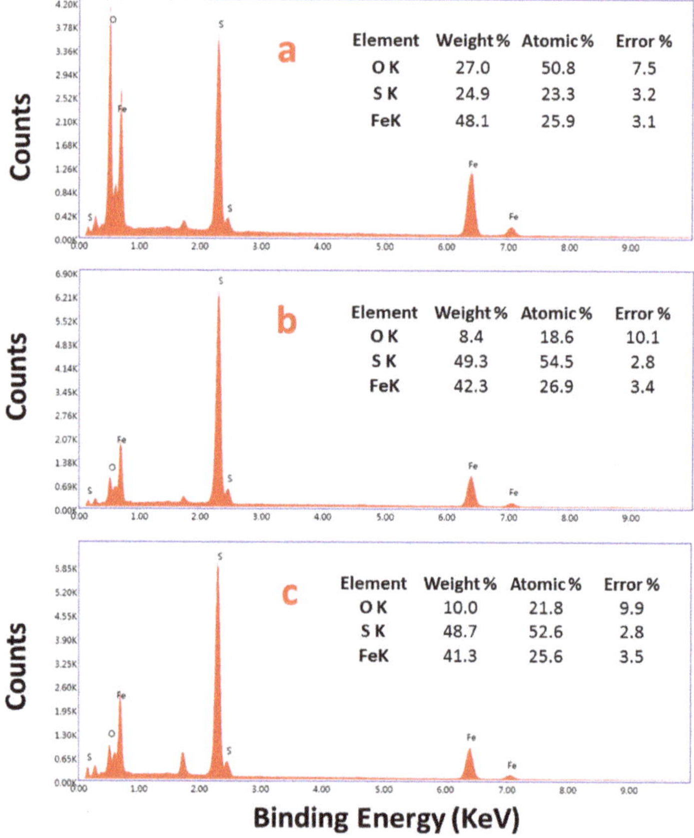

Figure 6. Energy-dispersive X-ray spectrum of the as-synthesized pyrite nanocrystals at 24 h with different concentrations of thiourea as the sulfur source: (**a**) 40 mmol/L, (**b**) 80 mmol/L, and (**c**) 160 mmol/L.

3.3. Characterization of Pyrite Prepared with a Mixture of Na₂S and S Powder as the Sulfur Source

To further investigate the influence of the type of sulfur source on the synthesis of pyrite, a mixture of Na_2S and S powder was used as the sulfur source to react with greigite precursor. Figure 2c displays the XRD patterns of the as-synthesized pyrite at 24 h with different concentrations of Na_2S and S powder as the sulfur source. As shown, the diffraction peaks of all obtained samples were very similar and could be well-indexed as pyrite. No other characteristic diffraction peaks could be found, indicating that the mixture of 5 mmol/L Na_2S and 5 mmol/L S powder was enough to transform greigite into pure pyrite, and the concentrations of Na_2S and S powder has no remarkable influence on the crystal phase of these products. Furthermore, with increasing concentration of Na_2S and S powder, the diffraction peak intensities of pyrite significantly increased in pace with the narrowing of the peak width, meaning an increase in the crystallinity of pyrite, which implied that a high concentration of mixture of Na_2S and S powder may be conducive to the growth of pyrite crystal particles.

The effect of the concentration of Na_2S and S powder on the morphologies and microstructures of the resulting pyrite with Na_2S and S powder as sulfur source was also investigated by SEM (shown in Figure 7). Compared with pyrite nanocrystals synthesized with thiourea as the sulfur source, the SEM images revealed that the as-synthesized iron pyrite crystals with Na_2S and S powder as the sulfur source had completely different morphologies, in which tetragonal, orthorhombic, cubic, and irregular pyrite crystal particles with sizes of 0.1–1 μm could be observed in all prepared samples. The co-existence of various morphologies and microstructures of pyrite indicated that the sulfur source had a significant influence on pyrite morphologies. When the concentrations of both Na_2S and S powder were increased from 5 to 15 mmol/L, no obvious morphological changes in pyrite were observed. The discrepancy of morphologies may be attributed to the different nucleation-growth process originating from the sulfur source.

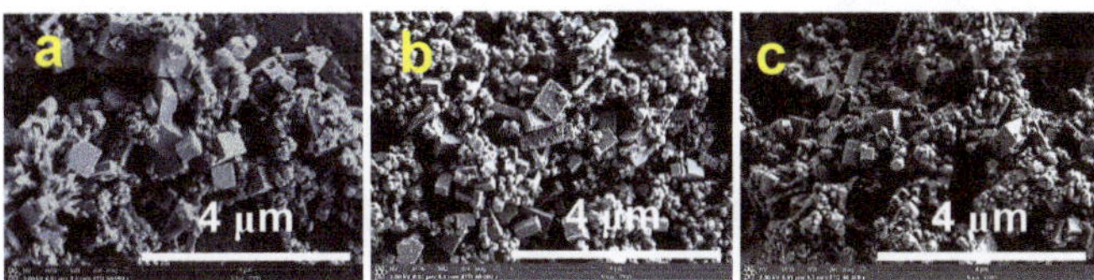

Figure 7. SEM images of the as-synthesized pyrite crystals at 24 h with Na_2S and S powder as the sulfur source: (**a**) 5 mmol/L Na_2S + 5 mmol/L S powder, (**b**) 10 mmol/L Na_2S + 10 mmol/L S powder, and (**c**) 15 mmol/L Na_2S + 15 mmol/L S powder.

3.4. Mechanism of Pyrite Formation

Based on the above results, it can be speculated that pyritization of greigite is a dissolution-precipitation process. The dissolution of greigite in water can produce ferrous (Fe^{2+}), S^{2-}, and zero-valence sulfur S [1]. Thiourea (NH_2CSNH_2) can react with H_2O to form small molecules of H_2S, CO_2, and NH_3 at high temperature. Dissolved H_2S in aqueous solution can be deprotonated to obtain HS^- and S^{2-}. S^{2-} can react with S to form aqueous polysulfide species S_{x+1}^{2-} at high temperature. A high concentration of S^{2-} is beneficial to the formation of pyrite. The Fe^{2+} species originating from dissolved greigite can further react with S_{x+1}^{2-} or H_2S to produce FeS_2 nuclei when the solution is supersaturated (with respect to pyrite). The consumption of Fe^{2+} in the solution would further facilitate the dissolution of greigite in water, leading to the growth of FeS_2 nuclei to form pyrite crystals. The reaction process and relative mechanism between greigite and thiourea as well as the mixture of Na_2S and S powder can be described as below [49]:

$$Fe_3S_{4(s)} \leftrightarrow 3Fe^{2+} + 3S^{2-} + S; \tag{1}$$

$$NH_2CSNH_2 + 2H_2O \rightarrow H_2S + CO_2 + 2NH_3; \tag{2}$$

$$H_2S \leftrightarrow H^+ + HS^-; \tag{3}$$

$$2HS^- \leftrightarrow 2H^+ + 2S^{2-}; \tag{4}$$

$$xS + S^{2-} \rightarrow S_{x+1}^{2-}; \tag{5}$$

$$Fe^{2+} + S_{x+1}^{2-} \rightarrow FeS_2 + (x-1)S; \tag{6}$$

$$Fe^{2+} + H_2O + H_2S_{(aq)}^- \leftrightarrow FeSH^+ + H_3O^+; \tag{7}$$

$$FeSH^+ + H_3O^+ + S^{2-} \leftrightarrow FeS_2 + H_2O + H_2. \tag{8}$$

The growth of nanocubic pyrite via a hydrothermal method with greigite as the iron precursor and thiourea as the sulfur source has not been reported to date. Furthermore, as deduced from our results, it can be concluded that the sulfur precursor plays a crucial role in controlling the crystal phase composition, morphology, and size of the as-prepared pyrite with greigite as the iron precursor. Using thiourea as the sulfur source, the obtained product had a narrower size distribution as well as smaller crystal sizes compared with that of mixture of Na_2S and S powder as the sulfur source. Additionally, the morphology and particle size of pyrite crystals greatly depended on the rates of dissolution of greigite, the formation pathway of pyrite, and the initial supersaturation with respect to pyrite. When the mixture of Na_2S and S powder was used as the sulfur source, pyrite was formed through the reaction between Fe^{2+} and polysulfide via Equation (6) [1]. If thiourea was used as the sulfur source, the formation of pyrite was achieved mainly through Equations (7) and (8) [49]. When initial supersaturation (with respect to pyrite) is low, large-sized pyrite crystals will form, owing to that only a few pyrite nuclei can be formed. In contrast, fine-grained pyrite crystals would be the dominant product at high initial supersaturation (with respect to pyrite) because of a significant increase in the number of pyrite nuclei [43]. Therefore, in our experimental system, a much smaller particle size and a narrower size distribution of pyrite crystals were obtained with thiourea as the sulfur source, compared with that of the mixture of Na_2S and S powder as the sulfur source, because of the significantly higher concentration of the sulfur source for thiourea, which enhanced supersaturation (with respect to pyrite) and consequently facilitated the formation of numerous nuclei of pyrite crystals [49].

4. Conclusions

In summary, nanocubic pyrite crystals with exposed (100) crystal faces and edge lengths of approximately 100–200 nm were fabricated via a facile hydrothermal method with greigite as the iron precursor and thiourea as the sulfur source. The Fe_3S_4 precursor was prepared by a refluxing method via mixing $FeSO_4 \cdot 7H_2O$ and $Na_2S \cdot 9H_2O$ at 100 °C for 3 h. All the greigite could be thoroughly transformed to pyrite within 24 h via the hydrothermal method with thiourea as a precursor, and further prolonging the hydrothermal times had no significant effect on the crystal phase composition and crystallization of product. By varying the hydrothermal time from 24 to 168 h, the morphologies and microstructures of nanocubic pyrite showed no noticeable changes. When the concentration of thiourea was 40 mmol/L, pyrite as well as hematite could be observed, which was ascribed to the low concentration of thiourea insufficient to completely convert the greigite into pyrite. Moreover, when the mixture of Na_2S and S powder was used as the sulfur source, tetragonal, orthorhombic, cubic, and irregular pyrite crystal particles with sizes of 100 nm–1 μm could be observed in prepared samples, indicating that the sulfur source had significant influence on the morphologies and microstructures of pyrite. The results obtained in this study may provide new insights for synthesizing nanocubic pyrite crystal with controllable morphology and help to better understand the formation mechanism of pyrite. Our findings can also provide new insights into the formation environments and pathways of nanocubic pyrite under hydrothermal conditions.

Author Contributions: Q.W. proposed the research direction and guided the project; X.N. performed the experiment and wrote this manuscript; S.L., M.Y., and P.Z. analyzed the data and discussed the results; Z.Q. and W.Y. provided some useful suggestions.

Funding: This work was financially supported by the Chinese Academy of Sciences ("Hundred Talents Program"), the National Natural Science Foundation of China (41173074, 41872046), Guizhou Provincial Science and Technology projects ([2018]1172, [2019]1460), the Opening Fund of State Key Laboratory of Ore Deposit Geochemistry (201602), and the Training Project of Zunyi Municipal Innovative Talents Team (201539).

Acknowledgments: In this section you can acknowledge any support given which is not covered by the author contribution or funding sections. This may include administrative and technical support, or donations in kind (e.g., materials used for experiments).

Conflicts of Interest: The authors declare no conflict of interest.

References

1. Rickard, D.; Luther, G.W. Chemistry of iron sulfides. *Chem. Rev.* **2007**, *107*, 514–562. [CrossRef]
2. Vaughan, D.J.; Corkhill, C.L. Mineralogy of Sulfides. *Elements* **2017**, *13*, 81–87. [CrossRef]
3. Huang, F.; Gao, W.; Gao, S.; Meng, L.; Zhang, Z.; Yan, Y.; Ren, Y.; Li, Y.; Liu, K.; Xing, M.; et al. Morphology Evolution of Nano-Micron Pyrite: A Review. *J. Nanosci. Nanotechnol.* **2017**, *17*, 5980–5995. [CrossRef]
4. Large, R.R.; Danyushevsky, L.; Hollit, C.; Maslennikov, V.; Meffre, S.; Gilbert, S.; Bull, S.; Scott, R.; Emsbo, P.; Thomas, H.; et al. Foster, Gold and Trace Element Zonation in Pyrite Using a Laser Imaging Technique: Implications for the Timing of Gold in Orogenic and Carlin-Style Sediment-Hosted Deposits. *Econ. Geol.* **2009**, *104*, 635–668. [CrossRef]
5. Reich, M.; Kesler, S.E.; Utsunomiya, S.; Palenik, C.S.; Chryssoulis, S.L.; Ewing, R.C. Solubility of gold in arsenian pyrite. *Geochim. Cosmochim. Acta* **2005**, *69*, 2781–2796. [CrossRef]
6. Deditius, A.P.; Reich, M.; Kesler, S.E.; Utsunomiya, S.; Chryssoulis, S.L.; Walshe, J.; Ewing, R.C. The coupled geochemistry of Au and As in pyrite from hydrothermal ore deposits. *Geochim. Cosmochim. Acta* **2014**, *140*, 644–670. [CrossRef]
7. Liu, W.; Wang, Y.; Ai, Z.; Zhang, L. Hydrothermal Synthesis of FeS_2 as a High-Efficiency Fenton Reagent to Degrade Alachlor via Superoxide-Mediated Fe(II)/Fe(III) Cycle. *ACS Appl. Mater. Interface* **2015**, *7*, 28534–28544. [CrossRef] [PubMed]
8. Zhang, P.; Yuan, S.H. Production of hydroxyl radicals from abiotic oxidation of pyrite by oxygen under circumneutral conditions in the presence of low-molecular-weight organic acids. *Geochim. Cosmochim. Acta* **2017**, *218*, 153–166. [CrossRef]
9. Lee, W.; Batchelor, B. Abiotic, reductive dechlorination of chlorinated ethylenes by iron-bearing soil minerals. 2. Green rust. *Environ. Sci. Technol.* **2002**, *36*, 5348–5354. [CrossRef] [PubMed]
10. Lee, W.; Batchelor, B. Abiotic reductive dechlorination of chlorinated ethylenes by iron-bearing soil minerals. 1. Pyrite and magnetite. *Environ. Sci. Technol.* **2002**, *36*, 5147–5154. [CrossRef]
11. Pham, H.T.; Kitsuneduka, M.; Hara, J.; Suto, K.; Inoue, C. Trichloroethylene transformation by natural mineral pyrite: The deciding role of oxygen. *Environ. Sci. Technol.* **2008**, *42*, 7470–7475. [CrossRef]
12. Shukla, S.; Xing, G.; Ge, H.; Prabhakar, R.R.; Mathew, S.; Su, Z.; Nalla, V.; Venkatesan, T.; Mathews, N.; Sritharan, T.; et al. Origin of Photocarrier Losses in Iron Pyrite (FeS_2) Nanocubes. *ACS Nano* **2016**, *10*, 4431–4440. [CrossRef] [PubMed]
13. Khalid, S.; Ahmed, E.; Khan, Y.; Riaz, K.N.; Malik, M.A. Nanocrystalline Pyrite for Photovoltaic Applications. *Chemistryselect* **2018**, *3*, 6488–6524. [CrossRef]
14. Cui, M.M.; Johannesson, K.H. Comparison of tungstate and tetrathiotungstate adsorption onto pyrite. *Chem. Geol.* **2017**, *464*, 57–68. [CrossRef]
15. Kantar, C.; Ari, C.; Keskin, S. Comparison of different chelating agents to enhance reductive Cr(VI) removal by pyrite treatment procedure. *Water Res.* **2015**, *76*, 66–75. [CrossRef]
16. Kirkeminde, A.; Ren, S. Thermodynamic control of iron pyrite nanocrystal synthesis with high photoactivity and stability. *J. Mater. Chem. A* **2013**, *1*, 49–54. [CrossRef]
17. Kuo, T.R.; Chen, W.T.; Liao, H.J.; Yang, Y.H.; Yen, H.C.; Liao, T.W.; Wen, C.Y.; Lee, Y.C.; Chen, C.C.; Wang, D.Y. Improving Hydrogen Evolution Activity of Earth-Abundant Cobalt-Doped Iron Pyrite Catalysts by Surface Modification with Phosphide. *Small* **2017**, *13*, 1603356. [CrossRef]
18. Zhu, Y.; Fan, X.; Suo, L.; Luo, C.; Gao, T.; Wang, C. Electrospun FeS_2@Carbon Fiber Electrode as a High Energy Density Cathode for Rechargeable Lithium Batteries. *ACS Nano* **2016**, *10*, 1529–1538. [CrossRef]
19. Igarashi, K.; Yamamura, Y.; Kuwabara, T. Natural synthesis of bioactive greigite by solid-gas reactions. *Geochim. Cosmochim. Acta* **2016**, *191*, 47–57. [CrossRef]

20. Yang, Z.; Liu, X.; Feng, X.; Cui, Y.; Yang, X. Hydrothermal synthesized micro/nano-sized pyrite used as cathode material to improve the electrochemical performance of thermal battery. *J. Appl. Electrochem.* **2014**, *44*, 1075–1080. [CrossRef]

21. Wang, D.; Wang, Q.; Wang, T. Shape controlled growth of pyrite FeS_2 crystallites via a polymer-assisted hydrothermal route. *Crystengcomm* **2010**, *12*, 3797–3805. [CrossRef]

22. Golsheikh, A.M.; Huang, N.M.; Lim, H.N.; Chia, C.H.; Harrison, I.; Muhamad, M.R. One-pot hydrothermal synthesis and characterization of FeS_2 (pyrite)/graphene nanocomposite. *Chem. Eng. J.* **2013**, *218*, 276–284. [CrossRef]

23. Wu, R.; Zheng, Y.F.; Zhang, X.G.; Sun, Y.F.; Xu, J.B.; Jian, J.K. Hydrothermal synthesis and crystal structure of pyrite. *J. Cryst. Growth* **2004**, *266*, 523–527. [CrossRef]

24. Zou, J.; Gao, J. Preparation of Nanosize Iron Pyrite FeS_2 and Its Properties. In *Materials Research, Pts 1 and 2*; Gu, Z.W., Han, Y.F., Pan, F.H., Wang, X.T., Weng, D., Zhou, S.X., Eds.; Trans Tech Publications Ltd.: Zurich, Switzerland, 2009; pp. 459–462.

25. Luo, S.; Nie, X.; Yang, M.; Fu, Y.; Zeng, P.; Wan, Q. Sorption of Differently Charged Gold Nanoparticles on Synthetic Pyrite. *Minerals* **2018**, *8*, 428. [CrossRef]

26. Liu, W.L.; Rui, X.H.; Tan, H.T.; Xu, C.; Yan, Q.Y.; Hng, H.H. Solvothermal synthesis of pyrite FeS_2 nanocubes and their superior high rate lithium storage properties. *RSC Adv.* **2014**, *4*, 48770–48776. [CrossRef]

27. Kar, S.; Chaudhuri, S. Solvothermal synthesis of nanocrystalline FeS_2 with different morphologies. *Chem. Phys. Lett.* **2004**, *398*, 22–26. [CrossRef]

28. Zhu, L.; Richardson, B.J.; Yu, Q. Controlled colloidal synthesis of iron pyrite FeS_2 nanorods and quasi-cubic nanocrystal agglomerates. *Nanoscale* **2014**, *6*, 1029–1037. [CrossRef] [PubMed]

29. Jiang, F.; Peckler, L.T.; Muscat, A.J. Phase Pure Pyrite FeS_2 Nanocubes Synthesized Using Oleylamine as Ligand, Solvent, and Reductant. *Cryst. Growth Des.* **2015**, *15*, 3565–3572. [CrossRef]

30. Yu, B.B.; Zhang, X.; Jiang, Y.; Liu, J.; Gu, L.; Hu, J.S.; Wan, L.J. Solvent-Induced Oriented Attachment Growth of Air-Stable PhasePure Pyrite FeS_2 Nanocrystals. *J. Am. Chem. Soc.* **2015**, *137*, 2211–2214. [CrossRef]

31. Macpherson, H.A.; Stoldt, C.R. Iron Pyrite Nanocubes: Size and Shape Considerations for Photovoltaic Application. *ACS Nano* **2012**, *6*, 8940–8949. [CrossRef]

32. Bai, Y.X.; Yeom, J.; Yang, M.; Cha, S.H.; Sun, K.; Kotov, N.A. Universal Synthesis of Single-Phase Pyrite FeS_2 Nanoparticles, Nanowires, and Nanosheets. *J. Phys. Chem. C* **2013**, *117*, 2567–2573. [CrossRef]

33. Morin, G.; Noël, V.; Menguy, N.; Brest, J.; Baptiste, B.; Tharaud, M.; Ona-Nguema, G.; Ikogou, M.; Viollier, E.; Juillot, F. Nickel accelerates pyrite nucleation at ambient temperature. *Geochem. Perspect. Lett.* **2017**, *5*, 6–11. [CrossRef]

34. Caban-Acevedo, M.; Liang, D.; Chew, K.S.; DeGrave, J.P.; Kaiser, N.S.; Jin, S. Synthesis, Characterization, and Variable Range Hopping Transport of Pyrite (FeS_2) Nanorods, Nanobelts, and Nanoplates. *ACS Nano* **2013**, *7*, 1731–1739. [CrossRef] [PubMed]

35. Wang, M.D.; Xing, C.C.; Cao, K.; Zhang, L.; Liu, J.B.; Meng, L. Template-directed synthesis of pyrite (FeS_2) nanorod arrays with an enhanced photoresponse. *J. Mater. Chem. A* **2014**, *2*, 9496–9505. [CrossRef]

36. Berry, N.; Cheng, M.; Perkins, C.L.; Limpinsel, M.; Hemminger, J.C.; Law, M. Atmospheric-Pressure Chemical Vapor Deposition of Iron Pyrite Thin Films. *Adv. Energy Mater.* **2012**, *2*, 1124–1135. [CrossRef]

37. Hsiao, S.-C.; Hsu, C.-M.; Chen, S.-Y.; Perng, Y.-H.; Chueh, Y.-L.; Chen, L.-J.; Chou, L.-H. Facile synthesis and characterization of high temperature phase FeS_2 pyrite nanocrystals. *Mater. Lett.* **2012**, *75*, 152–154. [CrossRef]

38. Yuan, B.; Luan, W.; Tu, S.-T. One-step synthesis of cubic FeS_2 and flower-like $FeSe_2$ particles by a solvothermal reduction process. *Dalton Trans.* **2012**, *41*, 772–776. [CrossRef]

39. Xu, J.; Xue, H.; Yang, X.; Wei, H.; Li, W.; Li, Z.; Zhang, W.; Lee, C.-S. Synthesis of Honeycomb-like Mesoporous Pyrite FeS_2 Microspheres as Efficient Counter Electrode in Quantum Dots Sensitized Solar Cells. *Small* **2014**, *10*, 4754–4759. [CrossRef]

40. Chakraborty, B.; Show, B.; Jana, S.; Mitra, B.C.; Maji, S.K.; Adhikary, B.; Mukherjee, N.; Mondal, A. Cathodic and anodic deposition of FeS_2 thin films and their application in electrochemical reduction and amperometric sensing of H_2O_2. *Electrochim. Acta* **2013**, *94*, 7–15. [CrossRef]

41. Khabbaz, M.; Entezari, M.H. Degradation of Diclofenac by sonosynthesis of pyrite nanoparticles. *J. Environ. Manag.* **2017**, *187*, 416–423. [CrossRef] [PubMed]

42. Xian, H.Y.; Zhu, J.X.; Liang, X.L.; He, H.P. Morphology controllable syntheses of micro- and nano-iron pyrite mono- and poly-crystals: A review. *RSC Adv.* **2016**, *6*, 31988–31999. [CrossRef]

43. Wang, Q.; Morse, J.W. Laboratory simulation of pyrite formation in anoxic sediments. *ACS Symp. Ser.* **1995**, *612*, 206–223. [CrossRef]

44. Li, Q.; Wei, Q.; Zuo, W.; Huang, L.; Luo, W.; An, Q.; Pelenovich, V.O.; Mai, L.; Zhang, Q. Greigite Fe_3S_4 as a new anode material for high-performance sodium-ion batteries. *Chem. Sci.* **2017**, *8*, 160–164. [CrossRef] [PubMed]

45. Blanchard, M.; Alfredsson, M.; Brodholt, J.; Wright, K.; Catlow, C.R.A. Arsenic incorporation into FeS_2 pyrite and its influence on dissolution: A DFT study. *Geochim. Cosmochim. Acta* **2007**, *71*, 624–630. [CrossRef]

46. Deditius, A.P.; Utsunomiya, S.; Renock, D.; Ewing, R.C.; Ramana, C.V.; Becker, U.; Kesler, S.E. A proposed new type of arsenian pyrite: Composition, nanostructure and geological significance. *Geochim. Cosmochim. Acta* **2008**, *72*, 2919–2933. [CrossRef]

47. Abraitis, P.K.; Pattrick, R.A.D.; Vaughan, D.J. Variations in the compositional, textural and electrical properties of natural pyrite: A review. *Int. J. Miner. Process.* **2004**, *74*, 41–59. [CrossRef]

48. Xian, H.Y.; Zhu, J.X.; Tang, H.M.; Liang, X.L.; He, H.P.; Xi, Y.F. Aggregative growth of quasi-octahedral iron pyrite mesocrystals in a polyol solution through oriented attachment. *Crystengcomm* **2016**, *18*, 8823–8828. [CrossRef]

49. Liang, Y.X.; Bai, P.P.; Zhou, J.; Wang, T.Q.; Luo, B.W.; Zheng, S.Q. An efficient precursor to synthesize various FeS_2 nanostructures via a simple hydrothermal synthesis method. *Crystengcomm* **2016**, *18*, 6262–6271. [CrossRef]

Article

Predominantly Ferruginous Conditions in South China during the Marinoan Glaciation: Insight from REE Geochemistry of the Syn-glacial Dolostone from the Nantuo Formation in Guizhou Province, China

Shangyi Gu [1,*], Yong Fu [1] and Jianxi Long [2]

[1] College of Resource and Environmental Engineering, Guizhou University, Guiyang 550025, China; byez1225@126.com

[2] Guizhou Geological Survey, Bureau of Geology and Mineral Exploration and Development of Guizhou Province, Guiyang 550081, China; with-163@163.com

* Correspondence: sygu@gzu.edu.cn; Tel.: +86-0851-83627126

Received: 23 April 2019; Accepted: 3 June 2019; Published: 5 June 2019

Abstract: The Neoproterozoic Era witnessed two low-latitude glaciations, which exerted a fundamental influence on ocean–atmosphere redox conditions and biogeochemical cycling. Climate models and palaeobiological evidence support the belief that open waters provided oases for life that survived snowball Earth glaciations, yet independent geochemical evidence for marine redox conditions during the Marinoan glaciation remains scarce owing to the apparent lack of primary marine precipitates. In this study, we explore variability in rare earth elements (REEs) and trace metal concentrations in dolostone samples of the Cryogenian Nantuo Formation taken from a drill core in South China. Petrological evidence suggests that the dolostone in the Nantuo Formation was formed in near-shore waters. All the examined dolostone samples featured significant enrichment of manganese (345–10,890 ppm, average 3488 ppm) and middle rare earth elements (MREEs) (Bell Shape Index: 1.43–2.16, average 1.76) after being normalized to Post-Archean Australian Shale (PAAS). Most dolostone samples showed slight to no negative Ce anomalies (Ce*/Ce 0.53–1.30, average 0.95), as well as positive Eu anomalies (Eu*/Eu 1.77–3.28, average 1.95). This finding suggests that the dolostone samples were deposited from suboxic to iron-enriched and anoxic waters. Although total REE concentrations correlated positively with Th concentrations in dolostone samples, MREE-enriched PAAS-normalized patterns preclude the conclusion that REEs were largely introduced by terrestrial contamination. Rather, we interpret the correlation between REEs and Th as an indication that the former were transported by colloids and nanoparticles in meltwaters. Taken together, we propose that anoxic and ferruginous water columns dominated in South China during the Marinoan glaciation with a thin oxic/suboxic layer restricted to coastal waters. The extreme anoxic and ferruginous conditions prevailing in the Cryogenian would have provided a baseline for subsequent transient Ediacaran ocean oxygenation and life evolution.

Keywords: Nantuo Formation; syn-glacial dolostone; rare earth elements; ferruginous conditions; South China

---

1. Introduction

The Neoproterozoic Era was characterized by two global Snowball Earth glaciations (the Sturtian and Marinoan glaciations), which extended to low latitudes [1]. The subsequent appearance of animal fossils and diversification of eukaryotes during the Ediacaran Period has been dubbed the Neoproterozoic Oxygenation Event [2–5]. It was initially proposed that ice covered the continents and oceans for several million years, which would have greatly reduced air sea-gas exchange [1].

As a result, atmospheric molecular oxygen would be exhausted by oxic weathering of the continents during the lengthy glaciations [6]. However, several lines of evidence have shown that this was not the case. Biomarkers and the benthic macro-algae fossil record have demonstrated that macroscopic phototrophs survived the Marinoan glaciation, with open water in coastal environments providing refuge for macro-algae [7–9]. Phototrophy, including that of eukaryotic algae, could also have persisted in meltwater and drainages on the sea glacier, producing a well oxygenated surface meltwater-based ecosystem in the equatorial zone of Snowball Earth [10–13]. Climate modeling allows for that an intermittent open water belt existed in low-latitude oceans during the Snowball Earth glaciation [14]. Redox-sensitive elements, iron speciation and nitrogen isotope data from diamictite support the existence of open marine waters and an oxygenated atmosphere during the Neoproterozoic Snowball Earth glaciations [6].

However, in contrast to the Sturtian glaciation, which was characterized by predominantly ferruginous conditions in marine chemistry, as indicated by the wide distribution of syn-glacial iron formation, few direct proxies have been seen to date for the redox condition of the Snowball Earth oceans [6]. A major problem in exploring seawater chemistry during the Marinoan glaciation has been the apparent lack of primary marine precipitates (such as carbonate), in which trace elements and stable isotopes (e.g., carbon and strontium) have been thought to provide a record of the seawater composition of ancient oceans [15,16]. Rare earth element (REE) compositions in carbonates are widely used to trace REE sources and oxygen levels in ancient ocean [17–22]. Here, we use rare earth and other trace elements found in bedded dolostone from the Nantuo Formation in South China to improve our knowledge of redox conditions during the Marinoan glaciation.

The dolostone samples analyzed in this study are derived from a drill core taken from the Nantuo Formation in the Yangtze Block of South China, specifically from a paleo-geographic slope environment in the Nanhua Basin (Figure 1). The sampled drill core is in Daotuo, about 14 km west of Songtao County, in eastern Guizhou Province. The Nantuo Formation (~654–635 Ma) is correlated with the Marinoan global glaciation, according to radiometric dating [23–25]. Its thickness varies from several metres, along the shallow shelf, to more than 2000 m, in the deep waters [25,26]. The Nantuo Formation is conformably underlain by the Datangpo Formation and is capped by dolostone 3–6 m thick at the base of the Doushantuo Formation [27].

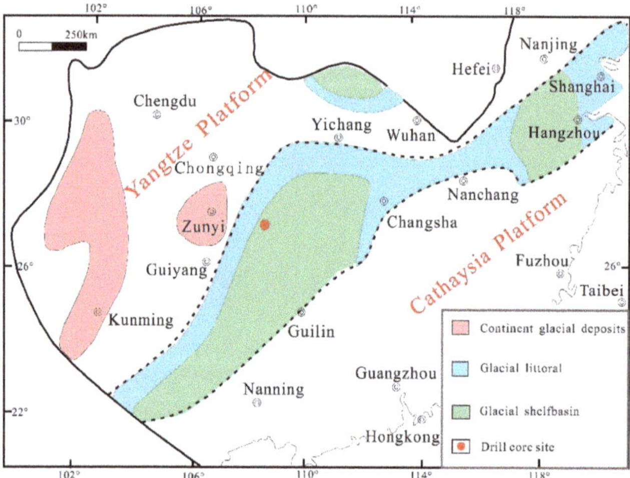

Figure 1. Paleogeographic map of the Nantuo Formation in South China (Modified after reference [27]). The red dot shows the locality of the drill core investigated in this study.

The thickness of the Nantuo Formation in the drill core is about 245.5 m, which is divided into four sections according to its lithofacies characteristics. The lower section is a coarsening-upward sequence composed of massive diamictites, representing the first episode of glaciation, when ice sheets reached the deep-water environment. The middle section is dominated by fine-grained siliciclastic lithofacies interbedded with a layer of laminated dolostone, which is a non-glacial marine deposit—an indication of glaciation's waning. The reappearance of massive diamictites and alternating coarse- and fine-grained lithofacies in the upper Nantuo Formation forms the third section, indicating a dynamic ice-grounding line. The fourth section, at the top of the Nantuo Formation, is characterized by alternating massive sandstone-stratified sandstone and laminated siltstone, signifying the end of glaciation. Figure 2 provides detailed information on the studied drill core.

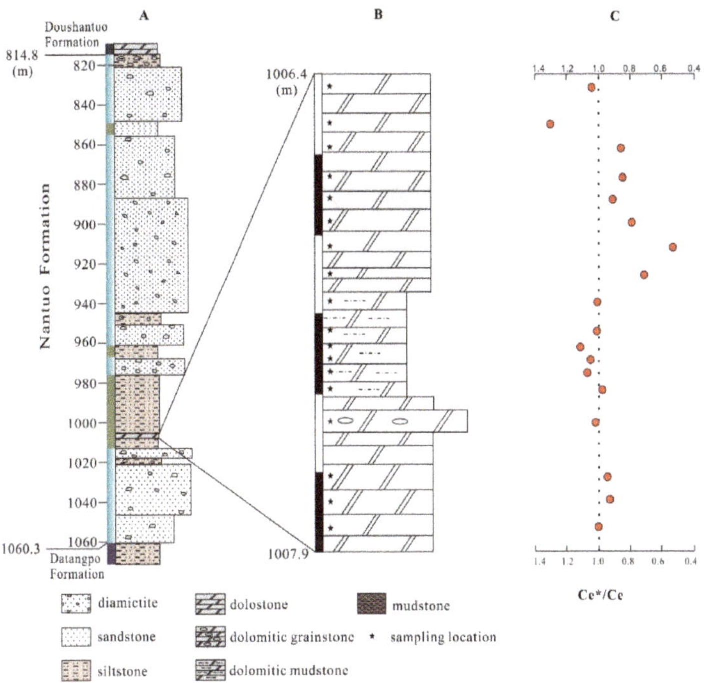

Figure 2. Stratigraphic log (**A,B**) and the composite Ce anomaly chemostratigraphy (**C**) of the Nantuo Formation from the Daotuo drill core in Guizhou Province, South China.

The dolostone bed intercalated in the mudstones is about 1.50 m thick and features micritic dolostones at its lower portion (ZK201-1–ZK201-4), calcareous mudstones at its middle (ZK201-5–ZK201-10), and micritic dolostones at its upper portion (ZK201-11–ZK201-18). Figure 3 depicts representative petrographic features, which can be seen under transmitted plane-polarized light. Massive grey dolomite with a micritic to microspar fabric was observed for all studied samples. Dolomitic peloids were found only in ZK201-4, where they varied in shape and size. These sedimentary structures indicate that the Nantuo Formation dolostones were deposited from a near-shore environment.

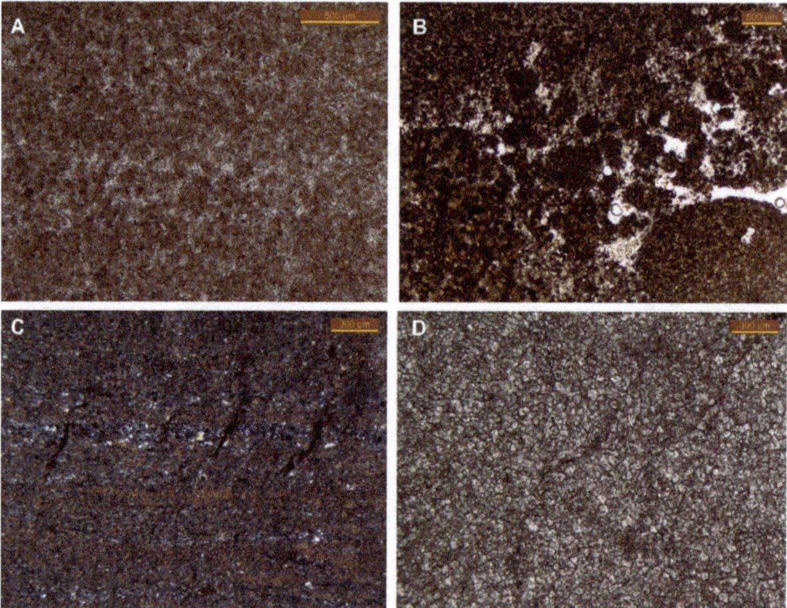

Figure 3. Transmitted plane-polarized light photomicrographs of petrographic thin sections showing micro-textures of Nantuo Formation dolostones from the Daotuo drill core in Guizhou Province, South China. (**A**) Micritic dolostone in sample ZK201-1. (**B**) Dolostone with peloidal fabrics in sample ZK201-4. (**C**) Calcareous mudstone in sample ZK201-8. (**D**) Micritic dolostone in sample ZK201-18.

2. Materials and Methods

Before sampling, a detailed petrographic analysis (including thin-section observation under polarizing microscope and staining with Alizarin Red S) was conducted on each sample. Scanning electron microscopy (SEM) using energy-dispersive (EDS) (SEM-EDS) analysis of selected samples confirmed that dolomite was the main chemical deposit in the samples. In all, 18 samples were obtained for chemical analysis by drilling with a 1 mm tungsten carbide micro-drill to ensure good spatial resolution. During sample drilling, visible veins and carbonate spars were avoided to ensure that samples represented their primitive deposits. About 500 mg was drilled for each sample. Because the samples were not pure dolostones, and included variable contents of siliciclastics and sulphides, the traditional $HF + HNO_3 + HClO_4$ dissolution protocol would inevitably introduce varying amounts of non-carbonate components. In this study, we adopted the dissolution protocol used by [16]: About 100 mg of a split of powder was caused to dissolve in 0.5 mol/L sub-boiling distilled acetic acid at room temperature for 4 h, and the resulting supernatant solution, after being centrifuged, was dried. Afterward, the dried solution was evaporated to dryness, refluxed with HNO_3, dissolved in 2% HNO_3 to 50 mL and then analyzed by Thermo Fisher inductively coupled plasma mass spectroscopy (ICP-MS) at the National Research Center for Geoanalysis at the Chinese Academy of Geological Sciences, Beijing, China. Reference materials GSR-1, GSR-2, and GSR-3, which were used to monitor the analytical reproducibility, obtained results matching those expected for the REE and other trace element contents with analytical errors less than 5%.

The predicted normalized concentrations of La*, Ce*, Eu*, as well as La, Ce and Eu anomalies, were calculated by Post-Archean Australian Shale [28] (PAAS) normalization according to [29,30], using the following equations:

$$La^* = Pr_N{}^*(Pr_N/Nd_N)^2 \tag{1}$$

$$Ce^* = Pr_N^*/(Pr_N/Nd_N) \tag{2}$$

$$Eu^* = Sm_N^*(Sm_N/Tb_N)^{1/2} \tag{3}$$

$$La/La^* = La_N/(Pr_N^3/Nd_N^2) \tag{4}$$

$$Ce/Ce^* = Ce_N/(Pr_N^2/Nd_N) \tag{5}$$

$$Eu/Eu^* = Eu_N/(Sm_N^{2}*Tb_N)^{1/3} \tag{6}$$

where REE_N represents the REE value after PAAS normalization. We used Equation (5) to calculate the Ce anomalies in this paper. Additionally, $(Nd/Yb)_N$ and $(Pr/Yb)_N$ were calculated as indicating relative light/heavy REE ratios (LREE/HREE). $(Pr/Tb)_N$ and $(Tb/Yb)_N$ are calculated as parameters indicative of light/middle REE (LREE/MREE) and middle/heavy REE (MREE/HREE), respectively. The bell-shaped index (BSI) is calculated using the following equation to represent the enrichment of the MREE [30]:

$$BSI = \frac{2(Sm_N + Gd_N + Dy_N)/3}{(La_N + Pr_N + Nd_N)/3 + (Ho_N + Er_N + Tm_N + Yb_N + Lu_N)/5}. \tag{7}$$

It should be noted that anomalies are calculated on a linear scale in the majority of the literature, with Ce/Ce* and Pr/Pr*, for example, expressed as follows [21,31]:

$$Ce/Ce^* = 2Ce_N/(Pr_N + La_N) \tag{8}$$

$$Pr/Pr^* = 2Pr_N/(Ce_N + Nd_N). \tag{9}$$

We inspected our data using both Equations (5) and (8) and found no critical differences.

3. Results

Tables 1 and 2 present the concentrations of the measured REEs and trace elements Th and Mn and calculated REE parameters, respectively. Figures 2C and 4 depict Ce anomaly variations and normalized patterns of REE distribution relative to the PAAS, respectively. All measured samples have consistently low REE concentrations of 3.30–29.77 ppm, high Mn concentrations of 345–10,890 ppm, and variable Th concentrations of 0.098–2.032 ppm. Despite differences in REE concentrations, these samples share similar REE patterns: (1) enrichment of MREE over both LREE ($[Pr/Tb]_N$ from 0.26 to 0.81) and HREE ($[Tb/Yb]_N$ from 1.30 to 2.99) with BSI from 1.43 to 2.16; (2) minor negative to slightly positive La anomalies with La/La* from 0.82 to 1.37; (3) slightly negative to slightly positive Ce anomalies, with Ce/Ce* from 0.53 to 1.30; (4) markedly positive Eu anomalies with Eu*/Eu from 1.77 to 3.28; (5) positive Gd anomalies with Gd/Gd* from 1.39 to 2.08; and (6) minor LREE depletion to moderate LREE enrichment, with $(Nd/Yb)_N$ from 0.55 to 2.09 and $(Pr/Yb)_N$ from 0.41 to 1.85.

Modern seawater composition is characterized by: (1) positive La anomaly; (2) negative Ce anomaly; (3) positive Gd anomaly; (4) supra-chondritic ratio (ppm) for Y/Ho (i.e. >40); and (5) LREE and MREE depletion, relative to HREE ($La_N/Yb_N < 1$; $Gd_N/Yb_N < 1$) [32]. The PAAS-normalized REE patterns for the carbonate samples are dramatically different from modern marine signatures as shown by the presence of little to no heavy REE enrichment ($[Nd/Yb]_N$ 0.55–2.09, average 1.21), convex shape (middle REE enriched, BSI 1.50–2.16, average 1.76), the presence of slightly positive Eu anomalies, an average of 1.95, and little to no negative Ce anomalies (Ce*/Ce 0.53–1.30, average 0.95). The PAAS-normalized REE patterns of the dolostones share similarities with the ferruginous water in modern acidic ferruginous lakes [32].

Table 1. Compositions of trace element Th, Mn, and rare earth elements (REE) from the carbonate rocks of the Nantuo Formation.

Sample No.	Stratigraphic Height (m)	Th	Mn	La	Ce	Pr	Nd	Sm	Eu	Gd	Tb	Dy	Ho	Er	Tm	Yb	Lu	TREE
ZK201-1	0.07	0.389	2476	2.180	5.412	0.722	3.335	0.750	0.230	1.230	0.160	0.814	0.145	0.391	0.037	0.236	0.031	15.67
ZK201-2	0.14	0.531	3081	2.768	6.881	1.005	4.730	1.088	0.300	1.728	0.212	1.098	0.199	0.530	0.062	0.324	0.047	20.97
ZK201-3	0.22	0.130	3054	0.957	2.219	0.289	1.219	0.184	0.130	0.416	0.044	0.258	0.043	0.110	0.013	0.091	0.011	5.98
ZK201-4	0.37	0.357	4577	1.326	3.093	0.373	1.582	0.217	0.149	0.467	0.055	0.292	0.053	0.120	0.017	0.093	0.012	7.85
ZK201-5	0.47	0.795	6025	2.597	5.636	0.805	3.856	0.803	0.229	1.321	0.172	0.958	0.179	0.517	0.063	0.380	0.053	17.57
ZK201-6	0.52	0.170	1774	0.536	1.246	0.154	0.697	0.071	0.043	0.241	0.024	0.143	0.025	0.059	0.007	0.044	0.006	3.30
ZK201-7	0.55	0.485	7507	2.673	5.134	0.586	2.414	0.396	0.158	0.697	0.085	0.482	0.100	0.266	0.029	0.176	0.024	13.22
ZK201-8	0.59	0.364	7367	1.568	3.371	0.369	1.546	0.221	0.176	0.525	0.059	0.365	0.072	0.219	0.026	0.154	0.025	8.70
ZK201-9	0.63	0.576	10890	3.736	7.690	1.025	4.774	0.980	0.530	1.502	0.200	1.027	0.219	0.642	0.074	0.415	0.056	22.87
ZK201-10	0.72	0.236	2052	1.078	2.993	0.488	2.769	0.862	0.223	1.256	0.162	0.921	0.165	0.442	0.054	0.344	0.047	11.80
ZK201-11	0.80	2.032	1010	5.637	8.626	1.610	7.398	1.617	0.395	2.027	0.241	1.176	0.185	0.462	0.057	0.295	0.039	29.77
ZK201-12	0.87	1.167	906	5.029	5.017	1.200	5.216	0.905	0.271	1.292	0.144	0.770	0.121	0.319	0.039	0.208	0.028	20.56
ZK201-13	0.95	0.571	3146	4.914	6.988	1.135	5.027	0.862	0.186	1.033	0.123	0.649	0.123	0.307	0.031	0.200	0.028	21.60
ZK201-14	1.01	0.398	2548	1.795	2.962	0.439	2.061	0.378	0.155	0.768	0.088	0.432	0.079	0.215	0.027	0.182	0.023	9.60
ZK201-15	1.07	0.098	345	0.767	1.670	0.219	0.850	0.098	0.062	0.259	0.027	0.157	0.024	0.060	0.007	0.067	0.007	4.27
ZK201-16	1.16	0.201	482	1.688	3.631	0.517	2.181	0.332	0.158	0.640	0.071	0.394	0.061	0.141	0.019	0.104	0.013	9.95
ZK201-17	1.23	0.319	4421	1.700	4.488	0.514	2.646	0.709	0.291	1.279	0.157	0.976	0.179	0.502	0.067	0.402	0.054	13.97
ZK201-18	1.33	0.205	1116	0.705	1.646	0.215	1.005	0.181	0.096	0.357	0.049	0.291	0.054	0.181	0.021	0.139	0.018	4.96

Table 2. Post-Archean Australian (PAAS)-normalized REE parameters calculated for the carbonates of the Nantuo Formation.

Sample No.	La/La*	Ce/Ce*	Pr/Pr*	Eu/Eu*	BSI	$(Dy/Sm)_N$	$(Nd/Yb)_N$	$(Pr/Yb)_N$	$(Pr/Tb)_N$	$(Tb/Yb)_N$
ZK201-1	1.01	1.00	0.98	1.37	2.06	1.29	1.18	0.98	0.39	2.48
ZK201-2	0.96	0.93	1.01	1.27	2.05	1.20	1.21	0.99	0.41	2.39
ZK201-3	0.93	0.94	1.02	3.02	1.81	1.66	1.11	1.01	0.57	1.78
ZK201-4	1.01	1.02	0.99	2.89	1.66	1.60	1.41	1.28	0.59	2.16
ZK201-5	1.16	0.97	0.99	1.27	1.72	1.42	0.84	0.68	0.41	1.65
ZK201-6	1.13	1.07	0.96	2.31	1.75	2.39	1.32	1.12	0.56	1.99
ZK201-7	1.22	1.04	0.98	1.77	1.48	1.44	1.14	1.06	0.60	1.76
ZK201-8	1.17	1.11	0.95	3.28	1.43	1.96	0.84	0.77	0.55	1.40
ZK201-9	1.24	1.01	0.98	2.44	1.60	1.24	0.96	0.79	0.45	1.77
ZK201-10	1.11	1.01	0.93	1.20	2.16	1.27	0.67	0.45	0.26	1.73
ZK201-11	1.16	0.71	1.12	1.22	2.04	0.86	2.08	1.74	0.58	2.99
ZK201-12	1.24	0.53	1.25	1.47	1.72	1.01	2.09	1.85	0.73	2.53
ZK201-13	1.33	0.79	1.09	1.10	1.54	0.89	2.09	1.81	0.81	2.25
ZK201-14	1.41	0.91	1.02	1.77	1.80	1.35	0.94	0.77	0.44	1.77
ZK201-15	0.82	0.85	1.08	2.59	1.63	1.89	1.05	1.04	0.71	1.48
ZK201-16	0.91	0.86	1.06	2.10	1.87	1.41	1.75	1.59	0.63	2.52
ZK201-17	1.37	1.30	0.87	1.80	1.87	1.63	0.55	0.41	0.28	1.43
ZK201-18	1.13	1.04	0.97	2.16	1.50	1.91	0.60	0.49	0.38	1.30

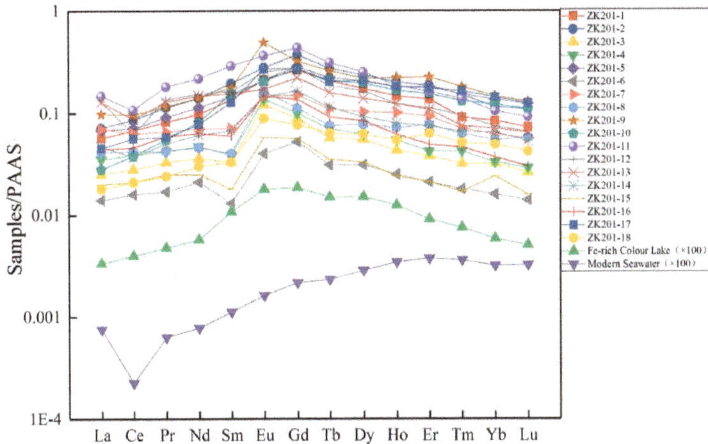

Figure 4. Comparisons of PAAS-normalized REE patterns among carbonate components from the Nantuo Formation Fe-rich Colour Lake [32], and modern oxic seawater [33].

4. Discussion

4.1. Primary Deposit Evaluation

It is essential to understand the origin of the studied carbonates in the Nantuo Formation. If the bedded dolostone was detrital in origin and derived from a pre-glacial dolostone through glacial erosion, the geochemical proxies of the carbonates would reflect the environment before the Marinoan glaciation. In the original Snowball Earth hypothesis, seawater would be undersaturated with carbonate minerals owing to the shutdown of alkalinity input from rivers, and dolomite could not be formed, given the very low Mg/Ca ratio and temperature at the freezing point [1,34]. However, Gernon et al. argued that Ca^{2+} and Mg^{2+} supersaturation caused by widespread hyaloclastite alteration allowed the deposition of dolostone during and after the global glaciations [35]. In fact, in the Nanhua Basin, pre-glaciation successions were dominated by sub-greenschist facies metamorphic siliciclastic rocks, volcanic rocks and marine siliciclastic rocks [36]. The reported dolostone was the Sturtian cap carbonate in the Nanhua Basin [37], and no carbonate clasts were found in diamictites from the Nantuo Formation. Stable carbon isotopic compositions ($\delta^{13}C_{carb}$) of the Sturtian cap carbonate ranged from −4.52‰ to +3.23‰, much higher than in analytical dolostone samples (average $\delta^{13}C_{carb}$ −7.1‰, data not shown in this article) in the Nantuo Formation. Accordingly, dolostone in the Nantuo Formation is unlikely to have been formed by re-deposition from pre-glacial dolostone. Instead, the bedded dolostone in the Nantuo Formation was deposited primarily during the Marinoan glaciation.

4.2. Evaluation of Contamination by Terrestrial Matter

Marine carbonates generally have low concentrations of REE, meaning that contamination by non-carbonates such as silicates and oxides can influence carbonate REE contents and patterns. To reduce contamination introduced by non-carbonates, analytical protocols called for the use of dilute acetic acid to dissolve the dolostone samples, according to reference [16]. Samples were then checked using geochemical contamination factors (high field strength elements (HFSE), e.g., Th) to detect silicate and oxide contamination [15,16].

REEs and other HFSEs, such as thorium, have very low concentrations in modern river waters and seawaters [38–41]. As a result, carbonate REE concentrations and patterns can be modified by terrestrial particulate matters (i.e., shale). The most widely used method for evaluating the contamination of shale is to investigate the correlation between REE concentrations and immobile elements such as Th, Zr, and Sc, which are abundant in shale (11.8 ppm, 210 ppm, and 16 ppm, respectively, in PAAS) but appear in significantly lower concentrations in low-temperature waters [41]. In addition, Y/Ho ratios in Phanerozoic carbonates have also been used to inspect shale contamination because pure marine carbonates have higher Y/Ho ratios than shale [16]. However, Wallace et al. argued that Y/Ho values in Neoproterozoic carbonates are lower than in modern sea waters and could be controlled by redox processes rather than resulting from silicate contamination [22]. Because Zr is mainly hosted by zircon which cannot be attacked by acetic acid and considering that Sc can be hosted by sulphide minerals [16], we preferred to use Th to track shale contamination. Terrestrial particulate matter contamination would systematically increase REE concentrations with Th contents. Good correlations exist between REE concentrations and the Th contents of the carbonate samples (Figure 5), suggesting that the carbonates were affected by contamination with the shale. However, if the shale contamination dominated REE concentrations of the carbonate samples, PAAS-normalized REE patterns should be flat, shale-like patterns with no element anomalies. This indicates that the shale contamination in the dolostones is negligible. Accordingly, we proposed that the PAAS-normalized REE patterns reflected the marine environment in which carbonates were deposited. An alternative and reasonable explanation for positive correlation of concentrations of REE and Th in the dolostone samples is the incorporation of near-shore colloidal and nano-particles related to Fe oxyhydroxide anoxic dissolution and scavenging. This is supported by the observed lack of a negative Ce anomaly in most of the studied carbonates. REE studies for meltwater and glacially-fed waters provide an analogue of the Cryogenian environment. The suggestion has been made that modern glacial meltwater could deliver significant amounts of particulate iron into surrounding coastal oceans in which iron oxide nanoparticulates occur as coatings or at the edges of sheet aluminosilicate minerals [42,43]. Nanoparticles and colloids serve as significant parts of REE and HFSE in modern meltwater and glacially-fed waters [44]. Large percentages of iron and manganese oxide nanoparticles and colloids and associated REE are removed by scavenging during coastal zone mixing [45] and subsequently incorporated into carbonate minerals as a result of reductive dissolution of iron and manganese oxyhydroxides once they enter anoxic water columns, as discussed in the following.

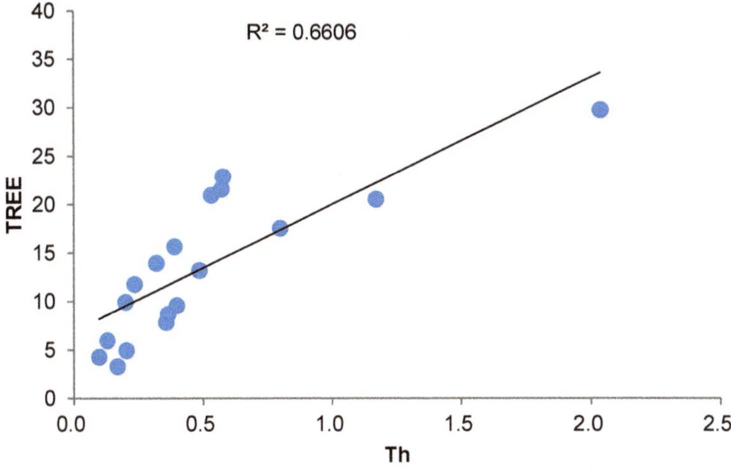

Figure 5. Positive correlation between the Th concentration and total REE concentration (TREE representing total REE concentration).

4.3. The Influence of Diagenesis

Owing to the large REE partition coefficients between the carbonate and seawater [46–49], early diagenetic models suggest that calcite REE patterns are less sensitive to post-deposition diagenetic alteration (including meteoric alteration and dolomitization) [38]. Well constrained variable diagenetic studies of REE distributions in ancient and modern limestones suggest that REE patterns, including Ce anomalies, are highly stable during meteoric processes, marine burial diagenesis, and dolomitization [47,50–53]. Taken together, carbonate REE patterns and Ce anomalies may record the seawater signatures of primary deposition.

4.4. Marine Redox Conditions during Nantuo Glaciation

The enrichment of MREE, the presence of weakly negative to slightly positive Ce anomalies, and the presence of pronounced positive Eu anomalies in the carbonates of the Nantuo Formation indicate an anoxic and iron-rich marine chemistry during the Nantuo glaciation. The reasons for the proposal are as follows: First, a lack of heavy REE enrichment is found in Precambrian anoxic marine basins to form iron formations [54]. Second, middle REE-enriched patterns are found in anoxic and/or ferruginous waters [32,55] and fossil biogenic apatite [56]. Because the analytical samples are carbonates, with no apatite observed under the microscope, middle REE enrichment patterns should be attributed to an anoxic and ferruginous environment, an interpretation supported by the elevated Mn concentrations in the samples (345–10,890 ppm), derived from anoxic dissolution of manganese oxides [55], a process likely to have occurred in oceans during the Marinoan glaciation. In modern anoxic waters with dissolved Mn and with no dissolved Fe present, PAAS-normalized REE showed linear patterns without MREE enrichment, whereas a distinctive MREE bulge-type pattern developed in anoxic waters with dissolved Mn and dissolved Fe [55].

Third, positive Eu anomalies in the REE patterns are interpreted as a suggestion of suboxic to anoxic ocean waters. A positive Eu anomaly is a typical feature of high-temperature hydrothermal alteration of ocean crust [57]. In contrast to the modern oxic marine environment, in which Ce negative anomalies are related to Ce (III) and oxidized and scavenged by Fe- and Mn-oxyhydroxide input from riverine or hydrothermal sources, the preservation of positive Eu anomaly signatures in shallow marine waters is compelling evidence for dominant marine anoxia in the Cryogenian oceans [58].

Fourth, negligible to positive Ce anomalies point to anoxic conditions, wherein the carbonates were deposited during the Marinoan glaciation. Ce is a redox sensitive element that is soluble in its trivalent form like other REEs, but less soluble in its tetravalent form. As a result, Ce is preferentially removed relative its neighboring La and Pr in modern oxic oceans causing a strong negative Ce anomaly to develop [19]. Under anoxic conditions, Ce anomaly is not well developed as a result of the reductive dissolution of Fe and Mn oxides [48]. Consequently, the Ce anomalies captured in ancient shallow-water carbonates could be a record of basin-scale redox conditions [22]. Pr/Pr* was used to distinguish real Ce anomalies from apparent ones, causing an overabundance of La relative to Ce [20]. If a sample has a Pr/Pr* value between 0.95 and 1.05, it is identified as a false Ce anomaly. Ce anomalies are much closer to unity, as indicated by samples ZK201-1–ZK201-10 and suggested shallow anoxic conditions (Figure 6.). Ce/Ce* values ranged from 0.53 to 0.87 (real Ce anomalies as identified by Pr/Pr*) in samples ZK201-11 to ZK201-16, indicating a transient oxygenation stage in shallow water. At the end of the carbonate deposition, the shallow water returned to anoxic conditions, as indicated by the slightly positive Ce anomaly seen in samples ZK201-17 and ZK201-18.

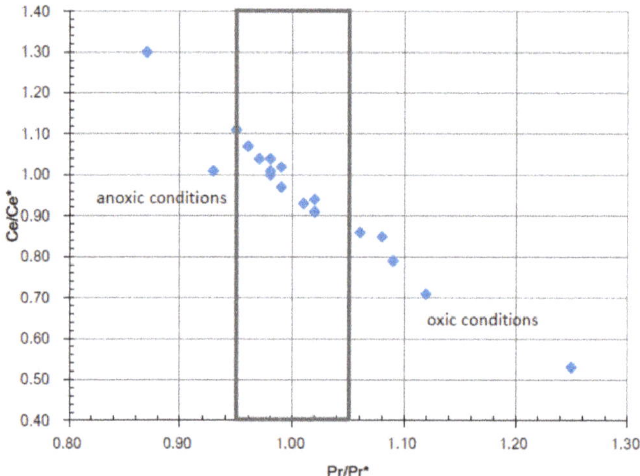

Figure 6. The cross-plot of Ce/Ce* vs. Pr/Pr* for carbonates from the Nantuo Formation.

Taken together, all this suggests that an almost completely anoxic water column dominated, with transient oxic conditions, during the Marinoan glaciation in the Nanhua Basin. By combining our Ce/Ce* values in the Nantuo Formation with those in the Sturtian iron formation [59] and interglacial carbonates [60] between these two glaciations, we may construct a redox picture in the Cryogenian oceans. Globally developed iron formations within the Sturtian glacial successions show no negative Ce anomalies (Ce/Ce* average 1.05), pointing to anoxic and ferruginous conditions during the older and lengthy Neoproterozoic glaciation [59]. Near-shore carbonates in interglacial Cryogenian reef complexes in Australia have negligible to negative Ce/Ce* (average 0.92), revealing a very shallow chemocline (several metres) overlying the prevailing ferruginous water [60]—findings that have been corroborated by new REE data from central Australia [61]. The extreme ocean anoxic conditions during the Cryogenian Period were probably caused by the two Snowball Earth glaciations. In contrast, cap carbonates in the Doushantuo Formation of the Nanhua Basin had lower Ce/Ce* than in the Nantuo Formation [62], indicating an important oxygenation increase in the immediate aftermath of the Marinoan glaciation. However, a multi-proxy palaeoredox study of Ediacaran successions showed that Ediacaran oxygenation was transient [63]. We propose that prevailing anoxic oceans during the Cryogenian Period exerted a primary influence on subsequent ocean oxygenation and evolution of life after the Marinoan glaciation.

5. Conclusions

Sedimentary carbonates from the lower units of the Nantuo Formation in eastern Guizhou Province correspond stratigraphically to the chemical sediments of the Marinoan ice age. These carbonates were geochemically analyzed for their REE and other trace elements to obtain a better understanding of the oceans at the time of the Marinoan glaciation. All the Nantuo Formation carbonates consistently showed similar REE patterns characterized by obvious MREE enrichment, slightly to no negative Ce anomalies, and positive Eu anomalies.

The combination of geochemical data and petrological evidence suggests that an anoxic and ferruginous water column dominated during the Marinoan glaciation with a thin oxic/suboxic layer restricted to coastal waters. Carbonates were precipitated from near shore environments that were affected by terrestrial particulate matters. Global Marinoan glaciation was probably the main cause of these anoxic and ferruginous conditions. Extreme ocean anoxia was dominant during the Cryogenian Period and constrained subsequent Ediacaran ocean oxygenation and the evolution of life.

Author Contributions: S.G. designed and supervised the project; Y.F. performed the experiments and analyzed the data; and J.L. performed the thin-section petrology of the samples. S.G. wrote this paper.

Funding: This research was funded by the National Natural Science Foundation of China (Nos. 41663005, 41762001).

Acknowledgments: We would like to thank the editors and anonymous reviewers for their helpful comments, which greatly improved the manuscript. Great thanks are also due to Yin Qin, Zhengze An, and Wenlang Qiao for their support of the field work and sample collection.

Conflicts of Interest: The authors declare no conflict of interest.

References

1. Hoffman, P.F.; Kaufman, A.J.; Halverson, G.P.; Schrag, D.P. A Neoproterozoic snowball earth. *Science* **1998**, *281*, 1342–1346. [CrossRef] [PubMed]
2. Xiao, S.; Knoll, A.H.; Yuan, X. Miaohe phyton, a possible brown alga from the terminal Proterozoic Doushantuo formation. *China J. Paleontol.* **1998**, *72*, 1072–1086. [CrossRef]
3. Narbonne, G.M. The Ediacara biota: Neoproterozoic origin of animals and their ecosystems. *Annu. Rev. Earth Planet. Sci.* **2005**, *33*, 421–442. [CrossRef]
4. Knoll, A.H.; Javaux, E.J.; Hewitt, D.; Cohen, P. Eukaryotic organisms in Proterozoic oceans. *Philosz. Trans. R. Soc. B* **2006**, *361*, 1023–1038. [CrossRef] [PubMed]
5. Och, L.M.; Shields-Zhou, G.A. The neoproterozoic oxygenation event: Environmental perturbations and biogeochemical cycling. *Earth Sci. Rev.* **2012**, *110*, 26–57. [CrossRef]
6. Johnson, B.W.; Poulton, S.W.; Goldblatt, C. Marine oxygen production and open water supported an active nitrogen cycle during the Marinoan Snowball Earth. *Nat. Commun.* **2017**, *8*, 1316. [CrossRef] [PubMed]
7. Corsetti, F.A.; Olcott, A.N.; Bakermans, C. The biotic response to Neoproterozoic Snowball Earth. Palaeogeogr. *Palaeoclimatol. Palaeoecol.* **2006**, *232*, 114–130. [CrossRef]
8. Parfrey, L.W.; Lahr, D.J.G.; Knoll, A.H.; Katz, L.A. Estimating the timing of early eukaryotic diversification with multigene molecular clocks. *Proc. Natl. Acad. Sci. USA* **2011**, *108*, 13624–13629. [CrossRef]
9. Ye, Q.; Tong, J.; Xiao, S.; Zhu, S.; An, Z.; Tian, L.; Hu, J. The survival of benthic macroscopic phototrophs on a Neoproterozoic snowball Earth. *Geology* **2015**, *43*, 507–510. [CrossRef]
10. Vincent, W.F.; Howard-Williams, C. Life on snowball Earth. *Science* **2000**, *287*, 2421. [CrossRef]
11. Vincent, W.F.; Gibson, J.A.E.; Pienitz, V.V.; Broady, P.A.; Hamilton, P.B.; Howard-Williams, C. Ice shelf microbial ecosystems in the high Arctic and implications for life on Snowball Earth. *Naturwissenschaften* **2000**, *87*, 137–141. [CrossRef] [PubMed]
12. Hoffman, P.F. Cryoconite pans on Snowball Earth: Supraglacial oases for Cryogenian eukaryotes? *Geobiology* **2016**, *14*, 531–542. [CrossRef] [PubMed]
13. Hawes, I.; Jungblut, A.D.; Matys, E.D.; Summons, R.E. The "Dirty Ice" of the McMurdo Ice Shelf: Analogues for biological oases during the Cryogenian. *Geobiology* **2018**, *16*, 369–377. [CrossRef] [PubMed]

14. Pierrehumbert, R.T.; Abbot, D.S.; Voigt, A.; Koll, D. Climate of the Neoproterozoic. *Annu. Rev. Earth Planet. Sci.* **2011**, *39*, 417–460. [CrossRef]

15. Frimmel, H.E. Trace element distribution in Neoproterozoic carbonates as palaeoenvironmental indicator. *Chem. Geol.* **2009**, *258*, 338–353. [CrossRef]

16. Zhao, Y.Y.; Zheng, Y.F.; Chen, F.K. Trace element and strontium isotope constraints on sedimentary environment of Ediacaran carbonates in southern Anhui, South China. *Chem. Geol.* **2009**, *265*, 345–362. [CrossRef]

17. Nothdurft, L.D.; Webb, G.E.; Kamber, B.S. Rare earth element geochemistry of Late Devonian reefal carbonates, Canning Basin, Western Australia: Confirmation of seawater REE proxy in ancient limestones. *Geochim. Cosmochim. Acta* **2004**, *68*, 263–283. [CrossRef]

18. Shields, G.A.; Webb, G.E. Has the REE composition of seawater changed over geological time? *Chem. Geol.* **2004**, *204*, 103–107. [CrossRef]

19. De Baar, H.J.W.; Schjif, J.; Byrne, R.H. Solution chemistry of the rare earth elements in seawater. *Eur. J. Solid State Inorg. Chem.* **1991**, *28*, 357–373.

20. Bau, M.; Dulski, P. Comparing yttrium and rare earths in hydrothermal fluids from the Mid-Atlantic Ridge: Implications for Y and REE behaviour during near-vent mixing and for the Y/Ho ratio of Proterozoic seawater. *Chem. Geol.* **1999**, *155*, 77–90. [CrossRef]

21. Kamber, B.S.; Webb, G.E. The geochemistry of late Archaean microbial carbonate: Implications for ocean chemistry and continental erosion history. *Geochim. Cosmochim. Acta* **2001**, *65*, 2509–2525. [CrossRef]

22. Wallace, M.W.; Hood, S.; Shuster, A.; Noah, A.G.; Planavsky, J.; Ree, C.P. Oxygenation history of the Neoproterozoic to early Phanerozoic and the rise of land plants. *Earth Planet. Sci. Lett.* **2017**, *466*, 12–19. [CrossRef]

23. Condon, D.; Zhu, M.; Bowring, S.A.; Wang, W.; Yang, A.; Jin, Y. U-Pb ages from the Neoproterozoic Doushantuo Formation, China. *Science* **2005**, *308*, 95–98. [CrossRef] [PubMed]

24. Zhang, S.; Jiang, G.; Han, Y. The age of the Nantuo Formation and Nantuo glaciation in South China. *Terra Nova* **2008**, *20*, 289–294. [CrossRef]

25. Zhang, Q.R.; Chu, X.L.; Feng, L.J. Neoproterozoic glacial records in the Yangtze region, China. In *The Geological Record of Neoproterozoic Glaciations*; Arnaud, E., Halverson, G.P., Shields-Zhou, G., Eds.; The Geological Society Publishing House: Bath, UK, 2011; pp. 357–366.

26. Jiang, G.; Shi, X.; Zhang, S.; Wang, Y.; Xiao, S. Stratigraphy and paleogeography of the Ediacaran Doushantuo Formation (ca. 635–551Ma) in South China. *Gondwana Res.* **2011**, *19*, 831–849. [CrossRef]

27. Lang, X.; Chen, J.; Cui, H.; Man, L.; Huang, K.J.; Fu, Y.; Zhou, C.; Shen, B. Cyclic cold climate during the Nantuo Glaciation: Evidence from the Cryogenian Nantuo Formation in the Yangtze Block, South China. *Precambrian Res.* **2018**, *310*, 243–255. [CrossRef]

28. McLennan, S.M. Rare earth elements in sedimentary rocks: Influence of provenance and sedimentary processes. In *Reviews in Mineralogy*; Lipin, B.R., McKay, G.A., Eds.; Mineralogical Society of America: Washington, DC, USA, 1989; pp. 170–200.

29. Lawrence, M.G.; Kamber, B.S. The behavior of the rare earth elements during estuarine mixing revisited. *Mar. Chem.* **2006**, *100*, 147–161. [CrossRef]

30. Tostevin, R.; Shields, G.A.; Tarbuck, G.M.; He, T.; Clarkson, M.O.; Wood, R.A. Effective use of cerium anomalies as a redox proxy in carbonate-dominated marine settings. *Chem. Geol.* **2016**, *438*, 146–162. [CrossRef]

31. Elderfield, H.; Upstillgoddard, R.; Sholkovitz, E.R. The rare-earth elements in rivers, estuaries, and coastal seas and their significance to the composition of ocean waters. *Geochim. Cosmochim. Acta* **1990**, *54*, 971–991. [CrossRef]

32. Johannesson, K.H.; Zhou, X. Origin of middle rare earth element enrichments in acid waters of a Canadian High Arctic lake. *Geochim. Cosmochim. Acta* **1999**, *63*, 153–165. [CrossRef]

33. Bolhar, R.; Kamber, B.S.; Moorbath, S.; Fedo, C.M.; Whitehouse, M.J. Characterisation of early Archaean chemical sediments by trace element signatures. *Earth Planet. Sci. Lett.* **2004**, *222*, 43–60. [CrossRef]

34. Jansen, M.F.; Nadeau, L.P. The effect of Southern Ocean surface buoyancy loss on the deep-ocean circulation and stratification. *J. Phys. Oceanogr.* **2016**, *46*, 3455–3470. [CrossRef]

35. Gernon, T.M.; Hincks, T.K.; Tyrrell, T.; Rohling, E.J.; Palmer, M.R. Snowball Earth ocean chemistry driven by extensive ridge volcanism during Rodinia breakup. *Nat. Geosci.* **2016**, *9*, 242–248. [CrossRef]

36. Wang, J.; Li, Z. History of Neoproterozoic rift basins in South China: Implications for Rodinia break-up. *Precambrian Res.* **2003**, *122*, 141–158. [CrossRef]

37. Yu, W.; Algeo, T.J.; Du, Y.; Zhou, Q.; Wang, P.; Xu, Y.; Yuan, L.; Pan, W. Newly discovered Sturtian cap carbonate in the Nanhua Basin, South China. *Precambrian Res.* **2017**, *293*, 112–130. [CrossRef]

38. Goldstein, S.J.; Jacobsen, S.B. Rare-earth elements in river waters. *Earth Planet. Sci. Lett.* **1988**, *89*, 35–47. [CrossRef]

39. Sholkovitz, E.R. The geochemistry of rare earth elements in the Amazon River estuary. *Geochim. Cosmochim. Acta* **1993**, *57*, 2181–2190. [CrossRef]

40. Sholkovitz, E.R. The aquatic chemistry of rare earth elements in rivers and estuaries. *Aquat. Geochem.* **1995**, *1*, 1–34. [CrossRef]

41. Gaillardet, J.; Viers, J.; Dupré, B. Trace elements in river water. In *Treatise on Geochemistry*; Drever, J.I., Holland, H.D., Turekian, K.K., Eds.; Elsevier: Amsterdam, The Netherlands, 2003; Volume 5, pp. 225–272.

42. Poulton, S.W.; Raiswell, R. Chemical and physical characteristics of iron oxides in riverine and glacial meltwater sediments. *Chem. Geol.* **2005**, *218*, 203–221. [CrossRef]

43. Bhatia, M.P.; Kujawinski, E.B.; Das, S.B.; Breier, C.F.; Henderson, P.B.; Charette, M.A. Greenland meltwater as a significant and potentially bioavailable source of iron to the ocean. *Nat. Geosci.* **2013**, *6*, 274–278. [CrossRef]

44. Tepe, N.; Bau, M. Importance of nanoparticles and colloids from volcanic ash for riverine transport of trace elements to the ocean: Evidence from glacial-fed rivers after the 2010 eruption of Eyjafjallajökull Volcano, Iceland. *Sci. Total Environ.* **2014**, *488–489*, 243–251. [CrossRef] [PubMed]

45. Tepe, N.; Bau, M. Distribution of rare earth elements and other high field strength elements in glacial meltwaters and sediments from the western Greenland Ice Sheet: Evidence for different sources of particles and nanoparticles. *Chem. Geol.* **2015**, *412*, 59–68. [CrossRef]

46. Zhong, S.; Mucci, A. Partitioning of rare earth elements (REEs) between calcite and seawater solutions at 25 °C and 1 atm, and high dissolved REE concentrations. *Geochim. Cosmochim. Acta* **1995**, *59*, 443–453. [CrossRef]

47. Webb, G.E.; Kamber, B.S. Rare earth elements in Holocene reefal microbialites: A new shallow seawater proxy. *Geochim. Cosmochim. Acta* **2000**, *64*, 1557–1565. [CrossRef]

48. Kamber, B.S.; Webb, G.E.; Gallagher, M. The rare earth element signal in Archaean microbial carbonate: Information on ocean redox and biogenicity. *J. Geol. Soc. Lond.* **2014**, *171*, 745–763. [CrossRef]

49. Zhao, M.Y.; Zheng, Y.F. Marine carbonate records of terrigenous input into Paleotethyan seawater: Geochemical constraints from Carboniferous limestones. *Geochim. Cosmochim. Acta* **2014**, *141*, 508–531. [CrossRef]

50. Della Porta, G.; Webb, G.E.; McDonald, I. REE patterns of microbial carbonate and cements from Sinemurian (Lower Jurassic) siliceous sponge mounds (Djebel Bou Dahar, High Atlas, Morocco). *Chem. Geol.* **2015**, *400*, 65–86. [CrossRef]

51. Banner, J.L.; Hanson, G.N. Calculation of simultaneous isotopic and trace element variations during water-rock interaction with applications to carbonate diagenesis. *Geochim. Cosmochim. Acta* **1990**, *54*, 3123–3137. [CrossRef]

52. Webb, G.E.; Nothdurft, L.D.; Kamber, B.S.; Kloprogge, J.T.; Zhao, J.X. Rare earth element geochemistry of scleractinian coral skeleton during meteoric diagenesis: A sequence through neomorphism of aragonite to calcite. *Sedimentology* **2009**, *56*, 1433–1463. [CrossRef]

53. Liu, X.M.; Hardisty, D.S.; Lyons, T.W.; Swart, P.K. Evaluating the fidelity of the cerium paleoredox tracer during variable carbonate diagenesis on the Great Bahamas Bank. *Geochim. Cosmochim. Acta* **2019**, *248*, 25–42. [CrossRef]

54. Bau, M.; Möller, P. Rare earth element systematics of the chemically precipitated component in Early Precambrian iron formations and the evolution of the terrestrial atmosphere lithosphere system. *Geochim. Cosmochim. Acta* **1993**, *57*, 2239–2249. [CrossRef]

55. Haley, B.A.; Klinkhammer, G.P.; McManus, J. Rare earth elements in pore waters of marine sediments. *Geochim. Cosmochim. Acta* **2004**, *68*, 1265–1279. [CrossRef]

56. Auer, G.; Reuter, G.; Hauzenberger, C.A.; Piller, W.E. The impact of transport processes on rare earth element patterns in marine authigenic and biogenic phosphates. *Geochim. Cosmochim. Acta* **2017**, *235*, 140–156. [CrossRef]

57. Bau, M. Rare-earth element mobility during hydrothermal and metamorphic fluid-rock interaction and the significance of the oxidation state of europium. *Chem. Geol.* **1991**, *93*, 219–230. [CrossRef]

58. Hood, A.V.S.; Wallace, M.W. Marine cements reveal the structure of an anoxic, ferruginous Neoproterozoic ocean. *J. Geol. Soc.* **2014**, *171*, 741–744. [CrossRef]

59. Lechte, M.; Wallace, M. Sub-ice shelf ironstone deposition during the Neoproterozoic Sturtian glaciation. *Geology* **2016**, *44*, 891–894. [CrossRef]

60. Hood, A.V.S.; Wallace, M.W. Extreme ocean anoxia during the Late Cryogenian recorded in reefal carbonates of Southern Australia. *Precambrian Res.* **2015**, *261*, 96–111. [CrossRef]

61. Verdel, C.; Phelps, B.; Welsh, K. Rare earth element and $^{87}Sr/^{86}Sr$ step-leaching geochemistry of central Australian Neoproterozoic carbonate. *Precambrian Res.* **2018**, *310*, 229–242. [CrossRef]

62. Hohl, S.V.; Becker, H.; Jiang, S.Y.; Ling, H.F.; Guo, Q.; Struck, U. Geochemistry of Ediacaran cap dolostones across the Yangtze Platform, South China: Implications for diagenetic modification and seawater chemistry in the aftermath of the Marinoan glaciation. *J. Geol. Soc.* **2017**, *174*, 893–912. [CrossRef]

63. Sahoo, S.K.; Planavsky, N.J.; Jiang, G.; Kendall, B.; Owens, J.D.; Wang, X.; Shi, X.; Anbar, A.D.; Lyons, T.W. Oceanic oxygenation events in the anoxic Ediacaran ocean. *Geobiology* **2016**, *14*, 457–468. [CrossRef]

 *minerals*

Article

Geochemical Alteration and Mineralogy of Coals under the Influence of Fault Motion: A Case Study of Qi'nan Colliery, China

Hewu Liu [1,2] **and Bo Jiang** [1,2,*]

[1] Key Laboratory of Coalbed Methane Resource & Reservoir Formation Process, Ministry of Education, China University of Mining and Technology, Xuzhou 221008, China
[2] School of Resources and Earth Science, China University of Mining and Technology, Xuzhou 221116, China
* Correspondence: jiangbo@cumt.edu.cn

Received: 30 April 2019; Accepted: 25 June 2019; Published: 27 June 2019

Abstract: Geochemical characteristics of rocks in fault zones have been extensively studied, while there are limited studies on coal occurring in fault zones of underground coal mine. In this study, five coal samples were carefully collected from a reverse fault zone in Qi'nan colliery. Systematical detection methods were employed to analyze the different chemical and physical characteristics of fault-related coal samples. Through comparative analysis, the following insights are obtained. Three subdivided fault zones were classified according to the deformation characteristics of coal samples. Frictional heat and strong ductile deformation generated by fault motion led to the dissociation of phenol and carboxyl groups in coal molecules, which sharply decreased the concentrations of elements Co and Mo bound to these functional groups in zone I. The modified pore-cleat system in zone I with higher pore volume and lower permeability allowed solutions containing enriched trace elements to migrate through zone I locally. Concentrations of HREE, MREE and related elements associated with the invasive solutions showed significant positive anomalies in zone I. Precipitation and smearing of clay minerals in zone I led to poorer connectivity. Disruption and delamination of laminar clay minerals by strong compression-shear stress significantly increased the adsorption sites for related elements, especially the HREE and MREE. Nano-scale clay minerals resulting from stress-induced scaly exfoliation also enhanced the retention capability of REE in zone I.

Keywords: minerals; trace elements; major elements; fault zone; tectonically deformed coal; differentiation mechanism

1. Introduction

Geochemical and geophysical characteristics of fault rocks are primarily controlled by fault zones or fault systems [1–5]. Fault zones in rocks are generally divided into fault core (comprised of ultracataclasite), damage zone and undeformed zone (comprised of protolith) [6,7]. Geochemical properties of fault rocks are altered in different ways. During faulting deformation, faults acting as major conduits for fluid migration lead to strong fluid–rock interactions [8]. Fluid–rock interactions in fault zones not only heal macro and micro fracture networks, but also lead to elemental variation and weathering of the fault core [9,10]. Most of the elements in fault core are depleted compared with those in the damage zone, which is ascribed to a stronger weathering process in the fault core [11]. Redistribution of elements and minerals in different subdivided zones could also result from the pressure solution of soluble minerals and diffusive mass transfer [12–14]. The formation of clay minerals in fault gouge is closely associated with the dehydration process and fluid–rock interactions in fault zones [9].

The permeability of fault rocks is both temporally and spatially heterogeneous. The permeability controls the migration of fluid in different subdivided zones, which is important for the chemical alteration of fault rocks [15–17]. A cyclic permeability evolution of fault zones occurs during the process of faulting deformation [8]. During faulting deformation, fractures and cataclasis within the fault core significantly increase the permeability of rocks, which is beneficial to the fluid migration [18,19]. After faulting deformation, the fault core sealed by mineral cementation and diffusive mass transfer has lower permeability [20–23]. Strong shear stress on fault planes leads to smearing of clay minerals into thin films (even nano-sized clay films), which decreases the permeability of fault-related coal as well [24,25]. As a summary, permeability evolution, fluid migration, and chemical alteration in fault zones are auxiliary to each other in a faulting deformation process.

Similarly, faults in coal seams are related to tectonic stress concentration, which could strongly alter intact coals into deformed ones [26]. In other words, fault structures control the local distribution of tectonically deformed coals (TDCs). TDCs are defined as coals superimposed by tectonic stress [27]. The chemical structure (specifically referring to the organic structure), physical structure and optical characteristics of TDCs are significantly affected by tectonic deformation [27–29]. Strong deformation (especially ductile deformation) on coal significantly increases the total volume and specific surface area of pore structures [30–36]. Moreover, it has been reported that there is early metamorphism of coal molecular structures in TDCs [37,38]. Side chains in TDCs including aliphatic structures and functional groups are disassociated by tectonic stress, which is followed by aromatic structures being collaged and rearranged (especially in ductile deformed coal) [39,40]. Furthermore, relevant studies have shown that coal molecular structures around faults are altered by strong deformation and frictional heat [41–43]. On the other hand, insufficient attention has been paid to the alteration of inorganic components in TDCs, particularly in fault-related coals. Thus, a study on deformed coal around fault structures could further advance the understanding of accumulation and dissipation mechanisms of elements under the influence of tectonic structures.

Elements in coal have long been investigated by geologists for preserved geological information, release of toxic elements during utilization process, and the potential economic significance of specific valuable elements [44–49]. The distribution of elements and minerals in coal seams is influenced by both syngenetic and epigenetic geological factors, for instance, the syngenetic inputs of terrigenous debris, mesotrophic anoxic conditions, volcanic ashes etc., and the epigenetic factors of igneous intrusion, circulating ground water etc. [47,50–56]. However, only few researchers have recognized that the characteristics of inorganic components in TDCs are influenced by tectonic stress as well. Actually, inorganic components in coal are closely associated with tectonic deformation [36,57–61]. For instance, fluids migrating through brecciated coals in fault zones probably lead to cleat mineralization and elemental enrichment [58,59,61]. Specifically, the concentrations of Ti, Sc, Ta, U and rare earth elements all increase as the coal deformation increases, while the concentration of Mo decreases with increasing deformation intensity, which can be mainly ascribed to the coal dynamic metamorphism accompanied by frictional thermal effects [62–65]. Analogously, concentrations of rare earth elements in fault-related boreholes are higher than those in unrelated boreholes in specific coal seams. Thus, it has been proposed that fault structures are beneficial to the accumulation of rare earth elements [66]. Therefore, the investigation of the geochemical alteration characteristics and mechanisms of deformed coals provides another perspective to understand the coal deformation mechanisms [62]. More importantly, coal and gas outbursts are considered to be closely related to strongly deformed coal, which is unpredictable during coal production [38]. Once the elemental variation characteristics and mechanisms in deformed coals are clarified, concentrations of elements could be applied as precise indicators of coal and gas outbursts [62,65,67].

In this study, a series of deformed coal samples located at different distances from a reverse fault are systematically studied from diverse perspectives. Based on the knowledge of geochemical characteristics of fault-related coal samples, differentiation mechanisms of trace elements in the fault zone are preliminarily clarified.

2. Geological Settings and Fault Zone Architecture

2.1. Geological Settings

Suxian coal mine, adjacent to Tanlu fault, is on the southeastern margin of the Northern China plate (Figure 1a). Multiple tectonic movements occurring in Suxian coal mine significantly altered the regional and primary structural features of coal deposited in this area. In the late Triassic (Indo-Chinese epoch), the collage and collision between Yangtze craton and North China craton led to the NS extrusion of Suxian coal mine, which induced the emergence of EW-trending folds and faults in the study area. The Yanshanian-Himalayan tectonic movement in early-middle Jurassic directly resulted in the formation of East Suzhou syncline, South Suzhou syncline and Xisipo thrust fault [27], which dominated the geological structure frame of Suxian coal mine (Figure 1b). South Suzhou syncline, with an axial direction of NE-20°, was cut into a half-baked syncline structure by Xisipo thrust fault. Under the influence of the cutting, the dip angle of east flank (about 45°) of South Suzhou syncline is greater than that of the west flank (about 25°). The main structural trace of Suxian coal mine is basically in the NNE direction (Figure 1b).

Qi'nan colliery, the selected sampling area, is at the turn end of South Suzhou syncline. Faults are the predominant structures that control the local distribution of TDCs in Qi'nan colliery. Two secondary folds, nearly along the EW direction, are develop in the central part of Qi'nan colliery. Samples were collected from a reverse fault that was adjacent to the axis of the Zhangxuewu syncline (Figure 1c). The EW-trending reverse fault was formed under the squeezing of tectonic stress in a NS direction originating from the Indo-China movement. This reverse fault was less affected by Yanshanian-Himalayan tectonic movement.

Furthermore, Yanshanian igneous intrusion in Huaibei coalfield belongs to the circum-Pacific tectonic magmatic belt. The Xierpu igneous rocks located in the northwest of Suzhou city were intruded up to the shallow part of through north Suzhou fault (Figure 1b). Faults developed in Zhuxianzhuang colliery are important magma transportation channels. Coal seam VI was influenced by the igneous intrusion as well (Figure 1d).

Shanxi Formation of lower Permian and lower Shihezi Formation of middle Permian are major coal-bearing strata in Qi'nan colliery (Figure 1d). Fault-related coal samples were collected in coal seam VI, lower Shihezi formation. The lower Shihezi formation, as a fluvial-dominated delta plain deposit, is mainly comprised of sandstone, mudstone and coal, which is conformable with the underlying Shanxi Formation and distributed regionally. The direct roof and floor of coal seam VI are mainly mudstone.

The hydrogeological resources of Qi'nan colliery are mainly the fourth aquifer in Cenozoic and the sandstone fissured aquifer of coal-bearing strata in Permian. The decreasing water level of the fourth aquifer after coal mining indicates that the source of water in coal is mainly the fourth aquifer. The sandstone-fissured aquifer could directly enter the coal seam through leaching and could be recharged by the fourth aquifer. The pH value of ground water penetrating into coal seams from an overlying aquifer is in the range of 7.2–7.9, which makes coal seam VI mildly alkaline as well [68].

2.2. Fault Zone Architecture

The dip direction and dip angle of the selected reverse fault are 358° and 35°, respectively (Figure 2). The direct roof and floor of coal seam VI are mainly grey mudstone, and the thick false roof and floor are mainly grey green mudstone. Roof and floor rocks dislocated by the fault are broken into cataclastic shapes near the fault plane followed by intact roof and floor rocks away from the fault plane. Field observation showed that there were dragged folds developing in hanging and foot walls near the fault plane (Figure 2). Coal, being more sensitive to tectonic stress, was broken into strongly deformed structures in dragged fold zones (zone I) with widths of ~2 meters in both hanging and foot walls. The influence of the fault on coal deformation became weak in the part with normal occurring roof and floor rocks. At a distance of ~4 meters from zone I in both hanging and foot walls, dense associated fractures cut most coal into mortar shapes (zone II). Coal beyond zone II showed relatively intact shapes (zone III). Besides, field observation showed that subdivided zones in the hanging wall were slightly broader than those in the footwall.

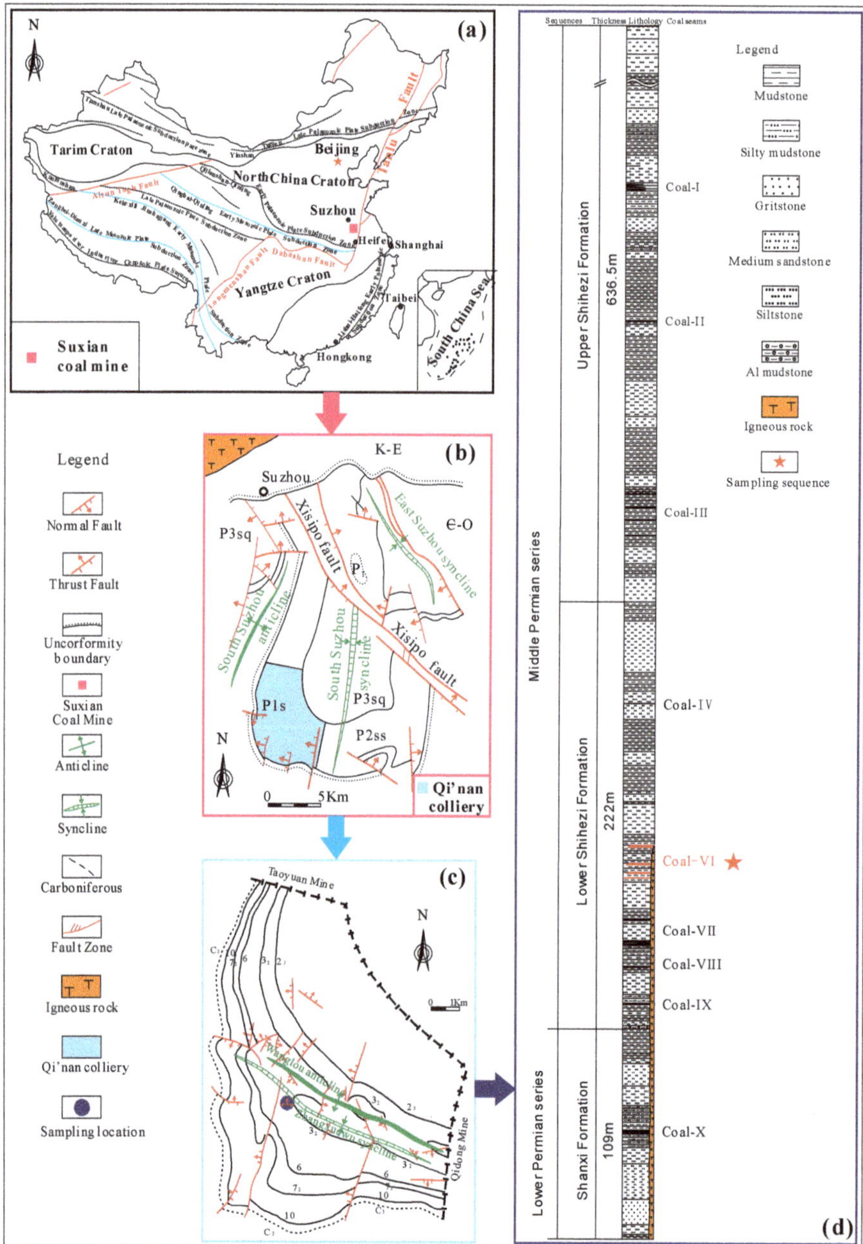

Figure 1. The location of sampling sites and sequence. (**a**) Map of China. (**b**) Structural outline of Suxian coal mine. (**c**) Structural outline of Qi'nan colliery and distribution of sampling location (modified from Jiang et al. [27]; Li et al. [69]). (**d**) Detailed stratigraphic column of coal-bearing strata in Qi'nan colliery.

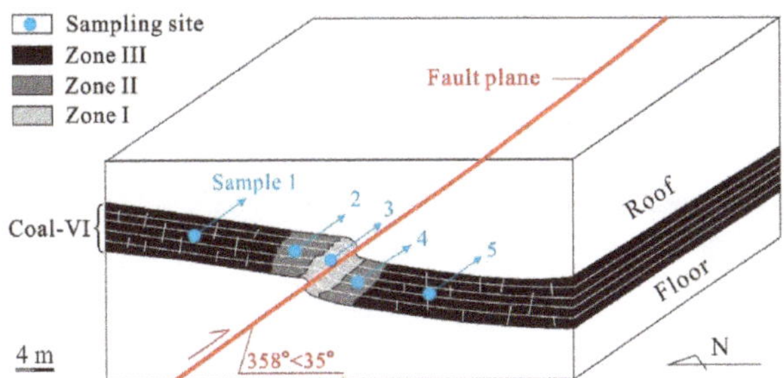

Figure 2. Schematic of fault zone architecture and distribution of sampling sites.

According to the guidelines of GB482-1995 Standard [70], five TDC samples with typical deformation characteristics in three subdivided zones were collected from the same sequence. Among them, samples 1, 2 and 3 were collected from the hanging wall, and samples 4 and 5 were collected from the foot wall. Samples 1 and 5 were collected in zone III; samples 2 and 4 were gathered in zone II; and sample 3 was collected in zone I (Figure 2).

Samples in zone III: Samples 1 and 5 in zone III are defined as cataclastic coal (Figure 3a,e). Maceral bands in these samples are clearly visible. Coal body with more integrated structures are mainly cut by straight and sparse cleats with flat surfaces.

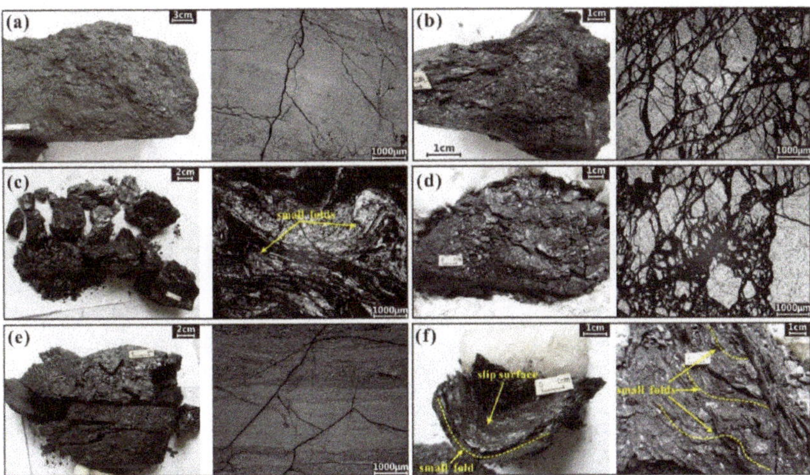

Figure 3. The macro and micro deformation characteristics of fault-related coal samples; (**a–e**) are images of micro and macro deformation characteristics of samples 1–5 sequentially (left images show macro deformation characteristics, right images show micro deformation characteristics). (**f**) Small folds and slip surface developed in sample 3.

Samples in zone II: Samples 2 and 4 in zone II are defined as mortar coal (Figure 3b,d). These samples are smashed into pieces by compression stress, and they show lower mechanical strength to the samples in zone III. The cleats developed in these two samples are denser compared with those of samples 1 and 5. Shapes of coal particles are angular or sub-angular.

Sample in zone I: Sample 3 is defined as wrinkle coal (Figure 3c). Coal developed in zone I is formed in a strong compression-shear stress environment. Sample 3 could easily be crumpled by hand into powder or grains. Small micro "folds" are well developed in zone I, indicating a strong ductile deformation (Figure 3c,f). Squeezing and slippage of the coal matrix led to the development of frictional mirror surfaces (Figure 3f).

3. Experiments

Bulk fault-related coal samples were firstly consolidated by injecting epoxy. Then, consolidated samples were polished perpendicular to marcel bands by using the sandpapers and Al_2O_3 colloidal solution. The micro deformation characteristics were observed through polished faces by using Polarizing Microscope Eclipse LV100N Pol (Nikon, Tokyo, Japan) in a reflected light. The polarizing microscope is equipped with illumination lamphouse (LV-LH50PC, Nikon, Tokyo, Japan).

Ultimate analysis of the samples was performed using Vario Macro Cube element analyzer (manufactured by Elementar, Hanau, Germany). Specifically, the detection of total sulfur contents was based on ASTM standard D3177-02 (2002). The proximate analysis of fault-related coal samples was in accordance with the ASTM Standards including ASTM D3174-12 (2012), ASTM D3175-17 (2017), ASTM D3173/D3173M-17a (2017), and ASTM D3175-17 (2017). Both ultimate analysis and proximate analysis were performed in Xuzhou Inspection Center of China National Administration of Coal Geology.

Concentrations of trace elements in various TDCs were analyzed by using inductively coupled plasma mass spectrometry (ICP-MS, Agilent, Santa Clara, CA, USA) following a two-step digestion method. Firstly, dry powder samples (200 mesh) were ashed (500 °C) in a muffle furnace for 4 h. Then, the ash samples were extracted with HNO_3, and the residue was dissolved by a mixture of $HF{:}HNO_3{:}HClO_4$ subsequently in a closed microwave digestion system with high pressure. Concentrations of trace elements in standard reference materials (SARM-19) and blank samples were also determined to calibrate the detection results. Concentrations of major element oxides in the high-temperature (815 °C) ash were determined by means of X-ray fluorescence spectrometry (XRF, Bruker, Karlsruhe, Germany) in the Advanced Analysis and Computation Center of China University of Mining and Technology.

Morphological characteristics of minerals in selected bulk TDC samples were observed by the FEI Quanta TM 250 scanning electron microscope equipped with an energy dispersive X-ray analyzer (SEM-EDX, FEI, Hillsboro, OR, USA) in the Advanced Analysis and Computation Center of China University of Mining and Technology. X-ray diffraction (XRD, Bruker, Karlsruhe, Germany) spectra for fault-related coal samples were acquired using Bruker D8 Advance instrument (Cu target, Kα radiation) equipped with 0.6-mm divergence slit and 8-mm anti-scatter slit systems. Operating conditions of the X-ray tube were U = 40 kV and I = 30 mA.

Characteristics of hydroxyl groups in organic and inorganic compounds of coal were analyzed by using Fourier transform infrared spectroscopy (FT-IR, Bruker, Karlsruhe, Germany). Powdered whole coal samples (0.9 mg, higher than 200 mesh) were initially ground with 80 mg of potassium bromide (KBr) for 20 min in an agate mortar. The powder mixture was pressed into a transparent disc for 10 min using a tablet machine. Pure ground KBr was used to obtain a reference spectrum. The discs were analyzed by VERTEX-70 FT-IR (Bruker, Germany), and the spectra were recorded in the range of 400 to 4000 cm^{-1} with a resolution of 4 cm^{-1}.

Characteristics of a pore-cleat system of fault-related coals were detected by using Macromerities 9500-type mercury porosimetry (Micromeritics, Drive Norcross, GA, USA) with a pore diameter ranging from 3 nm to 0.23 mm and pressure up to 413 MPa at the Key Laboratory of Coalbed Methane Resources and Reservoir Formation Process, China Ministry of Education. Pores with diverse pore sizes from millimeters to nanometers could be measured by this method [71].

4. Results

4.1. Coal Chemistry

As shown in Table 1, the fixed carbon (FC_d) values of the five samples were in the range of 51.81–55.78% and the volatile matter (V_{daf}) ranged from 34.27% to 35.88%, indicating high volatile bituminous coals according to the ASTM D388-18 standard. Samples with total sulfur content lower than 1% were classified as low grade based on the GB/T standard 15224.2-2004. The ash yields of samples 2 and 3 were slightly higher than others. Furthermore, the $R_{o,max}$ of sample 3 was a little higher than the other samples, indicating that sample 3 was of higher maturation degree (Table 1). The early-coalification phenomenon that occurred in sample 3 was mainly attributed to the dynamic metamorphism of intensive tectonic deformation [40–42].

Table 1. Proximate and ultimate analyses of fault-related coal samples.

Coal Samples	$R_{o,max}$ (%)	Proximate Analysis						Ultimate Analysis			
		M_{ad}	A_d	V_{daf}	Coking Index	FC_d	$S_{t.d}$	O_{daf}	C_{daf}	H_{daf}	N_{daf}
1	0.90	1.04	16.95	35.00	6	52.88	0.12	8.79	84.35	5.43	1.28
2	0.89	1.41	19.21	34.27	6	52.30	0.18	9.55	83.42	5.30	1.50
3	0.91	1.45	18.52	35.88	6	51.31	0.19	9.86	83.83	4.68	1.39
4	0.92	1.21	17.10	34.93	6	51.81	0.17	10.90	82.57	4.98	1.32
5	0.88	1.20	12.18	35.71	6	55.78	0.20	8.43	84.84	5.02	1.48

Note: $R_{o,max}$: Maximum reflectance of vitrinite; M_{ad}: Inherent moisture content with air-dried basis; A_d: Ash yield with dry basis; V_{daf}: Volatile matter yield with dry-ash-free basis; FC_d: Fixed carbon content with dry basis; ad: Air-dried basis; d: dry basis; daf: Dry-ash-free basis.

4.2. Minerals in the Reverse Fault Zone

XRD results showed that the minerals of five coal samples were mainly composed of clay minerals, carbonates and quartz (Figure 4). Clay minerals occupying the first place in samples mainly consisted of nacrite, kaolinite and dickite; and carbonates in the second place were mainly composed of calcite and dolomite. Compared with samples 2–4, quartz and carbonate were more developed in samples 1 and 5 respectively, while dickite could only be found in sample 3. Additionally, SEM-EDX results showed that sulfide and sulfate minerals could also be found in fault-related samples, for example, pyrite and sulphate.

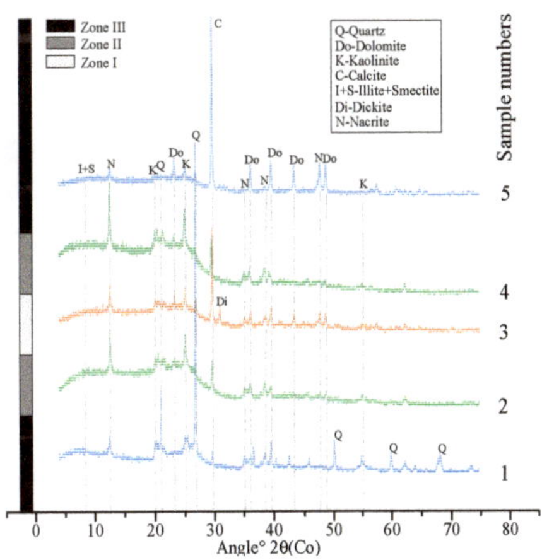

Figure 4. The XRD spectrums of fault-related coal samples.

In sample 3, clay minerals commonly precipitated in pore or cleat structures (Figure 6a–f). While for other samples, clay minerals usually occur as syngenetic precipitates in the cell of fusinite (Figure 5a,b). With a close distance to the fault structure, clay minerals filled in cells were obviously squeezed with the structural failure of fusinite. Besides, calcite and dolomite in the samples always occurred as cleat fillings (Figure 6g,i). Furthermore, pyrite in sample 3 was observed as cleat infillings, indicating the post-depositional mineralization process [72]. However, pyrite in other samples commonly occurred as syngenetic lumps or cell infillings. Therefore, different occurrence modes of minerals indicated that sample 3 was significantly affected by epigenetic factors compared with other samples.

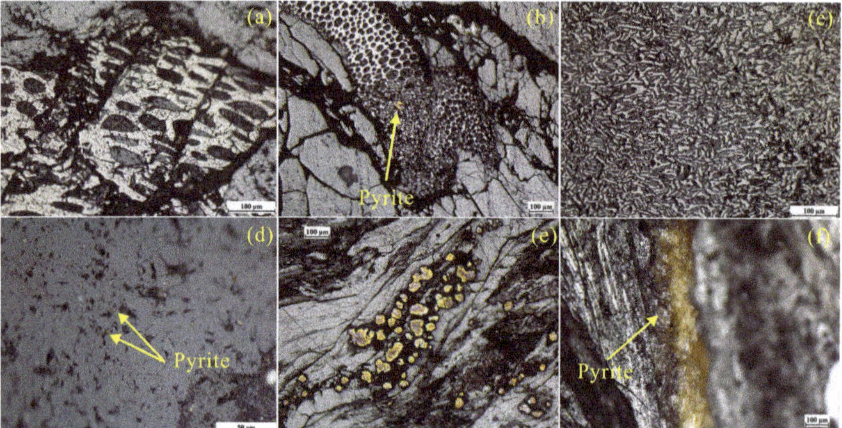

Figure 5. Minerals in fault-related coal samples. (**a**) Clay minerals filling in fusinite cell in sample 1; (**b**) Clay minerals filling in fusinite cell in sample 2; (**c**) Clay minerals filling in fusinite cell in sample 3; (**d**,**e**). Lumps pyrite in sample 4 and 5 respectively; (**f**) Pyrite filling in cleats in sample 3.

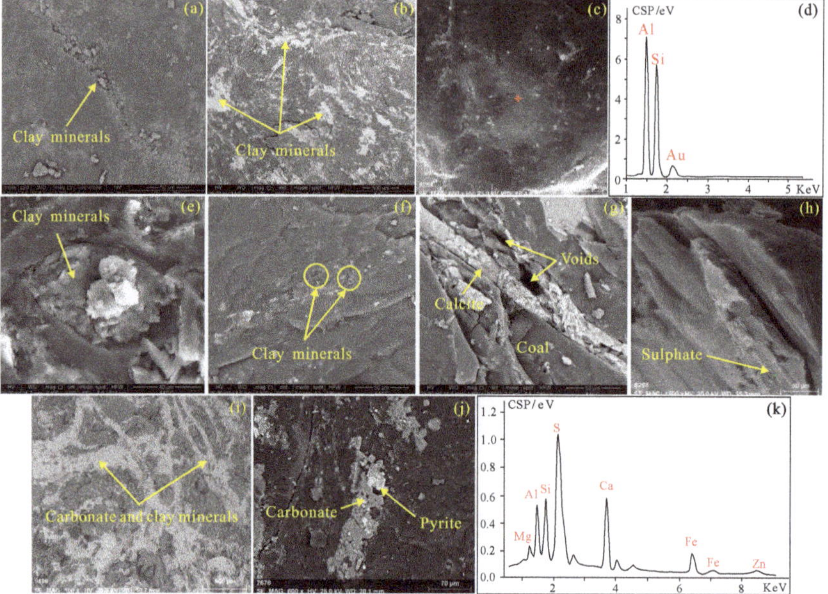

Figure 6. SEM-EDX images of minerals in fault-related coal samples. (**a**) Clay minerals occurring in cleat (sample 3); (**b**,**c**) clay minerals attached on the cleat surfaces filled with carbonates. (**d**) EDX spectrum of image c; (**e**,**f**) clay minerals filling in pore structures in sample 3; (**g**) distribution of calcite veins; (**h**) sulphate attached on cleat surface in sample; (**i**) the mix of carbonate and clay minerals filling in cleats; (**j**) the mix of carbonate, clay minerals and pyrite attached on cleat surface; (**k**) EDX spectrum of image j.

With the increase of coal deformation intensity, the phenomenon of mineral mix became more frequent. In sample 3, the sulphate usually coexisted with clay minerals or carbonates in cleats and fractures (Figure 6h). Besides, the mix between clay minerals and carbonates could also be observed in cleats and fractures (Figure 6i). Nevertheless, pyrite in sample 3 occurs in the cleat of coal matrix, sometimes even mixed with carbonates (Figure 6j,k). The phenomenon of mineral mix is primarily caused by the mechanical grinding of stronger tectonic stress.

More specifically, clay minerals in samples 1 and 5 were characterized by complete intact flakes. Particle sizes of clay minerals in samples 2 and 4 were slightly reduced under the influence of tectonic stress. Clay minerals appearing in zone I were broken into porphyritic shapes with smaller particle sizes compared with those in zone II (Figure 7). In zones I and II, clay minerals were less stacked compared with those in zone III. The same morphologies of clay minerals in TDCs were also found by Song et al. [62]. The assemblage characteristics of clay minerals was more complicated in sample 3, and dickite started to appear in sample 3 (Figure 4). The formation of dickite may be related to the thermal fluid migration associated with tectonic deformation [73].

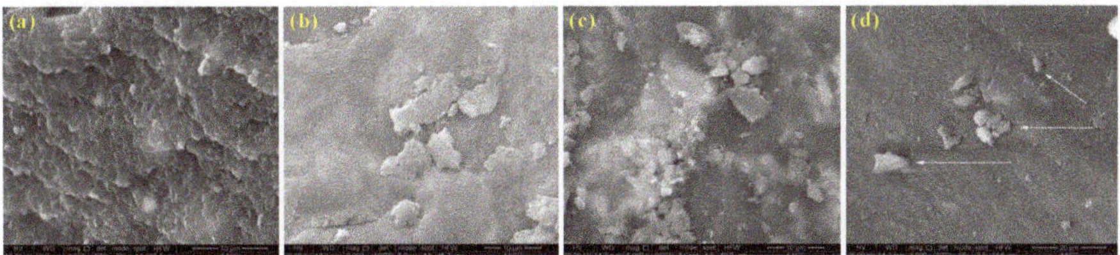

Figure 7. Morphological characteristics of clay minerals. (**a**) Clay minerals developed in zone III. (**b**) Clay minerals developed in zone II. (**c**) Clay minerals developed in zone I. (**d**) Sliding of clay minerals on frictional mirror surface of sample 3 (arrows refer to mobile direction of clay minerals).

4.3. Distribution of Oxides of Major Elements

Contents of SiO_2, CaO and Al_2O_3 were much higher than the contents of other oxides of major elements, which is consistent with the XRD result that clay minerals, carbonates and quartz were the main minerals developed in fault-related coals (Table 2, Figure 4). Higher SiO_2/Al_2O_3 ratios of the five samples (compared with theoretical ratio of kaolinite (1.18)) illustrated the existence of free SiO_2 [50]. Oxides of major elements in fault-related coal samples did not show obvious variation regularity (Figure 8).

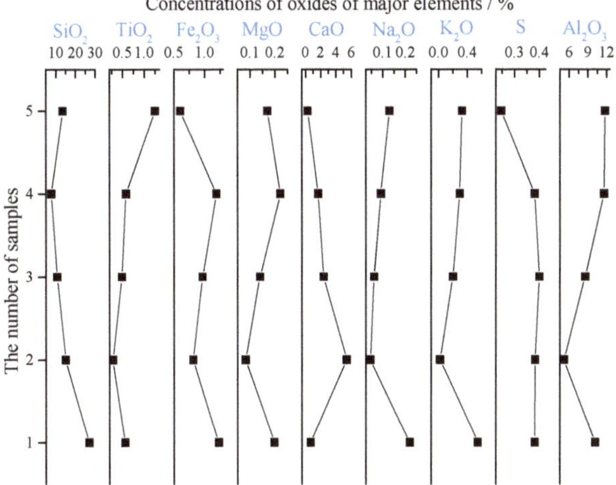

Figure 8. Distribution of oxides of major elements.

Table 2. Oxides of major elements (%) in fault-related coal samples.

Sample	1	2	3	4	5
SiO_2	26.92	14.95	10.7	7.68	13.55
TiO_2	1.25	0.583	0.491	0.283	0.566
Al_2O_3	11.77	11.6	8.59	5.18	10.2
Fe_2O_3	0.601	1.19	0.966	0.8113	1.23
MgO	0.17	0.222	0.14	0.083	0.198
CaO	0.288	1.66	2.39	5.406	0.667
Na_2O	0.13	0.092	0.062	0.045	0.22
K_2O	0.327	0.294	0.2	0.023	0.551
S	0.247	0.381	0.4	0.383	0.382

4.4. Distribution of Trace Elements

4.4.1. Distribution of Rare Earth Elements (REE)

REE in this research included La, Ce, Pr, Nd, Sm, Eu, Gd, Tb, Dy, Y, Ho, Er, Tm, Yb, and Lu. It is worth noting that Y with a very similar ion radius as Ho was included in REE as well [74]. REE in fault-related coals were subdivided into three groups, viz. light (LREE: La, Ce, Pr, Nd, and Sm), medium (MREE: Eu, Gd, Tb, Dy, and Y), and heavy (HREE: Ho, Er, Tm, Yb, and Lu) rare earth elements [75]. Concentrations of REE were normalized to the average value for upper continental crust as reported by Taylor and McLennan [76] (Figure 9).

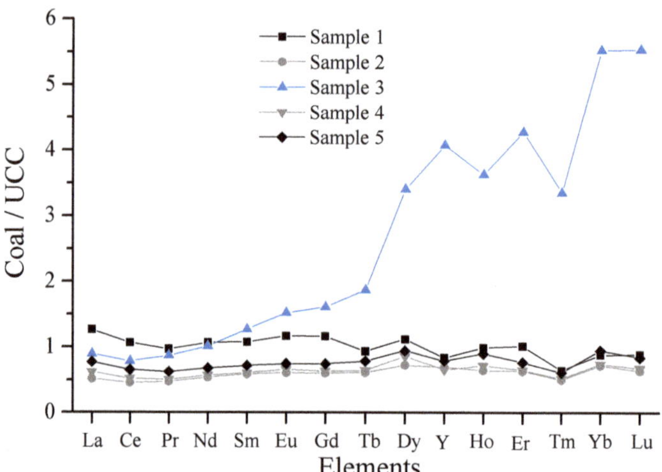

Figure 9. Distribution of normalized REE in five samples.

Interference of Ba on the anomalies of Eu should be clarified to ensure the reliability of data from ICP-MS analysis [74]. Fortunately, the ratios of Ba/Eu for the five samples were all lower than 1000 (Table 3), indicating that the interference of Ba could be ignored [77].

Table 3. Ratios of Ba/ Eu of five coal samples.

Samples	1	2	3	4	5
Ba/Eu	374.00	223.64	160.71	294.92	246.27

Calculated parameters of REE help to understand their distribution patterns. The La_N/Sm_N ratio of samples 2–5 was higher than the Gd_N/Yb_N ratio (Table 4), indicating that the fraction degree of elements La–Eu was higher than that of elements Gd–Lu in samples 2–5. Similarly, Ce_N/Yb_N of coals was much lower than the average value (4.6) [78], which indicated the enrichment of HREE and MREE compared with LREE. The distribution patterns of REE in samples 2–5 were H-type with La_N/Lu_N lower than 1. In sample 1, distribution pattern of REE belonged to L-type ($La_N/Lu_N > 1$). Samples with H-type REE were probably affected by natural waters circulating in the coal basins (especially for sample 3 with $La_N/Lu_N = 0.16$) [74,75], which is consistent with the above analysis and the hypothesis proposed by Yang et al [66]. Weak positive Eu anomalies (Eu_N/Eu_N^* ranging from 1.01 to 1.13) illustrated that coals in the fault zone were less affected by reducing conditions and high temperature (>250 degrees). Y_N/Ho_N in Chinese coal commonly shows very weak anomalies [74]. However, Y concentration in sample 3 showed obvious anomalies in comparison with others. Weak Ce anomalies (0.89–0.93) were observed in the five samples, indicating that the source origin of coals in the fault zone was terrigenous materials [74].

Table 4. Parameters calculation of normalized REE concentrations.

Samples	La_N/Sm_N	Gd_N/Yb_N	Ce_N/Yb_N	La_N/Lu_N	Gd_N/Lu_N	Ce_N/Ce_N^*	Eu_N/Eu_N^*	Y_N/Ho_N
1	1.18	1.31	1.21	1.42	1.31	0.93	1.13	0.84
2	0.88	0.84	0.63	0.81	0.94	0.91	1.02	1.07
3	0.70	0.29	0.14	0.16	0.29	0.87	1.04	1.12
4	1.02	0.86	0.70	0.91	0.93	0.90	1.07	0.91
5	1.07	0.78	0.69	0.91	0.88	0.92	1.01	0.88

Note: $Eu_N/Eu_N^* = Eu_N/(0.5Sm_N + 0.5Gd_N)$, where Eu_N refers to normalized concentration of element Eu.

Concentration coefficients (CCs) (which refer to ratios of concentrations of trace elements in fault-related coals to regional value) of trace elements in coal are useful parameters for estimating relative enrichment of those elements. Regional concentrations of trace elements provided by Zheng et al. [79,80] are used as a reference. According to the categorization scheme proposed by Dai et al. [74,81,82], CCs of fault-related coals were classified into four categories: enriched (5 < CC < 10), slightly enriched (2 < CC < 5), normal (0.5 < CC <2), and depleted (CC < 0.5). Average CCs of trace elements in samples 1 and 5 were used to represent CCs in zone III, and average CCs of samples 2 and 4 were applied to characterize CCs in zone II. Figure 10 shows that elements La–Gd were distributed in the three zones with normal value. Partial MREE (Tb, Dy and Y) and HREE were classified as enriched or slightly enriched types in Zone I. CCs of REE in zones II and III were all in the range of 0.5–2 without any exception.

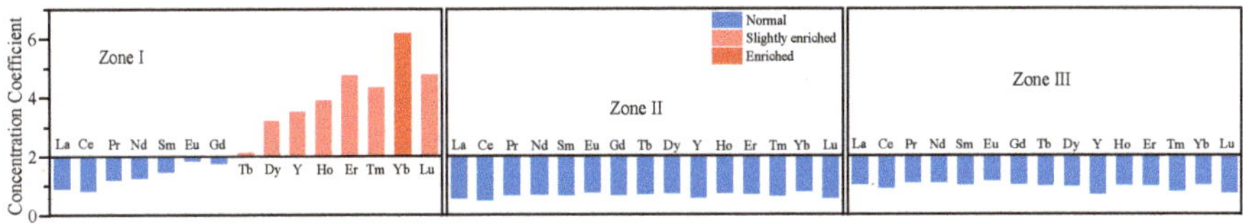

Figure 10. Distribution of CCs of REE in three subdivided zones.

Comparison of diverse TDCs showed that HREE and MREE were highly concentrated in sample 3. Concentrations of HREE and MREE in samples 2 and 4 were lower than those in samples 1 and 5 in zone II (Figure 11a, Table 5). Concentrations of LREE were higher in samples 1, 3 and 5, which is different from the distribution characteristics of HREE and MREE (Figure 11b).

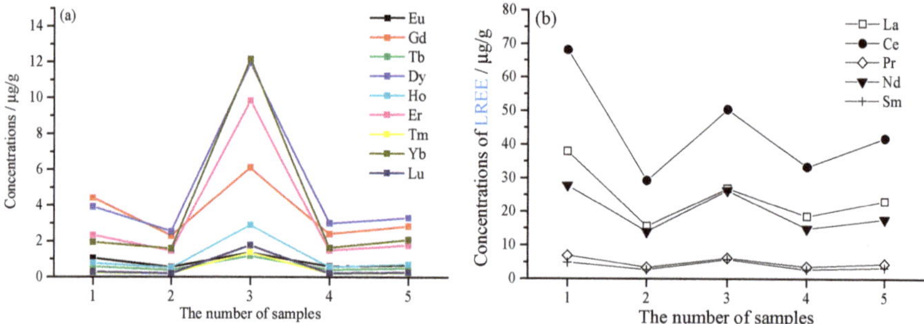

Figure 11. Distribution of REE in diverse fault-related coals. (**a**) Distribution of HREE and MREE in different fault-related coals. (**b**) Distribution of LREE in different fault-related coals.

Table 5. Concentrations of trace elements ($\mu g/g$) in fault-related coal samples.

Sample	1	2	3	4	5	Sample	1	2	3	4	5
Li	10	11	32	16	23	Ba	374	123	225	174	165
Be	1.6	1.1	6.6	1.2	1.6	La	38	16	27	19	23
Sc	7.3	4.0	19	6.1	8.0	Ce	68	29	50	33	42
Cr	39	18	84	28	30	Pr	6.9	3.4	6.2	3.6	4.4
Mn	9.8	6.9	43	28	19	Nd	28	14	26	15	18
Co	19	22	11	16	18	Sm	4.8	2.6	5.7	2.7	3.2
Ni	14	15	14	16	15	Eu	1.0	0.55	1.4	0.59	0.67
Cu	29	18	45	18	20	Gd	4.4	2.3	6.1	2.4	2.8
Zn	32	8.4	96	12	33	Tb	0.60	0.39	1.2	0.41	0.51
Ga	11	8.2	15	8.5	8.9	Dy	3.9	2.5	12	3.0	3.3
Rb	17	1.9	9.3	8.1	9.0	Ho	0.80	0.52	2.9	0.58	0.72
Sr	70	86	68	360	81	Er	2.3	1.5	9.8	1.5	1.8
Y	18	15	90	14	17	Tm	0.27	0.21	1.4	0.21	0.25
Nb	8.4	4.3	19	5.5	6.8	Yb	1.9	1.6	12	1.6	2.1
Mo	2.7	4.5	1.4	5.2	3.5	Lu	0.28	0.21	1.8	0.22	0.27
Cd	0.08	0.04	0.12	0.05	0.08	Ta	0.70	0.34	1.6	0.34	0.60
In	0.04	0.02	0.07	0.03	0.04	W	0.96	0.75	1.9	0.91	0.92
Cs	1.2	0.40	0.71	1.0	1.2	Tl	0.45	0.09	0.15	0.15	0.13
Pb	21	7.3	35	7.5	13	Th	5.7	3.0	15	3.9	6.6
Bi	0.22	0.19	0.36	0.16	0.22	U	2.1	1.1	8.0	1.5	1.5

4.4.2. Distribution of Other Trace Elements

In zone I, CCs of elements Li, Be, Sc, Cr, Zn, Nb, Pb, Ta, Th, and U ranged from 2 to 5 indicating a slight enrichment of those elements (Figure 12). Elements Li, Be, Sc, Cr, Nb, Ta, and U in zones II and III are within the global average values ($0.5 < CC < 2$). Elements Zn, Pb and Th were depleted ($CC < 0.5$) in zone II followed by normal CCs in zone III. CCs of elements Ni, Ga, In, Cs, Ba, W, and Tl were all at a normal level in the three zones, while CC of element Bi was lower than 0.5 in the three zones. Elements Cu and Rb with normal values in zones I and III were depleted in zone II. Element Sr with normal CCs in zones II and III was slightly enriched in zone II. Element Co which was slightly enriched in zones II and III showed normal concentration in zone I. Element Mo, depleted in zone I, showed constant normal distribution in zones II and III, while the distribution of element Cd was inverse in the three zones.

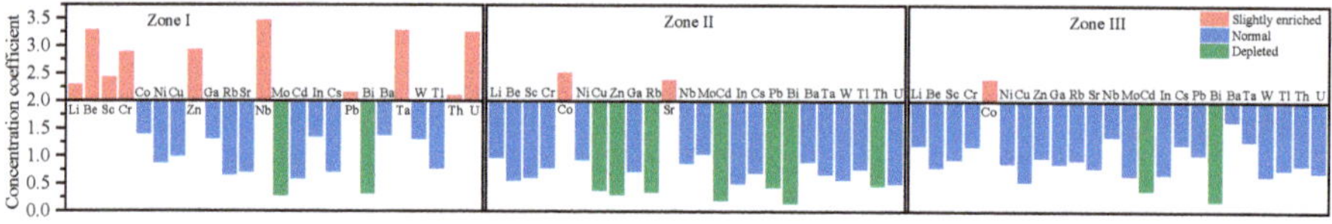

Figure 12. Distribution of CCs of REE in three subdivided zones.

For different types of TDCs, elements Li, Be, Sc, Cr, Cu, Mn, Zn, Ga, Nb, Cd, Bi, Ta, W, and U were significantly enriched in sample 3 (Figure 13a,b). Concentrations of elements Co and Mo were low in zone I, while they were relatively accumulated in zones II and III (Figure 13c). The distribution of elements Ni, Rb, Sr, Cs, Ba, and Tl belonged to disordered type (Figure 14).

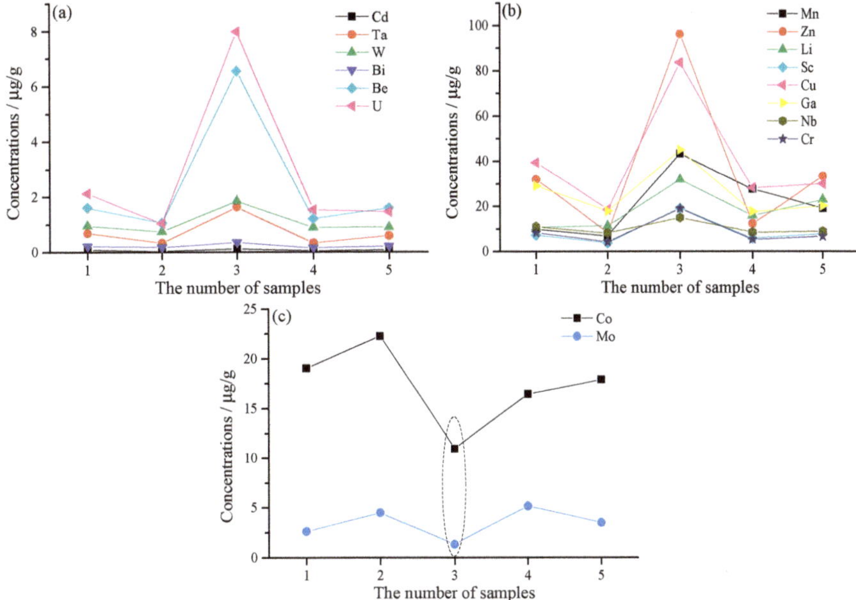

Figure 13. Distribution of accumulated and dissipated types of trace elements in the fault zone. (**a,b**) Distribution of accumulated type of trace elements in the fault zone. (**c**) Distribution of dissipated type of trace elements in the fault zone.

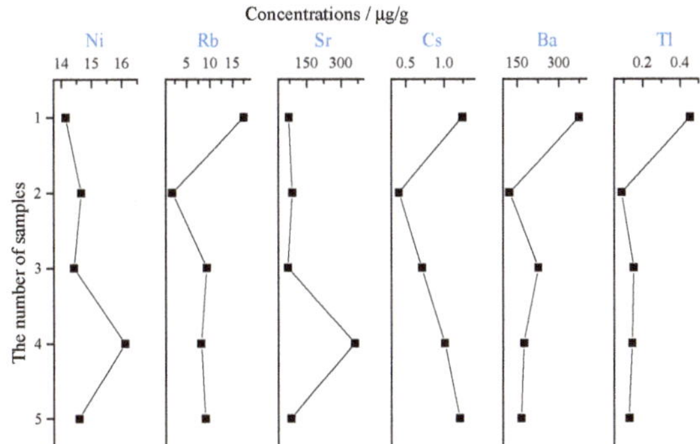

Figure 14. Distribution of disordered type of trace elements in the fault zone.

5. Discussion

5.1. Elemental Geochemical Associations

The key to understanding the differentiation mechanisms of trace elements is to determine their affinities. Concentrations of major element oxides, trace elements and ash yields were analyzed together with a hierarchical cluster analysis method (Figure 15). When 15 was adopted as the rescaled distance, all members could be classified into five groups.

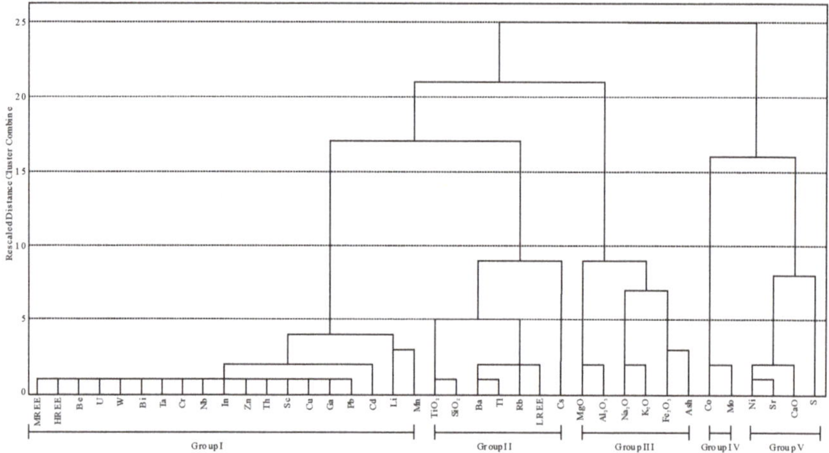

Figure 15. The family tree of hierarchical cluster analysis.

Group I included elements MREE, HREE, Be, U, W, Bi, Ta, Cr, Nb, In, Zn, Th, Sc, Cu, Ga, Pb, Cd, Li, and Mn. Among them, REE were found in the clay minerals attached on the cleat surfaces in sample 3 (Figure 16). Element Zn mainly occurred in pyrite according to the EDX spectrum (Figure 6j,k). Similarly, elements Cu and Pb were also closely correlated with sulfide minerals. Furthermore, the other elements that basically belong to lithophile elements also have aluminosilicate affinity to some extent as reported by Zhang et al. [83], Dai et al. [84], Tian et al. [85] and Finkelman et al. [86]. Elements in group I with significant positive anomalies were significantly influenced by thermal fluid or leached water as mentioned above.

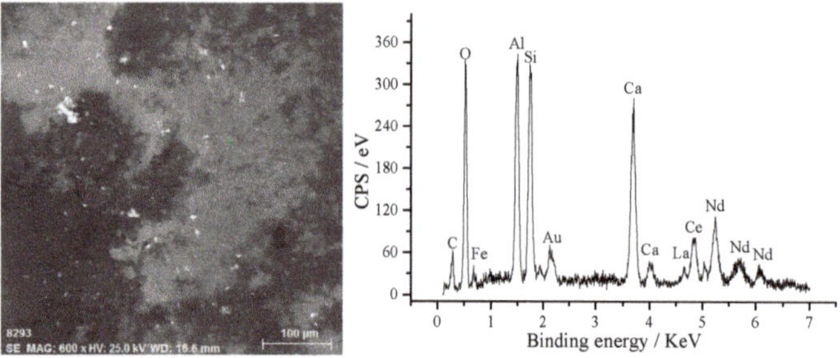

Figure 16. The SEM-EDX detection results of REE-containing minerals attached on cleat surfaces.

Group II included TiO_2, SiO_2, Ba, Tl, Rb, LREE, and Cs. Among them, the correlation coefficient between TiO_2 and Al_2O_3 was 0.782 and that between TiO_2 and SiO_2 is 0.988, indicating aluminosilicate affinity. Besides, the SEM-EDX results showed that LREE could be directly observed in clay minerals (Figure 16). In brief, elements in the group all showed aluminosilicate affinity.

Group III included MgO, Al_2O_3, Na_2O, K_2O, Fe_2O_3, and ash yield. Fe_2O_3 mainly occurred in dolomite and pyrite in those fault-related samples. Al_2O_3 is the major component of clay minerals. Besides, K_2O could be found in clay minerals, while, MgO was observed in both clay minerals and carbonates. Therefore, oxides of major elements in groups III were closely correlated with clay minerals or carbonates.

Group IV only included Co and Mo. Elements Co and Mo were strongly correlated with C_{daf} (r_{Cdaf} = 0.806 and 0.785 respectively), indicating a possible organic affinity. Ni, Sr, CaO, and S in group V were poorly correlated with ash yield or Al_2O_3. EDX data showed that element Ni was detected in the mix of sulphate minerals and carbonates, while Ni could not be detected in sole carbonate (Figure 6h, Table 6). Therefore, elements Ni and Sr were probably associated with sulphate minerals.

Table 6. EDX data for sulphate in sample 3.

Element	Concentration/wt.%	Element	Concentration/wt.%
O	18.07	Na	0.02
Al	2.06	Mg	0.01
C	49.68	K	0.02
Si	0.66	Ca	21.5
S	6.46	Fe	0.9
Ni	0.07		

5.2. Accumulation Mechanisms of HREE, MREE and Related Elements

5.2.1. Evidences of the Injection of Solutions

Local distribution of trace elements might be influenced by the invasion of solutions with higher concentrations of related elements [49]. The presence of dickite in sample 3 (Figure 4) might be related to the thermal fluid migration associated with tectonic deformation [73]. Similarly, in zone I, the developed fracture-fillings including carbonates, clay minerals and pyrite containing a higher concentration of MREE, HREE and related elements illustrated that those elements probably originated from an epigenetic process, such as the invasion hydrothermal fluids or leaching of meteoric water [50,87].

Furthermore, REE contains useful information on the epigenetic resources of related elements. For sample 3, H-type of REE ($La_N/Lu_N < 1$) with a significant Y positive anomaly was probably affected by invasive solutions [74,75]. Besides, Yb/La (=0.45) of sample 3 was much larger than common word hard coal (Table 7). Furthermore, the presence of higher U/Th and Yb/La in sample 3 was also caused by the elemental redistribution related to invasive solutions with higher concentrations of related elements [87,88]. The hydrothermal fluid or meteoric water leaching through the overlying clay partings caused the more active elements (U and Yb) to be deposited in coals [88].

Table 7. Calculated parameters of normalized REE concentrations.

Samples	1	2	3	4	5
U/Th	0.37	0.37	0.53	0.38	0.23
Yb/La	0.05	0.10	0.45	0.09	0.09

5.2.2. Migration

Higher CCs and concentrations of HREE and MREE of sample 3 compared to other samples indicate the significant enrichment of HREE and MREE in zone I (Figures 10 and 11a). Similarly, CCs and concentrations of related elements (Be, U, W, Bi, Ta, Cr, Nb, In, Zn, Th, Sc, Cu, Ga, Pb, Cd, Li and Mn) in sample 3 were higher than those in other samples (Figures 12 and 13a,b). Pore and cleat mineralization found in sample 3 (Figure 6a–f) can be considered another evidence of solutions movement [72,89]. Solutions migrating through coal seam carry different types of trace elements and play an important role in the enrichment of these elements [75,90,91]. Therefore, the accumulation of HREE, MREE and related elements in sample 3 was mainly associated with the invasive solutions.

The migration of solutions depends on the development of a pore-cleat system in fault-related coals. Coal samples collected around fault structures are altered by tectonic stress into TDCs with a modified pore-cleat system [32,34,53,59,61,62,92,93]. Clarifying different fluid paths in fault-related coal samples could provide useful information about the migration of solutions. Thus, total porosity data could be useful for determining the relationship between tectonic deformation and elemental enrichments in the studied samples. Pore volume and injection–ejection ratios of these coal samples could be calculated from the porosity results (Figure 17) [71,94]. As expected, the distribution of pore volume in fault-related coal samples was consistent with the distribution of HREE, MREE and related elements (Figures 11a and 13a,b), which confirms that the developed pore-cleat system in sample 3 was more beneficial for the movement of solutions and influx of specific trace elements.

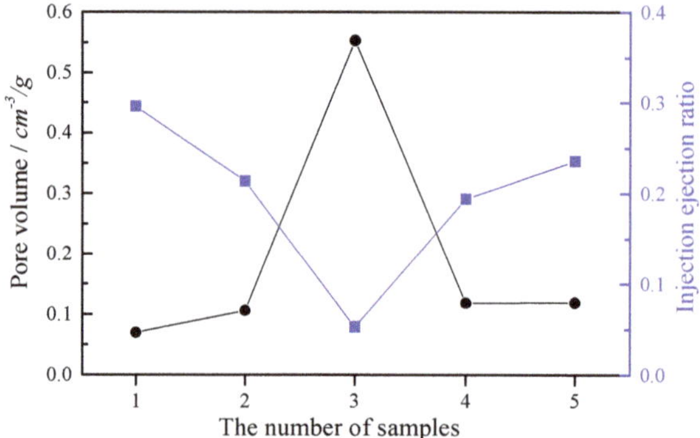

Figure 17. Distribution of pore volume and injection–ejection ratios of fault-related coal samples.

5.2.3. Accumulation

Sample 3 in zone I characterized by a low injection to ejection ratio had lower permeability (Figure 14) [71]. The lower permeability of sample 3 indicates that closed pore structures were likely formed under shear and compression stress or occluded by sealing minerals, viz. clay minerals (Figure 6e,f) in zone I [95,96]. During faulting deformation, many trace elements associated with clays and other minerals could be dissolved in acid solutions and transported through zone I [72,97], and then reprecipitated in a post-depositional pore-cleat system under alkaline condition (Figure 6a–f) [60,68]. Correspondingly, cleat-filling minerals retained and accumulated MREE; HREE and related elements occupied partial pore-cleat structures, and more closed pores and cleats were formed and the permeability of the conduit system in zone I became lower. Therefore, the solutions flowing into fault structures were limited in zone I, similar to the process occurring in fault zones of rocks [8,22,98]. The limitations of solutions in the local area could be used to explain the anomalous behavior that only occurred in zone I.

Pore-cleat mineralization could not fully explain the accumulation of HREE, MREE and related elements, because the concentration of Al_2O_3 (major constituent of clay minerals that is the host of these trace elements) in sample 3 was not the highest one. Clay minerals with special physical and chemical properties are sensitive to tectonic stress [99]. Similarly, decreasing particle sizes and less stacked laminar structures of clay minerals were also found in zones II and I, suggesting that strong mechanical grinding and dislocating of tectonic stress acted not only on the coal matrix, but also on minerals occurring in the coal (Figure 7a–c). Laminar clay minerals were preferentially disrupted along basal planes including layer-parallel extension and layer-parallel shortening [100], especially on the friction mirror surfaces (Figure 7d). Coating of clay minerals on the friction mirror surfaces further sealed the pore-cleat system of coal, which resulted in the reduction of permeability in zone I (Figure 6b,c) [25,101].

Studies on experimental and natural clay samples showed that stress promoted the breakdown and delamination of the clay mineral layers [24,100], which exposed the oxygen and hydroxyl planes beneficial to the sorption of elements [102]. It is necessary to evaluate whether similar chemical alterations (viz. hydroxyl exposure) of clay minerals occur in fault-related coals as well. FT-IR spectra over the range of 3600–3700 cm^{-1} provide useful information about hydroxyl groups of clay minerals in coal [103]. Adsorption bands at 3620 cm^{-1}, 3650 cm^{-1}, 3670 cm^{-1}, and 3695 cm^{-1} were assigned to the stretching vibration of inner hydroxyl, outer hydroxyl and inner-surface hydroxyl groups, respectively [104,105].

Figure 18 shows that concentrations of three types of hydroxyl groups in sample 3 were significantly higher than those of other samples. However, the concentration of Al_2O_3 in sample 3 was not the highest one, suggesting that the increase in number of hydroxyl groups in sample 3 was not due to the content variation of clay minerals (Figure 8). Nacrite, kaolinite and dickite found in fault-related coals all belong to the kaolinite group with similar lamellar structures [106–108], indicating that diverse assemblage characteristics of clay minerals in fault-related coals also did not lead to the increase in number of hydroxyl groups in zone I. To the best of our knowledge, an increase in number of hydroxyl groups of clay minerals is mainly due to the exposure of oxygen and hydroxyl planes by shear stress.

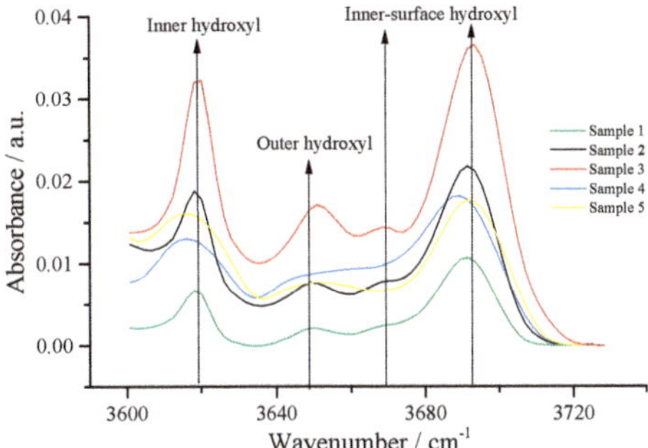

Figure 18. Characteristics of various types of hydroxyl groups of fault-related coal samples.

HREE, MREE and related elements that migrated into the coal seam in a dissolved form could be mostly adsorbed by clay minerals, which has been reported by previous studies and confirmed by relevant experimental studies [78,86,109–114]. Experimental studies have shown that the sorption of REE and related elements mainly depended on adsorption sites of clay minerals, namely, hydroxyl units on Al–O hexahedron or Si–O tetrahedron sheets [115–117]. An increase in the three types of hydroxyl groups supplied more edge and basal adsorption sites for trace elements in sample 3, which resulted in the accumulation of HREE, MREE and related elements in zone I. Besides, sorption experiments showed that HREE and MREE were more readily absorbed by clay minerals than LREE by clay minerals [118]. Therefore, HREE and MREE were more accumulated in sample 3 than LREE.

Furthermore, nano-scale clay minerals are known to develop in fault rocks [25,119]. Previous studies showed that scaly exfoliation of clay minerals usually led to the formation of nano-scale flakes under the influence of shear stress [24,120]. High-resolution transmission electron microscopy results showed that nano-scale clay minerals also developed in sample 3, which might result from compression-shear stress exfoliation (Figure 19). Chemical activity and specific surface of clay minerals were significantly enhanced due to the nano size, making them an important carrier of trace elements [121–123].

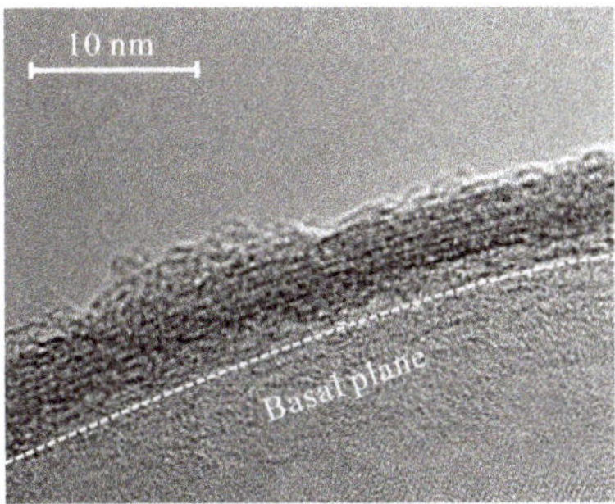

Figure 19. Nano-scale clay mineral in zone I.

5.3. Dissipation Mechanism of Elements Co and Mo

Extraction experiments showed that a large proportion of Co was associated with organic components (at least 20%) [86,124]. The concentration of Mo in lighter gravity fractions of sink-float experiments confirmed that Mo was at least partially associated with organic matter [86,125–127]. Co and Mo bonded to the organic compounds of coal which usually undergo ion-exchange with protons of oxygen functional groups, viz. carboxyl and phenolic groups [86,128,129]. Specific reactions between oxygen functional groups and ions are given in the following equations (According to Wang et al. [129]).

$$aCOAL\text{-}COO^- + M^{a+}(aq) \rightleftharpoons a(COAL\text{-}COO)M \tag{1}$$

$$aCOAL\text{-}O^- + M^{a+}(aq) \rightleftharpoons a(COAL\text{-}O)M \tag{2}$$

Coal molecular structures can be altered by tectonic stress, including the degradation of oxygen functional groups [28–30,130]. FT-IR is an effective method to characterize hydroxyl groups in coal chemical structures. Phenol (1100–1300 cm^{-1}) and carboxyl (1650–1700 cm^{-1}) can be distinguished in FT-IR spectrum ranging from 1100–1700 cm^{-1} [40]. The deconvolution process of FT-IR spectrum was performed using software Origin 7.5 (Figure 20a). Figure 20b shows that the molar content of phenol and carboxyl units significantly decreased in sample 3. Frictional heat generated by fault motion could lead to early-metamorphism of coal containing the degradation of oxygen functional groups [41,130,131]. According to recent research, strong ductile deformation in zone I could also change the molecular structures of coal by transformation into strain energy [39,40,132]. Besides, the higher $R_{o,max}$ of sample 3 also indicated that the maturation degree of sample 3 was advanced by the influence of stress-induced dynamic metamorphism, including the stress degradation of functional groups (Table 1). Therefore, oxygen functional groups (phenol and carboxyl), as a relatively unstable part of the coal molecular structure, were disassociated under the action of frictional heat and ductile deformation. Synchronized variation of phenol, carboxyl, Co and Mo confirmed that these two elements are probably associated with phenol and carboxyl groups. Dissociation of phenol and carboxyl groups led to dissolution of ion-exchanged Co and Mo into migrated solutions, which caused elements Co and Mo to be depleted in sample 3 (Figure 13c).

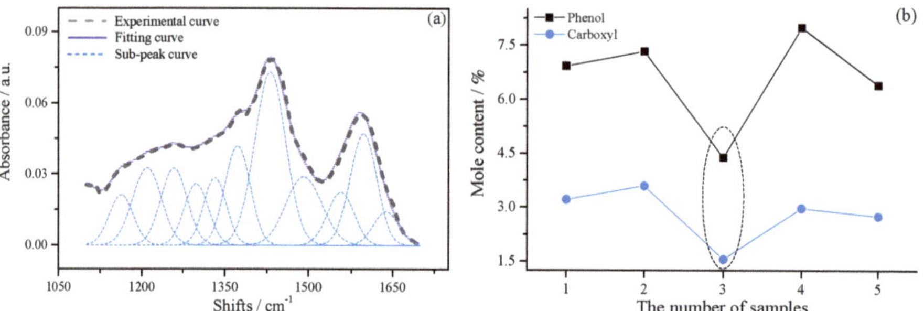

Figure 20. Distribution of phenol and carboxyl in fault-related coals. (**a**) Fitting of FT-IR spectrum in the range of 1100–1700 cm^{-1}. (**b**) Distribution of phenol and carboxyl in fault-related coals.

6. Conclusions

Five typical TDC samples were collected in the same sequence within the reverse fault zone. The geophysical and geochemical characteristics of fault-related coals were analyzed and compared by using appropriate detection and analytical methods. HREE, MREE and related elements (Be, U, W, Bi, Ta, Cr, Nb, In, Zn, Th, Sc, Cu, Ga, Pb, Cd, Li and Mn) commonly associated with invasive solutions showed significant positive anomalies in zone I. Precipitated occlusion and stress-induced smearing of clay minerals decreased the permeability of the pore-cleat system in zone I and then limited the thermal fluid in zone I, which was one of the major reasons for the appearance of significant positive anomalies only in zone I. Strong compression-shear tectonic stress led to disruption and delamination of clay minerals, which enhanced the their adsorption ability for HREE, MREE and related elements in zone I. Even nano-scale clay minerals with higher specific surface area and chemical activity were exfoliated by shear stress. Elements Co and Mo with strong organic affinity were sharply dissipated in zone I, which was ascribed to the disassociation of phenol and carboxyl groups resulting from the fault motion.

Author Contributions: H.L. undertook the experiments, prepared the manuscript and analyzed the data. B.J. conceived the project and designed the experiments. All authors discussed the results and commented on the manuscript.

Funding: This research was funded by the National Natural Science Foundation of China (No. 41672147, 41430317), and the Scientific Research Foundation of Key Laboratory of Coalbed Methane Resources and Reservoir Formation Process, Ministry of Education (China University of Mining and Technology) (No. 2017-004).

Acknowledgments: We would like to thank the reviewers and editors for their constructive comments and suggestions.

Conflicts of Interest: The authors declare no conflict of interest.

References

1. Roques, C.; Aquilina, L.; Bour, O.; Maréchal, J.; Dewandel, B.; Pauwels, H.; Labasque, T.; Vergnaud-Ayraud, V.; Hochreutener, R. Groundwater sources and geochemical processes in a crystalline fault aquifer. *J. Hydrol.* **2014**, *519*, 3110–3128. [CrossRef]

2. Duan, Q.; Yang, X.; Ma, S.; Chen, J.; Chen, J. Fluid–rock interactions in seismic faults: Implications from the structures and mineralogical and geochemical compositions of drilling cores from the rupture of the 2008 Wenchuan earthquake, China. *Tectonophysics* **2016**, *666*, 260–280. [CrossRef]

3. Kuo, L.; Huang, J.; Fang, J.; Si, J.; Song, S.; Li, H.; Yeh, E. Carbonaceous Materials in the Fault Zone of the Longmenshan Fault Belt: 2. Characterization of Fault Gouge from Deep Drilling and Implications for Fault Maturity. *Minerals* **2018**, *8*, 393. [CrossRef]

4. Gajst, H.; Weinberger, R.; Zhu, W.; Lyakhovsky, V.; Marco, S.; Shalev, E. Effects of pre-existing faults on compaction localization in porous sandstones. *Tectonophysics* **2018**, *747–748*, 1–15. [CrossRef]

5. Maciel, I.; Dettori, A.; Balsamo, F.; Bezerra, F.; Vieira, M.; Nogueira, F.; Salvioli-Mariani, E.; Sousa, J. Structural Control on Clay Mineral Authigenesis in Faulted Arkosic Sandstone of the Rio do Peixe Basin, Brazil. *Minerals* **2018**, *8*, 408. [CrossRef]

6. Wilson, J.E.; Chester, J.S.; Chester, F.M. Microfracture analysis of fault growth and wear processes, Punchbowl Fault, San Andreas system, California. *J. Struct. Geol.* **2003**, *25*, 1855–1873. [CrossRef]

7. Chester, J.S.; Chester, F.M.; Kronenberg, A.K. Fracture surface energy of the Punchbowl fault, San Andreas system. *Nature* **2005**, *437*, 133–136. [CrossRef] [PubMed]

8. Pei, Y.; Paton, D.A.; Knipe, R.J.; Wu, K. A review of fault sealing behaviour and its evaluation in siliciclastic rocks. *Earth Sci. Rev.* **2015**, *150*, 121–138. [CrossRef]

9. Buatier, M.D.; Chauvet, A.; Kanitpanyacharoen, W.; Wenk, H.R.; Ritz, J.F.; Jolivet, M. Origin and behavior of clay minerals in the Bogd fault gouge, Mongolia. *J. Struct. Geol.* **2012**, *34*, 77–90. [CrossRef]

10. Arancibia, G.; Fujita, K.; Hoshino, K.; Mitchell, T.M.; Cembrano, J.; Gomila, R.; Morata, D.; Faulkner, D.R.; Rempe, M. Hydrothermal alteration in an exhumed crustal fault zone: Testing geochemical mobility in the Caleta Coloso Fault, Atacama Fault System, Northern Chile. *Tectonophysics* **2014**, *623*, 147–168. [CrossRef]

11. Liu, B.; Liu, Y.; Wu, K.; Wang, X.; Pei, Y.; Guo, J. Geochemical characteristics of fault core and damage zones of the Hong-Che Fault Zone of the Junggar Basin (NW China) with implications for the fault sealing process. *J. Asian Earth Sci.* **2017**, *143*, 141–155. [CrossRef]

12. Richard, J.; Coulon, M.; Gaviglio, P. Mass transfer controlled by fracturing in micritic carbonate rocks. *Tectonophysics* **2002**, *350*, 17–33. [CrossRef]

13. Richard, J. Large-scale mass transfers related to pressure solution creep-faulting interactions in mudstones: Driving processes and impact of lithification degree. *Tectonophysics* **2014**, *612–613*, 40–55. [CrossRef]

14. Fossen, H.; Schultz, R.A.; Shipton, Z.K.; Mair, K. Deformation bands in sandstone: A review. *J. Geol. Soc. London.* **2007**, *164*, 755–769. [CrossRef]

15. Dockrill, B.; Shipton, Z.K. Structural controls on leakage from a natural CO_2 geologic storage site: Central Utah, U.S.A. *J. Struct. Geol.* **2010**, *32*, 1768–1782. [CrossRef]

16. Bauer, J.F.; Meier, S.; Philipp, S.L. Architecture, fracture system, mechanical properties and permeability structure of a fault zone in Lower Triassic sandstone, Upper Rhine Graben. *Tectonophysics* **2015**, *647–648*, 132–145. [CrossRef]

17. Frery, E.; Gratier, J.; Ellouz-Zimmerman, N.; Loiselet, C.; Braun, J.; Deschamps, P.; Blamart, D.; Hamelin, B.; Swennen, R. Evolution of fault permeability during episodic fluid circulation: Evidence for the effects of fluid–rock interactions from travertine studies (Utah–USA). *Tectonophysics* **2015**, *651–652*, 121–137. [CrossRef]

18. Indrevær, K.; Stunitz, H.; Bergh, S.G. On Palaeozoic–Mesozoic brittle normal faults along the SW Barents Sea margin: Fault processes and implications for basement permeability and margin evolution. *J. Geol. Soc. London.* **2014**, *171*, 831–846. [CrossRef]

19. Si, J.; Li, H.; Kuo, L.; Huang, J.; Song, S.; Pei, J.; Wang, H.; Song, L.; Fang, J.; Sheu, H. Carbonaceous Materials in the Longmenshan Fault Belt Zone: 3. Records of Seismic Slip from the Trench and Implications for Faulting Mechanisms. *Minerals* **2018**, *8*, 457. [CrossRef]

20. Yielding, G.; Freeman, B.; Needham, D.T. Quantitative Fault Seal Prediction. *AAPG Bull.* **1997**, *81*, 897–917.

21. Jones, R.M.; Hillis, R.R. An integrated, quantitative approach to assessing fault-seal risk. *AAPG Bull.* **2003**, *87*, 507–524. [CrossRef]

22. Tenthorey, E.; Cox, S.F.; Todd, H.F. Evolution of strength recovery and permeability during fluid–rock reaction in experimental fault zones. *Earth Planet. Sc. Lett.* **2003**, *206*, 161–172. [CrossRef]

23. Solum, J.G.; van der Pluijm, B.A.; Peacor, D.R. Neocrystallization, fabrics and age of clay minerals from an exposure of the Moab Fault, Utah. *J. Struct. Geol.* **2005**, *27*, 1563–1576. [CrossRef]

24. Sun, Y.; Jiang, S.Y.; Wei, Z.; Lu, X.C. Nano-Coating Texture on the Shear Slip Surface in Rocky Materials. *Adv. Mater. Res.* **2013**, *669*, 108–114. [CrossRef]

25. Viti, C.; Collettini, C.; Tesei, T. Pressure solution seams in carbonatic fault rocks: Mineralogy, micro/nanostructures and deformation mechanism. *Contrib. Mineral. Petr.* **2014**, *167*, 970. [CrossRef]

26. Frodsham, K.; Gayer, R.A. The impact of tectonic deformation upon coal seams in the South Wales coalfield, UK. *Int. J. Coal Geol.* **1999**, *38*, 297–332. [CrossRef]

27. Jiang, B.; Qu, Z.; Wang, G.G.X.; Li, M. Effects of structural deformation on formation of coalbed methane reservoirs in Huaibei coalfield, China. *Int. J. Coal Geol.* **2010**, *82*, 175–183. [CrossRef]

28. Ju, Y.; Li, X. New research progress on the ultrastructure of tectonically deformed coals. *Prog. Nat. Sci. Mater.* **2009**, *19*, 1455–1466. [CrossRef]

29. Li, W.; Jiang, B.; Moore, T.A.; Wang, G.; Liu, J.; Song, Y. Characterization of the Chemical Structure of Tectonically Deformed Coals. *Energ. Fuel.* **2017**, *31*, 6977–6985. [CrossRef]

30. Pan, J.; Zhu, H.; Hou, Q.; Wang, H.; Wang, S. Macromolecular and pore structures of Chinese tectonically deformed coal studied by atomic force microscopy. *Fuel* **2015**, *139*, 94–101. [CrossRef]

31. Song, Y.; Jiang, B.; Li, F.; Liu, J. Structure and fractal characteristic of micro- and meso-pores in low, middle-rank tectonic deformed coals by CO_2 and N_2 adsorption. *Micropor. Mesopor. Mat.* **2017**, *253*, 191–202.

32. Liu, Q.; Zhang, K.; Zhou, H.; Cheng, Y.; Zhang, H.; Wang, L. Experimental investigation into the damage-induced permeability and deformation relationship of tectonically deformed coal from Huainan coalfield, China. *J. Nat. Gas Sci. Eng.* **2018**, *60*, 202–213. [CrossRef]

33. Zhang, Z.; Cao, S.; Li, Y.; Guo, P.; Yang, H.; Yang, T. Effect of moisture content on methane adsorption- and desorption-induced deformation of tectonically deformed coal. *Adsorpt. Sci. Technol.* **2018**, *36*, 1648–1668. [CrossRef]

34. Li, F.; Jiang, B.; Cheng, G.; Song, Y.; Tang, Z. Structural and evolutionary characteristics of pores-microfractures and their influence on coalbed methane exploitation in high-rank brittle tectonically deformed coals of the Yangquan mining area, northeastern Qinshui basin, China. *J. Petrol. Sci. Eng.* **2019**, *174*, 1290–1302.

35. Xie, P.; Hower, J.C.; Liu, X. Petrographic characteristics of the brecciated coals from Panxian county, Guizhou, southwestern China. *Fuel* **2019**, *243*, 1–9. [CrossRef]

36. Karayiğit, A.İ.; Mastalerz, M.; Oskay, R.G.; Buzkan, İ. Bituminous coal seams from underground mines in the Zonguldak Basin (NW Turkey): Insights from mineralogy, coal petrography, Rock-Eval pyrolysis, and meso-and microporosity. *Int. J. Coal Geol.* **2018**, *199*, 91–112. [CrossRef]

37. Cao, D.; Li, X.; Zhang, S. Influence of tectonic stress on coalification: Stress degradation mechanism and stress polycondensation mechanism. *Sci. China Ser. D* **2007**, *50*, 43–54. [CrossRef]

38. Hou, Q.; Li, H.; Fan, J.; Ju, Y.; Wang, T.; Li, X.; Wu, Y. Structure and coalbed methane occurrence in tectonically deformed coals. *Sci. China Earth Sci.* **2012**, *55*, 1755–1763. [CrossRef]

39. Xu, R.; Li, H.; Hou, Q.; Li, X.; Yu, L. The effect of different deformation mechanisms on the chemical structure of anthracite coals. *Sci. China Earth Sci.* **2015**, *58*, 502–509. [CrossRef]

40. Song, Y.; Jiang, B.; Han, Y. Macromolecular response to tectonic deformation in low-rank tectonically deformed coals (TDCs). *Fuel* **2018**, *219*, 279–287. [CrossRef]

41. Bustin, R.M. Heating during thrust faulting in the Rocky mountains: Friction or fiction? *Tectonophysics* **1983**, *95*, 309–328. [CrossRef]

42. Cao, Y.X.; Mitchell, G.D.; Davis, A.; Wang, D.M. Deformation metamorphism of bituminous and anthracite coals from China. *Int. J. Coal Geol.* **2000**, *43*, 227–242. [CrossRef]

43. Cao, Y.X.; Davis, A.; Liu, R.X.; Liu, X.W.; Zhang, Y.G. The influence of tectonic deformation on some geochemical properties of coals—A possible indicator of outburst potential. *Int. J. Coal Geol.* **2003**, *53*, 69–79. [CrossRef]

44. Chen, J.; Liu, G.; Jiang, M.; Chou, C.; Li, H.; Wu, B.; Zheng, L.; Jiang, D. Geochemistry of environmentally sensitive trace elements in Permian coals from the Huainan coalfield, Anhui, China. *Int. J. Coal Geol.* **2011**, *88*, 41–54. [CrossRef]

45. Saikia, J.; Narzary, B.; Roy, S.; Bordoloi, M.; Saikia, P.; Saikia, B.K. Nanominerals, fullerene aggregates, and hazardous elements in coal and coal combustion-generated aerosols: An environmental and toxicological assessment. *Chemosphere* **2016**, *164*, 84–91. [CrossRef] [PubMed]

46. Dai, S.; Finkelman, R.B. Coal as a promising source of critical elements: Progress and future prospects. *Int. J. Coal Geol.* **2018**, *186*, 155–164. [CrossRef]

47. Stewart, B.W.; Capo, R.C.; Hedin, B.C.; Hedin, R.S. Rare earth element resources in coal mine drainage and treatment precipitates in the Appalachian Basin, USA. *Int. J. Coal Geol.* **2017**, *169*, 28–39. [CrossRef]

48. Dai, S.; Yan, X.; Ward, C.R.; Hower, J.C.; Zhao, L.; Wang, X.; Zhao, L.; Ren, D.; Finkelman, R.B. Valuable elements in Chinese coals: A review. *Int. Geol. Rev.* **2016**, *60*, 590–620. [CrossRef]

49. Dai, S.; Ren, D.; Chou, C.; Finkelman, R.B.; Seredin, V.V.; Zhou, Y. Geochemistry of trace elements in Chinese coals: A review of abundances, genetic types, impacts on human health, and industrial utilization. *Int. J. Coal Geol.* **2012**, *94*, 3–21. [CrossRef]

50. Dai, S.; Luo, Y.; Seredin, V.V.; Ward, C.R.; Hower, J.C.; Zhao, L.; Liu, S.; Zhao, C.; Tian, H.; Zou, J. Revisiting the late Permian coal from the Huayingshan, Sichuan, southwestern China: Enrichment and occurrence modes of minerals and trace elements. *Int. J. Coal Geol.* **2014**, *122*, 110–128. [CrossRef]

51. Bullock, L.; Parnell, J.; Perez, M.; Feldmann, J. Tellurium Enrichment in Jurassic Coal, Brora, Scotland. *Minerals* **2017**, *7*, 231. [CrossRef]

52. Yuan, Y.; Tang, S.; Zhang, S. Geochemical and Mineralogical Characteristics of the Middle Jurassic Coals from the Tongjialiang Mine in the Northern Datong Coalfield, Shanxi Province, China. *Minerals* **2019**, *9*, 184. [CrossRef]

53. Karayiğit, A.I.; Bircan, C.; Mastalerz, M.; Oskay, R.G.; Querol, X.; Lieberman, N.R.; Türkmen, I. Coal characteristics, elemental composition and modes of occurrence of some elements in the İsaalan coal (Balıkesir, NW Turkey). *Int. J. Coal Geol.* **2017**, *172*, 43–59. [CrossRef]

54. Dai, S.; Ji, D.; Ward, C.R.; French, D.; Hower, J.C.; Yan, X.; Wei, Q. Mississippian anthracites in Guangxi Province, southern China: Petrological, mineralogical, and rare earth element evidence for high-temperature solutions. *Int. J. Coal Geol.* **2018**, *197*, 84–114. [CrossRef]

55. Chen, J.; Liu, G.; Li, H.; Wu, B. Mineralogical and geochemical responses of coal to igneous intrusion in the Pansan Coal Mine of the Huainan coalfield, Anhui, China. *Int. J. Coal Geol.* **2014**, *124*, 11–35. [CrossRef]

56. Dai, S.; Wang, X.; Zhou, Y.; Hower, J.C.; Li, D.; Chen, W.; Zhu, X.; Zou, J. Chemical and mineralogical compositions of silicic, mafic, and alkali tonsteins in the late Permian coals from the Songzao Coalfield, Chongqing, Southwest China. *Chem. Geol.* **2011**, *282*, 29–44. [CrossRef]

57. Fowler, P.; Gayer, R.A. The association between tectonic deformation, inorganic composition and coal rank in the bituminous coals from the South Wales coalfield, United Kingdom. *Int. J. Coal Geol.* **1999**, *42*, 1–31. [CrossRef]

58. Dawson, G.K.W.; Golding, S.D.; Esterle, J.S.; Massarotto, P. Occurrence of minerals within fractures and matrix of selected Bowen and Ruhr Basin coals. *Int. J. Coal Geol.* **2012**, *94*, 150–166. [CrossRef]

59. Hower, J.C.; Williams, D.A.; Eble, C.F.; Sakulpitakphon, T.; Moecher, D.P. Brecciated and mineralized coals in Union County, Western Kentucky coal field. *Int. J. Coal Geol.* **2001**, *47*, 223–234. [CrossRef]

60. Permana, A.K.; Ward, C.R.; Li, Z.; Gurba, L.W. Distribution and origin of minerals in high-rank coals of the South Walker Creek area, Bowen Basin, Australia. *Int. J. Coal Geol.* **2013**, *116–117*, 185–207. [CrossRef]

61. Karayiğit, A.İ.; Mastalerz, M.; Oskay, R.G.; Gayer, R.A. Coal petrography, mineralogy, elemental compositions and palaeoenvironmental interpretation of Late Carboniferous coal seams in three wells from the Kozlu coalfield (Zonguldak Basin, NW Turkey). *Int. J. Coal Geol.* **2018**, *187*, 54–70. [CrossRef]

62. Song, Y.; Jiang, B.; Liu, H.; Li, F.; Shao, P.; Yan, G. Variations in stress-sensitive minerals and elements in the tectonic-deformation Early to Middle Permian coals from the Zhuxianzhuang mine, Anhui Province. *J. Geochem. Explor.* **2018**, *188*, 11–23.

63. Ghosh, S.; Rodrigues, S.; Varma, A.K.; Esterle, J.; Patra, S.; Dirghangi, S.S. Petrographic and Raman spectroscopic characterization of coal from Himalayan fold-thrust belts of Sikkim, India. *Int. J. Coal Geol.* **2018**, *196*, 246–259. [CrossRef]

64. Liu, X.; Song, D.; He, X.; Nie, B.; Wang, L. Insight into the macromolecular structural differences between hard coal and deformed soft coal. *Fuel* **2019**, *245*, 188–197. [CrossRef]

65. Li, Y.; Jiang, B.; Qu, Z. Controls on migration and aggregation for tectonically sensitive elements in tectonically deformed coal: An example from the Haizi mine, Huaibei coalfield, China. *Sci. China Earth Sci.* **2014**, *57*, 1180–1191. [CrossRef]

66. Yang, M.; Liu, G.; Sun, R.; Chou, C.; Zheng, L. Characterization of intrusive rocks and REE geochemistry of coals from the Zhuji Coal Mine, Huainan Coalfield, Anhui, China. *Int. J. Coal Geol.* **2012**, *94*, 283–295. [CrossRef]

67. Jiang, B.; Li, Y.; Qu, Z.; Li, M. Preliminary study on theory and method of structural geochemistry of gas outburst prediction. *J. China Coal Soc.* **2015**, *40*, 1408–1414, (In Chinese with English abstract).

68. Shen, H.Z. Hydrogeological characteristics of the fourth aquifer in Sunan coal mine. Master's Thesis, Anhui University of Science & Technology, Huainan, China, 2005. (In Chinese with English abstract).

69. Li, M. Structure evolution and deformation mechanism of tectonically deformed coal. Ph.D. Thesis, China University of Mining and Technology, Xuzhou, China, 2013. (In Chinese with English abstract).

70. China State Bureau of Technical Supervision. *Sampling of Coal in Seam (GB482-1995)*; Standards Press of China: Beijing, China, 1995. (In Chinese)

71. Liu, J.; Jiang, B.; Li, M.; Qu, Z.; Wang, L.; Li, L. Structural control on pore-fracture characteristics of coals from Xinjing coal mine, northeastern Qinshui basin, China. *Arab. J. Geosci.* **2015**, *8*, 4421–4431. [CrossRef]

72. Ward, C.R. Analysis and significance of mineral matter in coal seams. *Int. J. Coal Geol.* **2002**, *50*, 135–168. [CrossRef]

73. Buatier, M.D.; Trave, A.; Labaume, P.; Potdevin, J.L. Dickite related to fluid-sediment interaction and deformation in Pyrenean thrust fault zones. *Eur. J. Mineral.* **1997**, *9*, 875–888. [CrossRef]

74. Dai, S.; Graham, I.T.; Ward, C.R. A review of anomalous rare earth elements and yttrium in coal. *Int. J. Coal Geol.* **2016**, *159*, 82–95. [CrossRef]

75. Seredin, V.V.; Dai, S. Coal deposits as potential alternative sources for lanthanides and yttrium. *Int. J. Coal Geol.* **2012**, *94*, 67–93. [CrossRef]

76. Taylor, S.R.; McLennan, S.M. *The Continental Crust: Its Composition and Evolution*; Blackwell Scientific Publication: Carlton, Australia, 1985; Volume 94.

77. Yan, X.; Dai, S.; Graham, I.T.; He, X.; Shan, K.; Liu, X. Determination of Eu concentrations in coal, fly ash and sedimentary rocks using a cation exchange resin and inductively coupled plasma mass spectrometry (ICP-MS). *Int. J. Coal Geol.* **2018**, *191*, 152–156. [CrossRef]

78. Eskenazy, G.M. Aspects of the geochemistry of rare earth elements in coal: An experimental approach. *Int. J. Coal Geol.* **1999**, *38*, 285–295. [CrossRef]

79. Zheng, L.; Liu, G.; Chou, C.; Qi, C.; Zhang, Y. Geochemistry of rare earth elements in Permian coals from the Huaibei Coalfield, China. *J. Asian Earth Sci.* **2007**, *31*, 167–176. [CrossRef]

80. Zheng, L.; Liu, G.; Wang, L.; Chou, C. Composition and quality of coals in the Huaibei Coalfield, Anhui, China. *J. Geochem. Explor.* **2008**, *97*, 59–68. [CrossRef]

81. Dai, S.; Xie, P.; Jia, S.; Ward, C.R.; Hower, J.C.; Yan, X.; French, D. Enrichment of U-Re-V-Cr-Se and rare earth elements in the Late Permian coals of the Moxinpo Coalfield, Chongqing, China: Genetic implications from geochemical and mineralogical data. *Ore Geol. Rev.* **2017**, *80*, 1–17. [CrossRef]

82. Dai, S.; Seredin, V.V.; Ward, C.R.; Hower, J.C.; Xing, Y.; Zhang, W.; Song, W.; Wang, P. Enrichment of U-Se-Mo-Re-V in coals preserved within marine carbonate successions: Geochemical and mineralogical data from the Late Permian Guiding Coalfield, Guizhou, China. *Miner. Depos.* **2015**, *50*, 159–186. [CrossRef]

83. Zhang, J.; Ren, D.; Zheng, C.; Zeng, R.; Chou, C.; Liu, J. Trace element abundances in major minerals of Late Permian coals from southwestern Guizhou province, China. *Int. J. Coal Geol.* **2002**, *53*, 55–64. [CrossRef]

84. Dai, S.; Jiang, Y.; Ward, C.R.; Gu, L.; Seredin, V.V.; Liu, H.; Zhou, D.; Wang, X.; Sun, Y.; Zou, J.; et al. Mineralogical and geochemical compositions of the coal in the Guanbanwusu Mine, Inner Mongolia, China: Further evidence for the existence of an Al (Ga and REE) ore deposit in the Jungar Coalfield. *Int. J. Coal Geol.* **2012**, *98*, 10–40. [CrossRef]

85. Tian, C.; Zhang, J.; Zhao, Y.; Gupta, R. Understanding of mineralogy and residence of trace elements in coals via a novel method combining low temperature ashing and float-sink technique. *Int. J. Coal Geol.* **2014**, *131*, 162–171. [CrossRef]

86. Finkelman, R.B.; Palmer, C.A.; Wang, P. Quantification of the modes of occurrence of 42 elements in coal. *Int. J. Coal Geol.* **2018**, *185*, 138–160. [CrossRef]

87. Dai, S.; Li, T.; Seredin, V.V.; Ward, C.R.; Hower, J.C.; Zhou, Y.; Zhang, M.; Song, W.; Song, X.; Zhao, C. Origin of minerals and elements in the Late Permian coals, tonsteins, and host rocks of the Xinde Mine, Xuanwei, eastern Yunnan, China. *Int. J. Coal Geol.* **2014**, *121*, 53–78. [CrossRef]

88. Dai, S.; Zhang, W.; Ward, C.R.; Seredin, V.V.; Hower, J.C.; Li, X.; Song, W.; Wang, X.; Kang, H.; Zheng, L.; et al. Mineralogical and geochemical anomalies of late Permian coals from the Fusui Coalfield, Guangxi Province, southern China: Influences of terrigenous materials and hydrothermal fluids. *Int. J. Coal Geol.* **2013**, *105*, 60–84. [CrossRef]

89. Kolker, A.; Chou, C. Cleat-Filling Calcite in Illinois Basin Coals: Trace-Element Evidence for Meteoric Fluid Migration in a Coal Basin. *J. Geol.* **1994**, *102*, 111–116. [CrossRef]

90. Bouška, V.; Pesek, J.; Sykorova, I. Probable Modes of Occurrence of Chemical Elements in Coal. *Acta Mont. Ser. B.* **2000**, *117*, 53–90.

91. Dai, S.; Li, D.; Ren, D.; Tang, Y.; Shao, L.; Song, H. Geochemistry of the late Permian No. 30 coal seam, Zhijin Coalfield of Southwest China: Influence of a siliceous low-temperature hydrothermal fluid. *Appl. Geochem.* **2004**, *19*, 1315–1330. [CrossRef]

92. Ju, Y.; Luxbacher, K.; Li, X.; Wang, G.; Yan, Z.; Wei, M.; Yu, L. Micro-structural evolution and their effects on physical properties in different types of tectonically deformed coals. *Int. J. Coal Sci. Technol.* **2014**, *1*, 364–375. [CrossRef]

93. Li, B.; Zhuang, X.; Querol, X.; Moreno, N.; Córdoba, P.; Li, J.; Zhou, J.; Ma, X.; Liu, S.; Shangguan, Y. The mode of occurrence and origin of minerals in the Early Permian high-rank coals of the Jimunai depression, Xinjiang Uygur Autonomous Region, NW China. *Int. J. Coal Geol.* **2019**, *205*, 58–74. [CrossRef]

94. Li, M.; Jiang, B.; Lin, S.; Lan, F.; Wang, J. Structural Controls on Coalbed Methane Reservoirs in Faer Coal Mine, Southwest China. *J. Earth Sci. China* **2013**, *24*, 437–448. [CrossRef]

95. Niu, Q.; Pan, J.; Cao, L.; Ji, Z.; Wang, H.; Wang, K.; Wang, Z. The evolution and formation mechanisms of closed pores in coal. *Fuel* **2017**, *200*, 555–563. [CrossRef]

96. Chen, Y.; Qin, Y.; Wei, C.; Huang, L.; Shi, Q.; Wu, C.; Zhang, X. Porosity changes in progressively pulverized anthracite subsamples: Implications for the study of closed pore distribution in coals. *Fuel* **2018**, *225*, 612–622. [CrossRef]

97. Karacan, C.O.; Okandan, E. Fracture/cleat analysis of coals from Zonguldak Basin (northwestern Turkey) relative to the potential of coalbed methane production. *Int. J. Coal Geol.* **2000**, *44*, 109–125. [CrossRef]

98. Renard, F.; Gratier, J.; Jamtveit, B. Kinetics of crack-sealing, intergranular pressure solution, and compaction around active faults. *J. Struct. Geol.* **2000**, *22*, 1395–1407. [CrossRef]

99. Ju, Y.; Sun, Y.; Wan, Q.; Lu, S.; He, H.; Wu, J.; Zhang, W.; Wang, G.; Huang, C. Nanogeology: A Revolutionary Challenge in Geosciences. *Bull. Mineral. Petrol. Geochem.* **2016**, *35*, 1–20.

100. Vannucchi, P.; Maltman, A.; Bettelli, G.; Clennell, B. On the nature of scaly fabric and scaly clay. *J. Struct. Geol.* **2003**, *25*, 673–688. [CrossRef]

101. Guo, D.; Han, D.; Chen, Y.; Liu, F. Quantitative study of surface components of outburst coal with XPS method. *Coal Eng.* **1996**, *1*, 2–4, (In Chinese with English abstract).

102. Kristof, J.; Frost, R.L.; Kloprogge, J.T.; Horvath, E.; Mako, E. Detection of four different OH-groups in ground kaolinite with controlled-rate thermal analysis. *J. Therm. Anal. Calorim.* **2002**, *69*, 77–83. [CrossRef]

103. Strydom, C.A.; Bunt, J.R.; Schobert, H.H.; Raghoo, M. Changes to the organic functional groups of an inertinite rich medium rank bituminous coal during acid treatment processes. *Fuel Process. Technol.* **2011**, *92*, 764–770. [CrossRef]

104. Serratosa, J.M.; Hidalgo, J.M.; Vinas, J.M. Orientation of OH Bonds in Kaolinite. *Nature* **1962**, *195*, 486–487. [CrossRef]

105. Ledoux, R.L.; White, J.L. Infrared Study of the OH Groups in Expanded Kaolinite. *Science* **1964**, *143*, 244–246. [CrossRef] [PubMed]

106. Brindley, W.G.; Kao, C.C.; Harrison, J.L. Relation between Structural Disorder and Other Characteristics of Kaolinites and Dickites. *Clay Clay Miner.* **1986**, *34*, 239–249. [CrossRef]

107. Michalkova, A.; Szymczak, J.J.; Leszczynski, J. Adsorption of 2,4-Dinitrotoluene on Dickite: The Role of H-Bonding. *Struct. Chem.* **2005**, *16*, 325–337. [CrossRef]

108. Welch, M.D.; Crichton, W.A. Pressure-induced transformations in kaolinite. *Am. Mineral.* **2010**, *95*, 651–654. [CrossRef]

109. Ogasawara, M. Geochemistry of Rare-Earth Elements. *Min. Geol.* **2009**, *39*, 166–176.

110. Seredin, V.V. Rare earth element-bearing coals from the Russian Far East deposits. *Int. J. Coal Geol.* **1996**, *30*, 101–129. [CrossRef]

111. Bradbury, M.H.; Baeyens, B. Sorption of Eu on Na- and Ca-montmorillonites: Experimental investigations and modelling with cation exchange and surface complexation. *Geochim. Cosmochim. Acta* **2002**, *66*, 2325–2334. [CrossRef]

112. Moldoveanu, G.A.; Papangelakis, V.G. Recovery of rare earth elements adsorbed on clay minerals: I. Desorption mechanism. *Hydrometallurgy* **2012**, *117–118*, 71–78. [CrossRef]

113. Moldoveanu, G.A.; Papangelakis, V.G. Recovery of rare earth elements adsorbed on clay minerals: II. Leaching with ammonium sulfate. *Hydrometallurgy* **2013**, *131–132*, 158–166. [CrossRef]

114. Sen Gupta, S.; Bhattacharyya, K.G. Kinetics of adsorption of metal ions on inorganic materials: A review. *Adv. Colloid Interfac.* **2011**, *162*, 39–58. [CrossRef]

115. Tertre, E.; Berger, G.; Simoni, E.; Castet, S.; Giffaut, E.; Loubet, M.; Catalette, H. Europium retention onto clay minerals from 25 to 150 °C: Experimental measurements, spectroscopic features and sorption modelling. *Geochim. Cosmochim. Acta* **2006**, *70*, 4563–4578. [CrossRef]

116. Geckeis, H.; Lützenkirchen, J.; Polly, R.; Rabung, T.; Schmidt, M. Mineral–Water Interface Reactions of Actinides. *Chem. Rev.* **2012**, *113*, 1016–1062. [CrossRef] [PubMed]

117. Martin, L.A.; Wissocq, A.; Benedetti, M.F.; Latrille, C. Thallium (Tl) sorption onto illite and smectite: Implications for Tl mobility in the environment. *Geochim. Cosmochim. Acta* **2018**, *230*, 1–16. [CrossRef]

118. Coppin, F.; Berger, G.; Bauer, A.; Castet, S.; Loubet, M. Sorption of lanthanides on smectite and kaolinite. *Chem. Geol.* **2002**, *182*, 57–68. [CrossRef]

119. Schleicher, A.M.; van der Pluijm, B.A.; Warr, L.N. Nanocoatings of clay and creep of the San Andreas fault at Parkfield, California. *Geology* **2010**, *38*, 667–670. [CrossRef]

120. Sun, Y.; Shu, L.; Lu, X.; Liu, H.; Zhang, X.; Kosaka, K.; Lin, A. A comparative study of natural and experimental nano-sized grinding grain textures in rocks. *Chin. Sci. Bull.* **2008**, *53*, 1217–1221. [CrossRef]

121. Hochella, M.F.; Lower, S.K.; Maurice, P.A.; Penn, R.L.; Sahai, N.; Sparks, D.L.; Twining, B.S. Nanominerals, Mineral Nanoparticles, and Earth Systems. *Science* **2008**, *319*, 1631–1635. [CrossRef]

122. Silva, L.F.O.; Jasper, A.; Andrade, M.L.; Sampaio, C.H.; Dai, S.; Li, X.; Li, T.; Chen, W.; Wang, X.; Liu, H.; et al. Applied investigation on the interaction of hazardous elements binding on ultrafine and nanoparticles in Chinese anthracite-derived fly ash. *Sci. Total Environ.* **2012**, *419*, 250–264. [CrossRef]

123. Chen, T.; Xie, Q.; Liu, H.; Xie, J.; Zhou, Y. Nano-minerals and nano-mineral resources. *Earth Sci.* **2018**, *43*, 1439–1449, (In Chinese with English abstract).

124. Mastalerz, M. Modes of occurrence of trace elements in coal. *Int. J. Coal Geol.* **2001**, *46*, 66. [CrossRef]

125. Chatterjee, P.K. Coal mining disasters in New South Wales and Queensland between 1920–1979—A technical appraisal: Proc Australas Inst Min Metall, N281, March 1982, P37–43. *Int. J. Rock Mech. Min. Sci. Geomech. Abstr.* **1982**, *19*, 136. [CrossRef]

126. Finkelman, R.B. Trace elements in coal. *Biol. Trace Elem. Res.* **1999**, *67*, 197–204. [CrossRef] [PubMed]
127. Liu, J.; Yang, Z.; Yan, X.; Ji, D.; Yang, Y.; Hu, L. Modes of occurrence of highly-elevated trace elements in superhigh-organic-sulfur coals. *Fuel* **2015**, *156*, 190–197. [CrossRef]
128. Mizera, J.; Mizerová, G.; Machovič, V.; Borecká, L. Sorption of cesium, cobalt and europium on low-rank coal and chitosan. *Water Res.* **2007**, *41*, 620–626. [CrossRef] [PubMed]
129. Wang, Z.; Bai, Z.; Li, W.; Bai, J.; Guo, Z.; Chen, H. Effects of ion-exchanged calcium, barium and magnesium on cross-linking reactions during direct liquefaction of oxidized lignite. *Fuel Process. Technol.* **2012**, *94*, 34–39. [CrossRef]
130. O'Hara, K. Paleo-stress estimates on ancient seismogenic faults based on frictional heating of coal. *Geophys. Res. Lett.* **2004**, *31*. [CrossRef]
131. Kitamura, M.; Mukoyoshi, H.; Fulton, P.M.; Hirose, T. Coal maturation by frictional heat during rapid fault slip. *Geophys. Res. Lett.* **2012**, *39*. [CrossRef]
132. Liu, H.; Jiang, B.; Liu, J.; Song, Y. The evolutionary characteristics and mechanisms of coal chemical structure in micro deformed domains under sub-high temperatures and high pressures. *Fuel* **2018**, *222*, 258–268. [CrossRef]

Article

Metal-Bearing Nanoparticles Observed in Soils and Fault Gouges over the Shenjiayao Gold Deposit and Their Significance

Bimin Zhang [1,2], Zhixuan Han [1,2,*], Xueqiu Wang [1,2], Hanliang Liu [1,2], Hui Wu [1,2] and Hui Feng [3,*]

[1] Key Laboratory of Geochemical Exploration, Institute of Geophysical and Geochemical Exploration, CAGS, Langfang 065000, China
[2] UNESCO International Centre on Global-scale Geochemistry, Langfang 065000, China
[3] Beijing Institute of Geo-exploration Technology, Beijing 102209, China
* Correspondence: hanzhixuan@igge.cn (Z.H.); fenghuimail@163.com (H.F.); Tel.: +86-316-2267721 (Z.H.)

Received: 30 April 2019; Accepted: 3 July 2019; Published: 5 July 2019

Abstract: Mineral deposits concealed by thick cover sequences present special problems for geochemical exploration. A variety of penetrating geochemical methods have been developed in the last few decades to explore for buried deposits. The theoretical basis of the mechanism by which metals migrate upward from buried deposits through the cover to the surface is still not fully understood. One hypothesis is that metal particles or metal elements could be carried onto bubbles or micro-flow of geogas and migrate upward to the surface. After years of study, nano-scale metal-bearing particles have been widely observed in geogas samples from different kinds of concealed deposits. However, the occurrence of these metal-bearing particles carried by geogases in near-surface media, such as soil, has not been studied in detail. In this study, metal-bearing nanoparticles were observed in samples from soils and fault gouges over the Shenjiayao gold deposit. The results indicate that (1) the ore-forming elements in soils can only come from deep-seated ore bodies and they occur in nanoparticles in the study area; (2) there is an obvious relationship between metal nanoparticles in fault gouges and soils; (3) the metallic nanoparticles in fault gouges represent a transitional phase along the whole vertical migration process. In addition, the observation results show that the metal-bearing nanoparticles tend to be adsorbed on the surface of clay minerals, which provide theoretical support for using fine fraction soils as sampling media to carry out geochemical exploration in sediment-covered terrains. Based on the results and discussion, a simple migration model was built in this paper.

Keywords: metallic nanoparticles; migration mechanism; prospecting; Shenjiayao gold deposit

1. Introduction

As discoveries of world-class mineral deposits continue to decline, increased attention is being focused on geochemical exploration methods specifically designed for terrains covered by thick regolith [1,2]. These methods include partial extraction techniques, geogas analysis, electrogeochemistry, biogeochemistry, hydrochemistry, etc. [3–17]. Geogas is an effective method in the search for concealed deposits in covered terrains and has been used for mineral deposit exploration with satisfactory results [9,18–25]. This method is based on the assumption that the geogas could carry ultrafine metal-bearing particles or metal elements in the form of tiny bubbles or micro-flow and migrate upwards to the surface [9,12,18,26–29]. In order to prove the theory is plausible, the best way is to find ultrafine metal-bearing particles in geogas samples that definitely come from deep-seated ore bodies. After years of studies, ultrafine metal-bearing particles at the nano-scale have been widely observed in geogas samples from different kinds of concealed deposits [12,30–41], which provide the evidence for the vertical migration of elements through geogas media and also indicate that geogas is a proper sampling media for geochemical exploration in covered terrains. However, soil is the most common sampling media in prospecting activities. Many studies have proved that soils over concealed deposits contain anomalies that can reflect deep-seated ore bodies [2,9,10,14,15,36,37,42–52]. In some cases, the sediment cover over the deep-seated ore bodies is very simple and uniform. Often, the only source of anomalies in soils is from the deep-seated ore bodies [2,10,36,46,49]. In order to explain the anomaly formation mechanism, the key approach is to ascertain the occurrence and origin of ore-forming elements in soils over concealed deposits.

In this study, we observed the nanoparticles in soils and fault gouges over the Shenjiayao gold deposit in China, which is covered by several to tens of meters of loess. The loess land in China covers 632,000 km^2, occupying approximately 6% of the total area of the country. The objective of this study is to (a) characterize and compare the nanoparticles sampled from various media; (b) determine the origin of the metal-bearing nanoparticles; and (c) illuminate the migration mechanism of ore-forming elements from deep-seated ore bodies to the earth's surface and discuss the significance of nanoparticles in geogas for mineral prospecting.

2. Geological Setting

The Shenjiayao gold deposit (134,000 oz of gold) is situated in the Xiaoqinling-Xiong'ershan gold metallogenic belt in the Qinling Mountains of eastern Shananxi and western Henan provinces in central China [53]. The Xiaoshan area is a structural dome bounded by the regional Sanmenxia-Baofeng Fault to the north and the Jiaohe Fault to the south. The dome has a core of Archean basement rocks, which is surrounded by Proterozoic metamorphic rocks (Figure 1). The Xiaoqinling-Xiong'ershan region has an indicated resource of more than 400 tonnes of gold and is second in China [54]. Deposits in this region are commonly small but of high grade.

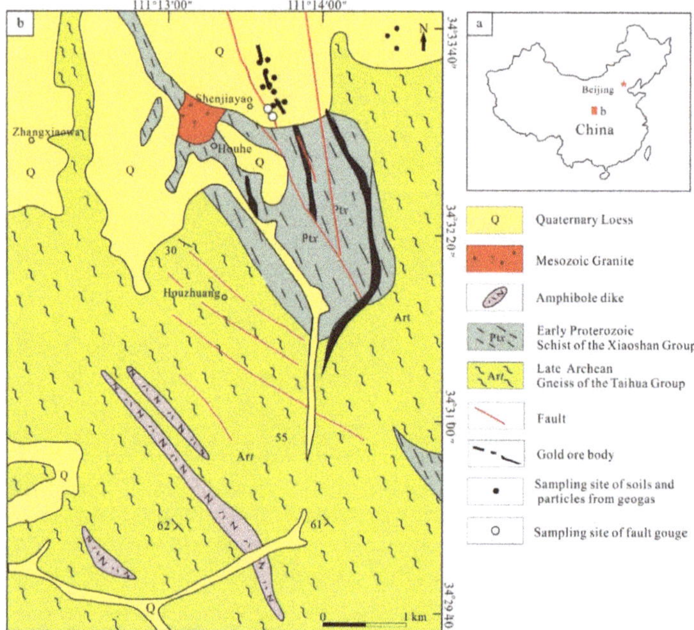

Figure 1. Geological map and sampling sites in the Shenjiayao gold deposit (modified after Chen et al. [55]).

The Shenjiayao deposit is a metamorphic hydrothermal gold deposit and is the largest deposit in the Xiaoshan area. The gold ores occur in quartz veins and in highly altered and fractured rocks. The faults in this area are primarily striking to the NNW–NW and the ore bodies are also primarily striking the same direction (Figure 1). Gneiss, plagiogranitic gneiss, and granodioritic rocks of the Taihua Group hosts the deposit. Principal ore minerals include: Native gold, electrum, pyrite, chalcopyrite, galena, sphalerite, arsenopyrite, and marmatite. Gangue minerals mainly include: Quartz, sericite, siderite, dolomite, calcite, chlorite, barite, and kaolinite [25,53].

The concealed ore bodies mainly locate in the contact zones of Archean plagioclase gneiss and Early Proterozoic schist. The bed rock is covered by loess in study area. The thickness of loess cover is 22–75 m.

3. Materials and Methods

3.1. Sample Collection and Preprocessing

Open stope method was adopted by Shenjiayao deposit for underground mining. To avoid mining-related contamination, the sampling sites were located away from the mine adits, roads, tailings lagoons, and the ore processing mill. Ten soil samples were collected in different sites above the concealed gold ore bodies. Two fault gouge samples were collected from a deep gully over the ore bodies in the loess-covered area, three soil samples were collected in the background area, and three ore samples were collected in the underground mine. The locations of sampling sites are shown in Figure 1. At each soil sampling site, one soil sample around 100 g was collected at a depth of 20–30 cm to avoid organic horizon. After field work, the soil samples were dried at room temperature, and sieved to <76 µm. About 10 g soil was scattered using an electromagnetic oscillation micrometer vibrating screen connected to a trap device and an air extractor (Figure 2). The trapping device contains 1 µm Millipore filter and carbon-coated aluminum TEM (transmission electron microscopy) grid to collect nanoparticles. During the oscillation of the vibrating screen and the air extractor, the gases in the sample carrying the micro-nanoparticles enter the collector through a 0.45 µm Millipore filter and metal-bearing particles are captured on a carbon-coated aluminum TEM grid. This process needs to continue about 3 minutes in order to adsorb appropriate particles on the grid. Following this procedure, a clean tweezer was used to pick up the TEM grids, then the grids with the attached nanoparticles were placed into a special grid sample box. The fault gouge samples and ore samples were ground by a ceramic mill to <76 µm. The fault gouge samples were separated nanoparticles in the same way as soil samples.

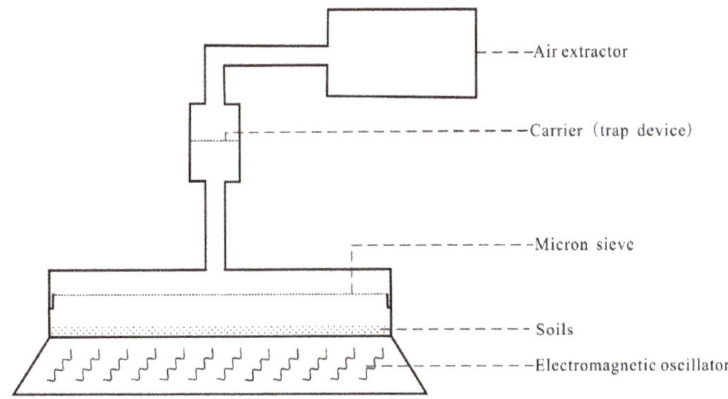

Figure 2. Schematic drawing of the device for separating nanoparticles for soils.

3.2. TEM Observation

After preprocessing, the TEM grids were analyzed by a transmission electron microscopy (TEM) at Beijing Center for Physical and Chemical Analysis to observe particle features (such as size, shape, structure, composition, and form of polymerization) (Beijing, China). The TEM (Tecnai G2 F30) had a spot resolution of 0.20 nm, lattice resolution of 0.1 nm, resolution of STEM HAADF of 0.17 nm, and minimum beam spot diameter of 0.8 nm. It was equipped with an X-ray energy dispersive spectrometer (EDS) at an accelerating voltage of 300 kV, which can measure the composition of the nanoparticles.

3.3. Chemical Analysis

At the same time, all samples were analyzed in the laboratory of Institute of Geophysical and Geochemical Exploration (CAGS) in Langfang, China. A 0.25 g sample was digested in a hot mixture of acids (HCl, HF, HNO_3, and $HClO_4$). Inductively coupled plasma-mass spectrometry (ICP-MS) was used for the determination of Ag, Cu, Pb, Sb, and Zn concentrations. In addition, a 10 g sample was digested in aqua regia and analyzed by graphite furnace atomic absorption spectrometry (GF-AAS) to obtain the Au concentration. Furthermore, a 0.5 g sample was subjected to an aqua regia digest and analyzed by hydride generation atomic fluorescence spectrometry (HG-AFS) to determine the As and Hg concentrations. Analytical accuracy and precision for the laboratory quality were strictly controlled by laboratory replicate samples and standard reference materials (SRMs).

4. Results

4.1. Nanoparticles from the Deposit

4.1.1. Nanoparticles in Soils over the Gold Deposit

Gold-bearing nanoparticles are very infrequent in the studied particles. An important reason is the abundance of Au in nature is too low, even in the gold mining area. Figure 3a,b shows an Au-bearing particle in soils over Shenjiayao gold deposit. The particle exhibits an irregular shape with a diameter of approximately 800 nm. The energy dispersive X-ray (EDX) results (Table 1; ID:1) show the particle contains Au (34.11%), Cu (12.16%), Fe (12.91%), Ni (5.54%), C (9.15%), and O (7.38%). Figure 3c is the high-resolution transmission electron microscope (HRTEM) image, which marks two circular regions. The gray black area (I) has a clear crystal face, which reveals a crystalline nature, while the gray white area (II) has no crystal face, which reveals an amorphous nature. The selected area electron diffraction (SAED) pattern (Figure 3d) shows regularly distributed diffraction spots of the gray black area. The scanning transmission electron microscope with high angle annular dark field (STEM-HAADF) image (Figure 3e) reveals that some smaller nanoparticles are attached on the surface of a big particle. The EDX was used to detect the gray area in the edge of the particle, which shows that the main components are C (20.15%), O (15.36%), and Fe (28.59%). The C:O:Fe atomic ratio is nearly 4:2:1. According to the above, we can infer that some nano-scale native gold, hematite, and Cu-, Ni-bearing particles are attached on the surface of a carbon-bearing particle.

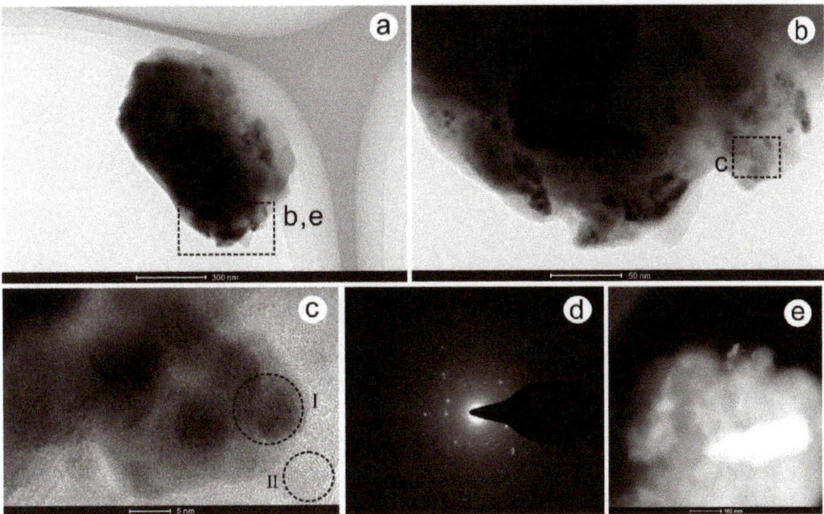

Figure 3. A gold-bearing nanoparticle in soil sample over the studied gold deposit; (**a**) TEM image; (**b**) TEM image; (**c**) high-resolution transmission electron microscope (HRTEM) image; (**d**) selected area electron diffraction (SAED) pattern; (**e**) scanning transmission electron microscope with high angle annular dark field (STEM-HAADF) image. Note that some small nanoparticles adsorb on the surface of a big particle.

Figure 4a shows a Cu–Zr–Au–C–O particle, approximately 1 μm in size and irregular in shape. It comprises two sub-particles. The EDX results (Table 1; ID:2) show that this particle contains Cu (24.15%), Zr (21.44%), Au (1.36%), C (36.71%), and O (10.22%). The magnified image (Figure 4b) and STEM-HAADF image (Figure 4e) reveals that some smaller nanoparticles are adsorbed on the surface of a big particle. The HRTEM image (Figure 4c) and SAED pattern (Figure 4d) shows that the smaller nanoparticles have clear crystal face and regularly distributed diffraction spots, indicative of a crystalline nature. This suggests that this particle is a carbon-bearing particle and some smaller Cu- and Zr-bearing nanoparticles are attached on its surface. It is worth noting that a small amount of gold occurs in this particle, too.

Table 1. Information about Au-, Pb-, Zn-, Cu-, and Fe-bearing nanoparticles in soils over gold deposits.

Particle ID	Size (nm)	Shape		Element																	
				C	O	Na	Mg	Al	Si	S	Cl	K	Ca	Fe	Ni	Cu	As	Zn	Zr	Au	Pb
1	600 × 900	irregular	wt %	9.15	7.38	-	-	3.07	5.34	-	-	10.33	-	12.91	5.54	12.16	-	-	-	34.11	-
			at %	30.69	18.59	-	-	4.59	7.66	-	-	10.65	-	9.31	3.81	7.71	-	-	-	6.98	-
2	600 × 1000	irregular	wt %	36.71	10.22	2.06	-	2.25	1.08	-	-	-	-	0.11	-	24.15	-	-	21.44	1.36	-
			at %	67.23	14.05	1.97	-	1.83	0.85	-	-	-	-	0.11	-	8.36	-	-	5.17	0.15	-
3	unmeasurable	aggregation	wt %	10.21	6.52	-	-	5.79	0.73	-	-	0.43	-	0.37	-	0.40	-	9.62	-	-	65.93
			at %	42.77	20.51	-	-	10.80	1.32	-	-	0.56	-	0.33	-	0.31	-	7.40	-	-	16.01
4	150 × 300	circular	wt %	-	18.35	-	-	3.71	4.10	-	-	0.75	-	6.57	-	-	-	64.90	-	-	-
			at %	-	44.10	-	-	5.28	5.61	-	-	0.74	-	4.52	-	-	-	38.18	-	-	-
5	unmeasurable	aggregation	wt %	-	8.46	-	-	25.98	2.37	2.58	24.27	-	-	0.65	-	33.83	-	1.86	-	-	-
			at %	-	18.15	-	-	33.05	2.89	2.76	23.49	-	-	0.40	-	18.28	-	0.98	-	-	-
6	200 × 400	irregular	wt %	-	23.04	-	-	7.96	15.28	0.33	0.19	2.33	-	28.79	-	0.52	-	17.15	-	-	-
			at %	-	46.58	-	-	8.60	16.58	0.32	0.16	1.82	-	15.70	-	0.25	-	7.99	-	-	-
7	1000 × 1000	Circular	wt %	-	21.59	-	-	2.10	-	-	-	-	-	76.31	-	-	-	-	-	-	-
			at %	-	48.30	-	-	2.79	-	-	-	-	-	48.91	-	-	-	-	-	-	-
8	900 × 900	irregular	wt %	-	23.25	-	-	2.13	1.46	-	-	-	-	71.33	-	0.28	-	-	-	-	-
			at %	-	50.24	-	-	2.73	1.80	-	-	-	-	44.15	-	0.15	-	-	-	-	-
9	400 × 600	irregular	wt %	-	21.87	-	-	3.50	1.92	0.33	-	-	-	67.79	-	1.23	-	-	-	-	-
			at %	-	48.43	-	-	4.60	1.02	0.37	-	-	-	43.01	-	0.69	-	-	-	-	-
10	unmeasurable	aggregation	wt %	-	25.38	-	-	4.37	3.66	0.45	-	-	-	59.85	-	0.22	-	2.56	-	-	-
			at %	-	52.30	-	-	5.34	4.29	0.46	-	-	-	35.34	-	0.11	-	1.29	-	-	-
11	100 × 200	irregular	wt %	-	29.70	-	-	4.35	11.83	-	-	1.45	0.25	45.08	-	-	-	0.59	-	-	-
			at %	-	54.23	-	-	4.71	12.31	-	-	1.08	0.18	23.58	-	-	-	0.26	-	-	-
12	unmeasurable	aggregation	wt %	-	26.06	-	-	4.25	3.78	-	-	-	-	65.73	-	0.18	-	-	-	-	-
			at %	-	52.53	-	-	5.08	4.34	-	-	-	-	37.96	-	0.09	-	-	-	-	-
13	400 × 500	irregular	wt %	-	30.62	-	0.59	6.05	9.93	0.61	-	-	-	51.31	-	-	-	-	-	-	-
			at %	-	54.95	-	0.70	6.44	10.15	0.55	-	-	-	26.38	-	-	-	-	-	-	-
14	unmeasurable	aggregation	wt %	14.26	22.89	-	-	3.91	1.24	0.15	-	0.24	-	53.09	-	-	-	-	-	-	-
			at %	31.12	37.49	-	-	3.80	1.16	0.12	-	0.16	-	24.92	-	-	-	-	-	-	-
15	unmeasurable	aggregation	wt %	57.01	14.29	-	-	3.23	1.60	-	-	-	-	22.17	-	-	-	-	-	-	-
			at %	76.26	14.35	-	-	1.92	0.91	-	-	-	-	6.38	-	-	-	-	-	-	-

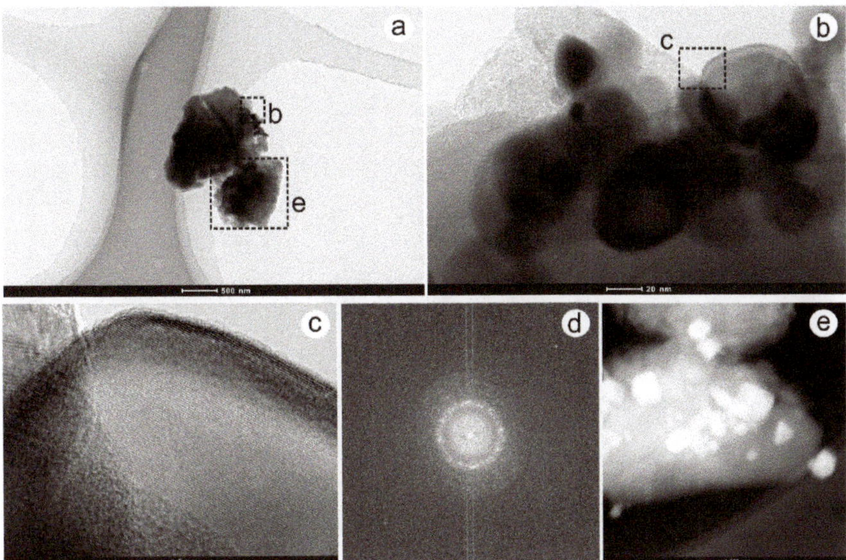

Figure 4. A Cu–Zr–Au–C–O particle in soils over gold deposit; (**a**) TEM image; (**b**) TEM image; (**c**) HRTEM image; (**d**) SAED pattern; (**e**) STEM-HAADF image. Some small nanoparticles adsorb on the surface of a big particle.

Numerous Pb-, Zn-, and Cu-bearing particles were observed in the soil samples over the studied gold deposit. Figure 5a shows a Pb- and Zn-bearing particle aggregation, approximately 50–100 nm in size and roughly circular in shape for every individual particle. The particle aggregation (Table 1; ID:3) contains Pb (65.93%), Zn (9.62%), C (10.21%), and O (6.52%). Figure 5b shows a Zn-bearing particle that exhibits a nearly circular shape with a diameter of approximately 150 nm. The particle (Table 1; ID:4) contains Zn (64.90%), O (18.35%), and Fe (6.57%). Figure 5c shows a Cu-bearing particle aggregation, approximately 200 nm in size and also roughly circular in shape for every individual particle. The particle aggregation (Table 1; ID:5) contains Cu (33.83%), Al (25.98%), Cl (24.27%), and Zn (1.86%). Figure 5d shows a Zn-bearing particle that exhibits an irregular shape with a diameter of approximately 250 nm. The particle (Table 1; ID:6) contains Zn (17.15%) and Fe (28.79%).

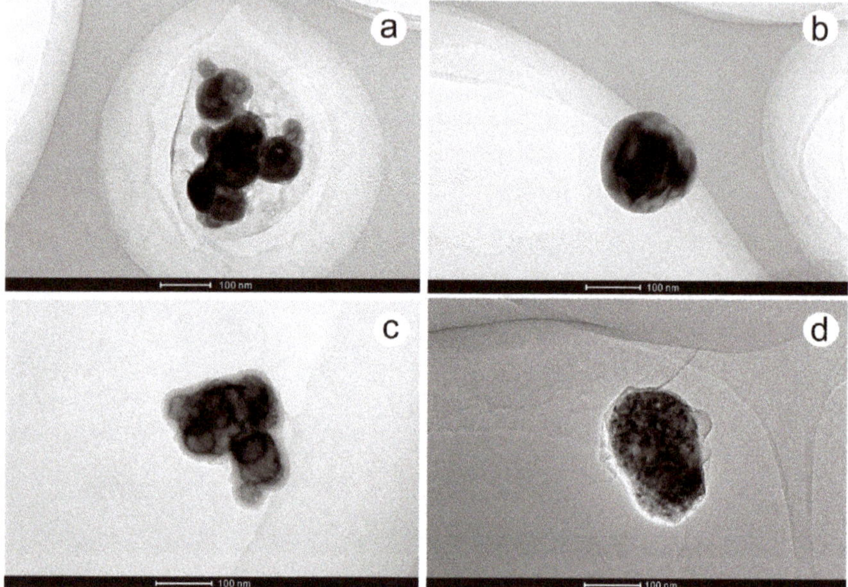

Figure 5. Pb-, Zn-, and Cu-bearing particles in soils over gold deposit; (**a**) TEM image of a Pb- and Zn-bearing particle aggregation; (**b**) TEM image of a Zn-bearing particle; (**c**) TEM image of a Cu-bearing particle aggregation; (**d**) TEM image of a Zn-bearing particle.

Fe-bearing particles are the most common metal-bearing particles in the studied soil samples and some of them are attached by clay minerals. Figure 6 shows some Fe-bearing particles. Most of them are adsorbed to clay minerals (Table 1, ID: 10, 11, 12, 13; Figure 6d–g) or carbon-bearing particles (Table 1, ID: 14, 15; Figure 6h,I) and some of them appear aggregations (ID: 10, 12, 14, 15; Figure 6d,f,h,i). Besides, numerous Ca-, Ba-, Ti-, and Na-bearing nanoparticles (Table 2, ID: 16–21; Figure 7a–f) occur in soils, most likely in the form of $CaCO_3$, $BaSO_4$, TiO_2, and NaCl. In addition, many quartz, amorphous carbon, organic matter, and clay nanoparticles were observed in soils, which indicates that these particles prevail in the studied soils.

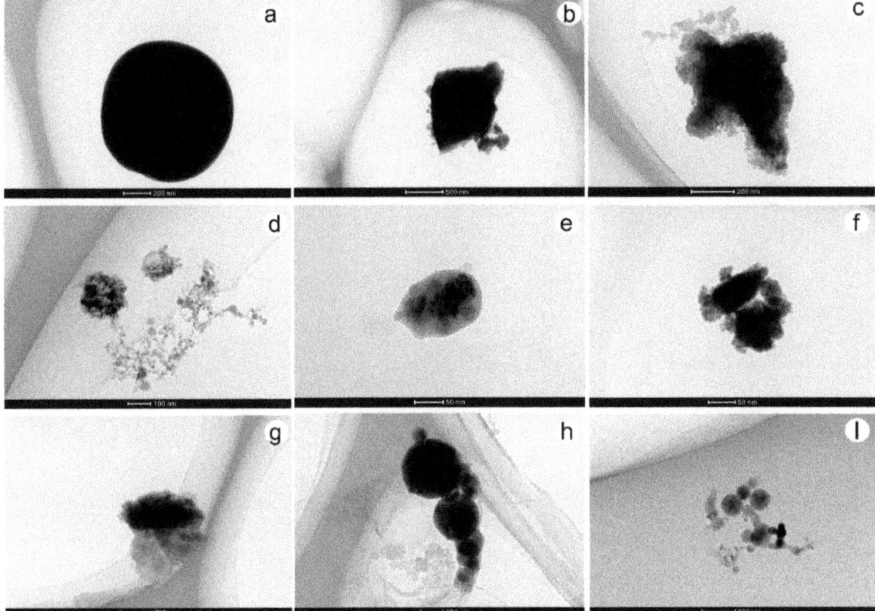

Figure 6. TEM images of Fe-bearing particles in soils over the studied gold deposit; (**a**) a Fe-bearing circular particle; (**b**) irregular Fe-bearing particles; (**c**) irregular Fe-bearing particles; (**d**) Fe-bearing particle aggregation; (**e**) a elliptic Fe-bearing particles; (**f**) Fe-bearing particle aggregation; (**g**) irregular Fe-bearing particles; (**h**) Fe-bearing particle aggregation; (**i**) Fe-bearing particle aggregation.

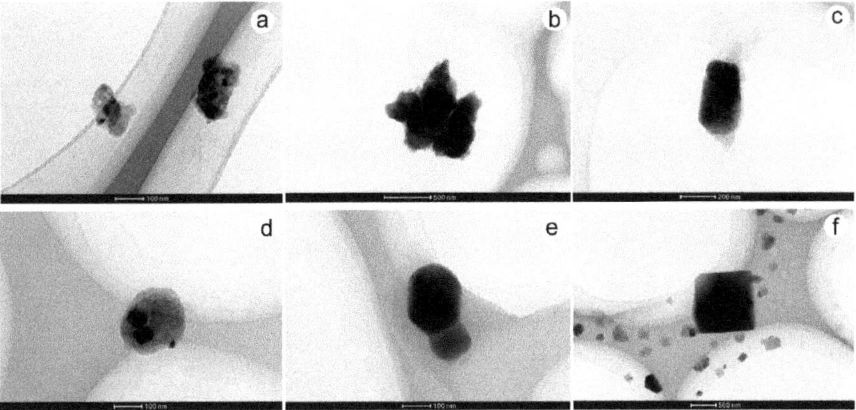

Figure 7. TEM images of Ca-, Ba-, Ti-, and Na-bearing nanoparticles in soils over the studied gold deposit; (**a**) Ca-bearing particles; (**b**) Ca-bearing particles; (**c**) Ba-bearing particles; (**d**) Ti-bearing particles; (**e**) Ti-bearing particles; (**f**) Na-bearing particles.

Table 2. Information about Ca-, Ba-, Ti-, and Na-bearing nanoparticles in soils over gold deposits.

Particle ID	Size (nm)	Shape		Element														
				C	O	Na	Mg	Al	Si	S	Cl	K	Ca	Ti	Ba	Fe	Cu	Zn
16	150 × 300	irregular	wt %	9.11	31.54	1.58	2.21	8.89	4.49	4.15	0.56	1.35	31.01	-	-	4.17	0.28	-
			at %	16.20	45.10	1.68	2.21	8.02	3.89	3.15	0.38	0.84	20.71	-	-	1.82	0.11	-
17	1000 × 1000	irregular	wt %	18.84	17.63	-	1.36	5.78	0.51	-	-	-	55.74	-	-	0.13	-	-
			at %	35.01	24.56	-	1.26	4.77	0.41	-	-	-	31.06	-	-	0.05	-	-
18	200 × 400	rectangle	wt %	-	-	1.32	-	4.34	0.36	15.82	-	-	0.78	-	76.97	-	-	-
			at %	-	-	4.38	-	12.26	0.98	37.63	-	-	1.49	-	42.74	-	-	-
19	200 × 200	sphere	wt %	-	24.71	-	1.14	7.26	8.32	-	-	-	3.17	48.72	-	2.54	1.06	1.10
			at %	-	45.49	-	1.38	7.93	8.72	-	-	-	2.33	29.96	-	1.34	0.49	0.49
20	200 × 200	hexagon	wt %	-	21.62	-	-	6.67	-	0.45	-	-	-	70.11	-	0.21	0.51	-
			at %	-	43.56	-	-	7.96	-	0.46	-	-	-	47.19	-	0.12	0.26	-
21	1000 × 1000	cube	wt %	-	2.33	76.80	-	-	-	-	19.33	1.21	0.33	-	-	-	-	-
			at %	-	3.58	82.07	-	-	-	-	13.39	0.76	0.20	-	-	-	-	-

4.1.2. Nanoparticles in Fault Gouges

Figure 8a shows an Au-bearing particle aggregation in fault gouges, approximately 200 nm in size and roughly circular in shape. The particle aggregation (Table 3; ID:22) contains Fe (87.44%), Mn (3.06%), Cu (1.10%), and Au (0.33%). Figure 8b shows another Au-bearing particle in fault gouges. The particle (Table 3; ID:23) contains Mn (54.24%), Fe (6.28%), and Au (0.23%). There are still four other particles contain Au in fault gouges, but the gold content of all these Au-bearing particles is very low.

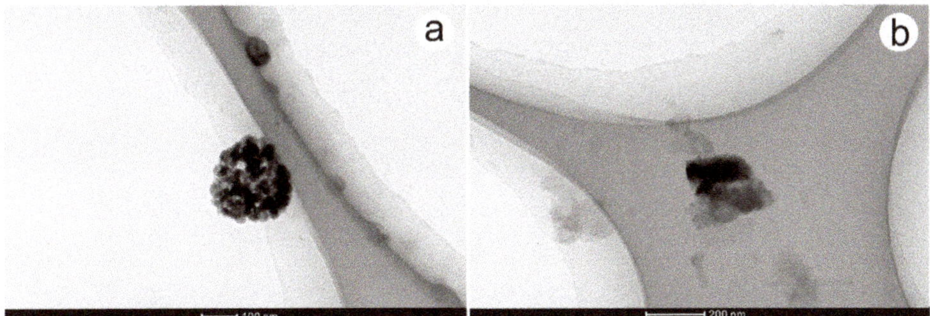

Figure 8. TEM images of Au-bearing nanoparticles in fault gouges; (**a**) Au-bearing particle aggregation; (**b**) a Au-bearing particle.

Numerous Cu-, Pb-, and Zn-bearing particles were observed in the fault gouge samples. Figure 9a shows a Cu-bearing particle in the fault gouges. The particle exhibits an irregular shape with a diameter of approximately 1000 nm. The EDX results (Table 3; ID:24 (I)) show the particle contains Cu (45.19%), S (16.73%), Si (11.97%), Al (8.46%), and O (14.83%) in area I (Figure 9b). The EDX results (Table 3; ID:24 (II)) show the particle contains Cu (3.10%), Si (31.28%), Al (20.90%), and O (42.12%) in area II (Figure 9b). It indicates that some copper sulfide or copper oxide nanoparticles are adsorbed by clay minerals, which also can be inferred from the STEM-HAADF image (Figure 9d). The HRTEM image (Figure 9c) shows that the small Cu-bearing nanoparticle has a clear crystal face, indicative of its crystalline nature. Figure 10a also shows a Cu-bearing particle (Table 3; ID: 25), which contains Cu (10.15%), Fe (14.92%), Al (19.68%), Si (15.41%), S (4.14%), and O (32.09%). It reveals some Cu-, Fe-bearing sulfide or oxide nanoparticles that are adsorbed to clay minerals. Figure 10b,c shows two Zn-bearing particles. The Zn content (Table 3; ID: 26, 27) is higher than 26% and S are higher than 17%. Figure 10d–f shows three Pb-bearing nanoparticles. The EDX results (Table 3; ID: 28, 29, 30) indicate that these three particles are composed of a lead oxide particle, a lead sulfide particle, and a native lead particle.

Table 3. Information about Au-, Pb-, Zn-, Cu-, and Fe-bearing nanoparticles in fault gouge over gold deposits.

Particle ID	Size (nm)	Shape		Element														
				C	O	Na	Al	Si	S	Mn	K	Ca	Ti	Fe	Cu	Zn	Au	Pb
22	300 × 300	irregular	wt %	-	6.92	-	-	1.15	-	3.06	-	-	-	87.44	1.10	-	0.33	-
			at %	-	20.47	-	-	1.94	-	2.64	-	-	-	74.06	0.82	-	0.08	-
23	300 × 300	irregular	wt %	-	23.07	-	-	7.38	-	54.24	5.31	2.26	-	6.28	1.23	-	0.23	-
			at %	-	47.80	-	-	8.71	-	32.72	4.50	1.87	-	3.72	0.64	-	0.04	-
24 (I)	1000 × 1000	irregular	wt %	-	14.83	1.66	8.46	11.97	16.73	-	-	1.16	-	-	45.19	-	-	-
			at %	-	30.88	2.40	10.45	14.20	17.40	-	-	0.96	-	-	23.70	-	-	-
24 (II)			wt %	-	42.12	-	20.90	31.28	-	-	-	2.61	-	-	3.10	-	-	-
			at %	-	56.80	-	16.71	24.03	-	-	-	1.40	-	-	1.05	-	-	-
25	900 × 900	irregular	wt %	-	32.09	-	19.68	15.41	4.14	-	1.00	2.61	-	14.92	10.15	-	-	-
			at %	-	51.03	-	18.56	13.96	3.29	-	0.65	1.65	-	6.80	4.06	-	-	-
26	100 × 100	irregular	wt %	-	16.99	-	4.29	8.50	17.56	-	0.33	0.15	-	24.24	0.61	26.45	-	-
			at %	-	36.08	-	5.40	10.28	18.61	-	0.28	0.13	-	14.75	0.33	13.75	-	-
27	300 × 300	irregular	wt %	30.88	4.80	-	-	1.04	22.92	-	1.01	0.56	-	5.18	0.39	33.21	-	-
			at %	60.22	7.03	-	-	0.87	16.74	-	0.61	0.33	-	2.17	0.14	11.90	-	-
28	400 × 500	irregular	wt %	-	16.10	-	7.64	2.38	-	-	-	1.24	-	0.55	0.62	0.60	-	70.87
			at %	-	56.66	-	15.94	4.78	-	-	-	1.74	-	0.56	0.55	0.52	-	19.26
29	300 × 300	irregular	wt %	-	30.51	3.03	4.90	2.50	8.90	-	7.84	0.80	-	22.76	0.24	-	-	16.76
			at %	-	56.82	3.92	5.41	2.66	8.27	-	5.97	0.59	-	12.14	0.11	-	-	2.41
30	200 × 300	irregular	wt %	-	1.99	-	5.42	0.95	-	-	0.96	-	-	0.83	1.85	0.28	-	87.71
			at %	-	14.56	-	23.50	3.97	-	-	2.86	-	-	1.75	3.40	0.50	-	49.47
31	300 × 500	irregular	wt %	-	32.12	-	10.52	11.78	0.28	-	0.45	-	-	42.52	0.60	1.02	-	-
			at %	-	55.23	-	10.73	11.54	0.24	-	0.32	-	-	20.95	0.26	0.43	-	-
32	500 × 1500	irregular	wt %	-	27.23	-	12.35	15.50	0.45	-	5.31	-	0.78	34.63	0.23	0.33	-	-
			at %	-	48.33	-	13.00	15.67	0.40	-	3.85	-	0.46	17.61	0.10	0.14	-	-
33	300 × 500	irregular	wt %	-	-	1.32	3.52	0.67	38.13	-	2.62	0.38	-	52.25	-	0.22	-	-
			at %	-	-	2.36	5.37	0.98	48.98	-	2.76	0.40	-	38.53	-	0.14	-	-

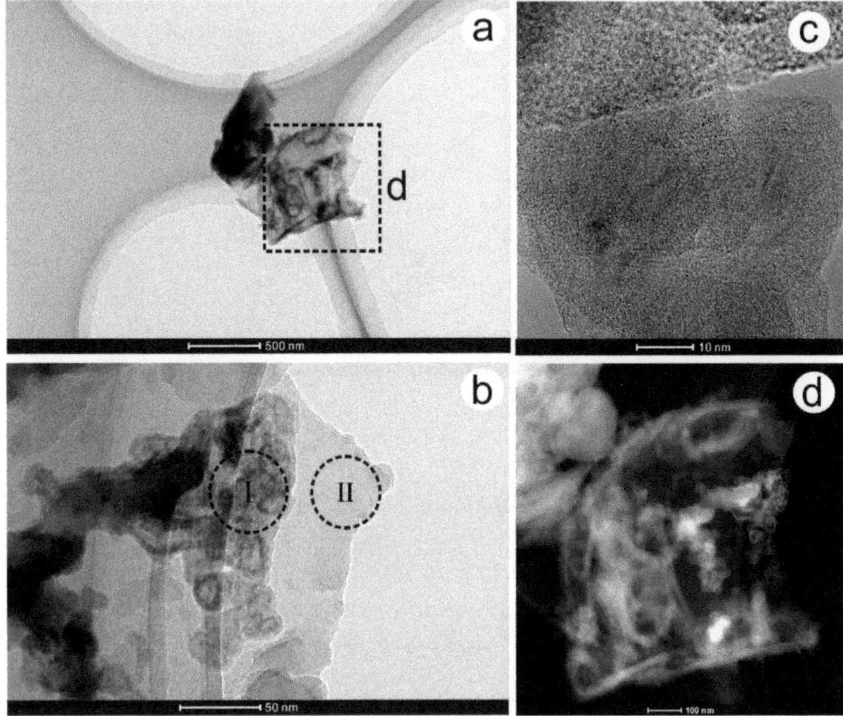

Figure 9. A Cu-bearing particle in the studied fault gouges; (**a**) TEM image; (**b**) TEM image; (**c**) HRTEM image; (**d**) STEM-HAADF image. Some small nanoparticles adsorb on the surface of a big particle.

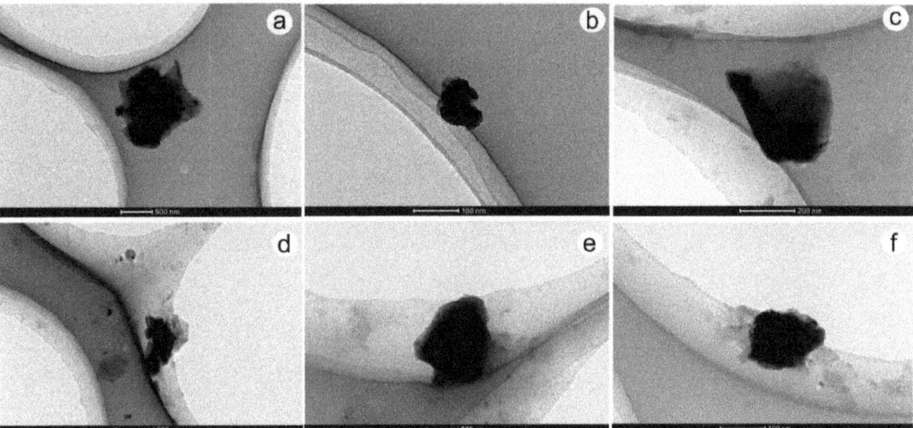

Figure 10. Cu-, Pb-, and Zn-bearing nanoparticles in fault gouges; (**a**) TEM image of a Cu-bearing particle; (**b**) TEM image of a Zn-bearing particle; (**c**) TEM image of a Zn-bearing particle; (**d**) TEM image of a Pb-bearing particle; (**e**) TEM image of a Pb-bearing particle; (**f**) TEM image of a Pb-bearing particle.

Fe-bearing particles also abound in the studied fault gouges. Most of Fe-bearing particles are adsorbed to clay minerals (Figure 11a,b; Table 3, ID: 31, 32). Figure 11c shows an Fe-bearing nanoparticle observed in fault gouges. The EDX results (Table 3; ID:33) show the particle contains Fe (52.25%) and S (38.13%), which indicates it is an independent pyrite particle. In addition, clay nanoparticles are also numerous in the studied gault gouges samples.

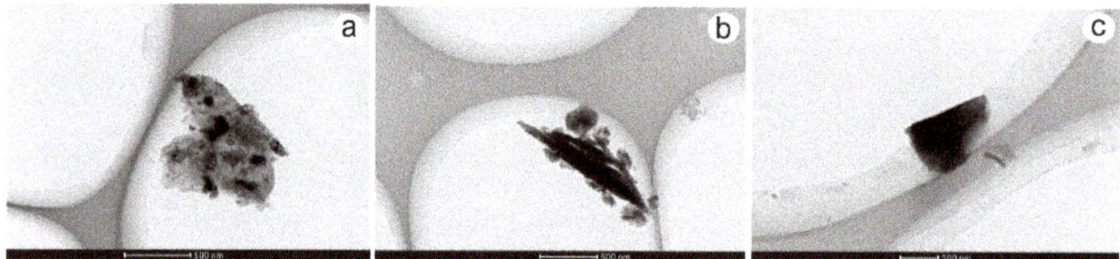

Figure 11. TEM images of Fe-bearing nanoparticles in the studied fault gouge samples. (**a**) Fe-bearing nanoparticles adsorbed to clay minerals; (**b**) Fe-bearing nanoparticles adsorbed to clay minerals; (**c**) a Fe-bearing nanoparticle.

4.2. Nanoparticles in the Geochemical Background Area

More than 50 nanoparticles were investigated in the background soil samples. Most of them are Fe-, Al-, Si-, Ca-, Ti-, and Ba-bearing nanoparticles. Although a few particles contain Cu, Pb, and Zn, the content of these ore-forming elements are usually lower than 1%. Hematite, clay minerals, quartz, and amorphous carbon particles are the most common nanoparticles observed in background soil samples.

4.3. The Concentration of Ore-Forming Elements in Soil Samples, Fault Gouge Samples, and Ore Samples

The ore-forming chemical element concentrations in various media are displayed in Table 4. The mean value of Au reaches up to 3520 ng/g in the ore samples. By contrast, the mean value of Au in fault gouge samples is 429 ng/g, in soil samples over the concealed ore bodies is 14.1 ng/g, and in soil samples from background area is 2.28 ng/g. The results show that Au concentrations are the highest in ores, which is followed by fault gouge material, soils over concealed ore bodies, and soils from background area. Silver, Cu, Pb, Zn, As, Sb, and Hg have the same distribution pattern as Au.

Table 4. Element content of different solid media from the study area.

Element	Unit	Soil over Concealed Ore Bodies ($n = 10$)			Soil from Background Area ($n = 3$)			Fault Gouge ($n = 2$)			Ore ($n = 3$)		
		Min	Max	Mean	Min	Max	Mean	Min	Max	Mean	Min	Max	Mean
Au	ng/g	3.18	35.0	14.1	1.67	3.01	2.28	137	722	429	1050	5760	3520
Ag	µg/g	0.06	1.34	0.41	0.06	0.27	0.14	4.92	14.4	9.7	89	439	291
Cu	µg/g	19.0	58.6	37.1	23.5	30.7	26.8	231	497	364	6945	10,980	8463
Pb	µg/g	30.3	430	114	25	32.7	28.6	1453	2617	2035	5521	8950	7231
Zn	µg/g	87.7	455	174	67.9	99.7	79.8	2776	8692	5734	4131	37,820	17,977
As	µg/g	14.2	93.6	48.3	13.5	24.7	17.3	363	2329	1346	7754	60,740	25,436
Sb	µg/g	1.27	4.00	2.03	1.02	1.24	1.12	12.2	60.5	36.3	294	773	610
Hg	µg/g	37.8	896	216	27.4	52.2	38.6	17.3	28.9	23.1	41,920	57,960	49,049

5. Discussion and Conclusions

5.1. Comparison of Metal-Bearing Nanoparticles in Soil and Fault Gouge Samples in the Studied Gold Mining Area

TEM analysis shows that nanoparticles are widespread in the soil and fault gouge samples collected in the studied gold mining area. Comparing the particles from these two different media, we found there were several common features. The nanoparticles, especially the metal-bearing nanoparticles, exhibit distinct features in shape, structure, component, and form of polymerization. The size of these particles ranges from several nanometers to more than 100 nanometer and their shape is ellipse, sphere, hexagon, schistose, or irregular. The metal-bearing nanoparticles tend to attach to the surface of clay minerals and amorphous carbon. In addition, numerous Cu-, Pb-, and Zn-bearing particles, as well as particles containing other metal compounds, occur in the soil and fault gouge samples collected at the deposit. Fe-bearing particles are the most common metal-bearing particles and a small number of Au-bearing particles also occur in samples. Most of the metal-bearing nanoparticles have internal ordered crystal structure.

At the same time, some differences between the metal-bearing particles in these two media have also been presented in this study. Firstly, nano-scale sulfides occur in the fault gouge samples, such as copper sulfide (Table 3; ID: 24, 25), lead sulfide (Table 3; ID: 29), zinc sulfide (Table 3; ID: 26, 27), and iron sulfide (Table 3; ID: 26, 29, 33). Metal-bearing particles mainly occur in the form of oxide in the studied soils, such as copper oxide (Table 1; ID: 1, 2, 5), lead oxide (Table 1; ID: 3), zinc oxide (Table 1; ID: 3, 4, 6), and iron oxide (Table 1; ID: 7–15). Secondly, the nanoparticles tend to form aggregate clusters in soils (Figure 5a,c; Figure 6c,d,f,h,i) and the nanoparticles in the fault gouge samples tend to occur as single particles. Thirdly, Au-, Cu-, Pb-, Zn-, and Fe-bearing particles in the fault gouges have better-defined crystal shape than in soils. The particles in the soil samples tend to exhibit a rounded or sub-rounded shape.

5.2. Comparison of Nanoparticles in Soil Samples in the Mining Area and in the Background Area

Fe-, Al-, Si-, Ca-, Ti-, and Ba-bearing nanoparticles commonly occur in the geochemical background of the studied gold mining area. Hematite, clay mineral, quartz, and amorphous carbon particles are the most common nanoparticles in samples. The main difference is that the Au-, Cu-, Pb-, and Zn-bearing nanoparticles are frequently observed in samples from the mining area and are very rare in samples from background area. Although a few particles contain Cu, Pb, and Zn, the contents of these ore-forming elements are very low.

5.3. Migration Mechanism of Metallic Nanoparticles from Mineralized Bodies to Earth's Surface

Understanding the mechanisms and their effectiveness in transferring ore-related metals upwards through the sedimentary cover are very important for the mineral exploration of areas with thick sediment cover [1]. Since element migration through sediment cover is a slow and complex process, the mechanism is still not fully understood, especially for allochthon cover, or for a thick sequence of various overlying post-mineralization rocks and regolith. It is generally thought that trace elements are transported from ore bodies to the earth's surface by one or more mechanisms [2]. The migration mechanisms include groundwater flow, capillary action, ionic diffusion, self-potential effect, vaporization, biological processes, and transportation by gases [1,2]. Because the movement of ions and particles through the sediment cover is upwards against gravity, a medium and a force are necessary to cause the upward transport of metals [1]. The transporting medium can include gas, water, and mineral particles and the transporting force can include pressure, concentration, electrical, and temperature gradients.

The migration mechanism of transportation by geogas has been proposed and studied for decades [3,9,12,18,56–58]. Nano-scale metal-bearing particles can be adsorbed onto the surfaces of gas bubbles and migrate with the bubbles upwards. The geogas may be derived from the atmosphere and driven to the surface by barometric pumping [15], be released from the ore minerals, or derived from mantle degassing. As mentioned in the first part of this paper, numerous metal-bearing nanoparticles have been observed in geogas samples [12,30–41]. However, answering the questions whether the element anomalies in soils are caused by particles transported by geogas and whether the ore-forming elements in soils occur in the form of nanoparticles need more research.

In this study, numerous Cu-, Pb-, and Zn-bearing particles, as well as a small number of particles containing Au, were identified in the soil samples collected in the studied gold mining area. Most of the metal-bearing nanoparticles have internal ordered crystal structure. Only a few particles contain Cu, Pb, and Zn, and the content of these ore-forming elements are very low in the studied nano-particles. The chemical analysis results show that the concentration of ore-forming elements in soils from the mining area are higher than in soils from the geochemical background area. Besides, the concentration of ore-forming elements in ore and fault gouge samples is much higher than in soil samples. It is assumed, therefore, that the higher concentrations of trace elements in soils from the mining area are due to the vertical migration of ore-forming elements with geogas after mineralization processes and very high concentrations of ore-forming elements in the fault gouge are, by contrast, mainly due to the migration of ore-forming elements with ore-forming fluids along the fault or fracture in or after the mineralization processes. Lu et al. [25] carried out a geogas prospecting experiment in the same study area. The experiment showed that clear Au anomalies occur in geogas samples over the concealed ore bodies. Besides, Au, Cu-, and Ag-bearing nanoparticles were also observed in geogas in Lu et al.'s study [25]. Because the soils in the mining area and in the background area are loess which are all from the same source and very homogeneous, the results clearly indicate that the ore-forming elements with high concentration in soils and geogas can only come from deep-seated ore bodies. Besides, the results also lead us to give a speculation that the ore-forming elements with high concentration in soils and geogas actually occur in the form of nanoparticles. At the same time, we can further conclude that geogas is a very important transfer medium for the vertical migration of metal-bearing nanoparticles. In addition, through comparison of the metallic nanoparticles from soils and fault gouges in the mining area, it can be inferred that the studied fault is a migration channel for ore-forming elements. It is assumed that the metallic nanoparticles in the fault gouge will further migrate into surface soils. The sulfides will turn into oxides. The crystal shape will be rounded. The nanoparticles will tend to aggregate in clusters. Based on the discussion above, a migration model of the ore-forming elements is shown in Figure 12. As reported by several studies, mineralization processes and late oxidation in deep orebodies formed metal-bearing nanoparticles [59–62]. Nanoscale particles have large specific surface areas that enable them to migrate with all kinds of geological fluids and be adsorbed by microbubbles and vertically migrate with air currents [33]. Faults, micro fractures, pores, and joints provide the migration channels for nanoparticles [40,63].When arriving at the surface, some of these metal-bearing particles would be trapped by soil geochemical barriers such as clays, oxide coatings, and colloids [33].

Besides, through this study, we found that many metal-bearing nanoparticles adsorb to the surface of clay minerals. Because clay minerals are very fine and occur in the fine fraction of soils, it provides a theoretical support for us to use fine fraction soil as sampling media to carry out geochemical exploration in covered terrains. Finally, based on the comparison of the nanoparticles from the mining area and the background area, we found that the metal-bearing nanoparticles are very different between these two areas. Many nanoparticles containing ore-forming elements were observed in soils collected in the mining area. It gives us an enlightenment to seek a new geochemical method through microscopic observation for mineral exploration in covered area.

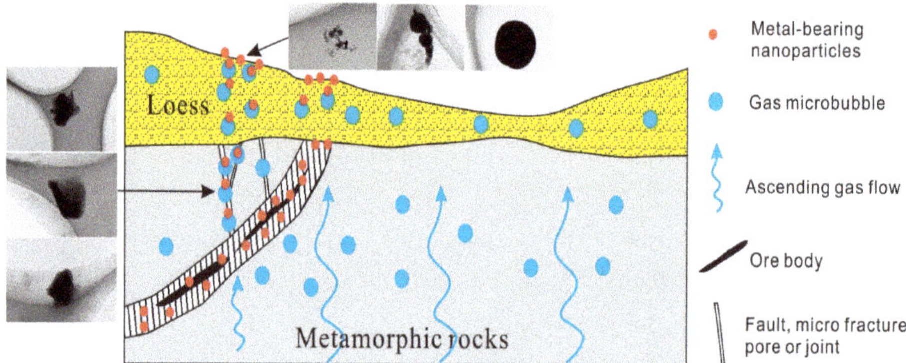

Figure 12. Schematic diagram of the migration model of the metal-bearing nanoparticles.

Author Contributions: X.W. and B.Z. conceived and designed the experiments; H.L. and H.W. collected the samples, Z.H. and H.F. performed the TEM experiments, B.Z. and Z.H. analyzed and wrote the paper, Z.H. and H.F. modified the manuscript.

Funding: This research was financially supported by the State Key Research & Development Project (2016YFC0600602), National Natural Science Foundation of China (41573044) and Special Scientific Research Fund of Public Welfare Profession of China (201511034) for the financial support.

Acknowledgments: Special thanks are given to related people and departments involved in this research. Sincere gratitude is extended to the editors and reviewers for the handling and reviewing of the manuscript. We also thank Gyozo Jordan for his comments and language modification. Constructive comments are of great help in improving the manuscript. Assistant Editor Heather Wu is acknowledged for handling the review process.

Conflicts of Interest: The authors declare no conflict of interest.

References

1. Anand, R.R.; Aspandiar, M.F.; Noble, R.R.P. A review of metal transfer mechanisms through transported cover with emphasis on the vadose zone within the Australian regolith. *Ore. Geol. Rev.* **2016**, *73*, 394–416. [CrossRef]

2. Wang, X.; Zhang, B.; Lin, X.; Xu, S.; Yao, W.; Ye, R. Geochemical challenges of diverse regolith-covered terrains for mineral exploration in China. *Ore. Geol. Rev.* **2016**, *73*, 417–431.

3. Malmqvist, L.; Kristiansson, K. Microflow of geogas—A possible formation mechanism for deep-sea nodules. *Mar. Geol.* **1981**, *40*, M1–M8. [CrossRef]

4. Malmqvist, L.; Kristiansson, K. Experiment evidence for an ascending micro-flow of geogas in the ground. *Earth. Plant. Sc. Lett.* **1984**, *70*, 407–416. [CrossRef]

5. Bradshaw, P.M.D.; Thomson, I.; Smee, B.W.; Larsson, J.O. The application of different analytical extractions and soil profile sampling in exploration geochemistry. *J. Geochem. Explor.* **1974**, *3*, 209–225. [CrossRef]

6. Clarke, J.R.; Meier, A.L. Enzyme leaching of surficial geochemical samples for defecting hydromorphic trace-element anomalies associated with precious metal mineralized bedrock buried beneath glacial overburden in northern Minneseta. In Proceedings of the Gold '90 Symposium—Gold '90, Salt Lake City, UT, USA, 26 February–1 March 1990; pp. 189–207.

7. Mann, A.W.; Birrell, R.D.; Gay, L.M.; Al, E. Partial extraction and mobile metal ions. In Proceedings of the 17th IGES, Townsville, Australia, 15–19 May 1995; pp. 31–34.

8. Clark, R.J. Innovative Enzyme Leach Provides. In Proceedings of the Exploration 97: Fourth Decennial International Conference on Mineral Exploration, Toronto, Canada, 14–18 September 1997; pp. 371–374.

9. Wang, X.; Cheng, Z.; Lu, Y.; Xu, L.; Xie, X. Nanoscale metals in Earthgas and mobile forms of metals in overburden in wide-spaced regional exploration for giant deposits in overburden terrains. *J. Geochem. Explor.* **1997**, *58*, 63–72. [CrossRef]

10. Mann, A.W.; Birrell, R.D.; Mann, A.T.; Humphreys, D.B.; Perdrix, J.L. Application of the mobile metal ion technique to routine geochemical exploration. *J. Geochem. Explor.* **1998**, *61*, 87–102. [CrossRef]

11. Cohen, D.R.; Shen, X.C.; Dunlop, A.C.; Rutherford, N.F. A comparison of selective extraction soil geochemistry and biogeochemistry in the cobar area, new south wales. *J. Geochem. Explor.* **1998**, *61*, 173–189. [CrossRef]

12. Chunhan, O.; Juchu, I.; Liangquan, E.; Fenggen, A. Experimental observation of the nano-scale particles in geogas matters and its geological significance. *Sci. China Ser. D Earth Sci.* **1998**, *4*, 325–329. [CrossRef]

13. Wang, X. Leaching of mobile forms of metals in overburden: Development and application. *J. Geochem. Explor.* **1998**, *61*, 39–55.

14. Kelley, D.L. The use of partial extraction geochemistry for copper exploration in northern Chile. *Geochem. Explor. Environ. Anal.* **2003**, *3*, 85–104. [CrossRef]

15. Cameron, E.M.; Hamilton, S.M.; Leybourne, M.I.; Hall, G.E.M.; McClenaghan, M.B. Finding deeply buried deposits using geochemistry. *Geochem. Explor. Environ. Anal.* **2004**, *4*, 7–32. [CrossRef]

16. Leybourne, M.I.; Cameron, E.M. Groundwater in geochemical exploration. *Geochem. Explor. Environ. Anal.* **2010**, *10*, 99–118. [CrossRef]

17. Lintern, M.; Anand, R.; Ryan, C.; Paterson, D. Natural gold particles in *Eucalyptus* leaves and their relevance to exploration for buried gold deposits. *Nat. Comun.* **2013**, *4*, 2614. [CrossRef] [PubMed]

18. Kristiansson, K.; Malmqvist, L. Trace elements in the geogas and their relation to bedrock composition. *Geoexploration* **1987**, *24*, 517–534. [CrossRef]

19. Tong, C.; Liang, X.; Li, J. The tentative geogas survey in the Dongji gold deposit. *Geophys. Geochem. Explor.* **1992**, *16*, 445–451.

20. Wang, X.; Xie, X.; Lu, Y. Dynamic collection of geogas and its preliminary application in search for concealed deposits. *Geophys. Geochem. Explor.* **1995**, *19*, 161–171.

21. Wang, M. Progress in the collection of geogas in China. *Geochem. Explor. Environ. Anal.* **2008**, *8*, 183–190. [CrossRef]

22. Wang, M.; Gao, Y.; Zhang, D.; Ren, T.; Liu, Y. Breakthrough in mineral exploration using geogas survey in the basin area of northern Qilian region and its significance. *Geophys. Geochem. Explor.* **2006**, *30*, 7–12.

23. Tong, C.; Li, J. Geogas prospecting and its mechanism in the search for deep-seated or concealed gold deposits. *Chin. J. Geophys.* **1999**, *42*, 135–142, 145–146.

24. Wang, Y.; Ye, R.; Zhang, B.; Yang, R.; Qi, F. The tentative geogas survey in the Jinwozi Concealed V210 gold deposit of Gobi area. *Geophys. Geochem. Explor.* **2012**, *36*, 263–266.

25. Lu, M.; Ye, R.; Wang, Z.; Wang, X. Geogas prospecting for buried deposits under loess overburden: Taking Shenjiayao gold deposit as an example. *J. Geochem. Explor.* **2019**, *197*, 122–129. [CrossRef]

26. Etiope, G.; Lombardi, S. Evidence for radon transport by carrier gas through faulted clays in Italy. *J Radioanal. Nucl. Chem.* **1995**, *193*, 291–300. [CrossRef]

27. Etiope, G.; Lombardi, S. Laboratory simulation of geogas microbubble flow. *Environ. Geol.* **1996**, *27*, 226–232. [CrossRef]

28. Wang, X.; Xie, X.; Cheng, Z.; Liu, D. Delineation of regional geochemical anomalies penetrating through thick cover in concealed terrains—A case history from the Olympic Dam deposit, Australia. *J. Geochem. Explor.* **1999**, *66*, 85–97. [CrossRef]

29. Wang, X. Conceptual model of deep-penetrating geochemical migration. *Geol. Bull. China* **2005**, *24*, 18–22.

30. Cao, J.; Hu, R.; Liang, Z.; Peng, Z. TEM observation of geogas-carried particles from the Changkeng concealed gold deposit, Guangdong Province, South China. *J. Geochem. Explor.* **2009**, *101*, 247–253. [CrossRef]

31. Cao, J. Particles carried by ascending gas flow at the Tongchanghe copper mine, Guizhou Province, China. *Sci. China* **2010**, *53*, 1647–1654. [CrossRef]

32. Cao, J. Simulation of adsorption of gold nanoparticles carried by gas ascending from the Earth's interior in alluvial cover of the middle lower reaches of the Yangtze River. *Geofluides* **2010**, *10*, 438–446. [CrossRef]

33. Wang, X.; Ye, R. Findings of nanoscale metal particles: Evidence for deep-penetrating geochemistry. *Acta Geoscientica Sin.* **2011**, *32*, 7–12.

34. Wei, X.; Cao, J.; Holub, R.F.; Hopke, P.K.; Zhao, S. TEM study of geogas-transported nanoparticles from the Fankou lead–zinc deposit, Guangdong Province, South China. *J. Geochem. Explor.* **2013**, *128*, 124–135. [CrossRef]

35. Wang, X.; Zhang, B.; Yao, W.; Liu, X. Geochemical exploration: From nanoscale to global-scale patterns. *Earth Sci. Front.* **2014**, *21*, 65–74.

36. Ye, R.; Zhang, B.; Wang, Y. Mechanism of the migration of gold in desert regolith cover over a concealed gold deposit. *Geochem. Explor. Env. Anal.* **2015**, *15*, 62–71. [CrossRef]

37. Zhang, B.; Wang, X.; Ye, R.; Zhou, J.; Liu, H.; Liu, D.; Han, Z.; Lin, X.; Wang, Z. Geochemical exploration for concealed deposits at the periphery of the Zijinshan copper–gold mine, southeastern China. *J. Geochem. Explor.* **2015**, *157*, 184–193. [CrossRef]

38. Wang, X.; Zhang, B.; Ye, R. Nanogeochemistry for mineral exploration through covers. *Bull. Miner. Petrol. Geochem.* **2016**, *35*, 43–51.

39. Cao, J.; Cheng, S.; Luo, S.; Liu, C. Study of particles in the ascending gas of ruptures caused by the 2008 Wenchuan earthquake. *Appl. Geochem.* **2017**, *82*, 38–46. [CrossRef]

40. Wang, X.; Zhang, B.; Ye, R. Nanoparticles observed by TEM from gold, copper-nickel and silver deposits and implications for mineral exploration in covered terrains. *J. Nanosci. Nanotechnol.* **2017**, *17*, 6014–6025. [CrossRef]

41. Jiang, T.; Cao, J.; Wu, Z.; Wu, Y.; Zeng, J.; Wang, Z. A TEM study of particles carried by ascending gas flows from the Bairendaba lead-zinc deposit, Inner Mongolia, China. *Ore Geol. Rev.* **2019**, *105*, 18–27. [CrossRef]

42. Turner, N.; Mills, D.; Fedikow, M.; Prince, P. The evaluation of geological exploration samples using multi-element mobile metal ion selective weak extraction and ICP-MS. In *Proceedings of Exploration 07: Fifth Decennial International Conference on Mineral Exploration*; Milkereit, B., Ed.; Namex Explorations Inc.: Westmount, QC, Canada, 2007; pp. 973–977.

43. Fabris, A.J.; Keeling, J.L.; Fidler, R.W. Surface geochemical expression of bedrock beneath thick sediment cover, Curnamona province, South Australia-2009. *Geochem. Explor. Environ. Anal.* **2009**, *9*, 237–246. [CrossRef]

44. Mokhtari, A.R.; Cohen, D.R.; Gatehouse, S.G. Geochemical effects of deeply buried Cu–Au mineralization on transported regolith in an arid terrain-2009. *Geochem. Explor. Environ. Anal.* **2009**, *9*, 227–236. [CrossRef]

45. Noble, R.R.P.; Stanley, C.R. Traditional and novel geochemical extractions applied to a Cu-Zn soil anomaly: A quantitative comparison of exploration accuracy and precision. *Geochem. Explor. Environ. Anal.* **2009**, *9*, 159–172. [CrossRef]

46. Cameron, E.M.; Leybourne, M.I.; Reich, M.; Palacios, C. Geochemical anomalies in northern Chile as a surface expression of the extended supergene metallogenesis of buried copper deposits. *Geochem. Explor. Environ. Anal.* **2010**, *10*, 157–169. [CrossRef]

47. Fedikow, M.A.F. Ligand-based partial extraction of near-surface soil samples. In Proceedings of the GeoCanada 2010–Working with the Earth, Calgary, AB, Canada, 10–14 May 2010; pp. 1–4.

48. Xie, X.; Lu, Y.; Yao, W.; Bai, J. Further study on deep penetrating geochemistry over the spence porphyry copper deposit, Chile. *Geosci. Front.* **2011**, *2*, 303–311. [CrossRef]

49. Wang, X.; Xu, S.; Zhang, B.; Zhao, S. Deep-penetrating geochemistry for sandstone-type uranium deposits in the Turpan–Hami basin, north-western China. *Appl. Geochem.* **2011**, *26*, 2238–2246. [CrossRef]

50. Morris, P.A. Fine fraction regolith chemistry from the East Wongatha area, Western Australia tracing bedrock and mineralization through thick cover. *Geochem. Explor. Environ. Anal.* **2013**, *13*, 21–40. [CrossRef]

51. Sylvester, G.C.; Mann, A.W.; Rate, A.W.; Wilson, C.A. Application of high-resolution mobile metal ion (MMI) soil geochemistry to archaeological investigations: An example from a Roman metal working site, Somerset, United Kingdom. *Geoarchaeology* **2017**, *32*, 563–574. [CrossRef]

52. Yeager, J.R.; Clark, J.R.; Mitchell, W.; Renshaw, R. Enzyme leach anomalies associated with deep Mississippi valley-type zinc ore bodies at the elmwood mine, Tennessee. *J. Geochem. Explor.* **1998**, *61*, 103–112. [CrossRef]

53. Mao, J.; Goldfarb, R.; Zhang, Z.; Xu, W.; Qiu, Y.; Deng, J. Gold deposits in the Xiaoqinling-Xiong'ershan region, Qinling Mountains, central China. *Miner. Deposita.* **2002**, *37*, 306–325. [CrossRef]

54. Qiu, Y.; Groves, D.I.; McNaughton, N.J.; Wang, L.; Zhou, T. Nature, age, and tectonic setting of granitoid-hosted, orogenic gold deposits of the Jiaodong peninsula, eastern North China craton, China. *Miner. Deposita.* **2002**, *37*, 283–305. [CrossRef]

55. Chen, Y.; Fu, S. Mineralization model and geological-geochemical features of the Shenjiayao gold deposit. *Geol. Prospect.* **1992**, *4*, 47–52.

56. Ren, T.; Liu, Y.; Wang, M. Nanosciences and concealed deposits. *Sci. Tech. Hera.* **1995**, *8*, 18–19.

57. Wang, X.; Xie, X.; Ye, S. Concepts for geochemical gold exploration based on the abundance and distribution of ultrafine gold. *J. Geochem. Explor.* **1995**, *55*, 93–101. [CrossRef]

58. Noble, R.R.P.; Lintern, M.J.; Gray, D.G.; Reid, N.; Anand, R.. Metal migration at the North Miitel Ni sulphide deposit in the southern Yilgarn Craton: Part 1, regolith and groundwater. *Geochem. Explor. Environ. Anal.* **2013**, *13*, 67–85. [CrossRef]

59. Palenik, C.S.; Utsunomiya, S.; Reich, M.; Kesler, S.; Wang, L.; Ewing, R.C. "Invisible" gold revealed: Direct imaging of gold nanoparticles in a Carlin-type deposit. *Am. Mineral.* **2004**, *89*, 1359–1366. [CrossRef]
60. Hochella, M.F., Jr. Nanoscience and technology: The next revolution in the earth sciences. *Earth Planet. Sci. Lett.* **2002**, *203*, 593–605. [CrossRef]
61. Deditius, A.P.; Utsunomiya, S.; Reich, M.; Kesler, S.E.; Ewing, R.C.; Hough, R.; Walshe, J. Trace metal nanoparticles in pyrite. *Ore Geol. Rev.* **2011**, *42*, 32–46. [CrossRef]
62. Hough, R.M.; Noble, R.R.P.; Reich, M. Natural gold nanoparticles. *Ore Geol. Rev.* **2011**, *42*, 55–61. [CrossRef]
63. Ju, Y.; Huang, C.; Sun, Y.; Wan, Q.; Lu, X.; Lu, S.; He, H.; Wang, X.; Zou, C.; Wu, J.; et al. Nanogeosciences: Research history, current status, and development trends. *J. Nanosci. Nanotechnol.* **2017**, *17*, 5930–5965. [CrossRef]

Article

Pyrite Morphology as an Indicator of Paleoredox Conditions and Shale Gas Content of the Longmaxi and Wufeng Shales in the Middle Yangtze Area, South China

Ziyi Liu [1,2], Dongxia Chen [1,2,*], Jinchuan Zhang [3], Xiuxiang Lü [1,2], Ziyi Wang [1,2], Wenhao Liao [1,2], Xuebin Shi [1,2], Jin Tang [1,2] and Guangjie Xie [1,2]

[1] State Key Laboratory of Petroleum Resources and Prospecting, China University of Petroleum, Beijing 102200, China
[2] College of Geosciences, China University of Petroleum, Beijing 102200, China
[3] Key Laboratory of Shale Gas Exploration and Evaluation (Ministry of Land and Resources), China University of Geosciences (Beijing), Beijing 100083, China
* Correspondence: lindachen@cup.edu.cn

Received: 26 April 2019; Accepted: 10 July 2019; Published: 12 July 2019

Abstract: Pyrite is the most common authigenic mineral preserved in many ancient sedimentary rocks. Pyrite also widely exists in the Longmaxi and Wufeng marine shales in the middle Yangtze area in South China. The Longmaxi and Wufeng shales were mainly discovered with 3 types of pyrites: pyrite framboids, euhedral pyrites and infilled framboids. Euhedral pyrites (Py4) and infilled framboids (Py5) belong to the diagenetic pyrites. Based on the formation mechanism of pyrites, the pyrites could be divided into syngenetic pyrites, early diagenetic pyrites, and late diagenetic pyrites. Under a scanning electron microscope (SEM), the syngenetic pyrites are mostly small framboids composed of small microcrystals, but the diagenetic pyrites are variable in shapes and the diagenetic framboids are variable in sizes with large microcrystals. Due to the deep burial stage, the pore space in the sediment was sharply reduced and the diameter of the late diagenetic framboids that formed in the pore space is similar to the diameter of the syngenetic framboids. However, the diameter of the syngenetic framboid microcrystals is suggested to range mainly from 0.3 μm to 0.4 μm, and that of the diagenetic framboid microcrystals is larger than 0.4 μm in the study area. According to the diameter of the pyrite framboids (D) and the diameter of the framboid microcrystals (d), the pyrite framboids could be divided into 3 sizes: syngenetic framboids (Py1, D < 5 μm, d ≤ 0.4 μm), early diagenetic framboids (Py2, D > 5 μm, d > 0.4 μm) and late diagenetic framboids (Py3, D < 5 μm, d > 0.4 μm). Additionally, the mean size and standard deviation/skewness values of the populations of pyrite framboids were used to distinguish the paleoredox conditions during the sedimentary stage. In the study area, most of the pyrite framboids are smaller than 5 μm, indicating the sedimentary water body was a euxinic environment. However, pyrite framboids larger than 5 μm in the shales indicated that the sedimentary water body transformed to an oxic-dysoxic environment with relatively low total organic carbon (TOC: 0.4–0.99%). Furthermore, the size of the framboid microcrystals could be used to estimate the gas content due to thermochemical sulfate reduction (TSR). The process of TSR occurs with oxidation of organic matter (OM) and depletes the H bond of the OM, which will influence the amount of alkane gas produced from the organic matter during the thermal evolution. Thus, syngenetic pyrites (d ranges from 0.35 μm to 0.37 μm) occupy the main proportion of pyrites in the Wufeng shales with high gas content (1.30–2.30 m^3/t), but the Longmaxi shales (d ranges from 0.35 μm to 0.72 μm) with a relatively low gas content (0.07–0.93 m^3/t) contain diagenetic pyrites. Because of TSR, the increasing size of the microcrystals may result in an increase in the value of $\delta^{13}C_1$ and a decrease in the value of $\delta^{13}C_1$-$\delta^{13}C_2$. Consequently, the size of pyrite framboids and microcrystals could be widely used for rapid evaluation of the paleoredox conditions and the gas content in shales.

Keywords: syngenetic framboids; diagenetic framboids; paleoredox conditions; shale gas content; middle Yangtze area; South China

---

1. Introduction

Longmaxi and Wufeng shales were recognized as the most important hydrocarbon generating shales in the middle Yangtze area in China [1–4]. Many studies have been published on the sedimentology of the Longmaxi and Wufeng formation, as well as on reservoir structure, the characteristics of shale gas and the way of gaseous accumulation [5–9], but less attention has been paid to the discussion of authigenic minerals such as pyrites. The only research on the pyrites in the Longmaxi and Wufeng shales has been limited to the pore structure of the pyrite framboids [10–12]. However, the pyrite morphology could be used to indicate paleoredox conditions. More recently, the size distribution of pyrite framboids has been successfully applied as an indicator for an anoxic–euxinic environment in ancient marine sediments [13–17], and the formation of diagenetic pyrite is related to hydrocarbon activity which will influence the hydrocarbons generated from the organic matter [18,19]. Thus, pyrites could be used as indicators of the paleoredox conditions in the middle Yangtze area and as the impact on the hydrocarbon of the Longmaxi and Wufeng shales.

As a widely distributed mineral in marine shales [13,20], pyrite can be divided into syngenetic pyrite and diagenetic pyrite. Syngenetic pyrites are formed at the oxic-anoxic interface and then drop to the sediment surface when buoyancy cannot keep them suspended (Figure 1a) [13,21,22]. The framboid is generally considered to be the unique structural feature of syngenetic pyrite [13,23,24]. Generally, pyrite framboids are not precipitated as pyrite directly, but are transformed through intermediate phases of iron sulfides [13,23,24]. The growth of syngenetic pyrite is restricted after burial and the original framboids are preserved (the diameter of a syngenetic framboid is smaller than 5 μm with small microcrystals). Therefore, syngenetic pyrite can be used to reflect the redox process of shale in a sedimentary water environment [13,25,26]. According to the different formation mechanisms of diagenetic pyrite, early diagenetic pyrites are formed in the sediments' porewater when the location of the oxic-anoxic interface falls down to the sediment (the diameter of an early diagenetic framboid is larger than 5 μm with a broad size distribution) (Figure 1b) [13,22,25,26]. With the burial of sediment, the microcrystals of the early diagenetic pyrites will over grow by the reaction between Fe and S from the porewater (Figure 1b). The formation of recrystallized pyrite (late diagenetic pyrite) and the overgrowth of early diagenetic pyrite are related to thermochemical sulfate reduction (TSR) [18,19]. The overall process of TSR can be summarized as the following simple reaction:

$$\text{Hydrocarbons} + SO_4^{2-} \rightarrow \text{alteredhydrocarbons} + \text{solidbitumen} + H_2S(HS^-) + HCO^-_3(CO_2) + H_2O \qquad (1)$$

The process of TSR occurs with oxidation of organic matter (OM) and depletes the H bond of the organic matter, which will influence the amount of alkane gas produced from the organic matter [18,19]. Further, TSR can result in the destruction of hydrocarbons [27,28] and deplete the content of hydrocarbon [29]. In the process of TSR, the size of late diagenetic pyrite is influenced by the supply of reactants, pore space and growth time. Therefore, the late diagenetic framboids with large microcrystals are often small in size [30]. Additionally, the euhedral pyrites or the pyritic masses may be derived from the overgrowth of diagenetic framboids (Figure 1c) [30,31]. The euhedral pyrites, pyritic masses, infilled framboids, overgrown framboids, polyframboid aggregates and stratiform pyrites are formed during the diagenetic stage with more positive $\delta^{34}S$ values [32–34].

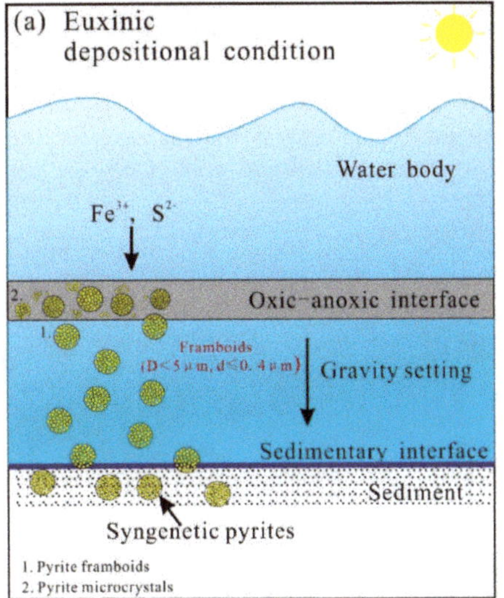
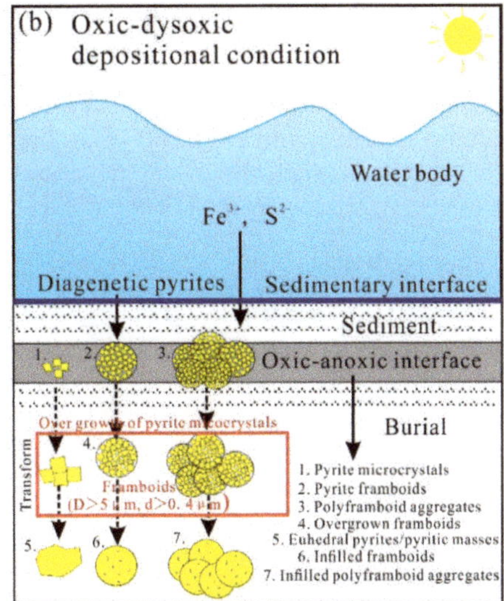
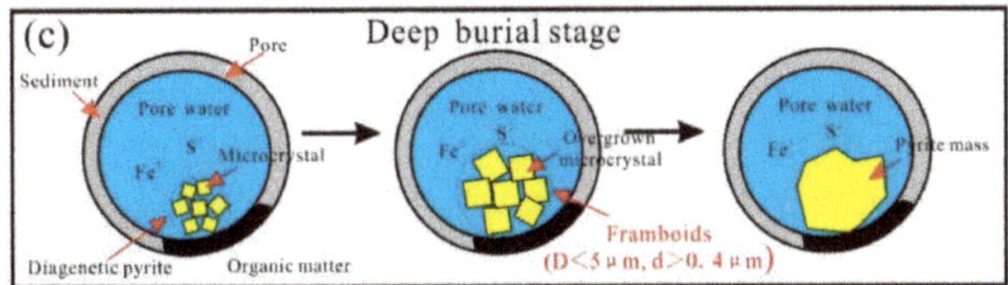

Figure 1. (**a**) Formation mechanism of syngenetic pyrites under the euxinic depositional conditions; (**b**) formation mechanism of early diagenetic pyrites under the oxic-dysoxic depositional conditions [22]; (**c**) formation mechanism of late diagenetic pyrites in the deep burial stage.

In general, the sulfur isotopic composition of pyrite is used to distinguish the diagenetic and syngenetic pyrite populations. The value of $\delta^{34}S$ in syngenetic pyrite is generally negative, while the value of $\delta^{34}S$ in diagenetic pyrite is positive [32,35]. Additionally, the average value of $\delta^{34}S$ measured from the pyrite of the marine shales is 4.15‰, which is similar to the average value of $\delta^{34}S$ measured from sulfate (2.19‰) in the south of China [36]. However, the $\delta^{34}S$ in pyrites of bulk rock is often measured through a mass spectrometer coupled to an elemental analyzer, and the measured sulfur isotope is derived from the mixed values of sulfur isotopes from sedimentary pyrite and diagenetic pyrite. In addition to the above experimental method, the diagenetic and syngenetic pyrite populations are usually distinguished by the smaller mean size of the framboids and standard deviation/skewness values of the populations [14,15,37,38]. Because the diagenetic framboids vary in size, the standard deviation/skewness values of diagenetic framboids are larger than those of syngenetic framboids. This method could be used in an area with apparently variable framboid sizes. For example, the diameter of framboids varies broadly, ranging from 3 μm to 25 μm in the Late Cretaceous Qingshankou formation, Songliao Basin, Northeast China [22]. The size of the pyrite framboids is classified as presenting two populations, framboids < 15 μm and framboids in the range 60–220 μm in the Toarcian Oceanic Anoxic Event, South Iberian Paleomargin [17]. However, the method does not adapt to the pyrites in the Longmaxi and Wufeng shales in the middle Yangtze area, South China. Apart from most of the sizes of the diagenetic framboids are similar to the sizes of syngenetic framboids, the arithmetic mean and standard deviations to predict a subset of framboids is indeed not accurate [39]. The standard deviation/skewness values of diagenetic framboids are similar to those of syngenetic framboids.

In this paper, the size of the framboid microcrystals has been used to distinguish diagenetic and syngenetic pyrite. Then, the size of the pyrite framboids could be used to analyze the paleoredox conditions of the water body in the study area. Furthermore, the relationship between pyrite morphology and shale gas content could be proposed by the size of the framboid microcrystals, which reflects the degree of pyrite recrystallization. The shale gas content could be influenced by the degree of pyrite recrystallization. Meanwhile, the carbon isotopes of alkanes will also be affected by the degree of pyrite recrystallization as a result of the TSR [40–45].

2. Geological Setting

The study area is mainly located in the Western Hunan and Hubei fold belt, which is a part of the middle Yangtze area in South China (Figure 2a) [46–49].

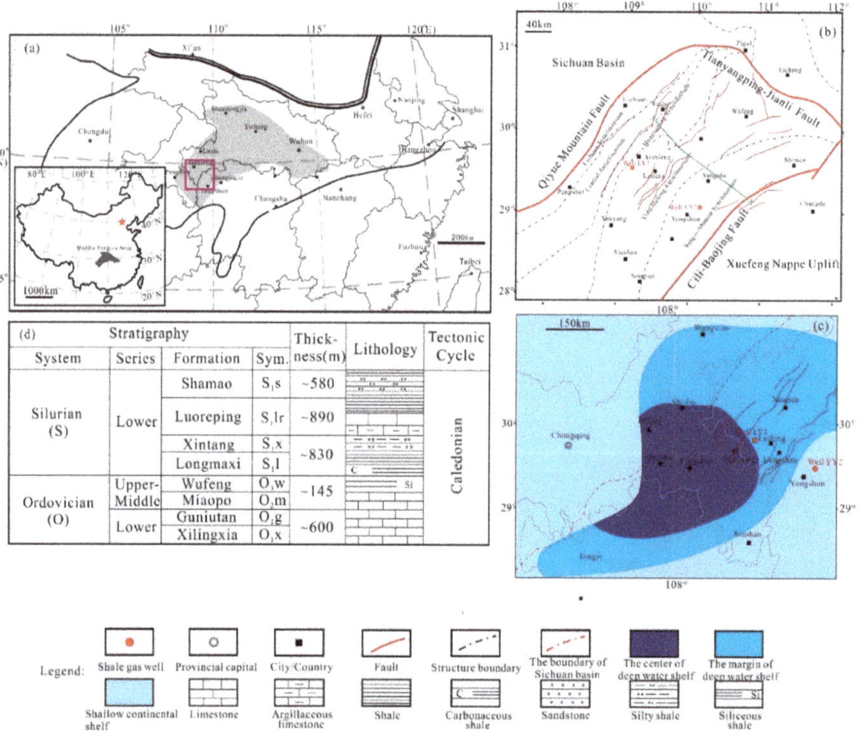

Figure 2. (**a**) The location of the study area in China; (**b**) the tectonic framework in the study area; (**c**) the sedimentary environment of the Longmaxi formation in the study area; (**d**) Simplified stratigraphic units in the middle Yangtze area and the Upper Ordovician Wufeng and the bottom of the lower Silurian Longmaxi formations. The wavy lines represent unconformities. Sym. = symbol.

For the tectonic setting, the study area is adjacent to the Sichuan basin in the west and the Jiangnan-Xuefeng nappe uplift in the east and consists of five tectonic units, namely, the Lichuan synclinorium, Yidu-Hefeng anticlinorium, Sangzhi-Shimen synclinorium, Huaguoping synclinorium and central anticlinorium (Figure 2b). Shale gas well LY1 was located at the syncline nucleus in the Yidu-Hefeng anticlinorium area, and shale gas well YY2 was located at the position of the syncline wing in the Sangzhi-Shimen synclinorium area. Both of the shale gas wells were located at positions far from faults (Figure 2b).

The studied strata in the the middle Yangtze area involve the Lower Silurian Longmaxi formation and the Upper Ordovician Wufeng formation (Figure 2d). In the Late Ordovician period, the study area formed a subsiding basin with frequent alternations of carbonate and mudstone. Then, the depth of the water in the southeastern and northern areas gradually deepened in the early Silurian period. A large set of black shales was formed laterally in mainly the shallow continental shelf and deep shelf areas. From the center of deep water shelf to the shallow continental shelf, the thickness of black shales was reduced. Subsequently, the whole area was uplifted, and the strata were denuded extensively, which was influenced by Caledonian tectonic movement in the late Silurian [50,51]. The lower sections of the Longmaxi and Wufeng Formation approximately 50–60 m thick are widely distributed over the study area [52]. Of the two wells involved in this paper, one well is located in the margin of deep-water shelf area and the other in the shallow continental shelf area (Figure 2c). As the most important hydrocarbon generating formations, the Wufeng formation is composed of black siliceous shales and the lower section of the Longmaxi formation consists of black carbonaceous shales and silty shales (Figure 2d). Additionally, the middle and upper sections of the Longmaxi formation are composed of gray-green shales, yellow-green shales and siltstones. The colors change gradually upward, indicating that the sedimentary environment of the Longmaxi formation varied from the deep shelf to the shallow shelf [52]. The Upper Ordovician Wufeng formation was in the deep water shelf sedimentary setting. The thermal maturity of the organic matter is higher than that in the North American shale and has equivalent vitrinite reflectance values greater than 2% [53–56].

3. Methods

For analysis of the mineralogical composition, 11 core samples were collected from well LY1 and 8 core samples were collected from well YY2. All of samples were ground in an agate mortar with more than 800-mesh sieve. The mineralogical composition of 17 samples was characterized using the XRD patterns of randomly oriented powders by using a Bruker D8 Discover X-ray diffractometer (Table 1). The operation and calculation followed the relevant oil-industry standard of China (SY/T 5163–2010). The sample with a particle size smaller than 10 μm was used to evaluate the total clay mineral content and the other nonclay mineral content [57].

Table 1. The mineralogical composition of 17 samples from well LY1 and well YY2 was derived based on the XRD analysis.

Well	Strata	Depth (m)	TOC (%)	Mineral Composition Content (%)				
				Quartz	Felspar	Carbonate	Pyrite	Clay Mineral
LY1	S_1l	903	0.99	32.7	11.6	9.2	3.2	43.3
LY1	S_1l	907	0.4	44.1	22.5	16.2	2.0	15.2
LY1	S_1l	908	1.13	34.3	13.8	11.8	3.9	36.2
LY1	S_1l	917	0.43	42.3	25.1	15.8	2.2	14.6
LY1	S_1l	921	0.92	40.8	17.1	13.4	2.9	25.8
LY1	S_1l	925	1.13	39.9	17.0	13.4	2.1	27.6
LY1	S_1l	929	1.06	40.5	18.8	11.8	2.5	26.4
LY1	S_1l	933	1.15	39.5	18.2	17.4	2.8	22.1
LY1	O_3w	943	2.33	44.6	13.2	13	3.8	25.4
LY1	O_3w	948	1.83	42.3	13.3	6.5	3.5	34.4
YY2	S_1l	1503	0.28	38.7	15.4	8.6	0.2	37.1
YY2	S_1l	1510	0.99	42	25	6.0	2.0	25.0
YY2	S_1l	1511	1.12	40.4	15.8	7.5	4.5	31.6
YY2	S_1l	1512	1.05	35	28	3.0	3.0	31.0
YY2	S_1l	1515	1.10	46	21	2.0	4.0	27.0
YY2	S_1l	1517	1.22	38.9	23	2.8	4.1	31.2
YY2	O_3w	1520	5.39	41	20	2.0	5.0	32.0

Notes: 1. S_1l means Longmaxi formation. 2. O_3w means Wufeng formation.

For pyrite morphological analyses, the chips of core samples were polished to 0.1 mm thickness using helium ion beams and cut into chips measuring 0.5 cm ×1 cm × 0.2 mm [58]. Then, the analyses were carried out under SU8010 cold field emission scanning electron microscope (SEM) equipped with low and high secondary-electron (SE) probes, black-scattered electrons (BSE) and an X-ray spectrometer (EDAX) at China University of Petroleum (Beijing). The pyrite morphological were obtained by SEM not optical methods. Then, the size of the pyrite framboids and the framboid microcrystals in the pictures were measured by the JMicroVision software (Version1.2.7, JMicroVision, Geneva, Switzerland), a software of image analysis similar to imageJ. Optical measurements underestimate the number of smaller framboids leading to larger arithmetic mean values. It is further interesting to note that the fit of the log-normal distribution is better for the SEM measurements than for the optical measurements [39].

The measured size of pyrite framboids are an approximation to within a 10% deviation because of the polished surface randomly intersecting the framboids [59].

The TOC data of 17 samples were measured by using a Leco CS-230 carbon analyzer in China University of Petroleum (Beijing, China). All samples were crushed and sieved with an 80-mesh sieve and then were blended with 10 vol% HCl for one hour to eliminate the inorganic carbon. After that, all samples were washed with distilled water to remove all traces of HCl. Finally, the TOC analysis was performed using a Leco CS-230 carbon analyzer with 99.5% oxygen as carrier gas under the temperature of 24 °C and relative humidity of 48% [60].

The gas content data of 17 samples were collected from desorption of the core samples from well LY1 and well YY2. The gas content was quantified after the core sample was recovered at the drilling site. The sample was placed as quickly as possible inside a hermetically sealed canister saturated with salt water, and the volumes of gas released inside the canister were periodically measured using a graduated cylinder at atmospheric pressure [61,62]. The gas desorption instruments were provided by the China University of Geosciences (Beijing, China).

The analysis of stable carbon isotopes was carried out at the Nuclear Industry Beijing Geological Research Analysis and Test Research Centre. Stable carbon isotope values were determined on a Finnigan Mat Delta Plus mass spectrometer interfaced with an HP 5890II chromato-graph. Individual hydrocarbon gas components (C_1–C_4) were separated on a gas chromatograph using a fused silica capillary column (PLOT Q 30 m × 0.32 mm). The GC oven was ramped from 35 °C to 80 °C at 8 °C/min, then to 260 °C at 5 °C/min, and maintained at the final temperature for 10 min. Stable carbon isotopic values are reported in the δ-notation in per mil (‰) relative to the Vienna Pee Dee Belemnite (VPDB). The measurement precision was estimated to be ±0.5‰ for δ13C [63–65].

4. Results

4.1. Pyrite Morphology in the Wufeng and Longmaxi Shales

As a widely distributed mineral in marine shales [13,20], pyrites appear with various shapes in Longmaxi and Wufeng shales in the middle Yangtze area, South China.

Core observation shows that the pyrites are not disorderly and irregularly distributed in the shales. Pyrites in the Longmaxi gray silty shales sometimes appear as stratiform pyrites, which can be seen as a layered distribution along the bedding surface of the shales (Figure 3a). Occasionally, the pyrites display a nodular form in the Longmaxi gray silty shales (Figure 3b). However, the pyrites in the Wufeng black shales are usually distributed in the form of thin strips along the bedding surface of shales (Figure 3c).

SEM analysis shows that the pyrites in the Wufeng and Longmaxi shales are mainly exhibits 3 types of pyrites: pyrite framboids, euhedral pyrites and infilled framboids. The pyrite framboids could be divided into 3 sizes: syngenetic framboids (Py1, D < 5 μm, d ≤ 0.4 μm) distributed in the organic matter enrichment zone (Figure 4a,c,e,f), early diagenetic framboids (Py2, D > 5 μm, d > 0.4 μm) (Figure 4c,e) and late diagenetic framboids (Py3, D < 5 μm, d > 0.4 μm) (Figure 4b). Euhedral pyrites (Py4) occurring in the pores between clay minerals and organic matter (Figure 4a,c,d) and infilled framboids (Py5) keeping the shape of sphericity or sub-sphericity (Figure 4f) belong to the diagenetic pyrites. Further, the pyrites in the Wufeng and Longmaxi shales are enriched in syngenetic framboids (Py1, D < 5 μm, d ≤ 0.4 μm) and diagenetic euhedral pyrites (Py4). However, the early diagenetic framboids (Py2, D > 5 μm, d > 0.4 μm), late diagenetic framboids (Py3, D < 5 μm, d > 0.4 μm) and diagenetic infilled framboids only exist in the Longmaxi shales.

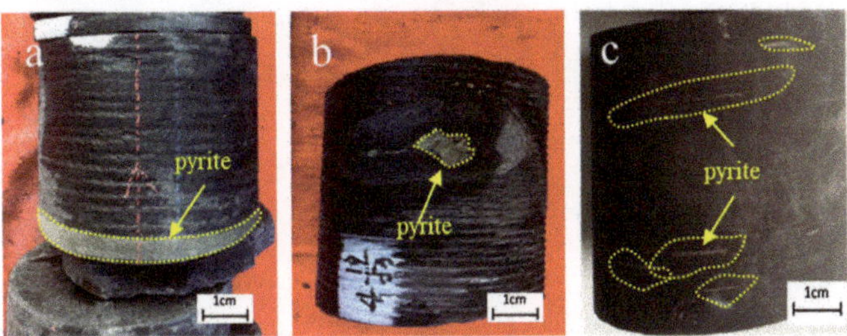

Figure 3. Core characteristics of the Longmaxi and Wufeng shales. (**a**) Stratiform pyrites in the gray silty shale of the Longmaxi formation, 921 m, well LY1; (**b**) separate nodular pyrites in the gray silty shale of the Longmaxi formation, 917 m, well LY1; (**c**) thin strips of pyrites in the Wufeng black shales, 1520 m, well YY2.

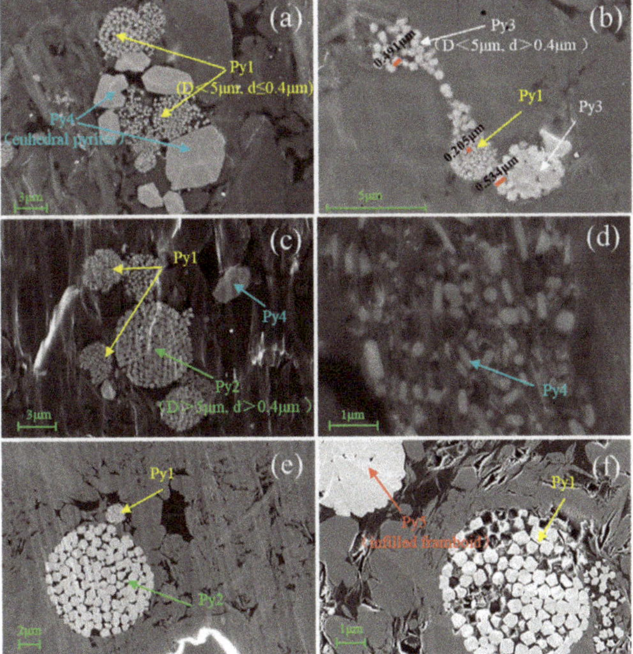

Figure 4. Pyrite micromorphology in the Longmaxi and Wufeng formations. (**a**) Syngenetic framboids (Py1) and small diagenetic euhedral pyrites (Py4) in the Wufeng black shale, 1520 m, well YY2; (**b**) pyrite framboids of similar size with different sized microcrystals between the syngenetic framboids (Py1) and late diagenetic framboids (Py3) in the Longmaxi black shale, 1510 m, well YY2; (**c**) in addition to Py1 and Py4, the early diagenetic framboid pyrites (Py2) are observed in the Longmaxi black-gray shale, 903 m, well LY1; (**d**) clay minerals and organic matter with euhedral pyrites in Lomgmaxi black shale, 925 m, well LY1; (**e**) early diagenetic framboid pyrites (Py2) in the Lomgmaxi shale, 921 m, well LY1; (**f**) infilled framboid (Py5) in the Longmaxi black shale, 1517 m, well YY2.

4.2. The size Distribution of Framboids and Microcrystals

As shown in Figure 5, the main peak of the diameter of the microcrystals ranges from 0.3 μm to 0.4 μm. Thus, the size of the syngenetic framboid microcrystals is suggested to range mainly from 0.3 μm to 0.4 μm, and the diagenetic framboid microcrystals are larger than 0.4 μm. The interval from 0.4 μm to 1.0 μm includes the diagenetic framboid microcrystals.

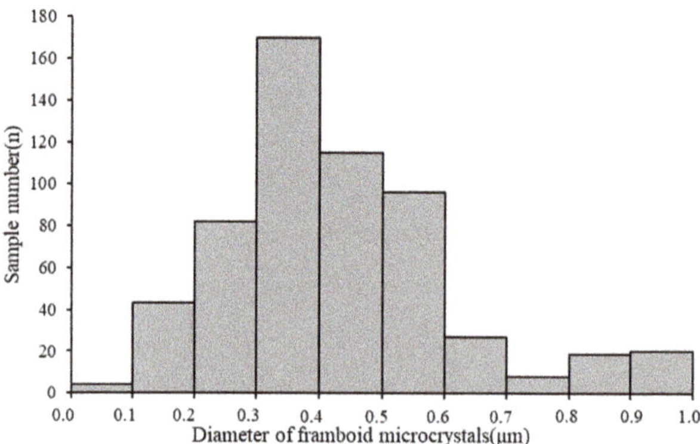

Figure 5. Distribution of the diameter of the framboid microcrystals in the Wufeng and Longmaxi formations.

According to the morphological characteristics of pyrite in the Wufeng and Longmaxi shales (Table 2), the statistics of the size distribution, such as standard deviation and skewness, have been calculated for 16 samples in the Wufeng and Longmaxi formations. The variation in standard deviation and skewness ranges from 0.62 to 2.94 and 0.03 to 3.86, respectively, with an average of 1.71 and 1.19, respectively. The average values of standard deviation and skewness of the Longmaxi framboids (1.73 and 1.35) are higher than those of the Wufeng framboids (1.62 and 0.51), which indicates that the pyrite framboids of the Longmaxi shales are more diverse and are larger than those of the Wufeng shales. Based on standard deviation and skewness, the pyrite framboids from euxinic and oxic-dysoxic environments have been plotted (Figure 6a,b). Most of the spots drop in the area of euxinic environments, except some spots from well LY1 that occur in oxic-dysoxic environments (Figure 6a,b). Meanwhile, the mean diameters of the framboids (Mean, D) range from 3.00 μm to 5.59 μm, with an average of 4.23 μm, and more than 70% of the pyrite framboids in black shales or black siliceous shales are smaller than 5.0 μm in mean size. Only a small number of the pyrite framboids have a mean size larger than 5.0 μm in the Longmaxi black-gray shales, black-gray silty shales and gray silty shales (Figures 7 and 8). The mean diameters of the microcrystals of the framboids (Mean, d) range from 0.35 μm to 0.72 μm, with an average of 0.45 μm, and the value of the Mean, d in the Wufeng black shales is lower than 0.4 μm.

Table 2. Morphological characteristics of pyrite in the Wufeng and Longmaxi shales.

Well	Strata	Depth (m)	Mean, D (μm)	Mean, d (μm)	Standard Deviation	Skewness	Framboids Measured	TOC (%)	Desorbed Gas Content (m³/t)
YY2	S1l	1503	3.57	0.7	-	-	1	0.28	0.07
YY2	S1l	1510	4.33	0.42	2.05	1.17	57	0.99	0.55
YY2	S1l	1511	4.08	0.35	1.56	0.64	72	1.12	1.04
YY2	S1l	1512	3.36	0.37	1.01	0.03	56	1.05	0.69
YY2	S1l	1515	3.00	0.38	0.92	0.14	60	1.1	0.93
YY2	S1l	1517	4.00	0.4	1.35	0.95	70	1.22	0.44
YY2	O_3w	1520	4.46	0.35	1.47	0.72	90	5.39	2.30
LY1	S1l	903	5.12	0.48	2.69	3.86	79	0.99	0.46
LY1	S1l	907	5.32	0.72	2.94	2.71	64	0.4	0.15
LY1	S1l	908	3.82	0.38	1.3	0.2	69	1.13	0.57
LY1	S1l	917	5.45	0.43	2.53	2.51	69	0.43	0.42
LY1	S1l	921	5.59	0.68	2.18	2.09	55	0.92	0.34
LY1	S1l	925	3.34	0.49	1.43	1.34	43	1.13	0.45
LY1	S1l	929	4.11	0.39	1.97	1.78	60	1.06	0.64
LY1	S1l	933	3.65	0.41	0.62	0.13	56	1.15	0.61
LY1	O_3w	943	4.61	0.37	1.84	0.28	92	2.33	1.30
LY1	O_3w	948	4.14	0.37	1.56	0.53	75	1.83	1.65

Notes: 1. Mean, D means the mean diameter of framboids. 2. Mean, d means the mean diameter of microcrystals composing the framboids. 3. The measurement precision was estimated to be 0.45% of measured value for TOC and 3% of measured value for gas content.

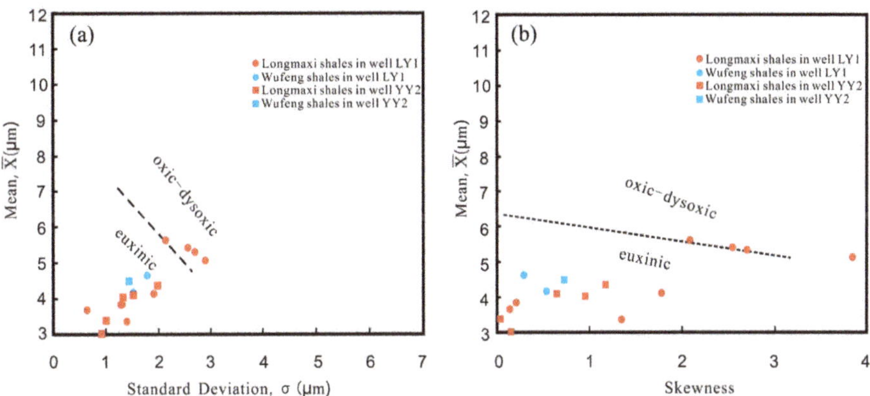

Figure 6. (**a**) Plot of the mean versus the standard deviation of the framboid size distribution of three wells. The dashed lines at the euxinic boundary and the oxic-dysoxic boundary are from Wei et al. (2016); (**b**) plot of the mean versus the skewness of the framboid size distribution of three wells. The dashed line is from Wei et al. (2016).

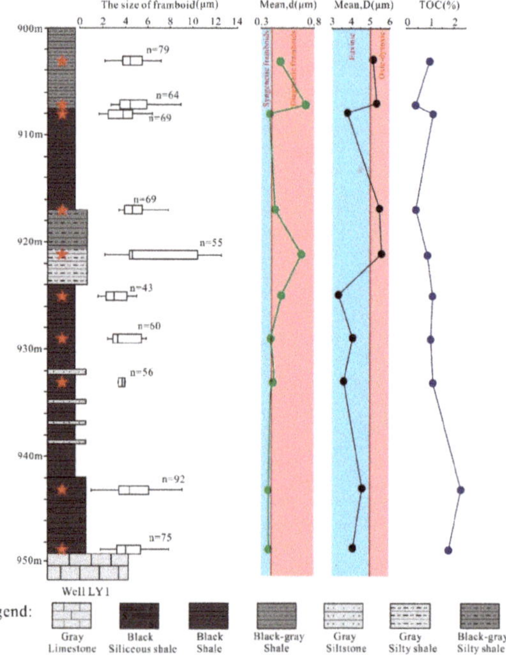

Figure 7. Size distribution of pyrite framboids, mean, D, mean, d, and TOC in the well LY1 shales.

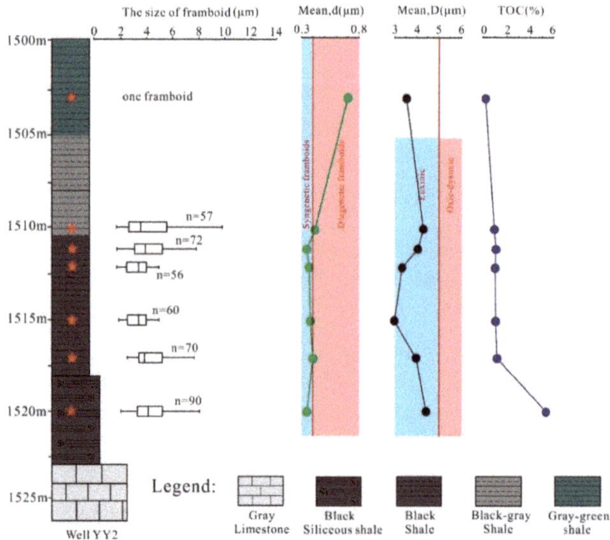

Figure 8. Size distribution of pyrite framboids, mean, D, mean, d, and TOC in the well YY2 shales.

Additionally, only one framboid could be observed in the Longmaxi gray-green shales at the depth of 1503 m with a low Mean, D and a relatively high Mean, d in well YY2 (Figure 7).

4.3. Characteristic of Shale Gas in the Wufeng and Longmaxi Shales

As shown in Table 2, the gas content tends to increase with increasing depth. Compared to the gas content in the Longmaxi shales (the gas content ranges from 0.07–1.04 m^3/t with an average of 0.53 m^3/t), relatively high gas content was discovered in the Wufeng shales (the gas content ranges from 1.30–2.30 m^3/t with an average of 1.75 m^3/t) in the middle Yangtze area. Though the plot of TOC to gas content shows a relatively positive correlation (Figure 9a), the plot of pyrite content to gas content shows an unobvious correlation (Figure 9b). According to Table 2 and Figure 10, the gas content of the area with Mean,d > 0.4 μm is lower than that of the area with Mean,d < 0.4 μm.

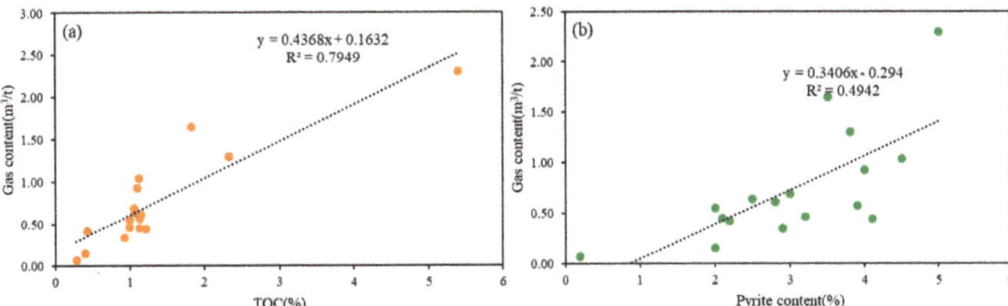

Figure 9. (**a**) Relationship between gas content and TOC values; (**b**) relationship between gas content and pyrite content for 17 shale samples from well YY2 and well LY1 in the middle Yangtze area, South China.

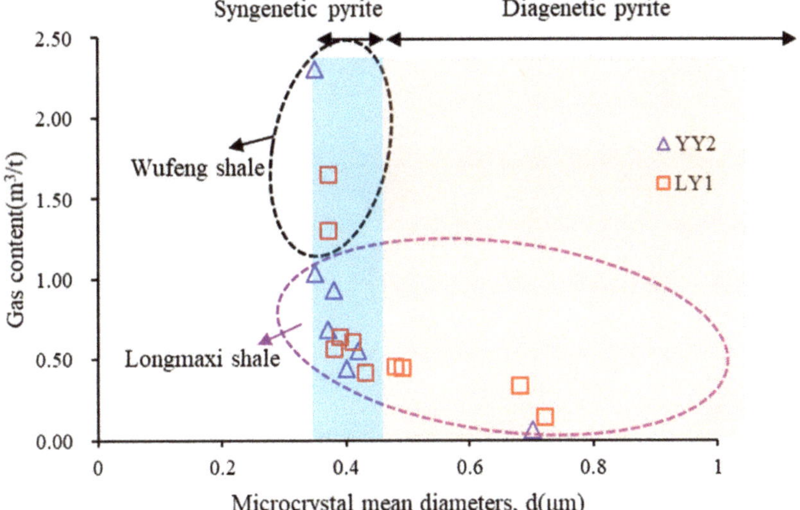

Figure 10. Different microcrystal diameters of pyrite framboids as indicators of the desorbed gas content.

For 10 gas samples from well YY2 and Well LY1, the $\delta^{13}C_1$ values range from −38.4‰ to −30.2‰, while the $\delta^{13}C_2$ values range from −41.5‰ to −35.5‰ (Table 3). Additionally, the $\delta^{13}C_1$(‰)-$\delta^{13}C_2$(‰) values range from 2.90‰ to 5.50‰. As the microcrystals continue to grow larger than 0.4 μm, the value of $\delta^{13}C_1$ and $\delta^{13}C_2$ increases, while the value of $\delta^{13}C_1$-$\delta^{13}C_2$ decreases (Table 3, Figure 11).

Table 3. Characteristics of microcrystalline mean diameters and the value of $\delta^{13}C$.

Well	Depth	Mean,d	$\delta^{13}C_1$(‰)VPDB	$\delta^{13}C_2$(‰)VPDB	$\delta^{13}C_1$(‰)-$\delta^{13}C_2$(‰)
YY2	1511	0.35	−38.0	−41.0	3.00
YY2	1512	0.37	−38.2	−41.1	2.90
YY2	1515	0.38	−38.3	−41.2	2.90
YY2	1520	0.35	−38.4	−41.3	2.90
LY1	907	0.72	−30.8	−35.8	5.00
LY1	917	0.43	−32.9	−38.4	5.50
LY1	921	0.68	−30.2	−35.5	5.30
LY1	925	0.49	−32.4	−37.7	5.30
LY1	933	0.41	−32.7	−38.1	5.40
LY1	943	0.37	−36.3	−41.5	5.20

Notes: 1. The measurement precision was estimated to be ±0.5‰ for $\delta^{13}C$.

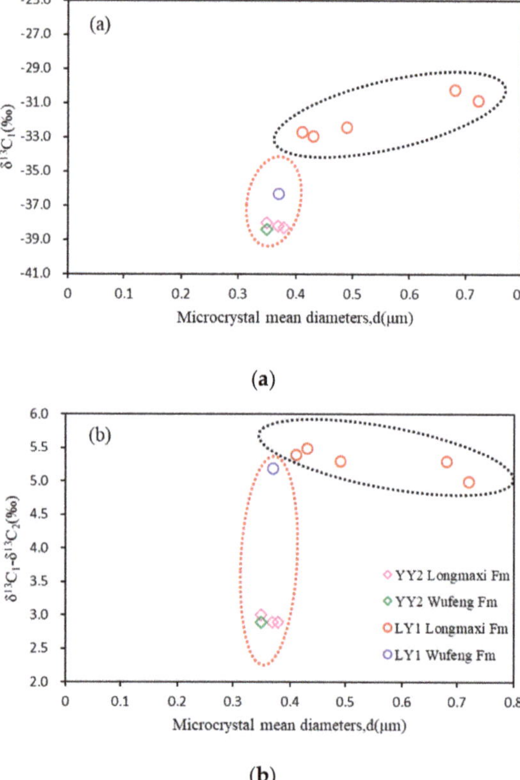

Figure 11. (a) Relationship between microcrystalline mean diameters and the value of the $\delta^{13}C_1$; (b) relationship between microcrystalline mean diameters and the value of the $\delta^{13}C_1$-$\delta^{13}C_2$.

5. Discussion

5.1. Pyrite Morphology as an Indicator of Paleoredox Conditions

Based on the formation mechanism of pyrite framboids, the pyrite framboids can be used as an indicator of paleoredox conditions [13–17,38,60]. Syngenetic framboids will not grow after pyrites are precipitated from the water and deposited in the sediments [13,66]. Nevertheless, the early diagenetic framboids that form under oxic-dysoxic conditions [13,22] have a longer growth time than the syngenetic framboids. This process results in the size of the early diagenetic framboids being large and variable. Hence, the framboids' mean size and standard deviation/skewness values of the populations can be used to estimate the paleoredox conditions in the water body [13–15,37,38].

The framboids' mean size and standard deviation/skewness values of the populations from the well LY1 and well YY2 are given in Table 2. As shown in Figure 6a,b, most of the spots drop in the area of euxinic environments. In the area of euxinic environments, pyrites are mainly syngenetic framboids with a narrow size distribution. However, some spots from well LY1 occur in oxic-dysoxic environments due to the samples of the spots include the early diagenetic framboids with a variable size. Meanwhile, most of the Mean, D values of the pyrite framboids in the two wells are smaller than 5 μm and reflect a euxinic environment with strongly reduction conditions in the study area (Figures 7 and 8) [25,26,67]. In addition, black-gray shales, black-gray silty shales and gray silty shales, all with relatively low TOC (0.4%–0.99%), reflect relatively weak reduction conditions in the sedimentary water body in well LY1. The shales with a Mean, D larger than 5 μm also indicate oxic-dysoxic conditions during the sedimentary process (Figure 7). In well YY2, the upper sections of the Longmaxi formation are composed of gray-green shales with low TOC (0.28%), which may suggest a sedimentary environment with oxidation conditions (Figure 8). During oxidation, it is difficult to form pyrites in the sediments [25,67], and only one pyrite framboid had been found under the SEM.

Although the Longmaxi and Wufeng formations are located in the deep-water shelf area and shallow continental shelf area, the redox environment of the sedimentary water is still different. According to the pyrite morphology, the Longmaxi and Wufeng formations with black shales are suggested to have formed under a euxinic environment, but the Longmaxi formation with black-gray shales, black-gray silty shales and gray silty shales is suggested to have formed under an oxic-dysoxic environment. The gray-green shales of the Longmaxi formation with only one or no pyrite framboids reflect an oxidizing environment. When a euxinic environment is transformed into an oxic-dysoxic environment, the sediment pore water is probably provided with more materials (e.g., Fe^{3+}, SO_4^{2-} and O_2) which benefits the process of TSR in the Longmaxi shales. However, the Wufeng formation is suggested to have formed under a euxinic environment with insufficient material supply (e.g., Fe^{3+}, SO_4^{2-} and O_2). The pore waters are short of Fe^{3+}, SO_4^{2-} and O_2, result in the lack of late diagenetic framboids in the Wufeng shales. The process of TSR during the diagenesis period is not obvious in the Wufeng formation.

5.2. Relationship between Pyrite Morphology and Gas Content

According to the observation of core samples and the corresponding data of the gas content, we found that the shales with diverse gas contents are accompanied by different pyrite morphologies in the middle Yangtze area, South China. A large amount of syngenetic framboids indicates a strongly reduction environment in the water column [15,22,25,68–72], which benefits the preservation of organic matter in the sediment. Thus, the shale that contains a number of syngenetic pyrites is usually the organic-rich shale. However, the relationship between TOC and pyrite content is not very good in the study area (Table 1) and the plot of pyrite content to gas content shows an unobvious correlation due to the existence of the late diagenetic pyrite (Figure 9b).

Redox reactions involving sulfate (SO_4^{2-}) and OM may occur during diagenetic processes [18,73,74]. In addition to sulfate reduction driven by bacterial sulfate reduction (BSR) at low temperatures during early diagenesis (i.e., $0 < T < 60$–$80\ ^\circ C$), sulfate reduction can also occur chemically (thermochemical sulfate reduction: TSR) at higher temperatures [18,73,74]. Thermochemical sulfate reduction (TSR) accompanied with the formation of diagenetic pyrite results in the oxidation of organic matter (OM) and the loss of functional groups [18,19], while H_2S produced from the TSR process in the source rock was depleted by iron in the shales [18,65,75]. Further, there is a significant negative correlation between the TOC content and $\delta^{34}S$ values of pyrite supporting the depletion of OM in the sections affected by

TSR [19]. The process of TSR could lead to the destruction of hydrocarbons [18,27,28] and the depletion of hydrocarbon content [29]. Meanwhile, TSR could give rise to highly aromatic, insoluble residues. These types of insoluble residues are usually pyrobitumens [18,27], and the pyrobitumens that are the final product of the organic thermal evolution cannot product alkanes again [76]. Additionally, the process of thermochemical sulfate reduction could result in depletion of the H bond of organic matter, which will influence the amount of alkane gas produced from the organic matter during the thermal evolution. Thus, with the increasing growth of late diagenetic pyrite and the overgrowth of the early diagenetic pyrites, the content of alkanes produced from OM became increasingly lower.

As shown in Figures 7 and 8, the sections of the Longmaxi formation with a Mean, d larger than 0.4 µm is coupled with relatively low shale gas contents. The plot of microcrystalline mean diameters to gas content reflects the negative relationship between the formation of diagenetic pyrites and the shale gas content in the Wufeng and Longmaxi shales (Figure 10). In contrast to the shales with diagenetic pyrites (d > 0.4 µm with a gas content range from 0.07 m^3/t to 0.61 m^3/t), shales with syngenetic framboids (d ≤ 0.4 µm) feature relatively high gas content (0.44–2.30 m^3/t) (Figure 10). Thus, syngenetic framboids (d ranges from 0.35 µm to 0.37 µm) occupy the main proportion of pyrites in the Wufeng shales with high gas content (1.30–2.30 m^3/t). However, the Longmaxi shales (d ranges from 0.35 µm to 0.72 µm) contain some diagenetic pyrites that occur with relatively low gas content (0.07–0.93 m^3/t) (Table 2, Figure 10). Consequently, the more diagenetic pyrite that develops as overgrowths of the framboid microcrystals, the lower the shale gas content is in the Longmaxi and Wufeng shales.

The process of thermochemical sulfate reduction companied with the formation of diagenetic pyrite could also influence the carbon isotope of alkanes [40–45]. As shown in Figure 11a, as the microcrystals continue to grow greater than 0.4 µm, the value of $\delta^{13}C_1$ increases. The thermochemical sulfate reduction (TSR) between sulfate and methane may result in the heavier carbon isotope of methane [40,41,44,45]. Furthermore, TSR could occur between sulfate and heavy hydrocarbon, which would result in the carbon isotope value of heavy hydrocarbon gas becoming heavier during the process [18,42,43]. $\delta^{13}C_2$ is more susceptible to TSR than $\delta^{13}C_1$, which leads to an increase in size of the microcrystals with a decreasing value of $\delta^{13}C1$-$\delta^{13}C_2$ (Table 3, Figure 11b).

6. Conclusions

(1) In the middle Yangtze area, the size of the pyrite framboid microcrystals is used to distinguish syngenetic pyrite and diagenetic pyrite due to their different formation mechanisms. The size of the syngenetic framboid microcrystals is suggested to range mainly from 0.3 µm to 0.4 µm, and that of the diagenetic framboid microcrystals is larger than 0.4 µm. Pyrite framboids in the study area are divided into syngenetic framboids (Py1, D < 5 µm, d ≤ 0.4 µm), early diagenetic framboids (Py2, D > 5 µm, d > 0.4 µm) and late diagenetic framboids (Py3, D < 5 µm, d > 0.4 µm).

(2) The sedimentary environment of the Longmaxi and Wufeng shales is mainly a euxinic environment with good reduction conditions. However, some part of the Longmaxi formation reflects an oxic-dysoxic environment with a lithology of black-gray shales, black-gray silty shales and gray silty shales. The gray-green shales of the Longmaxi formation with only one or no pyrite framboids reflect an oxidizing environment.

(3) The process of TSR accompanied with the formation of diagenetic pyrite results in oxidation of organic matter (OM) and depletion of the H bond of OM, which will influence the amount of alkane gas produced from the organic matter during the thermal evolution. The size of the framboid microcrystals could be used to estimate the shale gas content. Syngenetic pyrites (d ranges from 0.35 µm to 0.37 µm) occupy the main proportion of pyrites in the Wufeng shales with high gas content (1.30–2.30 m^3/t), but the Longmaxi shales (d ranges from 0.35 µm to 0.72 µm) with relatively low gas content (0.07–0.93 m^3/t) contain diagenetic pyrites in the middle Yangtze area, South China. Due to thermochemical sulfate reduction (TSR), the increasing size of the microcrystals may result in an increase in the value of $\delta^{13}C_1$ and a decrease in the value of $\delta^{13}C_1$-$\delta^{13}C_2$.

Author Contributions: Conceptualization, Z.L. and D.C.; Methodology, Z.L.; Software, J.T.; Validation, J.Z., X.L. and X.S.; Formal Analysis, G.X.; Investigation, Z.L.; Resources, J.Z.; D.C., Z.W.; Writing-Original Draft Preparation, Z.L.; Writing-Review & Editing, Z.L.; Visualization, Z.L.; Supervision, W.L.; Project Administration, D.C.; Funding Acquisition, D.C.

Funding: This work was supported by the Key Program of National Natural Science Foundation of China (41472110), The National Science and Technology Major Projects of China (2016ZX05034-001-05) and China Geological Survey (12120115007201).

References

1.   Qiu, X.S.; Yang, B.; Hu, M.Y. Characteristics of shale reservoirs and gas content of Wufeng-Longmaxi Formation in middle Yangtze region. *Nat. Gas Geosci.* **2013**, *24*, 1274–1283. (In Chinese)

2.   Nie, H.K.; Jin, Z.J.; Ma, X.; Liu, Z.B.; Lin, T.; Yang, Z.H. Graptolites zone and sedimentary characteristics of Upper Ordovician Wufeng Formation-Lower Silurian Longmaxi Formation in Sichuan Basin and its adjacent areas. *Acta Pet. Sin.* **2017**, *38*, 160–174. (In Chinese)

3.   Zhao, Z.; Li, R.; Feng, W.; Yu, Q.; Yang, H.; Zhu, L.X. Enrichment conditions and favorable zone prediction of Wufeng-Longmaxi shale gas reservoirs in the northern Yunnan-Guizhou provinces, China. *Nat. Gas Ind.* **2017**, *37*, 26–34. (In Chinese)

4.   Xiong, X.H.; Wang, J.; Xiong, G.Q.; Wang, Z.J.; Men, Y.P.; Zhou, X.L.; Zhou, Y.X.; Yang, X.; Deng, Q. Shale gas geological characteristics of Wufeng and Longmaxi formations in Northeast Chongqing and its exploration direction. *Acta Geol. Sin.* **2018**, *92*, 1948–1958. (In Chinese)

5.   Curtis, J.B. Fractured shale-gas systems. *AAPG Bull.* **2002**, *86*, 1921–1938.

6.   Chen, S.; Zhu, Y.; Wang, H.; Liu, H.; Wei, W.; Fang, J. Shale gas reservoir characterisation: A typical case in the southern Sichuan Basin of China. *Energy* **2011**, *36*, 6609–6616. [CrossRef]

7.   Milliken, K.L.; Day-Stirrat, R.J.; Papazis, P.K.; Dohse, C. Carbonate lithologies of the mississippian Barnett shale, Fort Worth Basin, Texas. *AAPG Mem.* **2012**, *97*, 290–321.

8.   Yang, C.; Zhang, J.C.; Tang, X.; Ding, J.H.; Zhao, Q.R.; Dang, W.; Chen, H.Y.; Su, Y.; Li, B.W.; Lu, D.F. Comparative study on micro-pore structure of marine, terrestrial, and transitional shales in key areas, China. *Int. J. Coal Geol.* **2017**, *171*, 76–92. [CrossRef]

9.   Liu, Y.; Zhang, J.C.; Zhang, P.; Liu, Z.Y.; Zhao, P.W.; Huang, H.; Tang, X.; Mo, X.X. Origin and enrichment factors of natural gas from the lower Silurian Songkan Formation in northern Guizhou province, south China. *Int. J. Coal Geol.* **2018**, *187*, 20–29. [CrossRef]

10. Zhao, J.H.; Jin, Z.K.; Jin, Z.J.; Wen, X.; Geng, Y.K. Origin of authigenic quartz in organic-rich shales of the Wufeng and Longmaxi Formations in the Sichuan Basin, South China: Implications for pore evolution. *J. Nat. Gas Sci. Eng.* **2017**, *38*, 21–28. [CrossRef]

11. Xu, H.; Zhou, W.; Zhang, R.; Liu, S.M.; Zhou, Q.M. Characterizations of pore, mineral and petrographic properties of marine shale using multiple techniques and their implications on gas storage capability for Sichuan Longmaxi gas shale field in China. *Fuel* **2019**, *241*, 360–371. [CrossRef]

12. Nie, H.K.; Sun, C.X.; Liu, G.X.; Du, W.; He, Z.L. Dissolution pore types of the Wufeng Formation and the Longmaxi Formation in the Sichuan Basin, south China: Implications for shale gas enrichment. *Mar. Pet. Geol.* **2019**, *101*, 243–251. [CrossRef]

13. Wilkin, R.T.; Barnes, H.L.; Brantley, S.L. The size distribution of framboidal pyrite in modern sediments: An indicator of redox conditions. *Geochim. Cosmochim. Acta* **1996**, *60*, 3897–3912. [CrossRef]

14. Wilkin, R.T.; Barnes, H.L. Pyrite formation in an anoxic estuarine basin. *Am. J. Sci.* **1997**, *297*, 620–650. [CrossRef]

15. Wignall, P.B.; Newton, R.; Brookfield, M.E. Pyrite framboid evidence for oxygenpoor deposition during the Permian-Triassic crisis in Kashmir. *Palaeogeogr. Palaeoclim. Palaeoecol.* **2005**, *216*, 183–188. [CrossRef]

16. Pisarzowska, A.; Berner, Z.A.; Racki, G. Geochemistry of Early Frasnian (Late Devonian) pyrite-ammonoid level in the Kostomłoty Basin, Poland, and a new proxy parameter for assessing the relative amount of syngenetic and diagenetic pyrite. *Sediment. Geol.* **2014**, *308*, 18–31. [CrossRef]

17. Gallego-Torres, D.; Reolid, M.; Nieto-Moreno, V.; Martínez-Casado, F.J. Pyrite framboid size distribution as a record for relative variations in sedimentation rate: An example on the Toarcian Oceanic Anoxic Event in Southiberian Palaeomargin. *Sediment. Geol.* **2015**, *330*, 59–73. [CrossRef]

18. Machel, H.G. Bacterial and thermochemical sulfate reduction in diagenetic settings—Old and new insights. *Sediment. Geol.* **2001**, *140*, 143–175. [CrossRef]

19. Ardakani, O.H.; Chappaz, A.; Sanei, H.; Mayer, B. Effect of thermal maturity on remobilization of molybdenum in black shales. *Earth Planet. Sci. Lett.* **2016**, *449*, 311–320. [CrossRef]

20. Wignall, P.B.; Newton, R. Pyrite framboid diameter as a measure of oxygen deficiency in ancient mudrocks. *Am. J. Sci.* **1998**, *298*, 537–552. [CrossRef]

21. Berner, R.A. Sedimentary pyrite formation. An update. *Geochim. Cosmochim. Acta* **1984**, *48*, 605–615. [CrossRef]

22. Wang, P.K.; Huang, Y.J.; Wang, C.S.; Feng, Z.H.; Huang, Q.H. Pyrite morphology in the first member of the Late Cretaceous Qingshankou Formation, Songliao Basin, Northeast China. *Palaeogeogr. Palaeoclimatol. Palaeoecol.* **2013**, *385*, 125–136. [CrossRef]

23. Love, L.G.; Amstutz, G.C. Review of microscopic pyrite from the Devonian Chattanooga Shale and Rammelsberg Banderz. *Fortschr Miner.* **1966**, *43*, 273–309.

24. Wang, X.J.; Zhang, S.Q.; Wei, M.J.; Mu, X.S.; Xu, T.W. Characteristics and formation mode of Es3 member pyrite in Wendong area of Dongpu depression. *Fault Block Oil Gas Field* **2015**, *22*, 178–183. (In Chinese)

25. Bond, D.P.G. Pyrite framboid study of marine Permian-Triassic boundary sections: A complex anoxic event and its relationship to contemporaneous mass extinction. *Geol. Soc. Am. Bull.* **2010**, *122*, 1265–1279. [CrossRef]

26. Wignall, P.B.; Bond, D.P.G.; Kuwahara, K.; Kakuwa, Y.; Newton, R.; Poulton, S.W. An 80 million year oceanic redox history from Permian to Jurassic pelagic sediments of the Mino-Tamba terrane, SW Japan, and the origin of four mass extinctions. *Glob. Planet. Chang.* **2010**, *71*, 109–123. [CrossRef]

27. Kelemen, S.R.; Walters, C.C.; Kwiatek, P.J.; Freund, H.; Afeworki, M.; Sansone, M.; Lamberti, W.A.; Pottorf, R.J.; Machel, H.G.; Peters, K.E.; et al. Characterization of solid bitumens originating from thermal chemical alteration and thermochemical sulfate reduction. *Geochim. Cosmochim. Acta* **2010**, *74*, 5305–5332. [CrossRef]

28. Hubert, E.K.; Walters, C.C.; Horn, W.C.; Zimmer, M.; Heines, M.M.; Lamberti, W.A.; Kliewer, C.; Pottorf, R.J.; Macleod, G. Sulfur isotopic analysis of bitumen and pyrite associated with thermal sulfate reduction in reservoir carbonates at the Big Piney-La Barge production complex. *Geochim. Cosmochim. Acta* **2014**, *134*, 210–220.

29. Ding, K.L.; Li, S.Y.; Yue, C.T.; Zhong, N.N. Review of thermochemical sulfate reduction. *J. Univ. Pet. China* **2005**, *29*, 150–155. (In Chinese)

30. Ye, Y.T.; Wu, C.D.; Zhai, L.N.; An, Z.Z. Pyrite morphology and episodic euxinia of the Ediacaran Doushantuo Formation in South China. *Sci. China* **2017**, *60*, 102–113. [CrossRef]

31. Lin, Z.Y.; Sun, X.M.; Peckmann, J.; Lu, Y.; Xu, L.; Strauss, H.; Zhou, H.Y.; Gong, J.L.; Lu, H.F.; Teichert, B.M.A. How sulfate-driven anaerobic oxidation of methane affects the sulfur isotopic composition of pyrite: A SIMS study from the South China Sea. *Chem. Geol.* **2016**, *440*, 26–41. [CrossRef]

32. Magnall, J.M.; Gleeson, S.A.; Stern, R.A.; Newton, R.J.; Poulton, S.W.; Paradis, S. Open system sulphate reduction in a diagenetic environment- Isotopic analysis of barite (δ^{34}S and δ^{18}O) and pyrite (δ^{34}S) from the Tom and Jason Late Devonian Zn-Pb-Ba deposits, Selwyn Basin, Canada. *Geochim. Cosmochim. Acta* **2016**, *180*, 146–163. [CrossRef]

33. Raiswell, R. Pyrite texture, isotopic composition and the availability of iron. *Am. J. Sci.* **1982**, *82*, 1244–1263. [CrossRef]

34. Sawlowicz, Z. Pyrite framboids and their development: A new conceptual mechanism. *Geol. Rundsch.* **1993**, *82*, 148–156. [CrossRef]

35. Canfield, D.E. The evolution of the earth surface sulfur reservoir. *Am. J. Sci.* **2004**, *304*, 839–861. [CrossRef]

36. Fu, X.D.; Qiu, N.S.; Qin, J.Z.; Ten, G.E.; Liu, W.H.; Wang, X.F. Content distribution and isotopic composition characteristics of sulfur in matrine source rocks in Middle-Upper Yangtze region. *Pet. Geol. Exp.* **2013**, *35*, 545–551. (In Chinese)

37. Wilkin, R.T.; Barnes, H.L. Formation processes of framboidal pyrite. *Geochim. Cosmochim. Acta* **1997**, *61*, 323–339. [CrossRef]

38. Wei, H.Y.; Wei, X.M.; Qiu, Z.; Song, H.Y.; Shi, G. Redox conditions across the G–L boundary in South China: Evidence from pyrite morphology and sulfur isotopic compositions. *Chem. Geol.* **2016**, *440*, 1–14. [CrossRef]

39. Rickard, D. Sedimentary pyrite framboid size-frequency distributions: A meta-analysis. *Palaeogeogr. Palaeoclimatol. Palaeoecol.* **2019**, *522*, 62–75. [CrossRef]

40. Cai, C.F.; Xie, Z.Y.; Worden, R.H.; Hu, G.Y.; Wang, L.S.; He, H. Methane-dominated thermochemical sulphate reduction in the Triassic Feixianguan Formation in East Sichuan Basin, China: Towards prediction of fatal H_2S concentrations. *Mar. Pet. Geol.* **2004**, *21*, 1265–1279. [CrossRef]

41. Cai, C.F.; Zhang, C.M.; He, H.; Tang, Y.J. Carbon isotope fractionation during methane-dominated TSR in East Sichuan Basin gasfields, China: A review. *Mar. Pet. Geol.* **2013**, *48*, 100–110. [CrossRef]

42. Hao, F.; Guo, T.; Zhu, Y.; Cai, X.; Zou, H.; Li, P. Evidence for multiple stages of oil cracking and thermochemical sulfate reduction in the Puguang gas field, Sichuan Basin, China. *AAPG Bull.* **2008**, *92*, 611–637. [CrossRef]

43. Xia, X.Y.; Ellis, G.S.; Ma, Q.S.; Tang, Y.C. Compositional and stable carbon isotopic fractionation during non-autocatalytic thermochemical sulfate reduction by gaseous hydrocarbons. *Geochim. Cosmochim. Acta* **2014**, *139*, 472–486. [CrossRef]

44. Jenden, P.D.; Titley, P.A.; Worden, R.H. Enrichment of nitrogen and ^{13}C of methane in natural gases from the Khuff Formation, Saudi Arabia, caused by the thermochemical sulfate reduction. *Org. Geochem.* **2015**, *82*, 54–68. [CrossRef]

45. Li, K.K.; George, S.C.; Cai, C.F.; Gong, S.; Sestak, S.; Armand, S.; Zhang, X.F. Fluid inclusion and stable isotopic studies of thermochemical sulfate reduction: Upper Permian and lower Triassic gasfields, northeast Sichuan Basin, China. *Geochim. Cosmochim. Acta* **2019**, *246*, 86–108. [CrossRef]

46. Hu, X. Oil and gas reservoir types in the western region of Hunan and Hubei and exploration orientation. *Oil Gas Geol* **2002**, *23*, 300–306. (In Chinese)

47. Long, Y. Lower Paleozoic shale gas exploration potential in the central Yangtze area, China. *Geol. Bull. China* **2011**, *30*, 344–348. (In Chinese)

48. Wang, J.; Li, X.; Huang, W. The shale gas exploration prospect assess of the Niutitang formation in western Hubei and Hunan-eastern Chongqing. *Geol. Sci. Technol. Inf.* **2014**, *33*, 98–103. (In Chinese)

49. Wang, Y.; Zhu, Y.; Chen, S.; Li, W. Characteristics of the nanoscale pore structure in Northwestern Hunan shale gas reservoirs using field emission scanning electron microscopy, high-pressure mercury intrusion, and gas adsorption. *Energy Fuels* **2014**, *28*, 945–955. [CrossRef]

50. Chen, K.; Zhang, J.C.; Tang, X.; Yu, J.D.; Liu, Y.; Yang, C. Lower Silurian Longmaxi formation shale adsorption capacity and main controlling factors analysis in western Hunan and Hubei areas. *Oil Gas Geol.* **2016**, *37*, 23–29. (In Chinese)

51. Yao, M.J.; Bao, H.Y.; Ding, Q.; Long, Y.K.; Liu, C.; Wang, Y.C. Stratigraphic and sedimentary characteristics of Wufeng-Longmaxi formation in the western Hunan-Hubei region. *Geol. Miner. Resour. South China* **2016**, *32*, 191–197. (In Chinese)

52. Hu, J.A.; Tang, S.H.; Zhang, S.H. Investigation of pore structure and fractal characteristics of the Lower Silurian Longmaxi shales in western Huan and Hubei Provinces in China. *J. Nat. Gas Sci. Eng.* **2016**, *28*, 522–535. [CrossRef]

53. Fan, E.; Tang, S.; Jiang, W.; Sun, C.; Zhang, C. Accumulation conditions and exploration potential of shale gas in Lower Silurian Longmaxi Formation. Northwestern Hunan. *Xi'an Shiyou Univ. Nat. Sci. Ed.* **2014**, *29*, 7–30.

54. Yang, F.; Ning, Z.; Liu, H. Fractal characteristics of shales from a shale gas reservoir in the Sichuan Basin, China. *Fuel* **2014**, *115*, 378–384. [CrossRef]

55. Yang, N.; Tang, S.; Zhang, S.; Jiang, W.; Zhu, W. Geological conditions and favorable exploration zones of shale gas in Longmaxi formation of North-western Hunan area. *Coal Sci. Technol.* **2014**, *42*, 104–108.

56. Wan, Y.; Pan, Z.J.; Tang, S.H.; Connell, L.D.; Down, D.D.; Camilleri, M. An experimental investigation of diffusivity and porosity anisotropy of a Chinese gas shale. *Nat. Gas Sci. Eng.* **2015**, *23*, 70–79. [CrossRef]

57. Chen, Q.; Zhang, J.; Tang, X.; Dang, W.; Li, Z.M.; Liu, C.; Zhang, X.Z. Pore Structure Characterization of the Lower Permian Marine Continental Transitional Black Shale in the Southern North China Basin, Central China. *Energy Fuels* **2016**, *30*, 10092–10105. [CrossRef]

58. Liu, Y.C.; Chen, D.X.; Qiu, N.S.; Wang, Y.; Fu, J.; Hu, Y.Y.; Jia, J.K.; Wu, H. Reservoir characteristics and methane adsorption capacity of the Upper Triassic continental shale in Western Sichuan Depression, China. *Aust. J. Earth Sci.* **2017**, *64*, 807–823. [CrossRef]

59. Cashman, K.V.; Marsh, B. Crystal size distribution (CSD) in rocks and the kinetics and dynamics of crystallization II: Makaopuhil lava lake. Contrib. *Miner. Pet.* **1988**, *99*, 292–305. [CrossRef]

60. Chen, G.; Gang, W.Z.; Liu, Y.Z.; Wang, N.; Jiang, C.; Sun, J.B. Organic matter enrichment of the Late Triassic Yanchang Formation (Ordos Basin, China) under dysoxic to oxic conditions: Insights from pyrite framboid size distributions. *J. Asian Earth Sci.* **2019**, *170*, 106–117. [CrossRef]

61. Mavor, M.; Nelson, C. *Coalbed Reservoir Gas-In-Place Analysis: GRI-97/0263*; Gas Research Institute: Chicago, IL, USA, 1997.

62. Ma, Y.; Zhong, N.N.; Li, D.H.; Pan, Z.J.; Cheng, L.J.; Liu, K.Y. Organic matter/clay mineral intergranular pores in the lower Cambrian lujiaping shale in the north-eastern part of the upper Yangtze area, China: A possible microscopic mechanism for gas preservation. *Int. J. Coal Geol.* **2015**, *137*, 38–54. [CrossRef]

63. Ni, Y.Y.; Dai, J.X.; Zhu, G.Y.; Zhang, S.C.; Zhang, D.J.; Su, J.; Tao, X.W.; Liao, F.R.; Wu, W.; Gong, D.Y.; et al. Stable hydrogen and carbon isotopic ratios of coal-derived and oil-derived gases: A case study in the tarim basin, NW China. *Int. J. Coal Geol.* **2013**, *116*, 302–313. [CrossRef]

64. Dai, J.X.; Gong, D.Y.; Ni, Y.Y.; Huang, S.P.; Wu, W. Stable carbon isotopes of coal-derived gases sourced from the Mesozoic coal measures in China. *Org. Geochem.* **2014**, *74*, 123–142. [CrossRef]

65. Dai, J.X.; Zou, C.N.; Liao, S.M.; Dong, D.Z.; Ni, Y.Y.; Huang, J.L.; Wu, W.; Gong, D.Y.; Huang, S.P.; Hu, G.Y. Geochemistry of the extremely high thermal maturity Longmaxi shale gas, southern Sichuan Basin. *Org. Geochem.* **2014**, *74*, 3–12. [CrossRef]

66. Fishman, N.S.; Hackley, P.C.; Lowers, H.A.; Hill, R.J.; Egenhoff, S.O.; Eberl, D.D.; Blum, A.E. The nature of porosity in organic-rich mudstones of the upper Jurassic Kimmeridge clay formation, North Sea, offshore United Kingdom. *Int. J. Coal Geol.* **2012**, *103*, 32–50. [CrossRef]

67. Raiswell, R.; Berner, R.A. Pyrite formation in euxinic and semi-euxinic sediments. *Am. J. Sci.* **1985**, *285*, 710–724. [CrossRef]

68. Liao, W.; Wang, Y.B.; Kershaw, S.; Weng, Z.T.; Yang, H. Shallow-marine dysoxia across the Permian–Triassic boundary: Evidence from pyrite framboids in the microbialite in South China. *Sediment. Geol.* **2010**, *232*, 77–83. [CrossRef]

69. Guan, C.; Zhou, C.; Wang, W.; Wan, B.; Yuan, X.; Chen, Z. Fluctuation of shelf basin redox conditions in the early Ediacaran: Evidence from Lantian Formation black shales in South China. *Precambrian Res.* **2014**, *245*, 1–12. [CrossRef]

70. Tian, L.; Tong, J.; Algeo, T.J.; Song, H.; Chu, D.; Shi, L.; Bottjer, D.J. Reconstruction of Early Triassic ocean redox conditions based on framboidal pyrite from the Nanpanjiang Basin, South China. *Palaeogeogr. Palaeoclim. Palaeoecol.* **2014**, *412*, 68–79. [CrossRef]

71. Takahashi, S.; Yamasaki, S.; Ogawa, K.; Kaiho, K.; Tsuchiya, N. Redox conditions in the end-Early Triassic Panthalassa. *Palaeogeogr. Palaeoclim. Palaeoecol.* **2015**, *432*, 15–28. [CrossRef]

72. Wei, H.; Algeo, T.J.; Yu, H.; Wang, J.; Guo, C.; Shi, G. Episodic euxinia in the Changhsingian (late Permian) of South China: Evidence from framboidal pyrite and geochemical data. *Sediment. Geol.* **2015**, *319*, 78–97. [CrossRef]

73. Orr, W.L. Geologic and geochemical controls on the distribution of hydrogen sulfide in natural gas. In *Advances in Organic Geochemistry*; Enadisma: Madrid, Spain, 1977; pp. 571–597.

74. Cross, M.M.; Manning, D.A.; Bottrell, S.H.; Worden, R.H. Thermochemical sulphate reduction (TSR): Experimental determination of reaction kinetics and implications of the observed reaction rates for petroleum reservoirs. *Org. Geochem.* **2004**, *35*, 393–404. [CrossRef]

75. Dixon, S.A.; Summers, D.M.; Surdam, R.C. Diagenesis and preservation of porosity in Norphlet Formation (Upper Jurassic), southern Alabama. *Am. Assoc. Pet. Geol. Bull.* **1989**, *73*, 707–728.

76. Chen, J.; Xiao, X.M. Evolution of nanoporosity in organic-rich shales during thermal maturation. *Fuel* **2014**, *129*, 173–181. [CrossRef]

 *minerals*

Review

Nanocrystalline Principal Slip Zones and Their Role in Controlling Crustal Fault Rheology

Berend A. Verberne [1,*], Oliver Plümper [2] and Christopher J. Spiers [2]

[1] Geological Survey of Japan, National Institute of Advanced Industrial Science and Technology, 1-1-1 Higashi, Tsukuba, Ibaraki 305-8567, Japan

[2] Department of Earth Sciences, Utrecht University, P.O. Box 80.021, 3508 TA Utrecht, The Netherlands; o.plumper@uu.nl (O.P.); c.j.spiers@uu.nl (C.J.S.)

* Correspondence: ba.verberne@aist.go.jp; Tel.: +81-29-861-5211

Received: 7 May 2019; Accepted: 25 May 2019; Published: 28 May 2019

Abstract: Principal slip zones (PSZs) are narrow (<10 cm) bands of localized shear deformation that occur in the cores of upper-crustal fault zones where they accommodate the bulk of fault displacement. Natural and experimentally-formed PSZs consistently show the presence of nanocrystallites in the <100 nm size range. Despite the presumed importance of such nanocrystalline (NC) fault rock in controlling fault mechanical behavior, their prevalence and potential role in controlling natural earthquake cycles remains insufficiently investigated. In this contribution, we summarize the physical properties of NC materials that may have a profound effect on fault rheology, and we review the structural characteristics of NC PSZs observed in natural faults and in experiments. Numerous literature reports show that such zones form in a wide range of faulted rock types, under a wide range of conditions pertaining to seismic and a-seismic upper-crustal fault slip, and frequently show an internal crystallographic preferred orientation (CPO) and partial amorphization, as well as forming glossy or "mirror-like" slip surfaces. Given the widespread occurrence of NC PSZs in upper-crustal faults, we suggest that they are of general significance. Specifically, the generally high rates of (diffusion) creep in NC fault rock may play a key role in controlling the depth limits to the seismogenic zone.

Keywords: nanograins; principal slip zone; crystallographic preferred orientation; amorphization; mirror-slip surface; faults; earthquakes; localization

---

1. Introduction

Nanocrystalline materials are widespread in the Earth's atmosphere, biosphere, and in the subsurface [1–5], including in principal slip zones (PSZs) within natural faults [6–8]. PSZs are zones of localized shear deformation that (have) accommodate(d) the bulk of displacement in the cores of upper-crustal faults [9,10], which suggests that the physical properties of the ultrafine(nano)-grained fault rock within PSZs plays an important role in controlling fault mechanical behavior or fault rheology. From observations on metals and ceramics it is well known that nanophase materials, characterized by grain sizes < 100 nm, frequently exhibit unusual deformation properties compared with coarser-grained counterparts [11–13]. The reason for this is the loss of cohesive energy between atoms comprising the grain as its size continues to decrease. In view of the generality of this nanograin size effect, it is important to consider the potential physical implications of nanogranular fault rock. Despite the emerging awareness on the importance of nanophase geomaterials in Earth sciences [1–8], their prevalence in upper-crustal faults and potential role in natural earthquake cycles remains insufficiently investigated.

In this paper, we aim to elucidate the significance of nanocrystalline PSZs in Earth's upper crust. We start with background on fault mechanics and upper-crustal seismogenesis, and summarize some key physical properties of nanophase materials which, when applied to fault rock, are expected to be of major importance in controlling fault strength and stability. We go on to review the micro- and nanostructural characteristics of natural and experimentally-formed nanocrystalline PSZs, and list reports from the literature of PSZs characterized by grains <100 nm in size. Our work demonstrates that nanocrystalline PSZs form under a wide range of conditions pertaining slow (a-seismic) and fast (co-seismic) upper-crustal fault slip. Also, we observe that they are frequently characterized by an internal crystallographic preferred orientation, and by the presence of amorphous materials and/or glossy fault plane interfaces known as "mirror-slip" surfaces. Given the abundant observations of nanocrystalline PSZs in field exposures of faults, as well as in experiments, we suggest that they are of general importance to upper-crustal fault deformation. The physical properties of nanocrystalline fault rock may play a key role in natural earthquake cycles, especially in controlling the depth distribution of upper-crustal seismicity.

2. Fault Zones, Earthquakes, and the Seismogenic Zone

The presence of long-lived, localized zones of shear deformation in the crust, or fault zones, implies that the fault rocks within are weaker than the surrounding country rocks and that their weakness is persistent [14,15]. The strength of the upper-crust is classically approximated using a Coulomb-type, brittle failure law, abruptly giving way to ductile deformation below ~15 to 20 km depth (Figure 1a) [16,17]. A brittle-to-ductile transition at ~15 to 20 km depth is consistent with geological and seismological observations of the base of the so-called "seismogenic zone", i.e., the depth interval in the upper-crust in which the bulk of upper-crustal earthquakes nucleate [18–25], suggesting that at greater depths earthquake rupture nucleation is inhibited by intrinsically stable, ductile or viscous flow in shear zones. A seismicity cut-off at shallower depths, typically observed at ~2–4 km, demarcates the upper limit of the seismogenic zone [24,26]. Field and laboratory studies of fault deformation suggest that within the seismogenic zone, "multi-mechanism" or "frictional-viscous" fault slip-involving coincident rate-sensitive (creep) and rate-insensitive (e.g., cataclasis) deformation mechanisms-plays an important role (Figure 1b) [27–37]. However, in general, the microphysical processes responsible for aseismic fault sliding above the seismogenic zone, and for seismogenic slip within, remain poorly understood for most fault rock types.

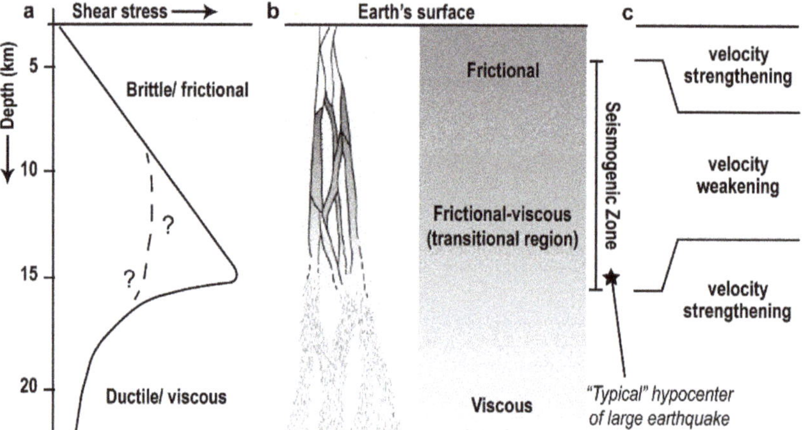

Figure 1. Schematic profile of a fault zone in Earth's upper-crust. (**a**) Fault strength vs. depth. (**b**) Fault zone sketch (from [15]), with a rough depth range indicating fault deformation regimes, the seismogenic zone, and (**c**) velocity dependence or intrinsic fault stability regimes.

In the case of earthquakes, sliding along faults is achieved by unstable, periodic slip events instead of by stable, continuous motion. This is similar to the jerky sliding motion that is frequently observed in laboratory rock friction experiments, known as "stick-slip" [38]. Regular stick-slip behavior can be easily envisioned using a spring-block model system, consisting of a rigid block or slider on a nominally flat surface, driven via a spring of a certain stiffness. When the spring is pulled at constant speed, an instability may develop depending on the frictional properties of the slider-surface contact, the mass of the block, the spring stiffness, and the loading rate, resulting in intermittent slider acceleration and stationary contact [39–42]. Ruina [41] showed that for regular stick-slip to occur the slider-surface contact must decrease in strength with increasing displacement rate, hence be "velocity-weakening". In the opposite case "velocity-strengthening" occurs, which leads to a state of stable sliding [41,43]. Thus, the seismogenic zone is believed to represent a depth interval in the upper-crust where shear deformation of fault zones leads to unstable, velocity-weakening behavior, as opposed to stable velocity-strengthening above and below (Figure 1c) [22,24,25,44].

Importantly, the velocity dependence of frictional strength is a material property of the sliding medium which constitutes the slider-surface contact. Applying this to natural faults, the sliding medium is represented by the granular wear product of cumulative slip along the fault, or "fault gouge", present in the fault core [45]. Field and drilling studies of active and inactive natural fault zones frequently demonstrate the presence of a mm- to cm-wide principal slip zone (PSZ) in the gouge-filled fault core that accommodates, or has accommodated, the bulk of displacement along the fault [9,46–52]. Tectonic loading of a faulted rock volume, as occurs continuously in numerous geologic settings (e.g., at tectonic plate boundaries), causes energy dissipation predominantly along the PSZ [10,53]. In the case of slow (aseismic) fault sliding, quasi-static deformation of fault rock is believed to be key [54–56], whereas at higher (seismic) slip rates frictional heat generated along the PSZ plays an increasingly important role [57], leading to dynamic fault rupture processes such as melting, decarbonation, and/or thermal pressurization [58–63].

3. The Physical Properties of Nanophase Materials

Material properties such as melting temperature or yield point frequently show drastic changes when the grain size decreases to the nanometer-realm (<100 nm) [11–13]. The reason for this is fundamental; a decreasing grain size implies a parabolic increase of the fraction of surface atoms (Figure 2a), which have a much lower average binding energy compared with atoms in the bulk phase. This means that when the grain size decreases to that of a few atoms or unit cells, it has major implications for thermodynamic stability and reactivity of the individual particles [64–66]. For example, the melting point of Au particles is observed to decrease from ~1300 K to 700 K as the grain size decreases from 20 nm to 5 nm [67] (Figure 2b). Observations on common rock-forming minerals are scarcer. However, in the case of calcite, which is the dominant constituent of limestone, the decomposition temperature decreases from ~1075 K to 950 K as the grain size decreases from 40 nm to 20 nm [68] (Figure 2b). Size-dependence of the melting or decomposition temperature of fault rock within a principal slip zone (PSZ) may have major implications for bulk fault rheology, for example at elevated temperatures due to frictional heating.

Another unique aspect relevant to nanostructured polycrystals and fault rock is their huge cumulative grain surface area, which naturally increases exponentially as the grain size continues to decrease (Figure 2c). This has major implications not only for chemical reactivity but also for the rheology of a material. For example, due to the short grain scale transport distances in nanostructured materials [69], grain boundary diffusion driven mechanisms [70–72] are generally fast, enabling superplastic deformation at much lower temperatures/higher strain rates than compared with in coarser-grained materials [73–75]. The high grain boundary density also plays an important role in dislocation-mediated plasticity. As the grain size decreases to the <100 nm size range, dislocations are emitted and adsorbed efficiently at grain boundaries, leading to a decrease of the material yield strength with decreasing grain size hence an "inverse Hall-Petch effect" [76–79] (Figure 2d). In this mechanism, dislocations traverse the crystallite within very brief time windows, achieving very large strains while leaving a micro-/nanostructure characterized by "strain-free" nanograins [80–82].

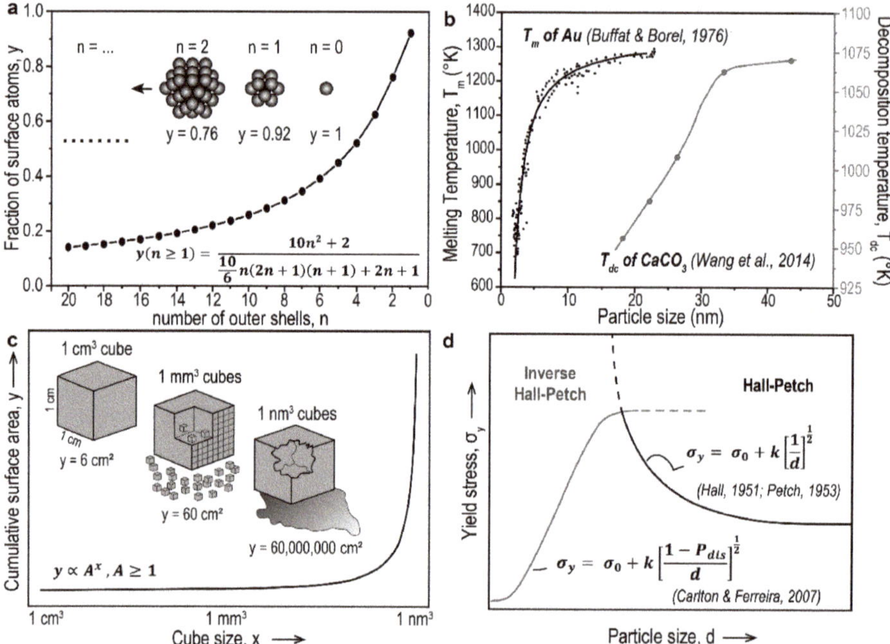

$$y(n \geq 1) = \frac{10n^2 + 2}{\frac{10}{6}n(2n+1)(n+1) + 2n + 1}$$

$$y \propto A^x, A \geq 1$$

$$\sigma_y = \sigma_0 + k\left[\frac{1}{d}\right]^{\frac{1}{2}}$$

$$\sigma_y = \sigma_0 + k\left[\frac{1 - P_{dis}}{d}\right]^{\frac{1}{2}}$$

Figure 2. Key characteristics of nanoparticles and nanostructured materials. (**a**) A simple model of hexagonal close-packed balls illustrating the effect of particle miniaturization. The fraction of surface particles increases near-exponentially with decreasing number of outer shells, or particle size (after [11,13]). (**b**) The melting temperature of Au [67] and the decomposition temperature of $CaCO_3$ [68] particles decrease sharply as particle size decreases within the nm-realm. (**c**) The cumulative surface area of polycrystals increases exponentially as the grain size continues to decrease. (**d**) The empirical relation between yield strength σ_y and grain size d, known as the Hall-Petch (HP) effect [76,77], reverses for very small d (after [78,79]). The expression for σ_y in the inverse-HP regime is the model by Carlton & Ferreira (2007) [79], where k is a constant and P_{dis} is the probability of a dislocation being absorbed by a grain boundary.

4. Nanocrystalline Principal Slip Zones in Natural Faults and in Experiments

Observations of natural and experimentally-formed principal slip zones (PSZs) showing the presence of <100 nm-sized grains are listed in respectively Tables 1 and 2. Below we summarize the micro- and nanostructural characteristics of nanocrystalline PSZs, highlighting seminal reports of field/drilling studies of natural faults and of laboratory studies. Much insight was obtained recently from studies of glossy or "mirror-like" fault slip surfaces (MSSs) formed in experimentally simulated faults composed of calcite fault gouge. These are described using a separate section. While we aspire to provide as complete an overview as possible, we may have overlooked some of the studies performed to date.

4.1. Nanocrystalline Principal Slip Zones in Exposures of Natural Faults

Faults that are exposed in an orientation normal to the fault plane display a cross-section through the damage zone that has developed upon repeated fault displacement, including the principal slip zone(s) (Figure 3a). Power and Tullis [83] used optical and transmission electron microscopy (TEM) to investigate sections prepared normal to the fault plane of rocks collected from the glossy fault trace of the Dixie Valley thrust fault (USA). In a zone just ~0.2 mm wide, the fault trace or PSZ is characterized by ultrafine grains down to 10 nm in size, and a uniform optical birefringence and extinction. This optical effect may be observed in (ultra)thin sections using crossed nicols in a polarizing light microscope (by rotating the microscope stage as shown in Supplementary Video S1) and is widely used as indicative of a crystallographic preferred orientation (CPO). Chester & Goldsby [84] also reported a nanocrystalline PSZ with a CPO, in fault core samples from the Punchbowl Fault (USA). Field investigation revealed visually distinct, 0.15 to 0.55 m thick layers of fine-grained fault gouge known as ultracataclasite, separated by what was identified as a "principal fracture surface" [46]. However, thin section analyses revealed that the ultracataclasite layers were separated by a zone of finite width (constituting a principal slip *zone*), characterized by a strong uniform birefringence, and the presence of grains down to 4 nm in size [7,84]. Other notable observations of naturally-formed nanocrystalline PSZs have been made from drilling of seismically-active fault zones, such as the Chelungpu Fault (Taiwan) [8,85] and the San Andreas Fault (USA) [86].

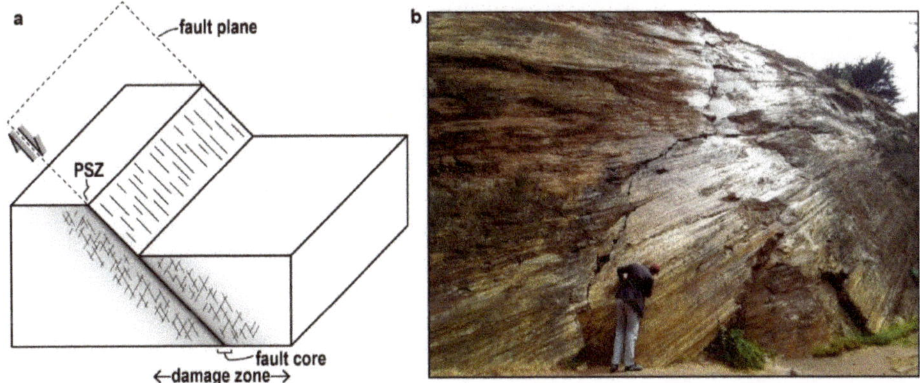

Figure 3. Principal slip zones (PSZs) in natural faults. (**a**) Sketch of a normal fault, highlighting the fault plane, damage zone, fault core, and PSZ (after [10,47]). (**b**). Striated, glossy surface of the Corona Heights Fault (USA) (courtesy of J. E. Samuelson).

Faults that are exposed parallel to the fault plane display the fault core, which is frequently characterized by slip-parallel striations and a relatively erosion-resistant, "glossy" or "well-polished" surface (Figure 3b) (Table 1). Such exposures have been reported as meter-scale outcrops in the field [87, 88], but also as cm-scale patches in drill core samples of active faults [85,86]. Siman-Tov et al. [87] coined the term "fault mirrors" for highly light-reflective fault surfaces cutting carbonate rocks in the Dead Sea transform region (Israel). They showed that the glossy fault plane is internally composed of a thin (<1 µm) veneer of calcite grains of a size down to ~50 nm. However, a glossy or mirror-like appearance has been described for fault surfaces composed of 0.1–1 µm-sized grains [85,89,90], and do not reveal much about the grain size within. As pointed out by Siman-Tov et al., the specular reflectivity occurs because the fault surface roughness has a wavelength shorter than that of visible light (400 nm) [91].

Another notable observation that has been frequently reported on using samples from naturally occurring nanocrystalline PSZs is the presence of (partly) amorphized material (Table 1). It may be observed as cm-thick veins in the field [92], or as thin coatings surrounding mineral clasts when observed using TEM [86,93,94]. Veins of glassy, amorphized rock known as pseudotachylytes, may form as a result of frictional melting along faults, pointing to high (co-seismic) slip rates [45,95,96]. For this reason, melt-origin pseudotachylytes are frequently used as field indicators of paleo-earthquake rupture [97,98]. However, the formation process is not implicit to the definition of pseudotachylytes, and they may form by other mechanisms than seismically-induced frictional melting [99]. Caution is necessary on the interpretation of field exposures of faults showing the presence of amorphous veins.

Table 1. List of reports, in chronological order, of ≤100 nm-sized grains in the cores (principal slip zones) of natural faults.

Location	Dominant Host Rock Mineralogy	d (nm)	CPO?	Glossy Surface?	Amorphous Material?	Source
Dixie Valley Thrust, USA	quartz	10	√	√		Power & Tullis [83]
Punchbowl Fault, USA	quartz, feldspar, clays	4	√			Chester & Goldsby [84]
Chelungpu Fault (TCDP borehole C)	quartz, clays	50				Ma et al. [8]
Nojima Fault Zone, Japan	quartz, feldspar	30				Keulen et al. [100]
Iida-Matsukawa Fault, Japan	quartz, feldspar	20			√	Ozawa & Takizawa [92]
San Andreas Fault (SAFOD main hole)	clays, quartz, feldspar	50		√	√	Janssen et al. [86]
Kfar Gladi Fault, Israel	calcite	50		√		Siman-Tov et al. [87]
Corona Heights Fault, USA	silica, quartz	10	√	√	√	Kirkpatrick et al. [88]
Gubbio Fault, Italy	calcite, clays	50		√		Bullock et al. [101]
Mt. Maggio fault, Italy	calcite	100				Collettini et al. [102]
Vado di Corno fault, Italy	calcite, dolomite	50		√		Demurtas et al. [103]
Capolivieri-Porto Azurro shear zone, Italy	tourmaline	10–100 [†]			√	Viti et al. [90]
Hsiaotungshi fault system (borehole), Taiwan	quartz, clays	50–100		√	√	Kuo et al. [85]
Maclure Glacier, USA	feldspar, quartz	10–100		√	√	Siman-Tov et al. [104]
Mt. Vettore Fault, Italy	calcite, clays	50		√		Smeraglia et al. [105]

"d" is the minimum grain size observed. [†] Fault mirrors (glossy surfaces) are also reported, however here they are composed mainly of >200 nm sized crystals. See the respective papers for details on localities, outcrops, and PSZ formation conditions.

4.2. Nanocrystalline Principal Slip Zones Formed in Fault-Slip Experiments

Laboratory experiments aiming to investigate upper-crustal fault deformation are carried out by imposing displacement along (initially) bare rock surfaces, or on a powdered sample layer representing a simulated fault gouge. The technology used to conduct fault-slip experiments varies greatly [106–108]. However, for simplicity, here we distinguish between two types of fault-slip tests; (i) low-velocity friction (LVF) tests, used to study slow fault-slip including the early (nucleation) stages of earthquake rupture, and (ii) high-velocity friction (HVF) tests, used to study dynamic earthquake rupture processes. Following Rowe & Griffith [98], slip rates (v) beyond ~10^{-4} m/s are "almost certainly dynamic", so we define HVF tests as using $v \geq 10^{-4}$ m/s, and LVF tests as using $v < 10^{-4}$ m/s. There are numerous other differences between LVF and HVF tests, in addition to the displacement rate used, that may affect micro- and nanostructural development along the simulated fault, or its recovery after an experiment. LVF tests typically achieve steady-state conditions of normal stress (σ_n) and temperature (T), but reaching cumulative displacements of maximally a few centimeters ($\sum x = 10^{-3}$–10^{-2} m). HVF tests may run for meters ($\sum x = 10^{1}$–10^{2} m), with frictional heat generated at the fault-slip interface playing an important role.

Despite the major differences between LVF and HVF tests, simulated fault samples recovered after an experiment typically show one or more, ultrafine-grained, shear plane-parallel bands of finite width, located in the sample interior ("Y-shears") or close to the loading piston interface ("boundary shears") [109,110] (Figure 4a–d). These shear bands mark a zone of abrupt grain size reduction with respect to the host rock (Figure 4b,d), and accommodated the bulk of the imposed shear displacement, i.e., representing experimentally-formed principal slip zones (PSZs). PSZ thicknesses may range from ~50–100 µm in samples recovered from LVF tests, to a few (tens of) microns in HVF deformed samples. Yund et al. [111] used TEM to investigate PSZs developed in simulated fault gouges of siliceous and carbonate compositions deformed in LVF and HVF rotary shear tests (Table 2). They reported grain sizes down to ~10–50 nm in all samples investigated, and the presence of amorphized materials, except in the carbonates. However, there are numerous (recent) reports of nanocrystalline PSZs formed in

LVF and HVF experiments using simulated fault samples composed of carbonates [58,93,112–119], many of which also showed the presence of amorphous materials (Table 2).

A crystallographic preferred orientation (CPO) was reported for nanocrystalline PSZs developed in LVF tests using simulated gouges composed of calcite [113,114,118] and quartz [120] (Table 2). The presence of a CPO may be inferred from uniform birefringence and extinction observed in thin sections (Supplementary Video S1), or else demonstrated using selected area diffraction in TEM. Electron backscatter diffraction is a powerful tool frequently used to quantify a CPO [121], however, to our knowledge this remains difficult to apply to extremely fine-grained aggregates such as those characterizing nanocrystalline PSZs (grain size << 100 nm). Recently, EBSD measurements were used to quantify a CPO characterizing a PSZ composed of 200–300 nm-sized, polygonal grains, formed in simulated calcite gouge sheared in HVF tests [122,123].

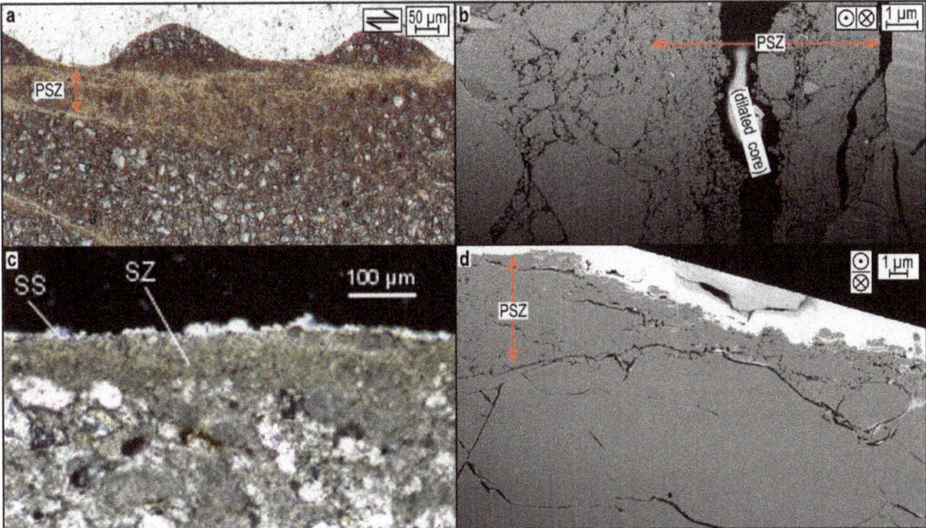

Figure 4. Principal slip zones (PSZs) in simulated calcite(-rich) fault gouge formed in LVF (**a**,**b**) and HVF experiments (**c**,**d**). (**a**) Plane polarized light micrograph of an ultra-thin section (parallel to the slip vector, sample CaCO₃-RT-dry of [113]). (**b**) Backscatter electron (BSE) micrograph prepared using a focused ion beam scanning electron microscope (FIB-SEM) (normal to the slip vector, sample lmst@150 °C of [124]). The central void is due to post-test dilation. (**c**) Cross-polarized light micrograph showing a narrow, fine-grained slip zone (SZ) bound by a slip surface (SS). Taken with publisher's permission from [93]. (**d**) BSE micrograph. Taken from [115].

Table 2. List of reports, in chronological order, of <100 nm-sized grains in experimentally-formed principal slip zones.

Dominant Sample Mineralogy	v_{max} (m/s)	σ_n (MPa)	$\sum x$ (m)	d (nm)	CPO?	Glossy Surface(s)?	Amorphous Material?	Source
quartz	10^{-5}	~135	10^{-5}	90			√[†]	Engelder [125]
quartz	$10^{-2.5}$	7–50	10^{-1}	10–15			√	Yund et al. [111]
quartz, feldspar	$10^{-2.5}$–10^{-6}	50–75	10^{-2}–10^{-1}	10–15			√	Yund et al. [111]
calcite	$10^{-2.5}$	15	10^{-2}	~50				Yund et al. [111]
dolomite	$10^{-2.5}$	75	10^{-1}	~50				Yund et al. [111]
calcite	10^{-2}–10^{0}	1.1–13.4	10^{0}–10^{1}	10		√		Han et al. [58]
siderite, magnetite	10^{0}	0.6–1.3	10^{1}	20–30		√		Han et al. [112]
antigorite	10^{0}	24.5	10^{0}	~50			√	Viti & Hirose [126]
quartz, feldspar	10^{-7}–10^{-8}	>10^{3}	10^{-3}	8		√	√	Pec et al. [127,128]
calcite	10^{-7}–10^{-5}	50	10^{-3}	5	√	√	√[††]	Verberne et al. [113,114]
dolomite	10^{0}	28.4	10^{-1}	10		√		Green II et al. [116]
quartz, feldspar	10^{-6}–10^{-5}	25	10^{-2}–10^{-1}	15–50			√	Hadizadeh et al. [129]
quartz, clays	10^{0}	1	10^{1}	10–50			√	Kuo et al. [130]
quartz, silica	10^{-6}	>10^{3}[†]	10^{-3}	5	√		√	Toy et al. [120]
calcite	10^{-1}–10^{0}	10	10^{-3}–10^{1}	5–10		√	√	Spagnuolo et al. [115]
calcite	10^{-1}	0.47–1.57	10^{1}	45		√		Siman–Tov et al. [117]
quartz	10^{-7}–10^{-3}[‡]	92–287	10^{-3}	10		√	√	Hayward et al. [131]
quartz, smectite	10^{-4}–10^{0}	5	10^{0}	10–50			√	Aretusini et al. [132]
quartz, muscovite	10^{-8}–10^{-5}	120	10^{-4}	10	√			Niemeijer [133]
quartz, muscovite	10^{-4}	120	10^{-4}	10	√			Niemeijer [133]
calcite	10^{-7}–10^{-5}	20–100	10^{-3}–10^{-4}	10				Mercuri et al. [119]
calcite	10^{-6}–10^{-5}	45	10^{-1}–10^{-2}	50	√		√	Delle Piane et al. [118]
quartz	10^{-4}–10^{-1}	2.5–5	10^{0}–10^{1}	10		√	√	Rowe et al. [134]

Grey shaded rows include data from LVF tests (here defined as tests employing $v < 10^{-4}$ m/s). v_{max} = max. displacement rate; σ_n = (effective) normal stress; $\sum x$ = accumulated displacement; d = min. grain size. Only the orders of magnitude of v_{max} and $\sum x$ are given. Glossy surfaces refer to the presence of "shiny" or "mirror-like" surfaces, regardless of continuity. For details see the respective literature. [†] Attributed to beam damage. [††] Attributed to unknown contamination. [‡] v_{max} here was estimated from unstable slip events.

Mirror-Slip Surfaces in Principal Slip Zones Developed in Calcite Gouge

"Glossy", "shiny", or "mirror-like" slip surfaces (MSSs) have been observed in experiments using a wide range of sample materials, characterized by a wide range of normal stresses (σ_n), displacement rates (v), and cumulative displacements ($\sum x$) (Table 2) (Figures 5 and 6). Recently, much attention has been given to MSSs developed in simulated faults composed of calcite, mainly because of their striking similarity with carbonate "fault mirrors" frequently observed in tectonically-active carbonate terrains [87,135], and the question whether they may be indicators of past seismic slip.

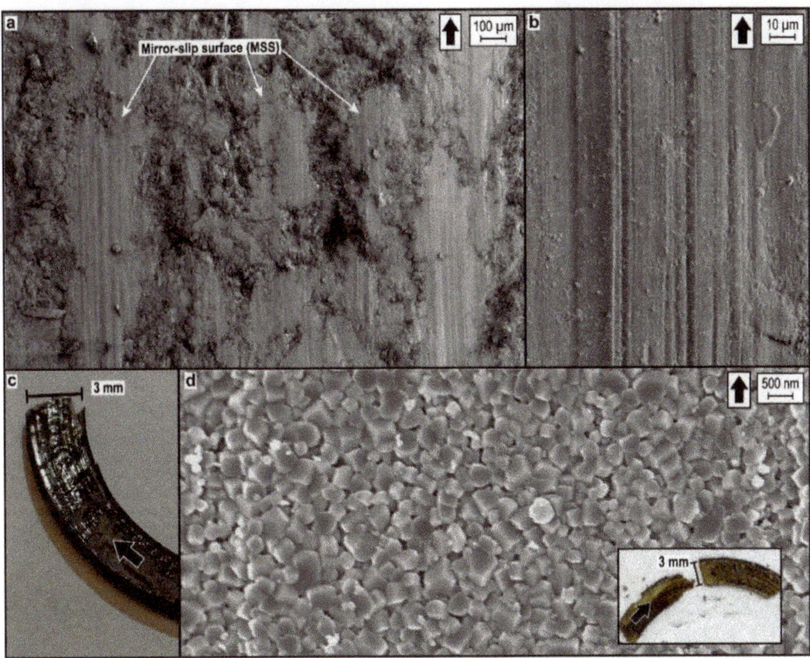

Figure 5. Principal slip zones with mirror-slip surfaces formed in simulated calcite gouge sheared in HVF tests ($v \geq 10^{-4}$ ms^{-1}). (**a,b**) Secondary electron (SE) micrographs. From [115]. (**c**) Sample fragment recovered from an experiment conducted using $0.1 \leq v \leq 100$ μms^{-1}, $\sigma_n^{\mathit{eff}} = 50$ MPa, $T = 550$ °C, $\sum x = 24.7$ mm (unpublished data). (**d**) Top view onto the PSZ developed in an experiment conducted at $v = 100$ μms^{-1} (sample CaCO$_3$-550-vhigh of [122]). Inset shows a photo of the (fragmented) sample recovered after the experiment.

In general, MSSs are characterized by extremely low surface roughness, especially in a direction parallel to the shear direction [136,137]. They have been reported as multiple, elongated patches aligned parallel to the shear direction (Figures 5a and 6a), or else as a single, continuous interface marking the PSZ boundary (Figure 5c). The number and extent of MSSs were shown to increase with increasing displacement and/or displacement rates, in HVF tests conducted at normal stresses up to 26 MPa [117,135,138], which led some authors to conclude that continuous MSSs may indeed serve as indicators of past seismic slip in natural faults cutting carbonates. However, a continuous MSS has also been observed in simulated calcite gouge sheared at $v = 10$ μms^{-1} (effective normal stress 20 MPa $\leq \sigma_n^{\mathit{eff}} \leq 100$ MPa, $T \approx 550$ °C, $\sum x = 12.4$ mm; Figure S2D of [122]). The role of effective normal stress, and of cumulative displacement, on the formation of (patchy) MSSs with progressive shear strain in LVF tests remains to be investigated. Pozzi et al. [139] recently reported on the microstructural development of MSS-bearing PSZs with progressive shear strain in HVF experiments on simulated calcite faults ($\sigma_n = 25$ MPa, v up to 1.4 m/s). Using polished sections prepared normal to the slip vector they observed that, after an initial stage of slip ($\Sigma x \approx 7$ cm), sharp discontinuities develop which are interpreted to represent MSSs. The matured PSZ is observed to consist of 200–300 nm sized, polygonal grains characterized by a crystallographic preferred orientation [123].

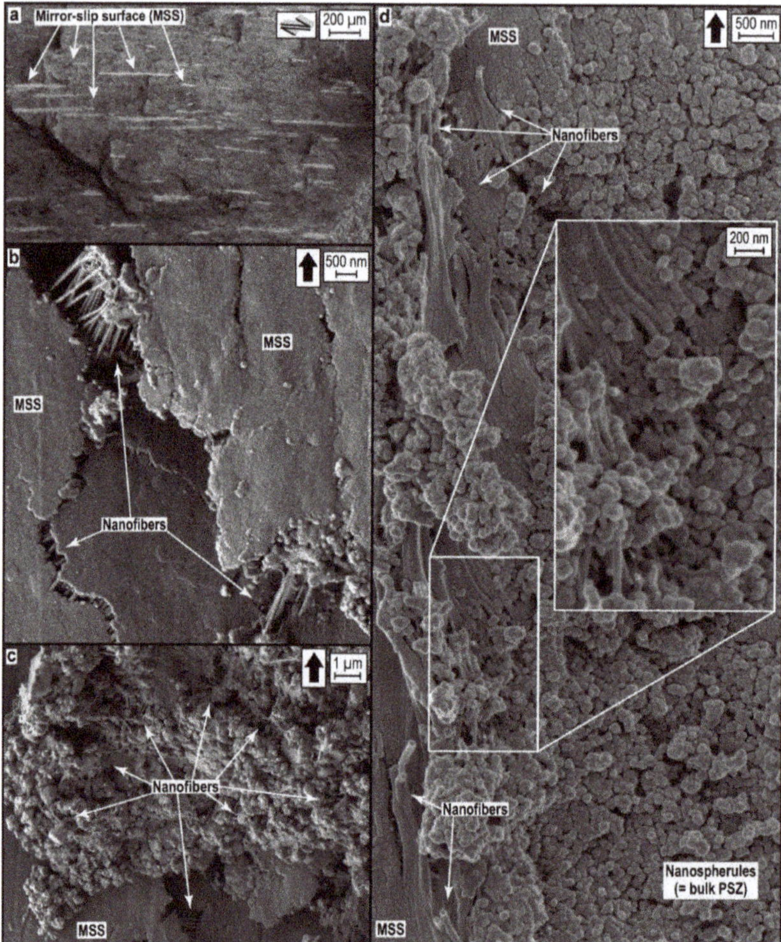

Figure 6. Principal slip zone with mirror-slip surfaces formed in simulated calcite gouge sheared in LVF tests ($v = 10^{-6}$ to 10^{-5} ms^{-1}, see [140]). Secondary electron micrographs. (**a**) Taken at an angle of 52° to the shear plane. The patches that are elongated parallel to the shear direction represent MSSs. (**b**) Stretched nanofibers and (**c**) nanofibers within the bulk PSZ. (**d**) Alignment of nanospherules at the edge of an MSS. (a) to (d) are top views onto the shear plane. The micrographs shown in (**a,d**) are from sample SEMB of [140], taken using a FEI Nova Nanolab FIB-SEM. The micrographs in (**b,c**) are of samples sheared using a gas-medium deformation apparatus installed at the Geological Survey of Japan (Tsukuba, Japan), and taken using a JEOL-7400F FEG-SEM.

Returning to patchy MSSs formed in LVF experiments using calcite gouge ($v = 10^{-6}$ m/s, $\sigma_n^{\it eff} = 50$ MPa, $\Sigma x \approx 5$–6 mm), individually these show remarkable micro- and nanostructural characteristics (Figure 6a–d) [140]. The PSZ itself comprises a porous, sheet-like volume of ~100 nm-sized spherical particles, with internal, 0.1 to 1 µm-thick, dense planar coatings comprising the MSSs. The MSS patches are observed at different topographic levels within the PSZ (Figure 6a) and are internally composed of ~100 nm wide fibers that show marked extension and plastic bending when stretched (Figure 6b–d). At locations where stretching led to nanofiber failure, necking structures are absent, suggestive of a low stress-sensitivity of the ductile strain rate (a low "n-value"), or superplastic behavior [140]. Nanofiber stretching in this way could only have occurred upon opening of the microcracks at room conditions after the experiment. The nanofibers are locally observed away from MSSs, i.e., within the porous volume constituting the broader PSZ (Figure 6c). Selected area diffraction patterns, taken using TEM, of a single nanofiber as well as of the spherical particles comprising the bulk PSZ, revealed a polycrystalline substructure composed of crystallites 5 to 20 nm in size, characterized by a CPO [113,114,140]. The uniform width of the nanofibers and the spherical nanograin aggregates or nanospherules, (both with diameter ~100 nm), combined with the alignment observed of nanospherules at the edge of some MSSs (Figure 6d), suggest that the nanofibers represent linear nanospherule chains.

5. Discussion

The compilation of literature observations reported above (Tables 1 and 2) demonstrates that nanocrystalline principal slip zones (NC PSZs) form in a wide range of rock types under a wide range of normal stresses and displacement rates pertaining to co-seismic and sub-seismic fault-slip in Earth's upper-crust. This suggests that NC PSZs play an important role in controlling fault sliding behavior, including earthquake rupture nucleation and dynamic propagation. Below we discuss possible formation mechanisms of the PSZ nanostructures observed, as well as a comparison between mirror-slip-surface-bearing PSZs developed in low-velocity friction (LVF; $v < 10^{-4}$ ms^{-1}) and in high-velocity friction (HVF; $v \geq 10^{-4}$ ms^{-1}) tests. We go on to discuss the role of NC PSZs in controlling upper-crustal fault strength and stability, and we consider their broader significance in the seismogenic zone.

5.1. Formation of PSZ Nanostructures, Amorphous Materials, and CPO

Under brittle conditions in a fault zone, grain size reduction occurs by cataclastic deformation involving intragranular fracture, comminution and intergranular friction [100,141–143]. However, below a certain critical grain size d_{crit} known as the grind limit, the stress required to initiate a fracture in compression becomes too high so that plastic yielding occurs [144]. For quartz, $d_{crit} \approx 0.9$ μm [100], whereas for calcite $d_{crit} \approx 0.85$ μm [145]. This means that the <100 nm-sized grains frequently observed in fault-slip experiments must point to a mechanism of grain size reduction involving plastic yielding. Using a model based on mode I Griffith failure [144,146] and low-temperature plasticity, Sammis and Ben-Zion [147] showed that in the case of quartz in a compressive regime, shock loading and subcritical crack growth may produce particles down to 3 nm in size. However, specifically for LVF experiments which employ displacement rates of μm/s and reach just millimeters of cumulative displacement over the timespan of a few hours, the formation of crystallites down to 5 nm in size combined with the presence of amorphous materials and a crystallographic preferred orientation (CPO) (Table 2), remains intriguing.

Focusing on simulated calcite faults (Figures 4–6), the internal polycrystalline substructure observed in PSZs nanograins formed in LVF [113,114,140] as well as HVF experiments [93] bears a striking similarity to microstructures found in shocked ductile metals [148,149]. As in metals, the high ductility of calcite [150,151] may therefore allow the observed ~5–20 nm substructure to form by progressive development of nano-cell walls from dense dislocation networks and tangles generated by low temperature crystal-plasticity (e.g., r(104) slip or e(108) twinning [152]). Following from this, we speculate that plastic deformation and/or fracturing and abrasion occurring at parent grain surfaces led to the detachment of ~100 nm sized nanocrystalline clusters or fragments from these micron-sized parent grains [148,149]. The nanograins produced in turn rounded, to form the rolling, grain-neighbor-swapping nanospherules comprising the porous nanogranular PSZ [140]. To further unravel the formation mechanism of nanocrystallites and nanospherules in calcite gouge is a challenging task which requires more elaborate experiments and micro-/nanostructural analyses.

Amorphous materials may form by melt quenching, mechanical deformation, chemical reactions or a coupling between the latter two. Here, we focus on those materials derived from sub-solidus derived processes. For a thought-provoking investigation into chemo-mechanical- vs. melt-derived amorphous solids the reader is referred to Pec et al. [127]. In general, solid-state amorphization is attributed to arise from (1) the introduction of externally-derived mechanical instabilities (e.g., dislocations), (2) externally-forced volume expansion at constant temperature or (3) thermal expansion during heating at constant pressure [65,153]. By contrast, mechano-chemical interactions may produce amorphous materials as a result of the reduction of chemical species. This is particularly relevant in carbonate fault rocks where decarbonation reactions release CO_2 that can subsequently be reduced to (amorphous) carbon phases. Several natural [154–156] and experimental [93,115,118,140] studies of carbonate fault rocks have reported the occurs of amorphous carbon, often intimately associated with nanogranular calcites. The exact chemical pathways for CO_2 reduction to amorphous carbon remain, however, debated [115,156,157]. Additionally, in natural systems fluids can facilitate the precipitation of amorphous solids [158], making it difficult to discriminate internally- from externally-controlled formation mechanisms. Sub-solidus-derived amorphous materials are also widely reported in silicate-dominated systems (Table 2). The detailed mechanism(s) of amorphization, and more generally the impact of differences in atomic order on mechanical properties [159], on fault rheology, remains subject of further study.

Aside from the formation of nanocrystallites and amorphization within a PSZ, this does not explain the development of an internal CPO. Pozzi et al. [123] suggested that grain-size insensitive (GSI) creep mechanisms (dislocation creep) may explain CPO formation in a PSZ composed of 200–300 nm-sized grains formed in calcite gouge sheared at $v = 1.4$ m/s, at $\sigma_n = 25$ MPa. However, in the case of CPO-bearing PSZs formed in LVF tests, the porous structure observed in the bulk PSZ (Figure 6) is suggestive of nanogranular flow, which is not known for generating, or retaining a pre-existing, CPO. One potential mechanism for CPO formation in (nano-)granular flow may be through oriented interface attachment (OA), which is widely reported as a mechanism by which nanocrystallites can rapidly coalesce to form single crystals in numerous nanomaterials [160–162], including in calcite [163]. The thermodynamic driving force for particle coalescence in an OA event originates from crystallographic orientation-dependent, interatomic Coulombic interactions arising from both the surface atoms, and of atoms within the interior of the approaching nanoparticles [164,165]. Particle coalescence leads to a reduction of total surface energy [120], which, in the case of calcite would lead to an alignment of the lowest energy (104) plane [166], consistent with observations of simulated calcite faults deformed in LVF tests [113,114,140].

5.2. MSS-Bearing PSZs as Indicators for Past Seismic Slip?

Microstructural studies of simulated dolomite and limestone faults sheared in HVF experiments suggest the following characteristics of "glossy", "shiny" or "mirror-like" slip surfaces:

(i)   They form only at high mechanical work input rates or power densities ($\dot{W} = \mu \cdot \sigma_n^{eff} \cdot v$, where μ is the coefficient of fault friction) [135,138].

(ii)   They are at least in part responsible for the strong dynamic weakening often seen in samples sheared at co-seismic slip rates [137].

(iii)   They are associated with dynamic recrystallization caused by heating at co-seismic slip rates [167].

However, the mirror-like surface patches formed in LVF experiments show very similar striated form and nanoscale topography [114,140] to those formed in HVF experiments, suggesting at least some degree of shared origin regardless of the areal extent or of shearing velocity. Moreover, a continuous MSS marking the PSZ has also been observed in simulated calcite gouge sheared at $0.1 \leq v \leq 100$ μms^{-1} (Figure 5c), which showed steady-state μ-values of ~0.5–0.6 (see [122,168] for data from experiments conducted under similar T-σ_n^{eff}-v conditions). These observations strongly suggest that MSSs are not related to any dynamic weakening mechanisms, and that their seismic origin remains debatable at best. Rather, as pointed out by Pozzi et al. [139], MSSs seem to demarcate a rheological discontinuity between an ultrafine-grained zone, which internally deforms via thermally-activated and grain-size dependent deformation mechanisms, i.e., the PSZ, and the adjacent, coarser-grained wall rock, which deforms by brittle processes (grain fracturing). The fact that patchy MSSs formed in LVF tests are found at different topographic levels throughout the PSZ (Figure 6a) indicate that they probably formed as isolated patches rather than as a single, through-going film. Furthermore, at some locations within the broader nanogranular PSZ volume, nanofibers are observed outside MSSs, or bridging between the MSS and the porous PSZ (Figure 6b,c). Combining these observations, we hypothesize that, with further increasing displacement, nanofibers within the bulk PSZ will ultimately align to form a single through-going MSS.

From rotary shear experiments on cylindrical cores of dolomite and limestone performed at $v = 0.002$–0.96 m/s and $\sigma_n^{eff} = 0.25$–6.9 MPa, Boneh et al. [138] showed that shiny striated slip-surface patches started to develop only at $\dot{W}$-values in excess of 30 kW/m^2. The cumulative area covered by these patches increased with increasing $\dot{W}$, ultimately producing a continuous, highly-reflective principal slip surface. Using experiments on simulated gouge prepared from dolostone, performed at $v = 0.001$–1.13 m/s and $\sigma_n^{eff} = 13$–26 MPa, Fondriest et al. [135] showed that shiny surfaces only developed at $\dot{W}$ values > 40 kW/m^2, covering an area of the sample that progressively increases with increasing displacement. By contrast, the shiny patches developed in simulated calcite gouge reported in LVF tests (Figure 6) [114,140] formed at $\dot{W} = \mu \cdot \sigma_n^{eff} \cdot v = 50$ MPa $\times (0.7 \pm 0.1) \times 10^{-6}$ m/s $= 35 \pm 5$ W/m^2, i.e., 2 to 5 orders of magnitude lower than considered necessary for them to form in HVF experiments. This demonstrates that such shiny striates surfaces do not exclusively form at the high-power densities (>30–40 kW/m^2) associated with HVF experiments and with co-seismic slip rates. The implication is that mirror-like PSZs cannot be used as field indicators of past co-seismic slip in carbonates rocks without additional geological or microstructural evidence. The role of normal stress and cumulative displacement achieved in controlling the continuity of MSSs should be investigated further.

The development of highly-reflective PSZs in HVF experiments performed by Smith et al. [167], using simulated calcite gouge ($v > 0.1$ m/s, $\sigma_n^{eff} = 2$–26 MPa, $\sum x > 1$ m) was shown to be associated with the presence of dynamically recrystallized grains characterized by a CPO, adjacent to the slipping zone, while the PSZ itself was composed of statically recrystallized grains. Dynamic recrystallization here refers to the growth of internal strain- or defect-free grains during shear, whereas static recrystallization refers to such growth upon piston arrest and cooling after the experiment. In the experiments by Smith et al. [167], recrystallization and CPO formation were attributed to the attainment of high temperatures (650–900 °C) reflecting heat dissipated from the PSZ during localized frictional slip at co-seismic rates [167,169]. Static recrystallization of PSZ grains was inferred to have played a role in experiments performed by Verberne et al. [122], on simulated calcite gouge sheared at $v = 100$ μm/s ($T = 550$ °C, $\sigma_n^{eff} = 50$ MPa, $\sum x = 10.4$ mm) (Figure 5d). Grain growth upon cooling after the experiment suggests that the grain size within the PSZ may have been smaller during shear. MSS-bearing PSZs developed in LVF tests ($v < 10^{-4}$ ms^{-1}) using calcite gouge showed no evidence for conventional dynamic or static recrystallization, either in or adjacent to the PSZ, nor is this likely to have occurred considering the low slip rates, temperatures, and $\dot{W}$-values applying to these tests. The implication is that the presence of a statically recrystallized PSZ, with adjacent dynamically recrystallized grains, may indeed offer a useful constraint to past high-velocity slip, at least in limestones [167,170].

5.3. The Role of Nanocrystalline PSZs in Controlling Fault Stability

As mentioned in Section 2 above, in the case of earthquakes, fault sliding is believed to occur by stick-slip motion [38], caused by potentially unstable "velocity(v)-weakening" properties of the fault sliding medium [39–41,43]. The physical processes responsible for v-weakening behavior of gouge-filled faults are only recently beginning to be elucidated. Based on observations from fault analogue experiments using powdered halite-muscovite mixtures, Niemeijer & Spiers [33,171] developed a micromechanical model for shear deformation of granular fault rock. They showed that competition between dilatant granular flow and compaction by water-assisted diffusive mass transfer leads to an increase in steady-state porosity with increasing shear rate, and v-weakening behavior [171]. However, any time-sensitive, Arrhenius-type deformation mechanism will, when in competition with time-insensitive dilatation or granular flow, produce v-weakening. This mechanism of competitive dilatation and compaction may well explain thermally-activated transitions in the v-dependence of friction seen in a wide range of fault rock types [168,172–174]. The extended "Chen-Niemeijer-Spiers" (CNS) model developed recently [175] is capable of quantitatively reproducing a wide range of laboratory fault gouge friction data using physically-based input parameters [176,177], therewith providing a powerful tool for numerical earthquake-cycle simulators [178].

To produce v-weakening in the CNS model, the rate of intergranular compaction ($\dot{\varepsilon}_{cp}$) and dilatation by granular flow ($\dot{\varepsilon}_{gr}$) within the deforming gouge zone must be within the same order of magnitude, i.e., $\left|\dot{\varepsilon}_{gr}\right| \approx \left|\dot{\varepsilon}_{cp}\right|$. Under conditions where either process dominates stable v-strengthening occurs. In the case of intergranular creep by water-assisted diffusive mass transfer, relevant for compaction of microgranular calcite up to 150 °C [179], the compactive strain rate $\dot{\varepsilon}_{cp}$ is described using [72]

$$\dot{\varepsilon}_{cp} = A \cdot \frac{\sigma \Omega}{RTd^3} \cdot DCS \cdot f(\varphi) \tag{1}$$

where A is a constant, σ is the (effective) axial stress, Ω is the molecular volume of the solid, D is the diffusion coefficient, C is the solubility of the solute, d is grain size, and $f(\varphi)$ is a porosity function (Table 3). In view of the inverse cubic dependence on d Equation (1), in the case of <100 nm sized particles in nanocrystalline PSZs, compaction by water-assisted diffusive mass transfer is expected to be very fast, even at relatively low temperatures.

Table 3. Values/expressions used for the terms appearing in Equation (1).

Term	Formula/Value	Source
A	$576/3\pi \approx 61$	Pluymakers & Spiers [180]
σ	50×10^6 Pa	Verberne et al. [140]
d	1×10^{-7} m	Figure 6
T	291 to 423 K	Verberne et al. [140]
Ω	3.69×10^{-5} m^3 mol^{-1}	Zhang et al. [179]
$f(\phi)$	$f(\phi) \approx 2\phi/(1-2\phi)^2 \approx 1.1$	Pluymakers & Spiers [180]
	Water-assisted diffusive mass transfer:	
D	$D = D_0 \exp\left(-\frac{Q}{RT}\right)$ $D = 1 \times 10^{-10}$ m^2·s^{-1} at $T = 298$ K $Q = 1.5 \times 10^4$ J·mol^{-1}	Nakashima [181]
C	$C = \sqrt{K_s}$ $\log K_s =$ $-171.9605 - 0.077993T + \frac{2839.319}{T} + \log T$	Plummer & Busenberg [182]
S	1 to 2×10^{-9} m	Verberne et al. [140]
	Solid-state grain boundary diffusion:	
DS	$DS = (DS)_0 \exp\left(-\frac{Q_d}{RT}\right)$ $(DS)_0 = 1.5 \times 10^{-9}$ m^3·s^{-1} $Q_d = 2.67 \times 10^5$ J·mol^{-1}	Farver & Yund [183]

In the case of solid-state grain boundary diffusion $C = 1$.

In Figure 7 we plot Equation (1) for the case of granular calcite, assuming grain sizes in the range from 5 nm to 1 μm. We also indicate the conditions of temperature (25–150 °C) and intergranular dilatation rate ($\left|\dot{\varepsilon}_{gr}\right| \approx 10^{-2}10^0$ s^{-1}) characterizing the PSZ developed in LVF experiments on simulated calcite faults (Figure 4a,b and Figure 6) [140]. This shows that for 100 to 500 nm-sized grains $\left|\dot{\varepsilon}_{gr}\right| \approx \left|\dot{\varepsilon}_{cp}\right|$ (Figure 7a), implying that under the conditions of normal stress and temperature used here (Table 3) v-weakening may be observed. Combined with the ~100 nm-sized nanospherules and -fibers observed in the broader

PSZ (Figure 6), this suggest that the mechanism of competitive dilation and compaction [33,171] can explain the thermally-activated transition from stable v-strengthening to unstable v-weakening seen at ~100 °C in experiments on simulated calcite(-rich) fault gouge [114,124,168,184,185]. For grains <100 nm size, this mechanism is not expected to be relevant when assuming intergranular creep by water-assisted diffusive mass transfer. However, equation 1 can also be used to assess creep rates assuming solid-state diffusion involving mass transfer through a grain boundary of thickness S [70]. This shows that for grain sizes down to 10 nm, $\dot{\varepsilon}_{cp} \approx 10^{-3}$ s^{-1} at temperatures > 650 °C (Figure 7b). Such high temperatures are common in HVF experiments, suggesting that the mechanism of competitive compaction and dilation may also play a role at co-seismic slip rates. This is consistent with claims that superplastic deformation of nanogranular fault rock controls (dynamic) fault rupture [93,116,186].

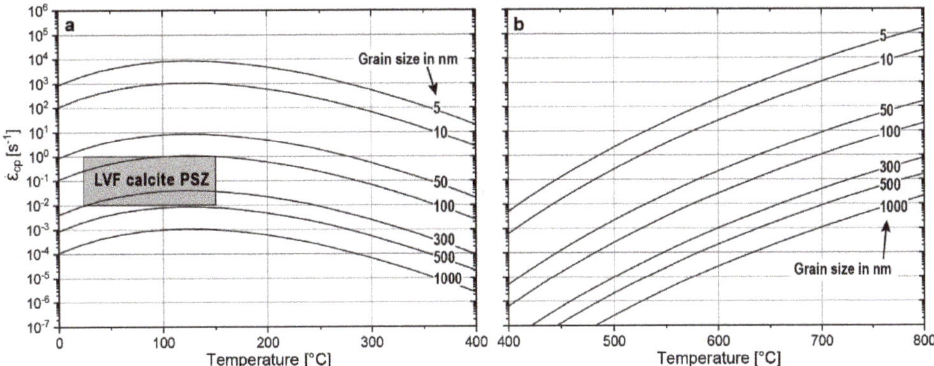

Figure 7. Intergranular compaction creep strain rates in granular calcite versus temperature for grain size *d*. (**a**). Water-assisted diffusive mass transfer. The grey shaded area indicates the conditions characterizing the PSZ developed in LVF experiments using simulated calcite gouge (Figure 4a,b and Figure 6) [140] (**b**). Solid-state grain boundary diffusion. For a list of parameters used see Table 3.

Notwithstanding all of the above, the micromechanical framework underlying the CNS model [171,175] is of course highly idealized. The model assumptions are reasonable at low slip rates, but break down when frictional heating and associated dynamic fault rupture processes come into play [57,61,62]. Another potentially problematical aspect is the knowledge and quantification of the relevant intergranular creep mechanisms. In view of the unusual deformation properties of nanocrystalline materials (Figure 2), extrapolation of data from compaction experiments using microcrystalline samples to the nanometer realm may be unreasonable. Parameter values and expressions such as listed in Table 3 have to be re-assessed in the case of nanocrystalline fault rock.

5.4. Implications for Natural Faulting in the Seismogenic Zone

In the above we have shown that the Chen-Niemeijer-Spiers (CNS) model describing competitive dilatant nanogranular flow and nanospherule/-fiber compaction may explain the transition from stable velocity strengthening to potentially unstable velocity weakening at temperatures ~80–100 °C seen in calcite fault rock [114,124,168,184,185]. This transition is consistent with the location of the upper seismogenic limit at shallow depths (~2–4 km) in tectonically-active limestone terrains, such as those characterizing the Mediterranean region [187–189]. Velocity weakening hence seismogenic fault slip on nanogranular PSZs becomes possible because (water-assisted) diffusive mass transfer is dramatically accelerated by the nanogranular nature of the slip zone rock that forms. Given the abundant observations of nanogranular fault surfaces in fault rocks of all types (Tables 1 and 2), and the fast diffusion rates in nanostructured materials [11,69,75], the mechanism of dilatation versus compaction, applied to sheared nanogranular fault rock, may be generally relevant in controlling the upper limit of the seismogenic zone.

In the CNS model framework, a comparison between fault rock creep rates and fault zone shear strain rates may yield clues on the depth to the limits seismogenic zone, i.e., taking the condition that $|\dot\varepsilon_{gr}| \approx |\dot\varepsilon_{cp}|$ may lead to unstable seismogenic fault-slip [33,140,171]. To illustrate this, we used Equation (1) for the case of calcite (Table 3) to plot grain size vs. depth curves for different $\dot\varepsilon_{cp}$, assuming a geothermal gradient of 30 °C/km and a density of 2700 kg/m^{-3} (Figure 8). Crustal fault zone shear strain rates remain poorly constrained, mainly because of the lack of observations on the width of the actively deforming zone [190]. Nonetheless, using the "commonly cited value" for upper-crustal fault zone shear strain rates of ~10^{-14} s^{-1} [190,191] for $\dot\varepsilon_{gr}$ in Equation (1) (Figure 8), shows that solid state creep may be relevant in grains <100 nm in size at ~3–4 km depth. This is close to the upper limit of the seismogenic zone in tectonically-active carbonate terrains [187–189]. However, importantly, in the above analysis using the compaction-dilation model, the processes controlling nanograin formation and the role of grain growth are ignored. Especially the latter potentially presents a major limitation, since at greater depths/temperatures static as well as dynamic recrystallization of PSZ grains are expected to play a role [147,192]. Future models aiming to describe the physical processes leading to dynamic fault rupture should take into account the progressive development of fault rocks with increasing shear strain, i.e., specifically, the competition between grain growth and grain size reduction the PSZ.

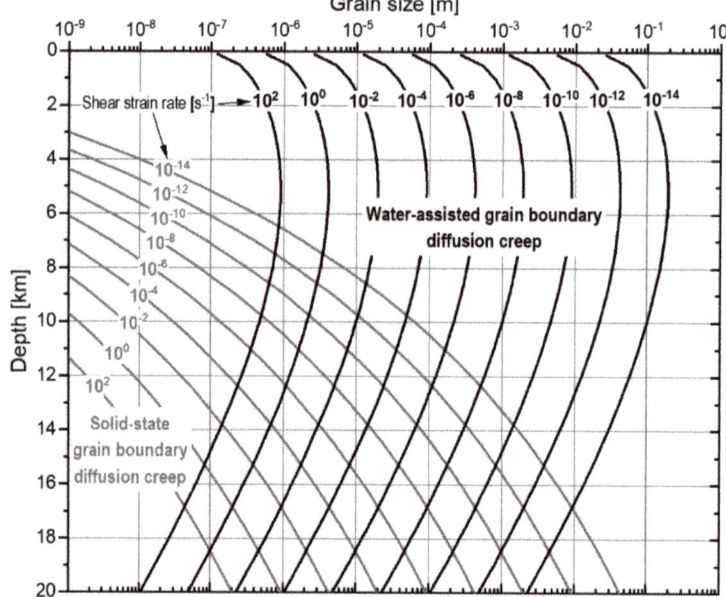

Figure 8. Water-assisted and solid-state grain boundary diffusion creep in calcite (Equation (1)), plotted as grain size vs. depth (taking 30 °C/km and a density of 2700 kg m^{-3}). For a list of parameters used see Table 3.

6. Conclusions

Nanocrystalline fault rock is consistently observed in natural and experimentally-formed principal slip zones (PSZs) and is frequently associated with the presence of a crystallographic preferred orientation (CPO), (partly) amorphized materials, and ultra-smooth interfaces known as "glossy", "shiny" or "mirror-like" slip surfaces (MSSs). Experiments conducted under a wide range of normal stresses, temperatures, and displacement rates demonstrate that these features can be produced over a wide range of conditions pertaining to upper-crustal fault-slip, covering co-seismic and sub-seismic displacement rates. Simple calculations using constitutive equations for compaction by water-assisted diffusive mass transfer, combined with existing models for velocity-weakening shear of gouge-filled faults, show that nanogranular fault rock plays a key role in controlling the depth to the upper-limit of the seismogenic zone. In view of the unusual deformation properties of nanocrystalline (NC) materials, an important task in Earth sciences is to improve insights on the rheology of NC PSZs, or of nanophased geomaterials in general.

Supplementary Materials: The following are available online at http://www.mdpi.com/2075-163X/9/6/328/s1, Video S1: unif_biref-[CaCO3-RT-dry].

Author Contributions: Conceptualization, B.A.V. and O.P.; methodology, B.A.V.; software, N/A; validation, B.A.V., O.P. and C.J.S.; formal analysis, B.A.V.; investigation, B.A.V.; resources, C.J.S.; data curation, B.A.V.; writing—original draft preparation, B.A.V.; writing—review & editing B.A.V., O.P., C.J.S.; visualization, B.A.V., O.P.; supervision, C.J.S.; project administration, B.A.V.; funding acquisition, B.A.V., C.J.S.

Funding: Part of this work was conducted within the framework of B.A.V.'s Ph.D. thesis, which was supported by grant 2011-75 awarded by the Netherlands Centre for Integrated Solid Earth Science (ISES). B.A.V. is now supported by JSPS KAKENHI grant #19K14823.

Acknowledgments: The authors thank André Niemeijer, Hans de Bresser, Jianye Chen, Colin Peach, Virginia Toy, and Martyn Drury for discussions at various stages of this work. Matthijs de Winter is thanked for preparing the FIB-section in sample lmst@150 °C (Figure 4b). Michael Hochella, Li-Wei Kuo, and Hiroko Kitajima are thanked for helping with literature. We thank two anonymous reviewers for helpful input that improved the paper significantly.

Conflicts of Interest: The authors declare no conflicts of interest.

References

1. Banfield, J.F.; Zhang, H.-Z. Nanoparticles in the environment. *Rev. Mineral. Geochem.* **2001**, *44*, 1–58. [CrossRef]
2. Hochella, M.F., Jr. Nanoscience and technology the next revolution in the Earth sciences. *Earth Planet. Sci. Lett.* **2002**, *203*, 593–605. [CrossRef]
3. Hochella, M.F., Jr.; Lower, S.K.; Maurice, P.A.; Penn, R.L.; Sahai, N.; Sparks, D.L.; Twining, B.S. Nanominerals, mineral nanoparticles, and Earth systems. *Science* **2008**, *319*, 1631–1635. [CrossRef] [PubMed]
4. Hochella, M.F., Jr.; Mogk, D.W.; Ranville, J.; Allen, I.C.; Luther, G.W.; Marr, L.C.; McGrail, B.P.; Muruyama, M.; Qafoku, N.P.; Rosso, K.M.; et al. Natural, incidental, and engineered nanomaterials and their impacts on the Earth system. *Science* **2019**, *363*, 10. [CrossRef] [PubMed]
5. Ju, Y.; Huang, C.; Sun, Y.; Wan, Q.; Lu, X.; Lu, S.; He, H.; Wang, X.; Zou, C.; Wu, J.; et al. Nanogeosciences: Research History, Current Status, and Development Trends. *J. Nanosci. Nanotechnol.* **2017**, *17*, 5930–5965. [CrossRef]
6. Wilson, B.; Dewers, T.; Reches, Z.; Brune, J. Particle size and energetics of gouge from earthquake rupture zones. *Nature* **2005**, *434*, 749–752. [CrossRef] [PubMed]
7. Chester, J.S.; Chester, F.M.; Kronenberg, A.K. Fracture surface energy of the Punchbowl fault, San Andreas system. *Nature* **2005**, *437*, 133–136. [CrossRef] [PubMed]
8. Ma, K.; Tanaka, H.; Song, S.; Wang, C.; Hung, J.; Tsai, Y.; Mori, J.; Song, Y.; Yeh, E.; Soh, W.; et al. Slip zone and energetics of a large earthquake from the Taiwan Chelungpu-fault Drilling Project. *Nature* **2006**, *444*, 473–476. [CrossRef]
9. Sibson, R.H. Earthquakes and rock deformation in crustal fault zones. *Annu. Rev. Earth Planet. Sci.* **1986**, *14*, 149–175. [CrossRef]
10. Sibson, R.H. Thickness of the seismic slip zone. *Bull. Seismol. Soc. Am.* **2003**, *93*, 1169–1178. [CrossRef]
11. Tjong, S.; Chen, H. Nanocrystalline materials and coatings. *Mater. Sci. Eng.* **2004**, *45*, 1–88. [CrossRef]
12. Meyers, M.; Mishra, A.; Benson, D. Mechanical properties of nanocrystalline materials. *Prog. Mater. Sci.* **2006**, *51*, 427–556. [CrossRef]
13. Rogers, B.; Adams, J.; Pennathur, S. *Nanotechnology: Understanding Small Systems*, 3rd ed.; CRC Press Taylor & Francis Group: Boca Raton, FL, USA, 2015; p. 407.

14. Rutter, E.; Holdsworth, R.; Knipe, R.; Strachan, R.; Magloughlin, J. The nature and tectonic significance of fault-zone weakening: An introduction. *Nat. Tecton. Signif. Fault Zone Weaken.* **2001**, *186*, 1–11. [CrossRef]

15. Holdsworth, R.E.; Stewart, M.; Imber, J.; Strachan, R.A. The strucure and rheological evolution of reactivated continental fault zones: A review and case study. In *Continental Reactivation and Reworking*; Miller, J., Holdsworth, R.E., Buick, I.S., Hand, M., Eds.; Geological Society: London, UK, 2001; Volume 184, pp. 115–137.

16. Brace, W.F.; Kohlstedt, D.L. Limits on lithospheric stress imposed by laboratory experiments. *J. Geophys. Res.* **1980**, *85*, 6248–6252. [CrossRef]

17. Kohlstedt, D.L.; Evans, B.; Mackwell, S.J. Strength of the lithosphere—Constraints imposed by laboratory experiments. *J. Geophys. Res. Solid Earth* **1995**, *100*, 17587–17602. [CrossRef]

18. Sibson, R.H. Fault zone models, heat-flow, and the depth distribution of earthquakes in the continental crust of the united states. *Bull. Seismol. Soc. Am.* **1982**, *72*, 151–163.

19. Sibson, R.H. Continental fault structure and the shallow earthquake source. *J. Geol. Soc.* **1983**, *140*, 741–767. [CrossRef]

20. Sibson, R.H. Roughness at the base of the seismogenic zone—Contributing factors. *J. Geophys. Res.* **1984**, *89*, 5791–5799. [CrossRef]

21. Meissner, R.; Strehlau, J. Limits of stresses in continental crusts and their relation to the depth-frequency distribution of shallow earthquakes. *Tectonics* **1982**, *1*, 73–89. [CrossRef]

22. Scholz, C.H. The brittle-plastic transition and the depth of seismic faulting. *Geol. Rundsch.* **1988**, *77*, 319–328. [CrossRef]

23. Shimamoto, T. The origin of S-C mylonites and a new fault-zone model. *J. Struct. Geol.* **1989**, *11*, 51–64. [CrossRef]

24. Scholz, C.H. *The Mechanics of Earthquakes and Faulting*, 2nd ed.; Cambridge University Press: Cambridge, UK, 2002; p. 471.

25. Fagereng, Å.; Toy, V.G.; Rowland, J.V. Geology of the earthquake source: An introduction. *Geol. Earthq. Sour.* **2011**, *359*, 1–16. [CrossRef]

26. Marone, C.; Scholz, C.H. The depth of seismic faulting and the upper transition from stable to unstable regimes. *Geophys. Res. Lett.* **1988**, *15*, 621–624. [CrossRef]

27. Chester, F.M.; Higgs, N.G. Multimechanism friction constitutive model for ultrafine quartz gouge at hypocentral conditions. *J. Geophys. Res.-Solid Earth* **1992**, *97*, 1859–1870. [CrossRef]

28. Reinen, L.A.; Tullis, T.E.; Weeks, J.D. Two-mechanism model for frictional sliding of serpentinite. *Geophys. Res. Lett.* **1992**, *19*, 1535–1538. [CrossRef]

29. Chester, F.M. Effects of temperature on friction—Constitutive equations and experiments with quartz gouge. *J. Geophys. Res.-Solid Earth* **1994**, *99*, 7247–7261. [CrossRef]

30. Stewart, M.; Holdsworth, R.; Strachan, R. Deformation processes and weakening mechanisms within the frictional-viscous transition zone of major crustal-scale faults: Insights from the Great Glen Fault Zone, Scotland. *J. Struct. Geol.* **2000**, *22*, 543–560. [CrossRef]

31. Bos, B.; Spiers, C. Fluid-assisted healing processes in gouge-bearing faults: Insights from experiments on a rock analogue system. *Pure Appl. Geophys.* **2002**, *159*, 2537–2566. [CrossRef]

32. Bos, B.; Spiers, C. Frictional-viscous flow of phyllosilicate-bearing fault rock: Microphysical model and implications for crustal strength profiles. *J. Geophys. Res.-Solid Earth* **2002**, *107*. [CrossRef]

33. Niemeijer, A.; Spiers, C. Velocity dependence of strength and healing behaviour in simulated phyllosilicate-bearing fault gouge. *Tectonophysics* **2006**, *427*, 231–253. [CrossRef]

34. Imber, J.; Holdsworth, R.E.; Smith, S.A.F.; Jefferies, S.P.; Collettini, C. Frictional-viscous flow, seismicity and the geology of weak faults: A review and future directions. In *Internal Structure of Fault Zones: Implications for Mechanical and Fluid-Flow Properties*; Wibberley, C.A.J., Kurz, W., Imber, J., Holdsworth, R.E., Collettini, C., Eds.; Geological Society: London, UK, 2008; Volume 299, pp. 151–173.

35. Noda, H.; Shimamoto, T. A rate- and state-dependent ductile flow law of polycrystalline halite under large shear strain and implications for transition to brittle deformation. *Geophys. Res. Lett.* **2010**, *37*. [CrossRef]

36. Noda, H.; Shimamoto, T. Transient behavior and stability analyses of halite shear zones with an empirical rate-and-state friction to flow law. *J. Struct. Geol.* **2012**, *38*, 234–242. [CrossRef]

37. Nicolas, A.; Fortin, J.; Regnet, J.B.; Verberne, B.A.; Plumper, O.; Dimanov, A.; Spiers, C.J.; Gueguen, Y. Brittle and semibrittle creep of Tavel limestone deformed at room temperature. *J. Geophys. Res. Solid Earth* **2017**, *122*, 4436–4459. [CrossRef]

38. Brace, W.F.; Byerlee, J.D. Stick-slip as a mechanism for earthquakes. *Science* **1966**, *153*, 990–992. [CrossRef] [PubMed]

39. Dieterich, J.H. Time-dependent friction and mechanics of stick-slip. *Pure Appl. Geophys.* **1978**, *116*, 790–806. [CrossRef]

40. Dieterich, J.H. Modeling of rock friction 1. Experimental results and constitutive equations. *J. Geophys. Res.* **1979**, *84*, 2161–2168. [CrossRef]

41. Ruina, A.L. Slip stability and state variable friction laws. *J. Geophys. Res.* **1983**, *88*, 359–370. [CrossRef]

42. Baumberger, T.; Heslot, F.; Perrin, B. Crossover from creep to inertial motion in friction dynamics. *Nature* **1994**, *367*, 544–546. [CrossRef]

43. Rice, J.R.; Ruina, A.L. Stability of steady frictional slipping. *J. Appl. Mech.* **1983**, *50*, 343–349. [CrossRef]

44. Scholz, C.H. Earthquakes and friction laws. *Nature* **1998**, *391*, 37–42. [CrossRef]

45. Sibson, R.H. Fault rocks and fault mechanisms. *J. Geol. Soc.* **1977**, *133*, 191–213. [CrossRef]

46. Chester, F.M.; Chester, J.S. Ultracataclasite structure and friction processes of the Punchbowl fault, San Andreas system, California. *Tectonophysics* **1998**, *295*, 199–221. [CrossRef]

47. Mitchell, T.; Faulkner, D. The nature and origin of off-fault damage surrounding strike-slip fault zones with a wide range of displacements: A field study from the Atacama fault system, northern Chile. *J. Struct. Geol.* **2009**, *31*, 802–816. [CrossRef]

48. Smith, S.A.F.; Billi, A.; Di Toro, G.; Spiess, R. Principal Slip Zones in Limestone: Microstructural Characterization and Implications for the Seismic Cycle (Tre Monti Fault, Central Apennines, Italy). *Pure Appl. Geophys.* **2011**, *168*, 2365–2393. [CrossRef]

49. Barth, N.; Boulton, C.; Carpenter, B.; Batt, G.; Toy, V. Slip localization on the southern Alpine Fault, New Zealand. *Tectonics* **2013**, *32*, 620–640. [CrossRef]

50. Li, H.; Wang, H.; Xu, Z.; Si, J.; Pei, J.; Li, T.; Huang, Y.; Song, S.; Kuo, L.; Sun, Z.; et al. Characteristics of the fault-related rocks, fault zones and the principal slip zone in the Wenchuan Earthquake Fault Scientific Drilling Project Hole-1 (WFSD-1). *Tectonophysics* **2013**, *584*, 23–42. [CrossRef]

51. Kuo, L.; Hsiao, H.; Song, S.; Sheu, H.; Suppe, J. Coseismic thickness of principal slip zone from the Taiwan Chelungpu fault Drilling Project-A (TCDP-A) and correlated fracture energy. *Tectonophysics* **2014**, *619*, 29–35. [CrossRef]

52. Shigematsu, N.; Kametaka, M.; Inada, N.; Miyawaki, M.; Miyakawa, A.; Kameda, J.; Togo, T.; Fujimoto, K. Evolution of the Median Tectonic Line fault zone, SW Japan, during exhumation. *Tectonophysics* **2017**, *696*, 52–69. [CrossRef]

53. Shipton, Z.; Evans, J.; Abercrombie, R.; Brodsky, E.; Abercrombie, R.; McGarr, A.; DiToro, G.; Kanamori, H. The missing sinks: Slip localization in faults, damage zones, and the seismic energy budget. *Earthq. Radiat. Energy Phys. Fault.* **2006**, *170*, 217. [CrossRef]

54. Lockner, D.A.; Byerlee, J.D.; Kuksenko, V.; Ponomarev, A.; Sidorin, A. Quasi-static fault growth and shear fracture energy in granite. *Nature* **1991**, *350*, 39–42. [CrossRef]

55. Marone, C. Laboratory-derived friction laws and their application to seismic faulting. *Annu. Rev. Earth Planet. Sci.* **1998**, *26*, 643–696. [CrossRef]

56. Ikari, M.J. Principal slip zones: Precursors but not recorders of earthquake slip. *Geology* **2015**, *43*, 955–958. [CrossRef]

57. Rice, J. Heating and weakening of faults during earthquake slip. *J. Geophys. Res. Solid Earth* **2006**, *111*. [CrossRef]

58. Han, R.; Shimamoto, T.; Hirose, T.; Ree, J.; Ando, J. Ultralow friction of carbonate faults caused by thermal decomposition. *Science* **2007**, *316*, 878–881. [CrossRef] [PubMed]

59. De Paola, N.; Hirose, T.; Mitchell, T.; Di Toro, G.; Viti, C.; Shimamoto, T. Fault lubrication and earthquake propagation in thermally unstable rocks. *Geology* **2011**, *39*, 35–38. [CrossRef]

60. Di Toro, G.; Hirose, T.; Nielsen, S.; Pennacchioni, G.; Shimamoto, T. Natural and experimental evidence of melt lubrication of faults during earthquakes. *Science* **2006**, *311*, 647–649. [CrossRef] [PubMed]

61. Di Toro, G.; Han, R.; Hirose, T.; De Paola, N.; Nielsen, S.; Mizoguchi, K.; Ferri, F.; Cocco, M.; Shimamoto, T. Fault lubrication during earthquakes. *Nature* **2011**, *471*, 494–498. [CrossRef]

62. Platt, J.; Brantut, N.; Rice, J. Strain localization driven by thermal decomposition during seismic shear. *J. Geophys. Res. Solid Earth* **2015**, *120*, 4405–4433. [CrossRef]

63. Chen, J.; Niemeijer, A.; Fokker, P. Vaporization of fault water during seismic slip. *J. Geophys. Res. Solid Earth* **2017**, *122*, 4237–4276. [CrossRef]
64. Couchman, P.R.; Jesser, W.A. Thermodynamic theory of size dependence of melting temperature in metals. *Nature* **1977**, *269*, 481–483. [CrossRef]
65. Fecht, H.J. Defect-induced melting and solid-state amorphization. *Nature* **1992**, *356*, 133–135. [CrossRef]
66. Qi, W.; Wang, M. Size effect on the cohesive energy of nanoparticle. *J. Mater. Sci. Lett.* **2002**, *21*, 1743–1745. [CrossRef]
67. Buffat, P.; Borel, J.P. Size effect on melting temperature of gold particles. *Phys. Rev. A* **1976**, *13*, 2287–2298. [CrossRef]
68. Wang, S.; Cui, Z.; Xia, X.; Xue, Y. Size-dependent decomposition temperature of nanoparticles: A theoretical and experimental study. *Physica B* **2014**, *454*, 175–178. [CrossRef]
69. Wurschum, R.; Herth, S.; Brossmann, U. Diffusion in nanocrystalline metals and alloys—A status report. *Adv. Eng. Mater.* **2003**, *5*, 365–372. [CrossRef]
70. Coble, R.L. A model for boundary diffusion controlled creep in polycrystalline materials. *J. Appl. Phys.* **1963**, *34*, 1679–1682. [CrossRef]
71. Ashby, M.F.; Verrall, R.A. Diffusion-accommodated flow and superplasticity. *Acta Metall. Mater.* **1973**, *21*, 149–163. [CrossRef]
72. Rutter, E.H. Kinetics of rock deformation by pressure solution. *Philos. Trans. R. Soc. A* **1976**, *283*, 203–219. [CrossRef]
73. Lu, L.; Sui, M.; Lu, K. Superplastic extensibility of nanocrystalline copper at room temperature. *Science* **2000**, *287*, 1463–1466. [CrossRef]
74. Mohamed, F.; Li, Y. Creep and superplasticity in nanocrystalline materials: Current understanding and future prospects. *Mater. Sci. Eng. A Struct.* **2001**, *298*, 1–15. [CrossRef]
75. Mohamed, F. Deformation mechanism maps for micro-grained, ultrafine-grained, and nano-grained materials. *Mater. Sci. Eng. A Struct.* **2011**, *528*, 1431–1435. [CrossRef]
76. Hall, E.O. The deformation and ageing of mild steel 3. Discussion of results. *Proc. Phys. Soc. Lond. B* **1951**, *64*, 747–753. [CrossRef]
77. Petch, N.J. The cleavage strength of polycrystals. *J. Iron Steel Inst.* **1953**, *174*, 25–28.
78. Louchet, F.; Weiss, J.; Richeton, T. Hall-Petch law revisited in terms of collective dislocation dynamics. *Phys. Rev. Lett.* **2006**, *97*. [CrossRef] [PubMed]
79. Carlton, C.E.; Ferreira, P.J. What is behind the inverse Hall-Petch effect in nanocrystalline materials? *Acta Mater.* **2007**, *55*, 3749–3756. [CrossRef]
80. Ma, E. Watching the nanograins roll. *Science* **2004**, *305*, 623–624. [CrossRef]
81. Shan, Z.; Stach, E.; Wiezorek, J.; Knapp, J.; Follstaedt, D.; Mao, S. Grain boundary-mediated plasticity in nanocrystalline nickel. *Science* **2004**, *305*, 654–657. [CrossRef]
82. Carlton, C.E.; Ferreira, P.J. Dislocation motion-induced strain in nanocrystalline materials: Overlooked considerations. *Mater. Sci. Eng. A Struct.* **2008**, *486*, 672–674. [CrossRef]
83. Power, W.L.; Tullis, T.E. The relationship between slickenside surfaces in fine-grained quartz and the seismic cycle. *J. Struct. Geol.* **1989**, *11*, 879–893. [CrossRef]
84. Chester, J.S.; Goldsby, D.L. *Microscale Characterization of Natural and Experimental Slip Surfaces Relevant to Earthquake Mechanics*; South California Earthquake Center: Los Angeles, CA, USA, 2003.
85. Kuo, L.; Song, S.; Suppe, J.; Yeh, E. Fault mirrors in seismically active fault zones: A fossil of small earthquakes at shallow depths. *Geophys. Res. Lett.* **2016**, *43*, 1950–1959. [CrossRef]
86. Janssen, C.; Wirth, R.; Rybacki, E.; Naumann, R.; Kemnitz, H.; Wenk, H.; Dresen, G. Amorphous material in SAFOD core samples (San Andreas Fault): Evidence for crush-origin pseudotachylytes? *Geophys. Res. Lett.* **2010**, *37*. [CrossRef]
87. Siman-Tov, S.; Aharonov, E.; Sagy, A.; Emmanuel, S. Nanograins form carbonate fault mirrors. *Geology* **2013**, *41*, 703–706. [CrossRef]
88. Kirkpatrick, J.; Rowe, C.; White, J.; Brodsky, E. Silica gel formation during fault slip: Evidence from the rock record. *Geology* **2013**, *41*, 1015–1018. [CrossRef]
89. Evans, J.; Prante, M.; Janecke, S.; Ault, A.; Newell, D. Hot faults: Iridescent slip surfaces with metallic luster document high-temperature ancient seismicity in the Wasatch fault zone, Utah, USA. *Geology* **2014**, *42*, 623–626. [CrossRef]

90. Viti, C.; Brogi, A.; Liotta, D.; Mugnaioli, E.; Spiess, R.; Dini, A.; Zucchi, M.; Vannuccini, G. Seismic slip recorded in tourmaline fault mirrors from Elba Island (Italy). *J. Struct. Geol.* **2016**, *86*, 1–12. [CrossRef]

91. Beckmann, P.; Spizzichino, A. *The Scattering of Electromagnetic Waves from Rough Surfaces*; Pergamon Press: Oxford, UK, 1963; Volume 4.

92. Ozawa, K.; Takizawa, S. Amorphous material formed by the mechanochemical effect in natural pseudotachylyte of crushing origin: A case study of the Iida-Matsukawa Fault, Nagano Prefecture, Central Japan. *J. Struct. Geol.* **2007**, *29*, 1855–1869. [CrossRef]

93. De Paola, N.; Holdsworth, R.; Viti, C.; Collettini, C.; Bullock, R. Can grain size sensitive flow lubricate faults during the initial stages of earthquake propagation? *Earth Planet. Sci. Lett.* **2015**, *431*, 48–58. [CrossRef]

94. Viti, C. Exploring fault rocks at the nanoscale. *J. Struct. Geol.* **2011**, *33*, 1715–1727. [CrossRef]

95. Sibson, R.H. Generation of pseudotachylyte by ancient seismic faulting. *Geophys. J. Int.* **1975**, *43*, 775–794. [CrossRef]

96. Spray, J.G. Pseudotachylyte controversy—Fact or friction? *Geology* **1995**, *23*, 1119–1122. [CrossRef]

97. Cowan, D. Do faults preserve a record of seismic slip? A field geologist's opinion. *J. Struct. Geol.* **1999**, *21*, 995–1001. [CrossRef]

98. Rowe, C.; Griffith, W. Do faults preserve a record of seismic slip: A second opinion. *J. Struct. Geol.* **2015**, *78*, 1–26. [CrossRef]

99. Wenk, H.R. Are pseudotachylytes products of fracture or fusion? *Geology* **1978**, *6*, 507–511. [CrossRef]

100. Keulen, N.; Heilbronner, R.; Stünitz, H.; Boullier, A.; Ito, H. Grain size distributions of fault rocks: A comparison between experimentally and naturally deformed granitoids. *J. Struct. Geol.* **2007**, *29*, 1282–1300. [CrossRef]

101. Bullock, R.; De Paola, N.; Holdsworth, R.; Trabucho-Alexandre, J. Lithological controls on the deformation mechanisms operating within carbonate-hosted faults during the seismic cycle. *J. Struct. Geol.* **2014**, *58*, 22–42. [CrossRef]

102. Collettini, C.; Carpenter, B.; Viti, C.; Cruciani, F.; Mollo, S.; Tesei, T.; Trippetta, F.; Valoroso, L.; Chiaraluce, L. Fault structure and slip localization in carbonate-bearing normal faults: An example from the Northern Apennines of Italy. *J. Struct. Geol.* **2014**, *67*, 154–166. [CrossRef]

103. Demurtas, M.; Fondriest, M.; Balsamo, F.; Clemenzi, L.; Storti, F.; Bistacchi, A.; Di Toro, G. Structure of a normal seismogenic fault zone in carbonates: The Vado di Como Fault, Campo Imperatore, Central Apennines (Italy). *J. Struct. Geol.* **2016**, *90*, 185–206. [CrossRef]

104. Siman-Tov, S.; Stock, G.; Brodsky, E.; White, J. The coating layer of glacial polish. *Geology* **2017**, *45*, 987–990. [CrossRef]

105. Smeraglia, L.; Billi, A.; Carminati, E.; Cavallo, A.; Doglioni, C. Field- to nano-scale evidence for weakening mechanisms along the fault of the 2016 Amatrice and Norcia earthquakes, Italy. *Tectonophysics* **2017**, *712*, 156–169. [CrossRef]

106. Tullis, T.E.; Tullis, J. Experimental rock deformation techniques. In *Mineral and Rock Deformation: Laboratory Studies*, 36th ed.; Hobbs, B.E., Heard, H.C., Eds.; AGU: Washington, DC, USA, 1986; pp. 297–324.

107. Niemeijer, A.; Di Toro, G.; Griffith, W.; Bistacchi, A.; Smith, S.; Nielsen, S. Inferring earthquake physics and chemistry using an integrated field and laboratory approach. *J. Struct. Geol.* **2012**, *39*, 2–36. [CrossRef]

108. Ma, S.; Shimamoto, T.; Lu, Y.; Togo, T.; Kitajima, H. A rotary-shear low to high-velocity friction apparatus in Beijing to study rock friction at plate to seismic slip rates. *Earthq. Sci.* **2014**, *27*, 469–497. [CrossRef]

109. Logan, J.M.; Friedman, M.; Higgs, N.; Dengo, C.A.; Shimamoto, T. Experimental studies of simulated gouge and their application to studies of natural fault zones. In Proceedings of the Conference VIII—Analysis of Actual Fault Zones in Bedrock, Menlo Park, CA, USA, 1–5 April 1979; pp. 305–343.

110. Logan, J.M.; Dengo, C.A.; Higgs, N.G.; Wang, Z.Z. Fabrics of experimental fault zones: Their development and relationship to mechanical behavior. In *Fault Mechanics and TRANSPORT Properties in Rocks*; Evans, B., Wong, T.-F., Eds.; Academic Press: London, UK, 1992; pp. 33–68.

111. Yund, R.A.; Blanpied, M.L.; Tullis, T.E.; Weeks, J.D. Amorphous material in high-strain experimental fault gouges. *J. Geophys. Res.* **1990**, *95*, 15589–15602. [CrossRef]

112. Han, R.; Shimamoto, T.; Ando, J.; Ree, J. Seismic slip record in carbonate-bearing fault zones: An insight from high-velocity friction experiments on siderite gouge. *Geology* **2007**, *35*, 1131–1134. [CrossRef]

113. Verberne, B.A.; de Bresser, J.H.P.; Niemeijer, A.R.; Spiers, C.J.; de Winter, D.A.M.; Plumper, O. Nanocrystalline slip zones in calcite fault gouge show intense crystallographic preferred orientation: Crystal plasticity at sub-seismic slip rates at 18–150 °C. *Geology* **2013**, *41*, 863–866. [CrossRef]

114. Verberne, B.A.; Spiers, C.J.; Niemeijer, A.R.; De Bresser, J.H.P.; De Winter, D.A.M.; Plumper, O. Frictional Properties and Microstructure of Calcite-Rich Fault Gouges Sheared at Sub-Seismic Sliding Velocities. *Pure Appl. Geophys.* **2014**, *171*, 2617–2640. [CrossRef]

115. Spagnuolo, E.; Plumper, O.; Violay, M.; Cavallo, A.; Di Toro, G. Fast-moving dislocations trigger flash weakening in carbonate-bearing faults during earthquakes. *Sci. Rep.* **2015**, *5*, 16112. [CrossRef] [PubMed]

116. Green, H.; Shi, F.; Bozhilov, K.; Xia, G.; Reches, Z. Phase transformation and nanometric flow cause extreme weakening during fault slip. *Nat. Geosci.* **2015**, *8*, U484–U491. [CrossRef]

117. Siman-Tov, S.; Aharonov, E.; Boneh, Y.; Reches, Z. Fault mirrors along carbonate faults: Formation and destruction during shear experiments. *Earth Planet. Sci. Lett.* **2015**, *430*, 367–376. [CrossRef]

118. Delle Piane, C.; Piazolo, S.; Timms, N.; Luzin, V.; Saunders, M.; Bourdet, J.; Giwelli, A.; Ben Clennell, M.; Kong, C.; Rickard, W.; et al. Generation of amorphous carbon and crystallographic texture during low-temperature subseismic slip in calcite fault gouge. *Geology* **2018**, *46*, 163–166. [CrossRef]

119. Mercuri, M.; Scuderi, M.; Tesei, T.; Carminati, E.; Collettini, C. Strength evolution of simulated carbonate-bearing faults: The role of normal stress and slip velocity. *J. Struct. Geol.* **2018**, *109*, 1–9. [CrossRef]

120. Toy, V.; Mitchell, T.; Druiventak, A.; Wirth, R. Crystallographic preferred orientations may develop in nanocrystalline materials on fault planes due to surface energy interactions. *Geochem. Geophys. Geosys.* **2015**, *16*, 2549–2563. [CrossRef]

121. Prior, D.; Boyle, A.; Brenker, F.; Cheadle, M.; Day, A.; Lopez, G.; Peruzzo, L.; Potts, G.; Reddy, S.; Spiess, R.; et al. The application of electron backscatter diffraction and orientation contrast imaging in the SEM to textural problems in rocks. *Am. Mineral.* **1999**, *84*, 1741–1759. [CrossRef]

122. Verberne, B.A.; Chen, J.; Niemeijer, A.R.; de Bresser, J.H.P.; Pennock, G.M.; Drury, M.R.; Spiers, C.J. Microscale cavitation as a mechanism for nucleating earthquakes at the base of the seismogenic zone. *Nat. Commun.* **2017**, *8*. [CrossRef] [PubMed]

123. Pozzi, G.; De Paola, N.; Holdsworth, R.; Bowen, L.; Nielsen, S.; Dempsey, E. Coseismic ultramylonites: An investigation of nanoscale viscous flow and fault weakening during seismic slip. *Earth Planet. Sci. Lett.* **2019**, *516*, 164–175. [CrossRef]

124. Verberne, B.A.; He, C.; Spiers, C.J. Frictional Properties of Sedimentary Rocks and Natural Fault Gouge from the Longmen Shan Fault Zone, Sichuan, China. *Bull. Seismol. Soc. Am.* **2010**, *100*, 2767–2790. [CrossRef]

125. Engelder, J.T. Quartz Fault Gouge: Its Generation and Effect on the Frictional Properties of Sandstone. Ph.D. Thesis, Texas A&M University, College Station, TX, USA, 1973.

126. Viti, C.; Hirose, T. Thermal decomposition of serpentine during coseismic faulting: Nanostructures and mineral reactions. *J. Struct. Geol.* **2010**, *32*, 1476–1484. [CrossRef]

127. Pec, M.; Stünitz, H.; Heilbronner, R.; Drury, M.; de Capitani, C. Origin of pseudotachylites in slow creep experiments. *Earth Planet. Sci. Lett.* **2012**, *355*, 299–310. [CrossRef]

128. Pec, M.; Stünitz, H.; Heilbronner, R.; Drury, M. Semi-brittle flow of granitoid fault rocks in experiments. *J. Geophys. Res. Solid Earth* **2016**, *121*, 1677–1705. [CrossRef]

129. Hadizadeh, J.; Tullis, T.; White, J.; Konkachbaev, A. Shear localization, velocity weakening behavior, and development of cataclastic foliation in experimental granite gouge. *J. Struct. Geol.* **2015**, *71*, 86–99. [CrossRef]

130. Kuo, L.; Song, Y.; Yang, C.; Song, S.; Wang, C.; Dong, J.; Suppe, J.; Shimamoto, T. Ultrafine spherical quartz formation during seismic fault slip: Natural and experimental evidence and its implications. *Tectonophysics* **2015**, *664*, 98–108. [CrossRef]

131. Hayward, K.; Cox, S.; Gerald, J.; Slagmolen, B.; Shaddock, D.; Forsyth, P.; Salmon, M.; Hawkins, R. Mechanical amorphization, flash heating, and frictional melting: Dramatic changes to fault surfaces during the first millisecond of earthquake slip. *Geology* **2016**, *44*, 1043–1046. [CrossRef]

132. Aretusini, S.; Mittempergher, S.; Plumper, O.; Spagnuolo, E.; Gualtieri, A.; Di Toro, G. Production of nanoparticles during experimental deformation of smectite and implications for seismic slip. *Earth Planet. Sci. Lett.* **2017**, *463*, 221–231. [CrossRef]

133. Niemeijer, A. Velocity-dependent slip weakening by the combined operation of pressure solution and foliation development. *Sci. Rep.* **2018**, *8*. [CrossRef] [PubMed]

134. Rowe, C.; Lamothe, K.; Rempe, M.; Andrews, M.; Mitchell, T.; Di Toro, G.; White, J.; Aretusini, S. Earthquake lubrication and healing explained by amorphous nanosilica. *Nat. Commun.* **2019**, *10*. [CrossRef] [PubMed]

135. Fondriest, M.; Smith, S.; Candela, T.; Nielsen, S.; Mair, K.; Di Toro, G. Mirror-like faults and power dissipation during earthquakes. *Geology* **2013**, *41*, 1175–1178. [CrossRef]

136. Tisato, N.; Di Toro, G.; De Rossi, N.; Quaresimin, M.; Candela, T. Experimental investigation of flash weakening in limestone. *J. Struct. Geol.* **2012**, *38*, 183–199. [CrossRef]

137. Chen, X.; Madden, A.; Bickmore, B.; Reches, Z. Dynamic weakening by nanoscale smoothing during high-velocity fault slip. *Geology* **2013**, *41*, 739–742. [CrossRef]

138. Boneh, Y.; Sagy, A.; Reches, Z. Frictional strength and wear-rate of carbonate faults during high-velocity, steady-state sliding. *Earth Planet. Sci. Lett.* **2013**, *381*, 127–137. [CrossRef]

139. Pozzi, G.; De Paola, N.; Nielsen, S.; Holdsworth, R.; Bowen, L. A new interpretation for the nature and significance of mirror-like surfaces in experimental carbonate-hosted seismic faults. *Geology* **2018**, *46*, 583–586. [CrossRef]

140. Verberne, B.A.; Plumper, O.; de Winter, D.A.M.; Spiers, C.J. Superplastic nanofibrous slip zones control seismogenic fault friction. *Science* **2014**, *346*, 1342–1344. [CrossRef]

141. Engelder, J.T. Cataclasis and generation of fault gouge. *Geol. Soc. Am. Bull.* **1974**, *85*, 1515–1522. [CrossRef]

142. Marone, C.; Scholz, C.H. Particle-size distribution and microstructures within simulated fault gouge. *J. Struct. Geol.* **1989**, *11*, 799–814. [CrossRef]

143. Stünitz, H.; Keulen, N.; Hirose, T.; Heilbronner, R. Grain size distribution and microstructures of experimentally sheared granitoid gouge at coseismic slip rates—Criteria to distinguish seismic and aseismic faults? *J. Struct. Geol.* **2010**, *32*, 59–69. [CrossRef]

144. Kendall, K. The impossibility of communiting small particles by compression. *Nature* **1978**, *272*, 710–711. [CrossRef]

145. Karihaloo, B.L. The impossibility of communiting small particles by compression. *Nature* **1979**, *279*, 169–170. [CrossRef]

146. Griffith, A. The phenomena of rupture and flow in solids. *Philos. Trans. R. Soc. Lond.* **1921**, *61*, 163–198. [CrossRef]

147. Sammis, C.; Ben-Zion, Y. Mechanics of grain-size reduction in fault zones. *J. Geophys. Res. Solid Earth* **2008**, *113*. [CrossRef]

148. Tao, N.; Wang, Z.; Tong, W.; Sui, M.; Lu, J.; Lu, K. An investigation of surface nanocrystallization mechanism in Fe induced by surface mechanical attrition treatment. *Acta Mater.* **2002**, *50*, 4603–4616. [CrossRef]

149. Yang, D.; Cizek, P.; Hodgson, P.; Wen, C. Microstructure evolution and nanograin formation during shear localization in cold-rolled titanium. *Acta Mater.* **2010**, *58*, 4536–4548. [CrossRef]

150. Kennedy, L.; White, J. Low-temperature recrystallization in calcite: Mechanisms and consequences. *Geology* **2001**, *29*, 1027–1030. [CrossRef]

151. De Bresser, J.H.P.; Evans, B.; Renner, J. On estimating the strength of calcite rocks under natural conditions. In *Deformation Mechanisms, Rheology and Tectonics: Current Status and Future Perspectives*; De Meer, S., Drury, M.R., de Bresser, J.H.P., Pennock, G.M., Eds.; Geological Society: London, UK, 2002; Volume 200, pp. 309–329.

152. De Bresser, J.H.P.; Spiers, C.J. Strength characteristics of the r, f, and c slip systems in calcite. *Tectonophysics* **1997**, *272*, 1–23. [CrossRef]

153. Wolf, D.; Okamoto, P.R.; Yip, S.; Lutsko, J.F.; Kluge, M. Thermodynamic parallels between solid-state amorphization and melting. *J. Mater. Res.* **1990**, *5*, 286–301. [CrossRef]

154. Oohashi, K.; Hirose, T.; Kobayashi, K.; Shimamoto, T. The occurrence of graphite-bearing fault rocks in the Atotsugawa fault system, Japan: Origins and implications for fault creep. *J. Struct. Geol.* **2012**, *38*, 39–50. [CrossRef]

155. Collettini, C.; Viti, C.; Tesei, T.; Mollo, S. Thermal decomposition along natural carbonate faults during earthquakes. *Geology* **2013**, *41*, 927–930. [CrossRef]

156. Ohl, M.; Plümper, O.; Chatzaras, V.; Wallis, D.; Vollmer, C.; Drury, M. Mechanisms of fault mirror formation in carbonate rocks. *EarthArXiv* **2019**. [CrossRef]

157. Oohashi, K.; Han, R.; Hirose, T.; Shimamoto, T.; Omura, K.; Matsuda, T. Carbon-forming reactions under a reducing atmosphere during seismic fault slip. *Geology* **2014**, *42*, 787–790. [CrossRef]

158. Luque, F.; Pasteris, J.; Wopenka, B.; Rodas, M.; Barrenechea, J. Natural fluid-deposited graphite: Mineralogical characteristics and mechanisms of formation. *Am. J. Sci.* **1998**, *298*, 471–498. [CrossRef]

159. Gao, G.; Mikulski, P.; Harrison, J. Molecular-scale tribology of amorphous carbon coatings: Effects of film thickness, adhesion, and long-range interactions. *J. Am. Chem. Soc.* **2002**, *124*, 7202–7209. [CrossRef]

160. Penn, R.; Banfield, J. Imperfect oriented attachment: Dislocation generation in defect-free nanocrystals. *Science* **1998**, *281*, 969–971. [CrossRef]

161. Niederberger, M.; Colfen, H. Oriented attachment and mesocrystals: Non-classical crystallization mechanisms based on nanoparticle assembly. *Phys. Chem. Chem. Phys.* **2006**, *8*, 3271–3287. [CrossRef]

162. De Yoreo, J.; Gilbert, P.; Sommerdijk, N.; Penn, R.; Whitelam, S.; Joester, D.; Zhang, H.; Rimer, J.; Navrotsky, A.; Banfield, J.; et al. Crystallization by particle attachment in synthetic, biogenic, and geologic environments. *Science* **2015**, *349*. [CrossRef] [PubMed]

163. Gehrke, N.; Colfen, H.; Pinna, N.; Antonietti, M.; Nassif, N. Superstructures of calcium carbonate crystals by oriented attachment. *Cryst. Growth Des.* **2005**, *5*, 1317–1319. [CrossRef]

164. Zhang, H.; Banfield, J. Energy Calculations Predict Nanoparticle Attachment Orientations and Asymmetric Crystal Formation. *J. Phys. Chem. Lett.* **2012**, *3*, 2882–2886. [CrossRef]

165. Zhang, H.; Banfield, J. Interatomic Coulombic interactions as the driving force for oriented attachment. *Crystengcomm* **2014**, *16*, 1568–1578. [CrossRef]

166. De Leeuw, N.; Parker, S. Atomistic simulation of the effect of molecular adsorption of water on the surface structure and energies of calcite surfaces. *J. Chem. Soc. Faraday Trans.* **1997**, *93*, 467–475. [CrossRef]

167. Smith, S.A.F.; Di Toro, G.; Kim, S.; Ree, J.; Nielsen, S.; Billi, A.; Spiess, R. Coseismic recrystallization during shallow earthquake slip. *Geology* **2013**, *41*, 63–66. [CrossRef]

168. Verberne, B.A.; Niemeijer, A.R.; De Bresser, J.H.P.; Spiers, C.J. Mechanical behavior and microstructure of simulated calcite fault gouge sheared at 20–600 °C: Implications for natural faults in limestones. *J. Geophys. Res. Solid Earth* **2015**, *120*, 8169–8196. [CrossRef]

169. Smith, S.A.F.; Nielsen, S.; Di Toro, G. Strain localization and the onset of dynamic weakening in calcite fault gouge. *Earth Planet. Sci. Lett.* **2015**, *413*, 25–36. [CrossRef]

170. Ree, J.; Ando, J.; Han, R.; Shimamoto, T. Coseismic microstructures of experimental fault zones in Carrara marble. *J. Struct. Geol.* **2014**, *66*, 75–83. [CrossRef]

171. Niemeijer, A.; Spiers, C. A microphysical model for strong velocity weakening in phyllosilicate-bearing fault gouges. *J. Geophys. Res. Solid Earth* **2007**, *112*. [CrossRef]

172. Hunfeld, L.; Niemeijer, A.; Spiers, C. Frictional Properties of Simulated Fault Gouges from the Seismogenic Groningen Gas Field Under In Situ P-T -Chemical Conditions. *J. Geophys. Res. Solid Earth* **2017**, *122*, 8969–8989. [CrossRef]

173. Den Hartog, S.; Spiers, C. Influence of subduction zone conditions and gouge composition on frictional slip stability of megathrust faults. *Tectonophysics* **2013**, *600*, 75–90. [CrossRef]

174. Blanpied, M.L.; Lockner, D.A.; Byerlee, J.D. Frictional slip of granite at hydrothermal conditions. *J. Geophys. Res. Solid Earth* **1995**, *100*, 13045–13064. [CrossRef]

175. Chen, J.; Spiers, C. Rate and state frictional and healing behavior of carbonate fault gouge explained using microphysical model. *J. Geophys. Res. Solid Earth* **2016**, *121*, 8642–8665. [CrossRef]

176. Chen, J.; Niemeijer, A.; Spiers, C. Microphysically Derived Expressions for Rate-and-State Friction Parameters, a, b, and D-c. *J. Geophys. Res. Solid Earth* **2017**, *122*, 9627–9657. [CrossRef]

177. Chen, J.; Niemeijer, A. Seismogenic Potential of a Gouge-filled Fault and the Criterion for Its Slip Stability: Constraints FRom a Microphysical Model. *J. Geophys. Res. Solid Earth* **2017**, *122*, 9658–9688. [CrossRef]

178. Van den Ende, M.P.A.; Chen, J.; Ampuero, J.P.; Niemeijer, A.R. A comparison between rate-and-state friction and microphysical models, based on numerical simulations of fault slip. *Tectonophysics* **2018**, *733*, 273–295. [CrossRef]

179. Zhang, X.; Spiers, C.; Peach, C. Compaction creep of wet granular calcite by pressure solution at 28 °C to 150 °C. *J. Geophys. Res. Solid Earth* **2010**, *115*. [CrossRef]

180. Pluymakers, A.M.H.; Spiers, C.J. Compaction creep of simulated anhydrite fault gouge by pressure solution: Theory v. experiments and implications for fault sealing. In *Rock Deformation from Field, Experiments and Theory: A Volume in Honour of Ernie Rutter*; Faulkner, D.R., Mariani, E., Mecklenburgh, J., Eds.; Geological Society, Special Publications: London, UK, 2014; Volume 409.

181. Nakashima, S. Diffusivity of ions in pore water as a quantitative basis for rock deformation estimates. *Tectonophysics* **1995**, *245*, 185–203. [CrossRef]

182. Plummer, L.N.; Busenberg, E. The solubilities of calcite, aragonite and vaterite in CO_2-H_2O solutions between 0 °C and 90 °C, and an evaluation of the aqueous model for the system $CaCO_3$-CO_2-H_2O. *Geochim. Cosmochim. Acta* **1982**, *46*, 1011–1040. [CrossRef]

183. Farver, J.; Yund, R. Volume and grain boundary diffusion of calcium in natural and hot-pressed calcite aggregates. *Contrib. Mineral. Petrol.* **1996**, *123*, 77–91. [CrossRef]

184. Chen, J.; Verberne, B.; Spiers, C. Effects of healing on the seismogenic potential of carbonate fault rocks: Experiments on samples from the Longmenshan Fault, Sichuan, China. *J. Geophys. Res.* **2015**, *120*, 5479–5506. [CrossRef]

185. Chen, J.; Verberne, B.; Spiers, C. Interseismic re-strengthening and stabilization of carbonate faults by "non-Dieterich" healing under hydrothermal conditions. *Earth Planet. Sci. Lett.* **2015**, *423*, 1–12. [CrossRef]

186. Schubnel, A.; Brunet, F.; Hilairet, N.; Gasc, J.; Wang, Y.; Green, H. Deep-Focus Earthquake Analogs Recorded at High Pressure and Temperature in the Laboratory. *Science* **2013**, *341*, 1377–1380. [CrossRef] [PubMed]

187. Bernard, P.; Lyon-Caen, H.; Briole, P.; Deschamps, A.; Boudin, F.; Makropoulos, K.; Papadimitriou, P.; Lemeille, F.; Patau, G.; Billiris, H.; et al. Seismicity, deformation and seismic hazard in the western rift of Corinth: New insights from the Corinth Rift Laboratory (CRL). *Tectonophysics* **2006**, *426*, 7–30. [CrossRef]

188. Valoroso, L.; Chiaraluce, L.; Piccinini, D.; Di Stefano, R.; Schaff, D.; Waldhauser, F. Radiography of a normal fault system by 64,000 high-precision earthquake locations: The 2009 L'Aquila (central Italy) case study. *J. Geophys. Res. -Solid Earth* **2013**, *118*, 1156–1176. [CrossRef]

189. Chiarabba, C.; De Gori, P.; Mele, F. Recent seismicity of Italy: Active tectonics of the central Mediterranean region and seismicity rate changes after the Mw 6.3 L'Aquila earthquake. *Tectonophysics* **2015**, *638*, 82–93. [CrossRef]

190. Fagereng, Å.; Biggs, J. New perspectives on 'geological strain rates' calculated from both naturally deformed and actively deforming rocks. *J. Struct. Geol.* **2018**, in press. [CrossRef]

191. Pfiffner, O.; Ramsay, J. Constraints on geological strain rates—Arguments from finite strain states of naturally deformed rocks. *J. Geophys. Res.* **1982**, *87*, 311–321. [CrossRef]

192. De Bresser, J.H.P.; Ter Heege, J.H.; Spiers, C.J. Grain size reduction by dynamic recrystallization: Can it result in major theological weakening? *Int. J. Earth Sci.* **2001**, *90*, 28–45. [CrossRef]

Article

Total Porosity Measured for Shale Gas Reservoir Samples: A Case from the Lower Silurian Longmaxi Formation in Southeast Chongqing, China

Fangwen Chen [1,2], Shuangfang Lu [1,*], Xue Ding [1], Hongqin Zhao [1] and Yiwen Ju [3,*]

[1] School of Geosciences, China University of Petroleum (East China), Qingdao 266580, China; cfwdqpi@163.com (F.C.); dingxue@upc.edu.cn (X.D.); 18354683115@163.com (H.Z.)

[2] State Key Laboratory of Oil and Gas Reservoir Geology and Exploitation, Chengdu University of Technology, Chengdu 610059, China

[3] Key Laboratory of Computational Geodynamics, College of Earth and Planetary Sciences, University of Chinese Academy of Sciences, Beijing, 100049, China

* Correspondence: lusfupc@163.com (S.L.); juyw03@163.com (Y.J.); Tel.: +86-532-86983191 (S.L.)

Received: 22 November 2018; Accepted: 21 December 2018; Published: 22 December 2018

Abstract: Measuring total porosity in shale gas reservoir samples remains a challenge because of the fine-grained texture, low porosity, ultra-low permeability, and high content of organic matter (OM) and clay mineral. The composition content porosimetry method, which is a new method for the evaluation of the porosity of shale samples, was used in this study to measure the total porosity of shale gas reservoir samples from the Lower Silurian Longmaxi Formation in Southeast Chongqing, China, based on the bulk and grain density values. The results from the composition content porosimetry method were compared with those of the Gas Research Institute method. The results showed that the composition content porosimetry porosity values of shale gas reservoir samples range between 2.05% and 5.87% with an average value of 4.04%. The composition content porosimetry porosity generally increases with increasing OM and clay content, and decreases with increasing quartz and feldspar content. The composition content porosimetry results are similar to the gas research institute results, and the differences between the two methods range from 0.05% to 1.52% with an average value of 0.85%.

Keywords: total porosity; shale gas reservoir; Longmaxi Formation; Southeast Chongqing

---

1. Introduction

Total porosity is the key parameter for resource evaluation and petroleum reserve calculation, especially for shale gas reservoirs. Porosity is defined as the percentage of pore or void volume of the porous sample in the bulk volume. A pore is the part of rock occupied by fluids. This definition is complicated in shale due to the presence of clay minerals and the variability in organic matter (OM). For shale samples, total porosity is the water content having molecular interaction with clay minerals, plus the free fluid (water, gaseous, and liquid hydrocarbons) in the relatively larger open pore and capillary spaces [1]. The total porosity measurement for shale gas reservoir samples is a particular problem because shale gas reservoir samples are organic-rich, low porosity and ultra-low permeability with strong heterogeneity [2–4]. Routine core analyses and traditional methods for measuring the porosity of shale gas reservoir samples have unreliable accuracy and suitability [5–7]. The pore systems in shale gas reservoirs are complex and diverse due to different sedimentation and diagenesis processes with different contents, types, and maturities of OM [8–11]. Pores in shale gas reservoir samples are much smaller than those in conventional carbonate and sandstone reservoirs [12–14]. The pore types include microfracture, microchannel, intraparticle pore, intercrystal pore, organopore, and fossil fragment pore [15]. At present, the techniques available for the measurement of micro-pore characteristics of shale samples are divided into two categories: the radiation method and the penetration fluids method [16].

The radiation methods include scanning electron microscopy (SEM), field emission scanning electron microscopy (FE-SEM), backscatter mode (BS), transmission electron microscopy (TEM), small-angle neutron scattering (SANS), ultra-small-angle neutron scattering (USANS), three-dimensional (3D) reconstruction technology, and computed tomography (CT). Most provide direct visual observation of microscopic features in shale samples [17–21]. 3D image reconstruction technology can be used to investigate shale microstructure and analyze the characteristics of pores. For the radiation methods, a higher resolution corresponds to a smaller sample size [22], and the smaller samples are less representative given shale's heterogeneity [16,19]. The penetration fluid methods include low temperature nitrogen adsorption/desorption (LTNA), mercury intrusion porosimetry (MIP), and nuclear magnetic resonance (NMR). The first two methods refer to injecting non-wetting fluid into a shale sample and recording the fluid volume and injection pressure. Then, the pore size distribution and specific surface area are calculated using several theoretical models [17,23–25]. Due to the differences in experimental environment (temperature and pressure) and injected fluid properties, LTNA and MIP methods have different detection ranges. The nanometer- to micrometer-scale pore systems in shale samples were evaluated by combining their results [17,23]. However, LTNA and MIP methods can only reflect interconnected pores, as the injected fluids cannot access isolated pores [16]. The result of the MIP method reflects the pore volumes connected by throats, corresponding to the pressure of injecting, which are much smaller than pore size [26,27]. NMR technology is used for evaluating the pore size distribution in saturated liquid samples [24], which can show the pore size and pore volume filling liquid regardless of the pores being connected or isolated.

Porosity measurement for shale is still challenging because of its fine-grained texture, low porosity, ultra-low permeability, and high content of OM and clay mineral. The Gas Research Institute (GRI) method is commonly used to evaluate the total porosity of shale gas reservoir samples. The GRI method uses the grain density and bulk density to calculate the total porosity of a shale gas reservoir sample [1,28]. There are several significant uncertainties in grain density measurement: (1) the crushing process is without standard, (2) the relative humidity (RH) of the measurement environment may change the gas saturation of the sample, (3) the pretreatment of hot toluene may dissolve macromolecular hydrocarbons and part of the solid bitumen [1], and (4) the matrix volume is underestimated because of the gas adsorbed by nanoscale pores.

A method using the results of quantitative X-ray diffraction (XRD) to calculate the total porosity of shale gas reservoir sample is introduced in this study. A systematic study of the method, including experimental uncertainty and the relationship between total porosity and content of various minerals, was performed on the Lower Silurian Longmaxi Shale in Southeast Chongqing, China. The results were compared with those of the GRI method on a series of parallel samples from the same depth.

2. Materials and Methods

2.1. Samples

Twenty-seven shale core samples from the Lower Silurian Longmaxi Shale of the Py1 well in Southeast Chongqing, China, were selected for measuring total porosity. Southeast Chongqing belongs to the Yangtze tectonic plate and is located in the Wuling Drape Zone and Western Hunan-Hubei Thrust Belt. The Xuefengshan Uplift and Sichuan Basin lie to the east and northwest of Southeast Chongqing, respectively (Figure 1). The sub-samples from these core samples were crushed to <425 μm (<40 mesh) powder to obtain representative homogeneous samples [29], which were used for the following experiments.

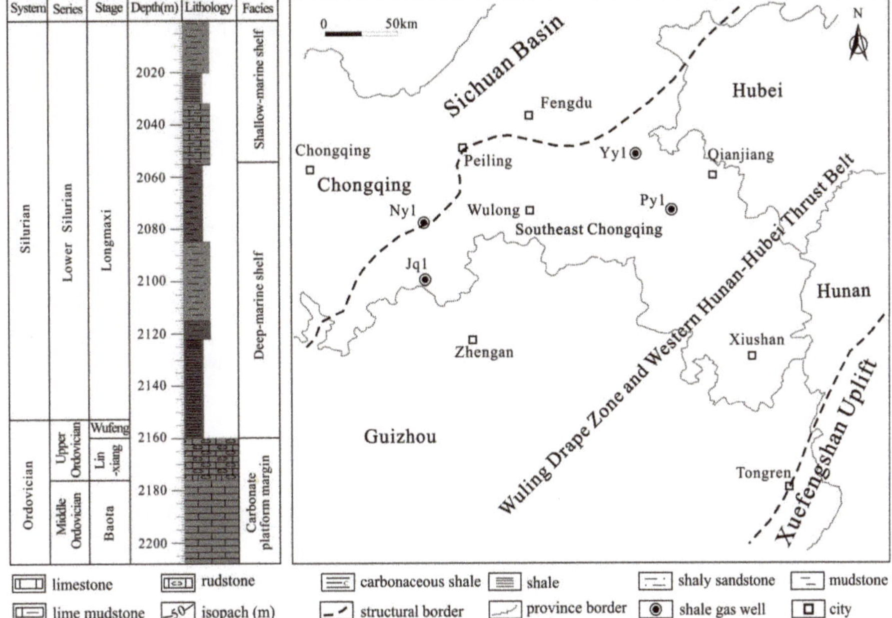

Figure 1. The stratigraphic column of the Py1 well and location of Southeast Chongqing, China.

2.2. TOC, Maturity and XRD Analyses

OM characterization, such as total organic carbon (TOC) content, type, and thermal maturity, were obtained using the Leco TOC and the Rock Eval II methodology. OM content is determined by TOC value divided by carbon element weight percentage in OM [18]. The mineralogical compositions of the samples were obtained based on XRD patterns measured on randomly oriented powder preparations using an Ultima IV X-ray diffractometer (Rigaku, Tokyo, Japan) at 40 kV and 30 mA with Cu Kα radiation (λ = 1.5406 for CuKα1). The crystalline mineral proportions were calculated based on the areas under the peaks corresponding to each mineral and corrected using Lorentz Polarization [30]. In this measurement, we recorded the content of non-clay minerals, the total content, and the relative content of clay minerals.

2.3. CCP Method

The composition content porosimetry (CCP) method uses the grain density and bulk density to calculate the total porosity of a shale gas reservoir sample. The bulk density (ρ_B) of the sample is calculated using Equation (1). The bulk volume was measured by mercury immersion at less than 6.6 Pa (50 μm Hg) using Archimedes' principle on a block sample of approximately 20 g without crushing. The bulk mass and grain mass were measured by balance set-up (Mettler Toledo AL104, readability 0.1 mg).

$$\rho_B = \frac{M_B}{V_B} \tag{1}$$

The grain density (ρ_G) of the sample is calculated using Equation (2):

$$\rho_G = M / \left(\frac{m_1}{\rho_1} + \frac{m_2}{\rho_2} + \ldots + \frac{m_{n-1}}{\rho_{n-1}} + \frac{m_n}{\rho_n} \right) \tag{2}$$

where M is the unit mass of shale gas reservoir sample; m_1, m_2, ... , m_{n-1}, and m_n are the masses of various mineral and organic matter components in terms of unit mass of shale gas reservoir sample; ρ_1, ρ_2, ... , ρ_{n-1}, and ρ_n are the true density values of various minerals and OM; and n is the number of components in the shale gas reservoir sample (Table 1) [31,32]. The true density of the illite-smectite mixed layer was calculated by using the true density values of illite and smectite according to their relative content in the illite-smectite mixed layer.

Table 1. The true density values of various minerals and organic matter (OM).

Density (g/cm^3)	Quartz	Orthoclase	Plagioclase	Calcite	Dolomite	Pyrite
Max true density	2.65	2.57	2.76	2.80	3.20	5.20
Min true density	2.22	2.54	2.61	2.60	2.86	4.90
Average true density	2.44	2.56	2.69	2.70	3.03	5.05

Density (g/cm^3)	Barite	Marcasite	Smectite	Chlorite	Illite	OM
Max true density	4.60	4.90	2.70	3.60	2.90	1.20
Min true density	4.00	4.80	2.00	3.60	2.60	1.20
Average true density	4.30	4.85	2.35	3.60	2.75	1.20

The porosity (Φ_{CCP}) of a shale gas reservoir sample measured by CCP is calculated using Equation (3):

$$\Phi_{CCP} = \frac{V_B - V_G}{V_B} \times 100\% = \frac{M/\rho_B - M/\rho_G}{M/\rho_B} \times 100\% = \frac{\rho_G - \rho_B}{\rho_G} \times 100\% \tag{3}$$

The total porosity measured by CCP is the total porosity, including the contributions of microfractures, microchannels, intercrystal pores, intracrystalline pores, and organopores.

2.4. GRI Method

In the GRI method, the bulk density was calculated using the bulk mass and the bulk volume, and the grain density was calculated using the grain mass and the grain volume. The bulk volume was measured by mercury immersion at less than 6.6 Pa (50 µmHg) using Archimedes' principle on a block sample of approximately 20 g without crushing. The grain volume was measured using He pycnometry and Boyle's law on the crushed sample, which was pretreated in an oven for 12–16 h at 80 °C to drive off any pore fluids such as water, oil and adsorbed gas. The bulk mass and grain mass were measured by balance set-up (Mettler Toledo AL104, readability 0.1 mg).

3. Results

3.1. Composition of Sample

The TOC content of the samples ranged from 0.11% to 4.12% with an average value of 1.58% (Table 2). The δC^{13} of organic carbon ranged from –32.04‰ to –28.78‰, which indicates that the kerogen is type I–II$_1$. Vitrinite reflectance (R_o) ranged from 1.91% to 3.09% with an average value of 2.66%, which indicates that the organic matter is in the stage of high thermal maturity [33]. The mineralogical compositions of the samples are composed of illite and smectite (14.0–49%), quartz (20.3–50.3%), orthoclase + plagioclase (5.9–18.3%) and carbonate minerals (2.3–31.5%, including calcite and dolomite). The content of smectite in illite-smectite mixed-layer ranged from 5% to 15%.

Table 2. The results of total organic content (TOC), maturity, and mineralogical composition of the samples.

Sample No.	TOC (%)	Ro (%)	Quartz	Orthoclase	Plagioclase	Calcite	Dolomite	Aragonite	Pyrite	Barite	Clay	I	C	I/S	%S
1	0.17	2.54	22.2	1.2	6.2	24.2	3.1	0.5	1.4	22.2	41.0	49	17	34	5
2	0.11	2.47	25.7	1.1	5.4	27.3	4.9	1.0	1.9	25.7	32.6	36	18	46	5
3	0.13	2.73	28.3	1.5	5.6	21.7	6.7	2.1	1.9	28.3	32.1	39	20	41	5
4	1.48	2.70	31.1	1.6	8.2	7.8	/	3.0	2.8	31.1	44.0	38	15	47	5
5	1.51	2.83	26.6	2.1	4.7	6.1	3.6	4.1	2.3	26.6	49.0	45	13	42	5
6	1.40	2.78	26.6	1.8	8.6	2.6	3.3	3.7	3.4	26.6	48.6	35	16	49	5
7	0.46	3.09	42.0	2.1	17.2	9.0	10.7	1.4	2.2	42.0	14.9	50	11	39	5
8	0.78	2.75	38.5	5.0	12.3	10.4	6.7	2.9	2.1	38.5	21.3	40	10	50	5
9	1.12	2.48	35.6	3.7	12.8	6.5	3.2	3.5	2.4	35.6	31.2	47	5	48	10
10	0.84	2.55	34.2	3.7	12.2	5.1	3.6	2.2	3.7	34.2	34.5	48	12	40	5
11	0.99	2.31	34.9	3.6	12.0	6.4	4.5	2.3	1.4	34.9	33.9	34	14	52	5
12	1.21	2.81	34.2	4.1	12.7	8.9	6.2	3.2	2.1	34.2	27.4	39	11	50	5
13	1.25	1.91	40.1	4.7	12.6	3.8	4.2	3.2	1.5	40.1	28.7	45	15	40	10
14	1.37	2.90	35.0	3.4	11.0	5.9	2.8	2.7	3.5	35.0	34.3	40	10	50	10
15	1.55	/[1]	34.8	3.3	11.9	5.9	2.0	3.5	1.8	34.8	35.3	48	14	38	10
16	1.12	2.91	39.3	3.9	12.1	3.8	1.9	2.4	4.4	39.3	31.1	40	13	47	10
17	1.21	/	41.2	5.6	14.0	3.4	3.0	2.0	2.0	41.2	27.6	47	12	41	10
18	1.66	2.66	35.2	4.0	11.7	5.3	2.5	1.8	1.7	35.2	36.1	44	13	43	10
19	1.51	2.59	40.2	3.5	11.0	4.1	3.4	1.7	2.9	40.2	31.7	50	11	39	15
20	2.38	2.76	39.6	4.0	6.6	14.7	4.7	5.0	2.4	39.6	20.6	56	7	37	5
21	2.59	2.50	40.3	3.6	7.8	7.2	5.7	5.0	2.1	40.3	25.7	59	9	32	10
22	3.32	2.81	43.4	2.7	6.6	8.5	5.6	5.1	2.5	43.4	22.3	43	14	43	5
23	3.90	2.68	44.8	2.8	7.6	4.8	2.1	3.6	2.9	44.8	27.5	65	9	26	10
24	3.81	2.87	50.5	3.0	6.6	5.5	2.2	4.2	4.8	50.5	19.4	45	8	47	10
25	4.12	2.57	34.3	1.5	10.7	2.3	4.4	/	3.1	34.3	39.6	35	14	51	5
26	2.52	2.79	33.4	5.1	9.8	0.0	6.6	2.2	1.7	33.4	38.7	36	7	57	5
27	0.27	2.49	35.2	2.6	8.4	1.1	2.1	1.1	1.4	35.2	47.8	46	8	46	10

[1] / represents no data. I represents illite. C represents chlorite. I/S represents illite/smectite. %S represents the weight percentage of smectite in the illite/smectite.

3.2. Sample Density

The CCP grain density values of the samples were calculated using Equation (2). The bulk density, GRI grain density, and CCP grain density values ranged from 2.525 to 2.701 g/cm^3, 2.636 to 2.787 g/cm^3, and 2.625 to 2.824 g/cm^3 with average values of 2.622 g/cm^3, 2.728 g/cm^3, and 2.731 g/cm^3, respectively (Figure 2).

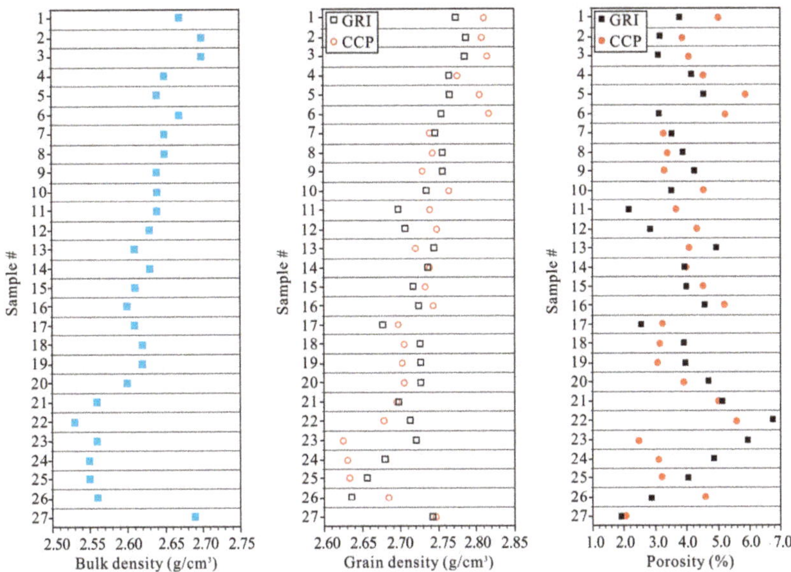

Figure 2. The bulk density, grain density and porosity values of the samples using the composition content porosimetry (CCP) and the Gas Research Institute (GRI) methods.

3.3. Porosity Values from CCP and GRI Methods

The CCP porosity values of the 27 shale gas reservoir samples were calculated using Equation (3). The CCP porosity and GRI porosity values of these samples ranged from 2.05% to 5.87% and 1.90% to 6.75% with average values of 4.04% and 3.91%, respectively (Figure 2). The difference in the maximum and minimum GRI porosity ranged from 0.01 to 1.08% with a mean of 0.40%.

4. Discussion

4.1. Relationship between Compositions and CCP Porosity

The CCP porosity values generally increased with increasing clay and OM content, except for a few samples (Figure 3a,b), and decreased with increasing quartz and feldspar content (Figure 4c,d). In samples with low OM content (<3%), the CCP porosity increased appreciably with total clay content (Figure 3a). CCP porosity also increased with OM content when the total clay content was less than 30% (Figure 3b). The OM and clay content significantly affect pore volume in shale samples. OM is the contributor of organic pores derived from hydrocarbon generation process via consuming OM. For high-maturity shale, the more the OM, the more the organic pores. Clay interparticle pores are among the major contributors to pore volumes with widths <100 nm [25]. Clay swelling may produce natural fractures in the shale sample and create additional pore volume [34]. In samples with low feldspar content (<10%), the total porosity decreased with carbonate content (Figure 3e). The organopores and clay intercrystal pores are important contributors to the total porosity due to the high thermal maturity of the Longmaxi shale [33,35]. Therefore, when the content of one of the two compositions is low, the porosity values increase obviously with the other composition content. The quartz and feldspar content increases with decreasing clay and OM content. Hence, the porosity values decrease with increasing quartz and feldspar content (Figure 3c,d).

4.2. Comparison of the Results of CCP and GRI Methods

To evaluate the precision of the CCP method, the CCP results were compared with those of the GRI method. The differences between the CCP and GRI results ranged from less than 0.01% to 3.27%. The trends in CCP and GRI porosity values were nearly the same for sample 1 to sample 27 (Figure 2). Several trends were identified in the comparison of CCP and GRI results, including grain density and porosity. The grain density values measured by the CCP and GRI methods declined with increasing OM and quartz content (Figure 4a,c), and increased with increasing carbonate content (Figure 4e). In the sample with high OM content, CCP grain density was lower than GRI grain density (Figure 4a). This behavior was the opposite in the sample with high total clay mineral content, low quartz content, and high carbonate content (Figure 4b,c,e). Grain density did not demonstrate any clear relationship with the feldspar content (Figure 4d). The CCP porosity values were larger than the GRI results in the sample with low OM and quartz content and high total clay mineral content, and were smaller than the GRI results in other samples with high OM and quartz content and low total clay mineral content (Figure 5a–c). The CCP porosity values did not exhibit a clear relationship between the feldspar content or carbonate content (Figure 5d,e).

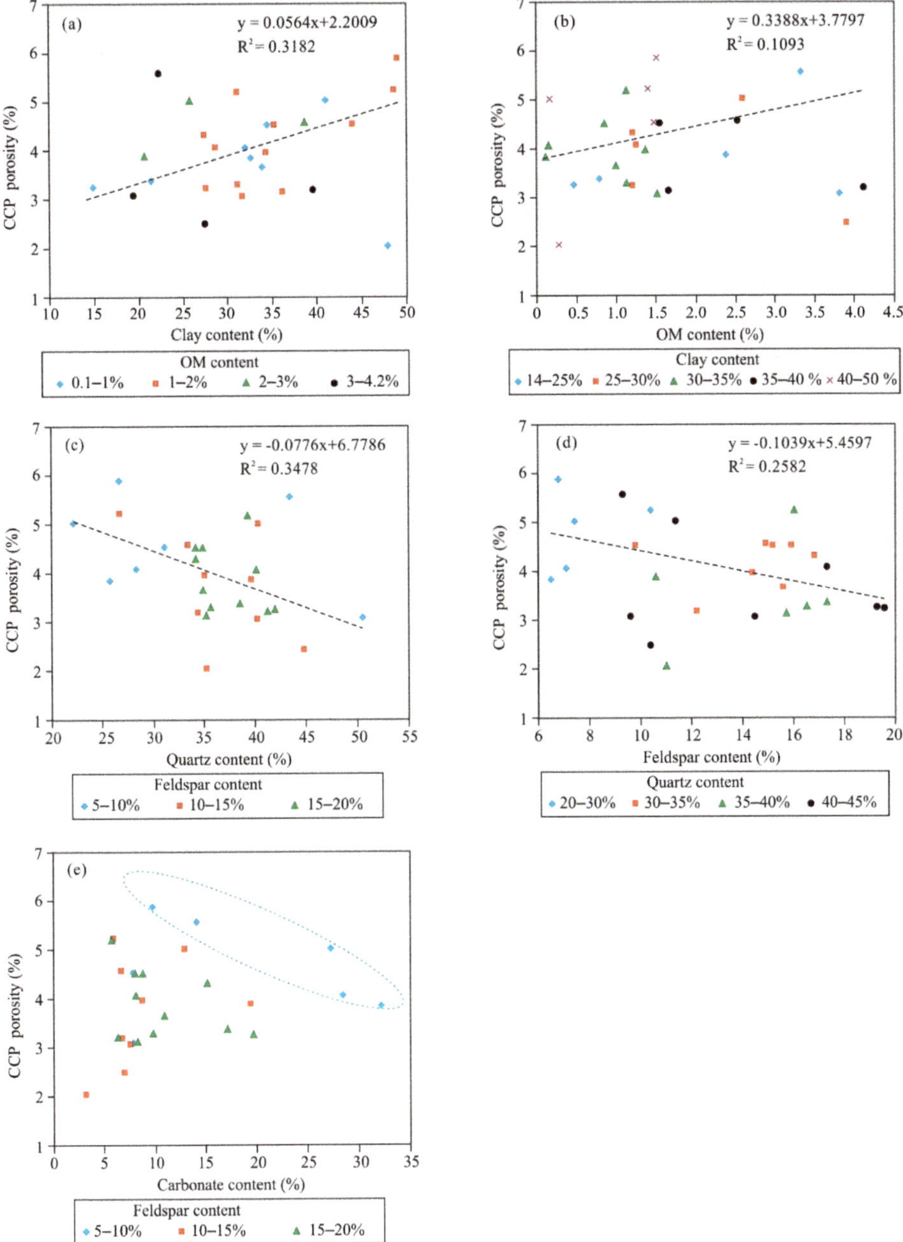

Figure 3. The relationships between CCP porosity values and composition: (**a**) clay content, (**b**) organic matter content, (**c**) quartz content, (**d**) feldspar content, (**e**) carbonate content.

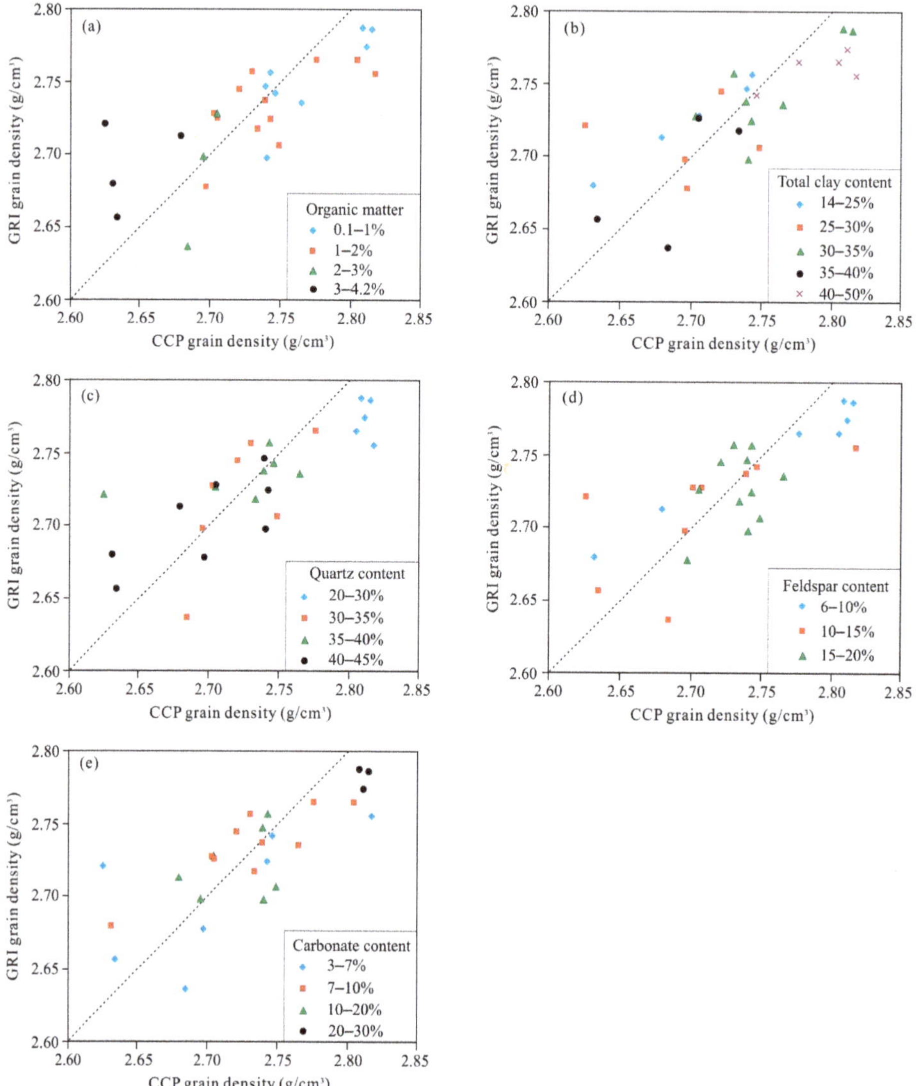

Figure 4. Comparison of the CCP grain density and GRI grain density values (variance $\sigma^2 = 0.0012$): (**a**) clay content, (**b**) organic matter content, (**c**) quartz content, (**d**) feldspar content, (**e**) carbonate content.

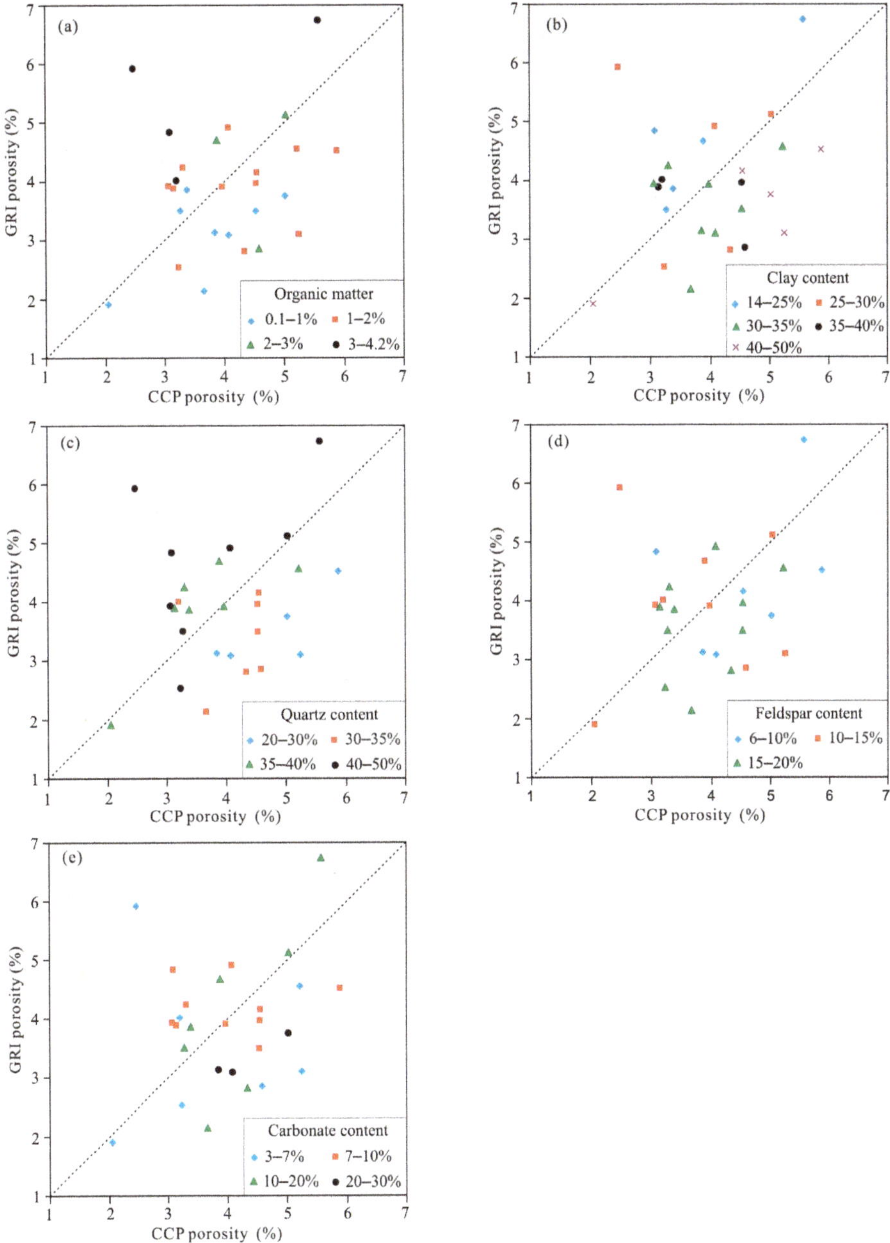

Figure 5. Comparison of the CCP porosity and GRI porosity values (variance $\sigma^2 = 0.0143$): (**a**) clay content, (**b**) organic matter content, (**c**) quartz content, (**d**) feldspar content, (**e**) carbonate content.

4.3. Advantages of the CCP Method

The benefits of the CCP method include: (1) no standard core plug (2.5 cm diameter) is required; (2) it is based on some conventional measurements such as XRD analysis and organic carbon content analysis; and (3) the total porosity of shale gas reservoir samples can be measured, including the unconnected pores. Volumetric measurements based on gas are not only affected by leaks and temperature variation but also overestimate pore volume due to gas adsorbed by the organic-rich shale gas reservoir sample. The bulk density calculated from the bulk mass and bulk volume can be well controlled in the CCP methodology. The content of various minerals measured by XRD are less affected by different measurement environments. The CCP experiment is operable and credible, which makes it applicable to measuring porosity in shale gas reservoir samples. In addition, compared with other penetration fluid methods, such as GRI porosity, LTNA, MIP, and MNR methods, the CCP method is less affected by fluid (gas or water) adsorption. For the shale reservoir samples, the porosities derived from the penetration fluid methods are apparent porosities because of adsorption of OM and clay, as well as the permeability behaviors [36–38].

4.4. Disadvantages of the CCP Method

There are two disadvantages of the CCP method. The first is that XRD analysis only measures common minerals such as quartz, feldspars, and carbonates. It ignores uncommon minerals in the samples because of their low content. The other disadvantage is that the true density values of various minerals used in this paper are the average values of their maximum and minimum true density values. Although the differences between the average true density values and the actual density values are less than 0.35 g/cm^3, the precision of the CCP result will be affected. The largest differences are due to the true density values of smectite and barite.

5. Conclusions

The results suggest that the CCP porosity values of shale gas reservoir samples range between 2.05% and 5.87% with an average value of 4.04%. The CCP method does not require a standard core plug and is based on conventional experiments. The CCP porosity generally increases with increasing OM and clay content, and decreases with increasing quartz and feldspar content. The CCP results were similar to the GRI results, and the differences between the two methods range from 0.05% to 1.52% with an average value of 0.85%. The disadvantages of the CCP method include ignoring content of uncommon minerals and the uncertainty in the true density values of various minerals.

Author Contributions: Conceptualization, F.C. and S.L.; methodology, F.C. and Y.J.; software, X.D.; validation, F.C., S.L., X.D., H.Z. and Y.J.; formal analysis, F.C. and H.Z.; investigation, F.C.; resources, F.C.; data curation, F.C. and H.Z.; writing—original draft preparation, F.C.; writing—review and editing, S.L. and Y.J.; visualization, X.D.; supervision, S.L.; project administration, S.L.; funding acquisition, F.C. and Y.J.

Funding: This research was funded by the Open Fund of State Key Laboratory of Oil and Gas Reservoir Geology and Exploitation (Chengdu University of Technology) (grant number PLC20180202), the National Natural Science Foundation of China (grant number 41530315), the National Science and Technology Major Project of China (grant number 2016ZX05061), and the Fundamental Research Funds for the Central Universities (grant number 18CX02071A).

Conflicts of Interest: The authors declare no conflict of interest.

References

1. Kuila, U.; McCarty, D.K.; Arkadiusz, D.; Fischer, T.B.; Prasad, M. Total porosity measurement in gas shales by the water immersion porosimetry (WIP) method. *Fuel* **2014**, *117*, 1115–1129. [CrossRef]
2. Pan, Z.J.; Ma, Y.; Connell, L.D.; Down, D.I.; Camilleri, M. Measuring anisotropic permeability using a cubic shale sample in a triaxial cell. *J. Nat. Gas Sci. Eng.* **2015**, *26*, 336–344. [CrossRef]
3. Ma, Y.; Pan, Z.J.; Zhong, N.N.; Connell, L.D.; Down, D.I.; Lin, W.L.; Zhang, Y. Experimental study of anisotropic gas permeability and its relationship with fracture structure of Longmaxi Shales, Sichuan Basin, China. *Fuel* **2016**, *180*, 106–115. [CrossRef]
4. Wang, L.; Wang, S.H.; Zhang, R.L.; Wang, C.; Xiong, Y.; Zheng, X.S.; Li, S.R.; Jin, K.; Rui, Z.H. Review of multi-scale and multi-physical simulation technologies for shale and tight gas reservoirs. *J. Nat. Gas Sci. Eng.* **2017**, *37*, 560–578. [CrossRef]

5. Jiang, F.J.; Chen, D.; Wang, Z.F.; Xu, Z.Y.; Chen, J.; Liu, L.; Huyan, Y.Y.; Liu, Y. Pore characteristic analysis of a lacustrine shale: A case study in the Ordos Basin, NW China. *Mar. Petrol. Geol.* **2016**, *73*, 554–571. [CrossRef]

6. Sun, J.M.; Dong, X.; Wang, J.J.; Schmitt, D.R.; Xu, C.L.; Mohammed, T.; Chen, D.W. Measurement of total porosity for gas shales by gas injection porosimetry (GIP) method. *Fuel* **2016**, *186*, 694–707. [CrossRef]

7. Zolfaghari, A.; Dehghanpour, H.; Xu, M.X. Water sorption behaviour of gas shales: II Pore size distribution. *Int. J. Coal Geol.* **2017**, *179*, 187–195. [CrossRef]

8. Chen, F.W.; Lu, S.F.; Ding, X. Organoporosity evaluation of shale: A case study of the Lower Silurian Longmaxi Shale in Southeast Chongqing, China. *Sci. World J.* **2014**, 1–9. [CrossRef]

9. Tuo, J.C.; Wu, C.J.; Zhang, M.F. Organic matter properties and shale gas potential of Paleozoic shales in Sichuan Basin, China. *J. Nat. Gas Sci. Eng.* **2016**, *28*, 434–446. [CrossRef]

10. Zhang, H.; Zhu, Y.M.; Wang, Y.; Kang, W.; Chen, S.B. Comparison of organic matter occurrence and organic nanopore structure within marine and terrestrial shale. *J. Nat. Gas Sci. Eng.* **2016**, *32*, 356–363. [CrossRef]

11. Chen, F.W.; Ding, X.; Lu, S.F. Organic porosity evaluation of lower cambrian Niutitang shale in Qiannan depression, China. *Petrol. Sci. Technol.* **2016**, *34*, 1083–1090. [CrossRef]

12. Nelson, P.H. Pore-throat sizes in sandstones, tight sandstones, and shales. *AAPG Bull.* **2009**, *93*, 329–340. [CrossRef]

13. Shi, M.; Yu, B.S.; Xue, Z.P.; Wu, J.S.; Yuan, Y. Pore characteristics of organic-rich shales with high thermal maturity: A case study of the Longmaxi gas shale reservoirs from well Yuye-1 in southeastern Chongqing, China. *J. Nat. Gas Sci. Eng.* **2015**, *26*, 948–959. [CrossRef]

14. Chen, F.W.; Ding, X.; Lu, S.F.; Ju, Y.W. The pore characterization of the lower silurian Longmaxi shale in the southeast Chongqing, China using FE-SEM, LTNA and MIP methods. *J. Nanosci. Nanotechnol.* **2017**, *17*, 6482–6488. [CrossRef]

15. Lu, S.F.; Chen, F.W.; Ju, Y.W. Type and quantitative evaluation of micropores in Longmaxi shale of southeast Chongqing, China. *J. Nanosci. Nanotechnol.* **2017**, *17*, 7035–7043. [CrossRef]

16. Chen, F.W.; Lu, S.F.; Ding, X.; He, X.P.; Xing, H.L. The splicing of backscattered scanning electron microscopy method used on evaluation of microscopic pore characteristics in shale sample and compared with results from other methods. *J. Petrol. Sci. Eng.* **2018**, *160*, 207–218. [CrossRef]

17. Clarkson, C.R.; Solano, N.; Bustin, R.M.; Bustin, A.M.M.; Chalmers, G.R.L.; Hec, L.; Melnichenko, Y.B.; Radlin'ski, A.P.; Blach, T.P. Pore structure characterization of north American shale gas reservoirs using USANS/SANS, gas adsorption, and mercury intrusion. *Fuel* **2019**, *103*, 606–616. [CrossRef]

18. Chen, F.W.; Lu, S.F.; Ding, X. Pore types and quantitative evaluation of pore volumes in the Longmaxi shale of southeast Chongqing, China. *Acta Geol. Sin.-Engl.* **2018**, *92*, 342–353. [CrossRef]

19. Slatt, R.M.; O'Brien, N.R. Pore types in the Barnett and Woodford gas shales: Contribution to understanding gas storage and migration pathways in fine-grained rocks. *AAPG Bull.* **2011**, *95*, 2017–2030. [CrossRef]

20. Curtis, M.E.; Cardott, B.J.; Sondergeld, C.H.; Rai, C.S. Development of organic porosity in the Woodford shale with increasing thermal maturity. *Int. J. Coal Geol.* **2012**, *103*, 26–31. [CrossRef]

21. Curtis, M.E.; Sondergeld, C.H.; Ambrose, R.J.; Rai, C.S. Microstructural investigation of gas shales in two and three dimensions using nanometer-scale resolution imaging. *AAPG Bull.* **2012**, *96*, 665–677. [CrossRef]

22. Long, H.; Swennen, R.; Foubert, A.; Dierick, M.; Jacobs, P. 3D quantification of mineral components and porosity distribution in Westphalian C sandstone by microfocus X-ray computed tomography. *Sediment. Geol.* **2009**, *220*, 116–125. [CrossRef]

23. Okolo, G.N.; Everson, R.C.; Neomagus, H.W.J.P.; Roberts, M.J.; Sakurovs, R. Comparing the porosity and surface areas of coal as measured by gas adsorption, mercury intrusion and SAXS techniques. *Fuel* **2015**, *141*, 293–304. [CrossRef]

24. Li, J.; Wang, W.; Cao, Q.; Shi, Y.; Yan, X.; Tian, S. Impact of hydrocarbon expulsion efficiency of continental shale upon shale oil accumulations in eastern China. *Mar. Pet. Geol.* **2015**, *59*, 467–479. [CrossRef]

25. Chen, F.W.; Zhao, H.Q.; Lu, S.F.; Ding, X.; Ju, Y.W. The effects of composition, laminar structure and burial depth on connected pore characteristics in a shale oil reservoir, the Raoyang Sag of the Bohai Bay Basin, China. *Mar. Pet. Geol.* **2019**, *101*, 290–302. [CrossRef]

26. Kuila, U.; Prasad, M. Specific surface area and pore-size distribution in clays and shales. *Geophys. Prospect.* **2013**, *61*, 341–362. [CrossRef]

27. Lin, B.T.; Chen, M.A.; Jin, Y.; Pang, H.W. Modeling pore size distribution of southern Sichuan shale gas reservoirs. *J. Nat. Gas Sci. Eng.* **2015**, *26*, 883–894. [CrossRef]

28. Luffel, D.L.; Guidry, F.K.; Curtis, J.B. Evaluation of devonian shale with new core and log analysis methods. *J. Petrol. Technol.* **1992**, *44*, 1192–1198. [CrossRef]
29. Omotoso, O.; McCarty, D.K.; Hillier, S.; Kleeberg, R. Some successful approaches to quantitative mineral analysis as revealed by the 3rd Reynolds Cup contest. *Clays Clay Miner.* **2006**, *54*, 748–760. [CrossRef]
30. Chalmers, G.R.L.; Bustin, R.M. Lower Cretaceous gas shales in northeastern British Columbia, Part I: Geological controls on methane sorption capacity. *Bull. Can. Petrol. Geol.* **2008**, *56*, 1–21. [CrossRef]
31. Pan, Z.L. *Crystallography and Mineralogy*, 3rd ed.; Geological Press: Bejing, China, 2006; pp. 130–174.
32. Okiongbo, K.S.; Aplin, A.C.; Larter, S.R. Changes in type II kerogen density as a function of maturity: Evidence from the Kimmeridge Clay Formation. *Energy Fuels* **2005**, *19*, 2495–2499. [CrossRef]
33. Löhr, S.C.; Baruch, E.T.; Hall, P.A.; Kennedy, M.J. Is organic pore development in gas shales influenced by the primary porosity and structure of thermally immature organic matter? *Org. Geochem.* **2015**, *87*, 119–132. [CrossRef]
34. Zolfaghari, A.; Dehghanpour, H.; Holyk, J. Water sorption behaviour of gas shales: I. Role of clays. *Int. J. Coal Geol.* **2017**, *179*, 130–138. [CrossRef]
35. Guo, H.J.; Jia, W.L.; Peng, P.A.; Zeng, J.; He, R.L. Evolution of organic matter and nanometer-scale pores in an artificially matured shale undergoing two distinct types of pyrolysis: A study of the Yanchang Shale with Type II kerogen. *Org. Geochem.* **2017**, *105*, 56–66. [CrossRef]
36. Jia, B.; Tsau, J.S.; Barati, R. A workflow to estimate shale gas permeability variations during the production process. *Fuel* **2018**, *220*, 879–889. [CrossRef]
37. Cui, X.; Bustin, A.M.; Bustin, R.M. Measurements of gas permeability and diffusivity of tight reservoir rocks: Different approaches and their applications. *Geofluids* **2009**, *9*, 208–223. [CrossRef]
38. Jia, B.; Tsau, J.S.; Barati, R. Different flow behaviors of low-pressure and high-pressure carbon dioxide in shales. *SPE J.* **2018**, *23*, 1452–1468. [CrossRef]

Article

Nano-Scale Pore Structure and Fractal Dimension of Longmaxi Shale in the Upper Yangtze Region, South China: A Case Study of the Laifeng–Xianfeng Block Using HIM and N$_2$ Adsorption

Cheng Huang, Yiwen Ju *, Hongjian Zhu, Yu Qi, Kun Yu, Ying Sun and Liting Ju

Key Laboratory of Computational Geodynamics, College of Earth and Planetary Sciences, University of Chinese Academy of Sciences, Beijing 100049, China; huangcheng150@126.com (C.H.); zhj8641@163.com (H.Z.); qiuqiuyu911@163.com (Y.Q.); yukun@cumt.edu.cn (K.Y.); sunyinglytwy2008@126.com (Y.S.); jlt@ucas.ac.cn (L.J.)
* Correspondence: juyw03@163.com

Received: 30 April 2019; Accepted: 8 June 2019; Published: 12 June 2019

Abstract: This paper tries to determine the key evaluation parameters of shale reservoirs in the complex tectonic provinces outside the Sichuan Basin in South China, and also to target the sweet spots of shale reservoirs accurately. The pore-structure characteristics of the Lower Silurian Longmaxi shale gas reservoirs in Well LD1 of the Laifeng–Xianfeng Block, Upper Yangtze region, were evaluated. N$_2$ adsorption and helium ion microscope (HIM) were used to investigate the pore features including pore volume, pore surface area, and pore size distribution. The calculated results show good hydrocarbon storage capacity and development potential of the shale samples. Meanwhile, the reservoir space and migration pathways may be affected by the small pore size. As the main carrier of pores in shale, organic matter contributes significantly to the pore volume and surface area. Samples with higher total organic carbon (TOC) content generally have higher porosity. Based on the Frenkel–Halsey–Hill equation (FHH model), two different fractal dimensions, D_1 and D_2, were observed through the N$_2$ adsorption experiment. By analyzing the data, we found that large pores usually have large values of fractal dimension, owing to their complex pore structure and rough surface. In addition, there exists a good positive correlation between fractal dimension and pore volume as well as pore surface area. The fractal dimension can be taken as a visual indicator that represents the degree of development of the pore structure in shale.

Keywords: Longmaxi Formation shale; pore structure; N$_2$ adsorption; helium ion microscope; fractal dimension

---

1. Introduction

According to the trends of the global petroleum industry, hydrocarbon exploration is mainly focused on marine deep water, onshore deep layer, and unconventional oil and gas [1]. Natural gas is well known as the cleanest fossil fuel by global countries, compared with oil and coal. To reduce exploration risk and determine economic feasibility, considerable efforts have been undertaken to improve the knowledge of gas storage and transport mechanisms.

Shale was previously thought to be either impossible or uneconomic in regard to having industrial capacity. However, the combination of horizontal drilling and hydraulic fracturing can extract huge quantities of natural gas from impermeable shale formations, which has made it one of the landmark achievements in the 21st century [2]. A shale gas reservoir is characterized as a self-contained source reservoir system. Abundant gas can be stored as free gas in intergranular porosities and natural fractures, as adsorbed gas in organic matter and clay particle surfaces, or dissolved in kerogen and bitumen [3,4]. Thus, a series of studies have been published over the past decade about the pore-structure characteristics of shale, including its shape, size, porosity, and connectivity [5–10].

From the Precambrian to Tertiary periods, organic-rich shale deposited in marine, transitional marine, or lacustrine settings, and is widely distributed in China [11]. Related researches have expounded the depositional environment, geochemical and reservoir characteristics, gas concentration, and prospective resource potential of the three different types of shale in China [12–16]. The Upper Yangtze Platform, where the Sichuan Basin is located, is one of the largest conventional natural gas provinces of China. Beyond that, the reported unconventional reserves, mainly in the lower Silurian and lower Cambrian shale formations, are significantly higher than the total reserves of conventional petroleum [17]. Among them, the Wufeng–Longmaxi formation, especially the bottom of the Longmaxi formation, is the leading target of shale gas development in the Sichuan Basin, with a large amount of natural gas produced from this over-mature marine shale [18–21]. Many scholars have been making studies on the Longmaxi formation in the Sichuan Basin, focusing on its south and east parts, i.e., southeastern Sichuan province and southeastern Chongqing province. However, there are few researches about Longmaxi shale in the western Hubei Province [22], which is outside the eastern margin of the Sichuan Basin, though a series of shale gas exploration wells were drilled in recent years. The study of this region can provide experience and guidance for the exploration and exploitation of complex tectonic deformation areas of marine shale in south China.

Many high-resolution techniques have been applied to study the pore structures in shale with low porosity and low permeability, large parts of which are nanoscale pores (pore diameter of less than 100 nm). Based on previous studies, all of the following methods have achieved good results: Scanning electron microscope (SEM), Nano-CT, helium ion microscope (HIM), mercury intrusion, gas adsorption, nuclear magnetic resonance (NMR), and small-angle and ultra-small-angle neutron scattering (SANS and USANS) [8,23–29]. In particular, SEM and N_2 adsorption are the most frequently used techniques, and both meet the demands of direct observation and quantitative analysis [30]. However, in this study, we choose HIM and N_2 adsorption to characterize the pore structures from several nanometers to a few microns. HIM has better capacity for identifying small pores with higher magnification and depth of field. This is not just for the sake of full-scale characterization, but also for mutual authentication of the results by using the two methods. Moreover, to get more quantitative features and to characterize the pore geometry more intuitively, fractal theory was applied in the data processing of this paper.

Fractal theory, which is used to describe the geometric and structural properties of a solid surface [31], is an important tool for evaluating surface roughness. The fractal behavior is associated with power law behavior for a number of features, as a function of the feature size on the pore–rock interface [32]. It has already been used to investigate either the permeability or surface appearance of coal and shale samples [33,34]. The fractal characteristics of shale and coal collected from China have been calculated and analyzed through some experiments like mercury porosimetry, N_2 gas adsorption, NMR, small-angle X-ray scattering (SAXS), and SEM digital images [35–40].

2. Geological Setting

Shale gas resources in China are mainly distributed in the southern Paleozoic marine shale, with significant geologic challenges such as strong structural deformation, high tectonic stress, big burial depth, slow drilling speed in hard formations, and complex surface conditions [41–43]. Located in the northwest side of the Yangtze platform, the Sichuan Basin is a typical superimposed basin with multiple periods of geologic structures [44]. Both conventional gas and unconventional gas with huge resource potential have been found within the Sichuan Basin. Most of the hydrocarbon-source rocks are from the Lower Palaeozoic system or of even older strata [44,45]. In recent years, many research results about the macro- and micro-structures of shale have emerged, aiming at the Ordovician Wufeng Formation–Lower Silurian Longmaxi Formation, and the Lower Cambrian Qiongzhusi Formation and Niutitang Formation in the Weiyuan and Fuling shale gas fields [17,45–48].

Our research area, the Laifeng–Xianfeng Block, is located at the thick fault–fold zone in western Hubei Province, in the Upper Yangtze region (Figure 1). This fault–fold zone borders the Sichuan Basin to the west, Qinling–Dabie Orogen to the north, and the Jiangnan–Xuefeng thrust uplift belt to the east. During the period of Late Ordovician–Early Silurian, this area was a clastic continental shelf sedimentary environment, showing a trend where water gradually became shallower and the depositional thickness got thinner, from the west to the east. After that, mainly because of the Yanshan movement, the fold and fault structure formed along the northeast–southwest direction [49,50].

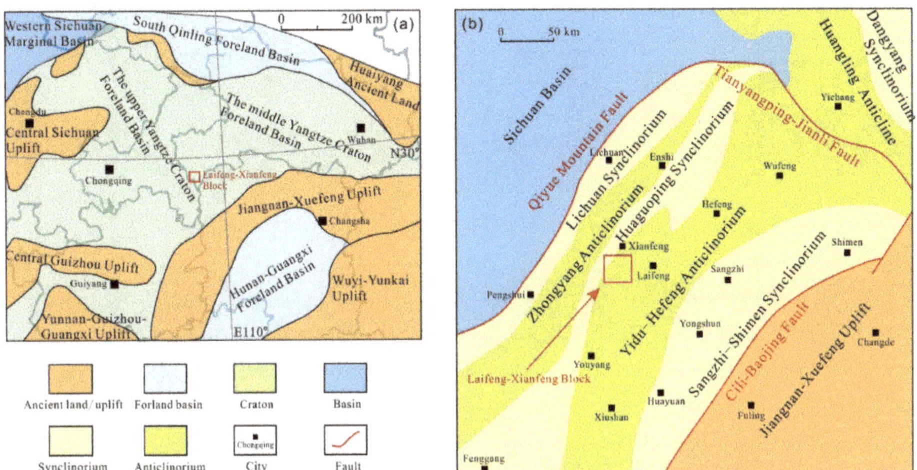

Figure 1. The geologic aspects around the research area. (**a**) Palaeogeographic pattern around the Laifeng–Xianfeng Block in the Sinian-Lower Paleozoic system (modified after [51]). (**b**) Regional tectonic pattern around the Laifeng–Xianfeng Block (modified after [49]).

The Laifeng–Xianfeng Block is the key area of shale gas exploration in the Middle and Upper Yangtze region, classified as the deformation zone outside basin [52]. It mainly shows the tectonic framework of wide-spaced anticlines, due to the influence of later tectonic movement [42,49,53]. The main exploration target strata of this area are the Upper Ordovician Wufeng Formation and the Lower Silurian Longmaxi Formation. Several shale gas production areas with economic value in China are mainly located around the Sichuan Basin. In addition, a large number of high-quality shale gas reservoirs have been discovered in southeast Chongqing province in recent years, just next to our research area. Though not far to the east, the Laifeng–Xianfeng Block has entirely different geological features from the Fuling shale gas field, the most successful commercial gas production area in China. The Longmaxi Formation, which belongs to the marginal area of deep-sea shelf facies of Silurian and with current burial depth of less than 2500 m, was taken as the research object of this paper.

Well LD1 is located in Enshi, in southwestern Hubei province (Figure 1). The drilling stratum of the Longmaxi Formation is 51 m thick, with abundant graptolite fossil and considerable desorption capacity. Horizontal bedding in low-energy environments and thin-plate pyrite in anoxic reductive sedimentary environments are widely developed in this formation. From top to bottom, the Longmaxi Formation can be roughly divided into three sections according to the lithology (Figure 2): Gray-black clay shale, dark gray massive pelitic siltstone, and gray-black carbonaceous clay shale interbedded with black carbonaceous siliceous shale.

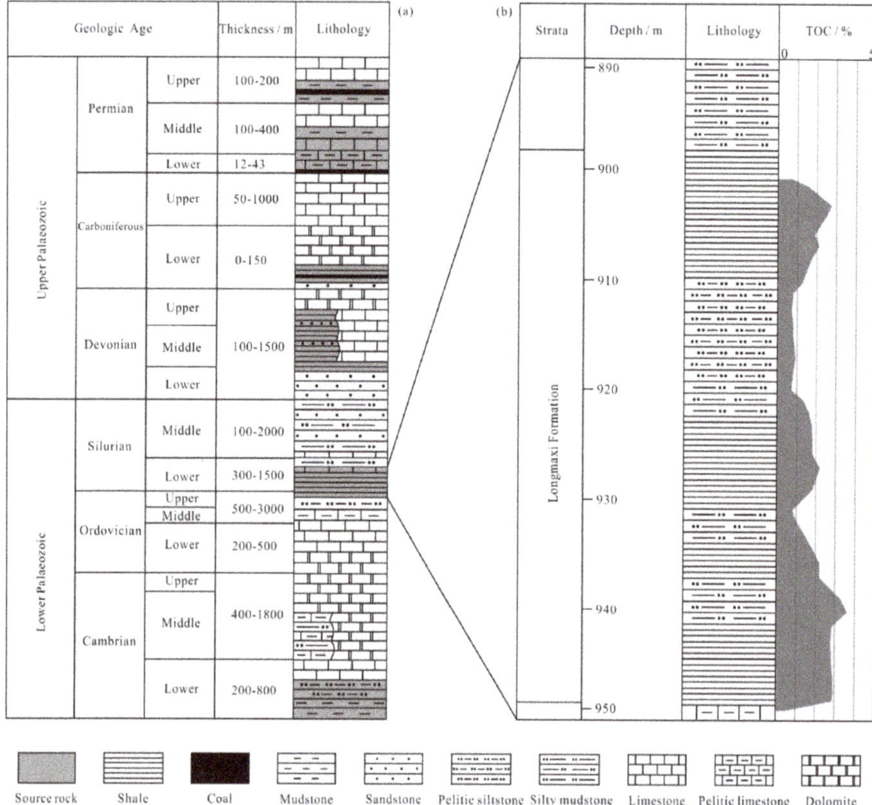

Figure 2. Synthetical stratum histogram of the research area. (**a**) The simplified stratigraphic column and the location of several high-quality source rocks of South China during the Paleozoic period (modified after [12,51]). (**b**) The lithological column and the change curve of measured total organic carbon (TOC) values of the Longmaxi Formation in Well LD1 (modified after [54]).

3. Samples and Methods

In this study, a total of five core samples were collected from Well LD1. The basic information and mineral composition characteristics of these samples are listed in Table 1. The sets of samples were taken at intervals, with the depth ranging from 904 to 944 m. Each of them was cut into several parts for analyzing the mineral composition, organic matter features, and pore structures.

Table 1. The depth, TOC content, and mineral composition (all in wt %) of all the samples in Well LD1.

Sample ID	Depth/m	TOC	Quartz	K-Feldspar	Anorthose	Calcite	Dolomite	Pyrite	Clay
L-1	904.8	2.14	46.6	1.5	6.9	1.2	/	6.0	37.8
L-2	915.3	0.80	48.9	3.9	16.7	4.7	6.5	2.0	17.3
L-3	933.2	1.21	48.5	4.4	12.4	5.2	4.1	2.1	23.3
L-4	943.4	2.37	48.9	3.0	13.7	2.7	2.4	3.9	25.4
L-5	943.8	1.98	47.5	0.4	1.3	7.4	2.4	2.9	38.1

3.1. Mineral and Organic Composition

All the samples were tested for mineral composition by X-ray diffraction analyses using a Rigaku Smartlab Multifunction X-ray Diffractometer (Rigaku, University of Chinese Academy of Sciences,

Beijing). Crushed samples (180–250 μm) were mixed with ethanol, hand ground in a mortar and pestle, and then smear mounted on glass slides for X-ray diffraction analyses. A semi-quantitative estimation of the mineral contents of samples was determined using Reitveld analyses [55], which fits a polynomial curve to the diffractograms [6].

The organic geochemical characteristics were determined by a Rock-Eval II apparatus with a total organic carbon (TOC) module. The value of TOC is calculated from the amount of CO_2 evolved during hydrocarbon generation and during oxidation at 650 °C.

3.2. Helium Ion Microscope (HIM)

To obtain more subtle features, and to verify the results of N_2 adsorption, two of these samples were tested by a helium ion microscope. This kind of instrument is a good addition to a scanning electron microscope, with its excellent resolution for imaging nanoscale pores [25]. Same as the preparation process for scanning electron microscope (SEM), the small block sample was polished to create a level surface using dry emery paper combined with argon ions. However, it is not necessary to spray carbon or metal on the surface to enhance electrical conductivity. This is to exclude the impact of carbon or metal particles on the observation of pores, since some of the pores are as small as a few nanometers wide and are easily covered by nanoparticles. HIM has higher resolution than SEM, which can make nanopores in shales more clearly observed. However, without the sprayed carbon or metal coating, the samples have extremely poor conductivity. As a result, the gray value in the image generally represents the electrical conductivity of the material. Namely, the bright area means organic matter, pyrite, and most clay minerals, while the dark area in the image usually represents quartz, feldspar, and calcite, which are different from SEM images. The shale samples were analyzed using an Orion NanoFab FIB-HIM produced by Carl Zeiss (Institute of Geology and Geophysics, Beijing). The instrument was operated at an acceleration voltage of 35 kV, a beam current set at 0.4–0.55 pA, and a scan dwell time of 1 μs.

3.3. Ultra-Low N_2 Adsorption

We conducted ultra-low N_2 adsorption analyses to measure the pore diameters between 0.3 and 200 nm [9]. Before the experiments, crushed samples (180–250 μm) needed to be oven dried and degassed at 110 °C for 24 h. A Quantachrome Autosorb iQ instrument (Quantachrome Ins, Changzhou University, Changzhou) was used to obtain the N_2 adsorption/desorption curves at 77.35 K. On the basis of the classification of the International Union of Pure and Applied Chemistry (IUPAC), the pores were split into three categories according to the diameter: Micropores (<2 nm), mesopores (between 2 and 50 nm), and macropores (>50 nm). The adsorption data at ultra-low pressure can provide abundant information about pore characteristics within the range of micropores (<2 nm). Previous studies have proposed combining CO_2 adsorption and N_2 adsorption experiments to characterize the pore structure on a wide range of pore diameters. CO_2 adsorption can be used to investigate micropore volume, and N_2 adsorption can be used to investigate pore volume from mesopores to macropores [8,56]. In this study, for a lower detection line, a molecular pump coupled with a diaphragm roughing pump was used for outgassing to make sure that the vacuum degree can reach $<1 \times 10^{-7}$ Pa [57]. Thus, based on the Barrett–Joyner–Halenda (BJH) and Brunauer–Emmett–Teller (BET) methods, pore volume and pore surface area within a wide range of sizes (from micropores to macropores) were calculated through analysis of N_2 adsorption data.

4. Results and Analysis

4.1. Mineral Constituents and TOC Content

The mineral constituents, TOC content, and depth of all the five samples collected from Well LD1 are shown in Table 1. The results suggest that all the shales samples have similar mineral composition that take quartz and clay as the basis. Among them, quartz, K-feldspar, and anorthose, usually classified as the siliceous part, account for about 60%. Although there exist some differences between the samples regarding the percentage of clay minerals (from 17.3% to 38.1%), these still have the second highest content of all the minerals. The TOC contents, calculated by rock pyrolysis, vary from 0.8% to 2.37%, with a mean value of 1.70%. Considering the exploration and development horizon of shale gas, the value means that there is certain potential. Seeing that organic matter is less dense than the whole rock and the aforesaid TOC content is expressed as a mass fraction, the volume percent of organic matter will be larger. The value of Ro is generally greater than 2.0% [49], which indicates that the organic matter of the shale is in the stage of high- to over-maturity. This corresponds to the late diagenetic epoch of Longmaxi shale, which means that most interparticle pores may be pressed or filled to be smaller or even disappear in the long-term subsidence and reservoir diagenesis process.

4.2. Nano-Scale Pore Structure by N_2 Adsorption Isotherms

4.2.1. N_2 Adsorption/Desorption Curves

Previous researches usually measured the N_2 adsorption isotherms for the relative pressure (P/P_o), starting from 0.01 [34,35]. Wang and Ju [57] classified this as conventional low-pressure N_2 physisorption, and the relative pressure range of 10^{-7}–10^{-2} was named "ultra-low pressure". In this study, we used the same experimental procedure as [57]. It can be seen from the local enlarged view of the adsorption curves that the initial relative pressure can achieve the order of 10^{-7} (Figure 3). Although the part of relative pressure that is less than 10^{-2} is very narrow in the ordinary coordinate, in fact, the data points are very dense. They account for about one third of the total adsorption data.

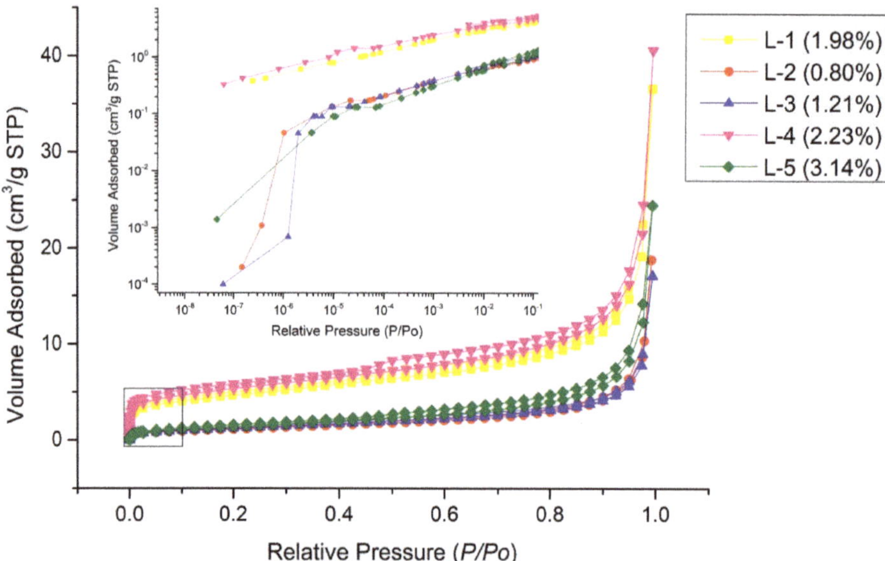

Figure 3. Low-pressure N_2 adsorption isotherms of samples, and a partial enlargement of their low relative pressure section (the area enclosed by a box). P and P_o means equilibrium pressure and gas saturation pressure, respectively. Notice that the logarithmic X-axis was used to characterize the differences more clearly between each sample.

All the shale samples have similar adsorption/desorption curves through data analysis (Figure 3). They are more likely to adjust to the isotherms of type H3 and type H4, according to the IUPAC classification, which means that the pores filled with N_2 are slit-shaped or ink-bottle-like. In addition, the volume adsorptions of N_2 show a clear gap between the five samples. They show no correlation with depth, but correlate with TOC content of the layer. Samples with higher TOC contents (L-1 = 2.14%, L-4 = 2.37%) have larger adsorption capacity. Meanwhile, L-2, L-3, and L-5, with lower TOC contents of 0.8%, 1.21%, and 1.98%, respectively, have relatively small adsorption capacities.

4.2.2. Pore Volume and Surface Area

In general, small pores have large specific surface area. The results listed in Table 2 show that all the samples have huge pore surface area, which means great hydrocarbon adsorption potential. Corresponding to the high adsorption amounts of the adsorption curves, L-1 and L-4 have the largest pore volume and surface area calculated from the Barrett–Joyner–Halenda (BJH) and Brunauer–Emmett–Teller (BET) methods, respectively. In terms of pore volume, these two samples are far ahead of the others, showing good reservoir space. In respect to pore surface area, L-1 and L-4 can have values more than twice those of L-2, which has the smallest specific surface area. In addition, the pore volume and surface area of each sample shows general corresponding relations. That is, a sample with large pore volume also has large surface area. With the same volume, the specific surface areas of micropores and mesopores will be much more than those of macropores. Thus, this kind of corresponding relation illustrates that although they are collected from different depths of the Longmaxi Formation, the series of samples have similar pore size proportions (Figure 4).

Table 2. Pore parameters of samples calculated from N_2 adsorption data.

Sample ID	V_{BJH} (cm^3/g)	S_{BET} (m^2/g)	Average Pore Diameter (nm)
L-1	0.052	16.62	13.7
L-2	0.029	4.286	27.2
L-3	0.026	4.782	22.2
L-4	0.058	18.96	13.3
L-5	0.038	5.492	27.6

V_{BJH} = pore volume; S_{BET} = pore surface area; BJH = Barrett–Joyner–Halenda; BET = Brunauer–Emmett–Teller.

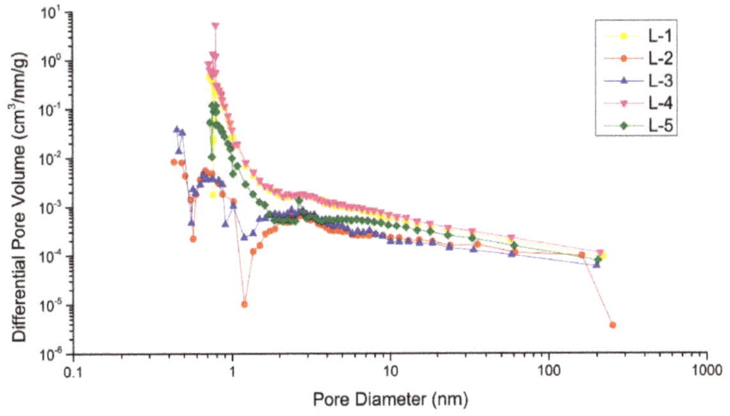

Figure 4. Pore size distribution, defined by differential pore volume analyses.

The calculated pore volume and surface area clarify the overall pore features of each sample. Meanwhile, the scatter plots of cumulative volume percent and surface area percent can, as shown below, highlight the contributions of pores in different diameter ranges. A sharp increase of the three samples with higher pore volume was observed in the micropore range, especially at pore diameters of less than 1 nm, (Figure 5). The percentage of micropores to pore volume of L-1 and L-4 can even achieve 50–60%. This may also be the reason that the average pore diameter of these two samples is significantly lower than that of the other three samples (Table 2). Furthermore, even in the samples with relatively underdeveloped micropores, L-2 and L-3, micropores can occupy a tenth of the total volume. Considering that their diameters are much smaller than those of the mesopores and macropores, this volume fraction means a huge advantage of number. Another thing to note is that L-2 and L-3 have higher volume proportions of macropores (more than 60% in the diagram). However, this does not mean that they have bigger volumes than the other samples in the range of macropores, due to their limited total pore volume. Undeveloped micropores lead to this result. This, on the other hand, reflects the importance of micropores on pore volume.

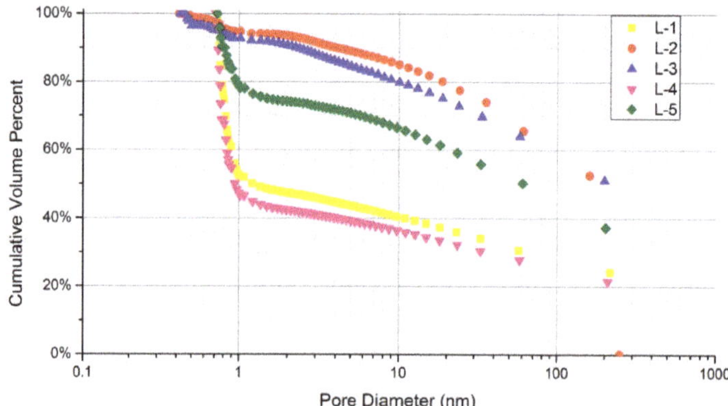

Figure 5. Pore size distribution by percent cumulative volume for the five selected samples in Table 1. Notice that we use logarithmic coordinates on the X-axis.

The cumulative percentage of pore surface area reveals more about the impact of micropores and mesopores. The scatter plot illustrates that for all the samples, micropores and mesopores provide more than 90% of the surface area (Figure 6). Comparisons between the samples also show the same trend as the pore volume, as mentioned above. That is, the higher the proportion of micropores, the larger surface area samples will have. L-1 and L-4, which can achieve total surface areas of 16 m^2/g or more, have the most prominent feature that micropores alone account for more than 95% of the surface area, while the macropores' surface area is close to zero. Unlike the pore volume features, the surface area fractions of macropores of all the samples are much the same, whereas samples with the smallest surface area (L-2, L-3) have large proportions of mesopores.

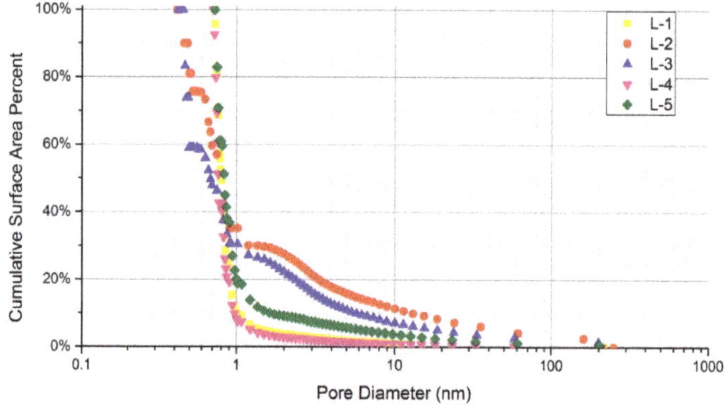

Figure 6. Pore size distribution by percent cumulative surface area for the five selected samples in Table 1. Notice that we use logarithmic coordinates on the X-axis.

4.3. Fractal Characteristics

Fractal theory has been widely applied to several porous materials. Fractal dimension is a ratio providing a statistical index of complexity, comparing how details in a pattern change with the scale at which it is measured. In the research of shale pore characteristics, several experimental methods can be used to calculate the fractal dimension. Among all the methods, the fractal dimensions estimated by N_2 adsorption have proven to be a useful and reliable petrophysical parameter to depict and quantify the irregular surfaces of pores and microstructures [36,58]. Based on the gas adsorption isotherms, the most common and effective method to obtain the fractal dimension is the Frenkel–Halsey–Hill equation (FHH model) [26,31,34,59]. Relevant researches on pores of coal and shale have shown its feasibility [35,37].

In this study, we used the aforementioned N_2 adsorption data to calculate the fractal dimension. The FHH equation is simplified as follows:

$$\ln(V) = (D-3)\ln\left(\ln\left(\frac{P_0}{P}\right)\right) + C$$

where V is the volume of N_2 adsorbed at each equilibrium pressure P, P_0 is the gas saturation pressure, D is the fractal dimension, and C is a constant. Thus, to get the value of the fractal dimension, we need to fit the slope of a straight line onto the plot of $\ln(V)$ versus $\ln(\ln(P_0/P))$, which can be obtained from the N_2 adsorption data.

We made the piecewise linear fitting of each scatter plot according to the different slopes of points, and calculated the root mean square error to show the fitting degree. At the segment of low relative pressure of all the samples, i.e., the black points at the far right part of the plot (Figure 7), the slope of fitting to a straight line is less than −1, which means that the calculated value of the fractal dimension will be smaller than 2. However, a valid fractal dimension of shale pores should be between 2 and 3 [31]. In addition, the fitting degrees of several samples in this segment were very low (the correlation coefficient can be as low as 0.49), especially at the points with the lowest relative pressure. Thus, we masked this part of the data, and followed the usual practice with the relative pressure from 0.01 to 1 [34,35]. Referring to previous studies, the two fitting straight lines based on the remaining points were bounded at a relative pressure of 0.5, where the x value is about −0.37 (Figure 8). The final fitting results are good, such that even the worst correlation coefficient is more than 0.99, and the calculated fractal dimensions are all within the range of definition.

We use D_1 and D_2 to represent the values of the fractal dimension at the P/P_0 intervals of 0–0.5 and 0.5–1, respectively. The corresponding slope plus three is the value of the fractal dimension. The results of fitted equations, correlation coefficients, and fractal dimensions calculated from the FHH model of all the samples are shown in Table 3. The fractal dimensions in the first segment (D_1) vary between 2.32 to 2.618, while the fractal dimensions in the second segment (D_2) are slightly larger, changing from 2.49 to 2.655. There is little difference in the fractal dimensions between each sample, which suggests that they have similar fractal pore structures.

As for the fractal dimension, a large value usually indicates a more complex pore structure or rougher surface. The comparison of fractal dimensions of different samples suggests that both L-1 and L-4 have the maximum values of D_1 and D_2, which means that their pore structures are more complex. The results obtained in the previous section also show that L-1 and L-4 have the largest percentages of pore volume and surface area of micropores and mesopores among all the samples. This suggests that the structures of these small pores are complicated, and as their numbers increase, the complexity of the microstructure will be reflected in the growth of the fractal dimension.

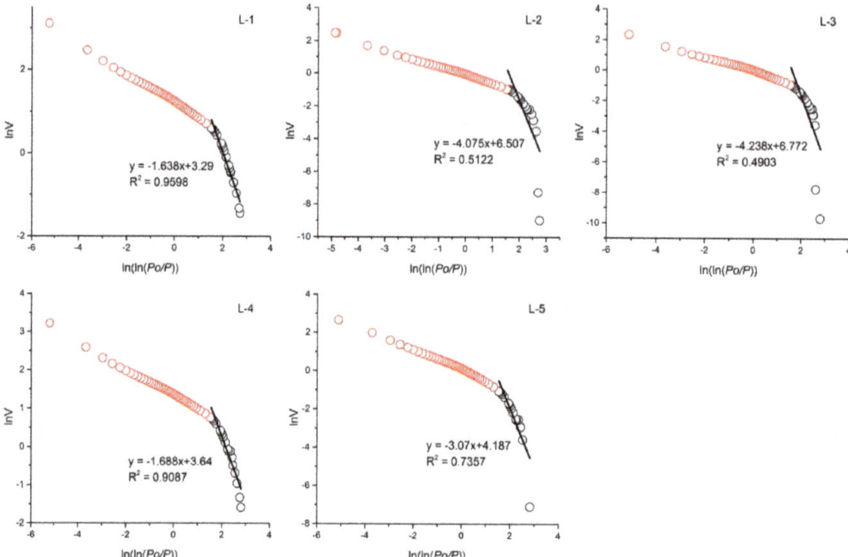

Figure 7. Fractal dimension calculation results with $\ln V$ versus $\ln(\ln(P_o/P))$ from N_2 adsorption isotherms. All the data points on the adsorption curve are shown in this figure. The fractal diameter calculated from the slope of the fitting line, which was made from the black data points of each diagram, is invalid and meaningless. The red data points were used to make the fitting line. The relevant results were shown in the next figure. R^2 is the correlation coefficient.

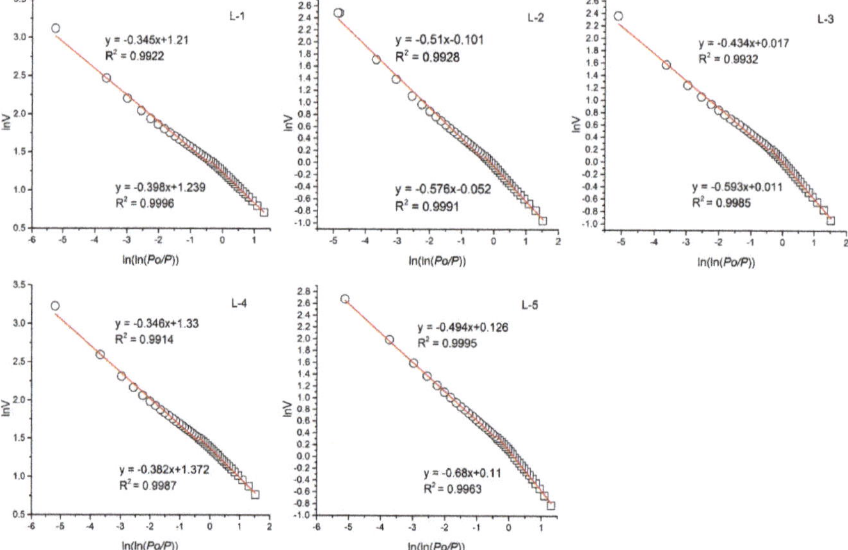

Figure 8. Fractal dimension calculation results with $\ln V$ versus $\ln(\ln(P_o/P))$ from N_2 adsorption isotherms (P/P_o from 0.01 to 1). Notice that we use different shapes of points to represent the fitting data of two parts. The hollow rectangles denote the fractal dimension D_1, and the hollow circles indicate the fractal dimension D_2. R^2 is the correlation coefficient.

Table 3. Fractal dimensions obtained from Frenkel–Halsey–Hill (FHH) model.

Sample ID	Region 1 ($P/P_o = 0$–0.5)			Region 2 ($P/P_o = 0.5$–1)		
	Equation 1	R^2	D_1	Equation 2	R^2	D_2
L-1	y = −0.398x + 1.239	0.9996	2.602	y = −0.345x + 1.21	0.9922	2.655
L-2	y = −0.576x − 0.052	0.9991	2.424	y = −0.51x − 0.101	0.9928	2.49
L-3	y = −0.593x + 0.011	0.9985	2.407	y = −0.434x + 0.017	0.9932	2.566
L-4	y = −0.382x + 1.372	0.9987	2.618	y = −0.346x + 1.33	0.9914	2.654
L-5	y = −0.68x + 0.11	0.9963	2.32	y = −0.494x + 0.126	0.9995	2.506

4.4. Pore Structure from HIM

The HIM image processing was performed on the samples of L-1 and L-2, which were selected as the representatives of developed and relatively undeveloped pores, respectively, based on analysis of N_2 adsorption. As the reference, the pores with diameter of 2 and 50 nm are shown in Figures 9 and 10. Notice that the red circle has a diameter of 50 nm, while the red dot represents a diameter of 2 nm. These are the two cut-off points for the rule of thirds of pore size by the IUPAC. By this means, we can make direct descriptions of the pore sizes of the two samples. Moreover, the average pore diameters determined by the N_2 adsorption experiments in the previous section are directly supported by the HIM images.

The dominant parts under HIM are quartz, clay minerals, and organic matters. However, with relatively low content, the pyrite can be identified easily under HIM due to its framboid shape and high brightness. The gray level of organic matter is close to clay minerals in some images, but it is easy to distinguish them. Dense pores, especially micropores and mesopores, are usually developed in organic matter. There will be bright sides at the edges of organic matters, which make their outlines more clear. Organic matter pores, rather than mineral-hosted pores, are considered to be the dominant contributors to total porosity and hydrocarbon storage in many organic-rich unconventional reservoirs [60–62]. Meanwhile, other types of pores are rarely found under HIM, especially those with diameters below those of macropores. Thus, we focused on the development of organic matter pores of the two samples in this section.

The HIM images of L-1 show a large volume of micropores and fine mesopores developing with associated organic matters, which are usually mixed with clay minerals. These pores mainly show round or oval shapes within organic matters. However, very few slit-like pores also appear at the grain boundaries between organic matters and minerals (Figure 9b), possibly because of the shrinkage of organic matter during the hydrocarbon generation process. Sometimes, these kinds of small cracks can extend into a long fracture by connecting with each other (Figure 9d). Except for when they are mixed with clay minerals, organic matters can also fill into the space between the mineral particles, which are usually characterized by their flat boundaries, such as pyrite and quartz (Figure 9f). These organic pores can be considered to be formed after filling, i.e., secondary pores. It is interesting to note that no pore development was observed in a few organic matters (Figure 9g). Some macropores are formed by the dissolution of minerals such as calcite, which may create large pore spaces (Figure 9h). Generally speaking, the shale sample L-1 has a well-developed pore structure, and is dominated by micropores and fine mesopores.

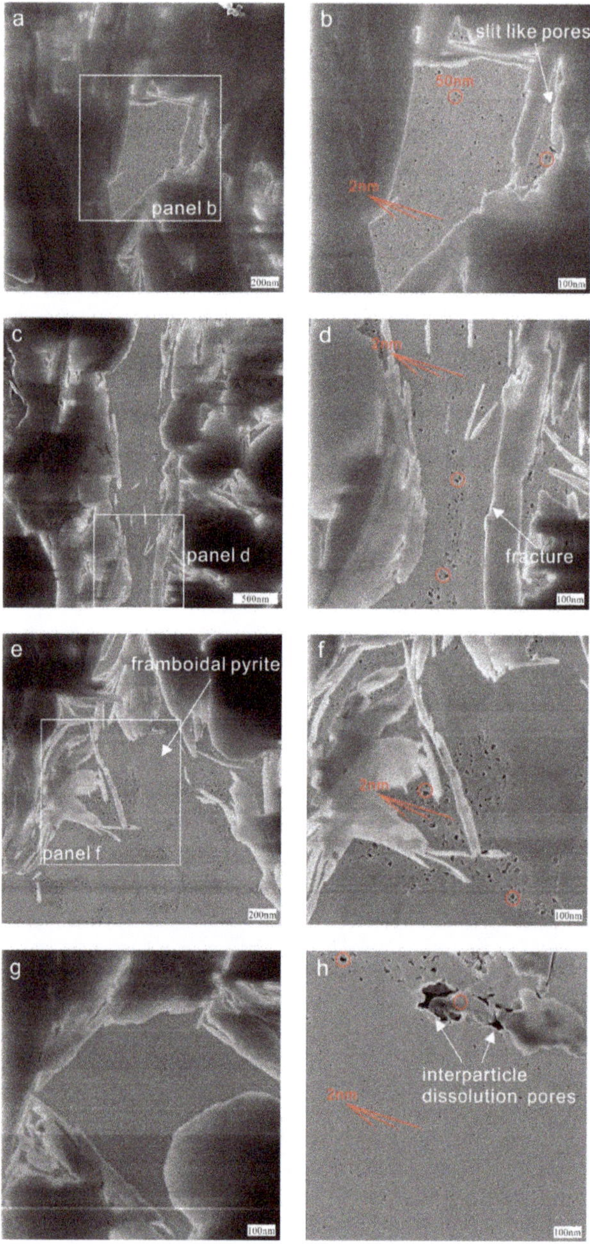

Figure 9. Helium ion microscope (HIM) images of sample L-1, showing abundant porosity with a wide range in the size of micropores and fine mesopores. (**a**,**b**): Densely developed organic pores with a few slit-like pores forming between organic matter and clay minerals; (**c**,**d**): Banded organic matter formed by the compaction process. Some directionally distributed pores and fractures are developed in the organic matter; (**e**,**f**): The organic matter between the particles of framboidal pyrite, and the pores within it; (**g**): Organic matter without pore development; (**h**): Well-developed pores less than ten nanometers in diameter. Some large pores were also formed by dissolution of mineral particles.

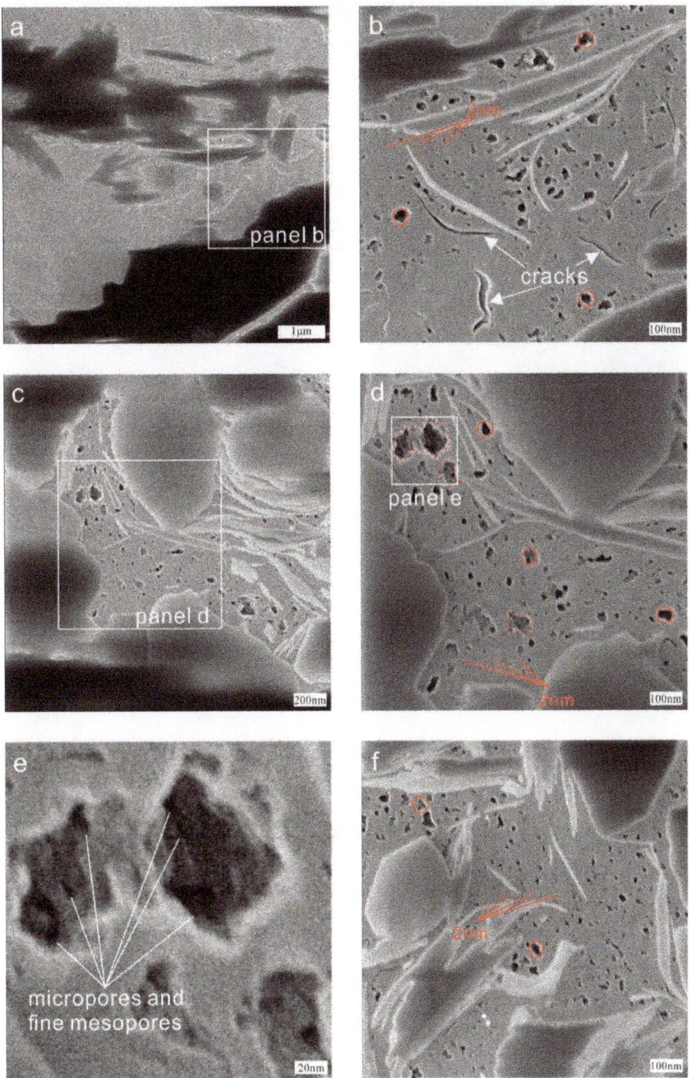

Figure 10. Helium ion microscope (HIM) images of sample L-2, showing abundant porosity with a wide range in the size of coarse mesopores and macropores. (**a,b**): Many cracks caused by shrinkage can be obtained within the organic matter; (**c,d**): The organic matter mixed with clay minerals and brittle minerals. There are some small pores developing at the walls of larger pores (circled by a red dashed line); (**e**): Many micropores and fine mesopores on the inner walls of macropores; (**f**): Well-developed organic pores, most of which are coarse mesopores.

The phenomenon of organic matters mixing with clay minerals or brittle minerals is more obvious in sample L-2. Compared with L-1, the HIM images of L-2 show limited porosity, with most in the range of coarse mesopores and macropores. Not only are there larger pores, but many cracks caused by shrinkage can also be observed (Figure 10b). Similar to the observations in the L-1 sample, the pore shapes of L-2 are mainly round or nearly round. As some micropores and fine mesopores are found from the walls of large pores (Figure 10d) due to the great depth of field of HIM, it can be proven that micropores may cause irregular surfaces in the walls of larger pores [6]. This is interpreted as being because densely small pores produced in a certain area will connect to become larger pores during the process of hydrocarbon generation. To sum up, the pore size of L-2 is larger than that in L-1, which is the same conclusion as in the N_2 adsorption experiment.

5. Discussion

The observation of HIM images does not correspond very well to the division of developed and undeveloped pore structures based on analysis of N_2 adsorption. For instance, sample L-2 has larger pore size according to the HIM images (Figure 10), while its pore volume and surface area are both the lowest among all the five samples. Furthermore, it was found that between the sample L-1 and L-2, with undeveloped and developed macropores respectively, the gap in the total volume is not as large as the difference in surface area. Considering that there are only slight differences in mineral components between the five samples, we assume that the pore structure is least affected by mineral components. Furthermore, the pores observed in the HIM images are mainly related to organic matter. Thus, the results can be explained by the effect of organic matter content in one way.

The diagram of pore volume and surface area versus TOC content is shown below (Figure 11). There is a moderate-to-good positive correlation between the TOC contents and pore volumes as well as pore surface areas of samples. One problem is that the trend is less obvious, due to limited sample quantity. The figure suggests that, as the main carrier of pores, the amount of organic matter is the major factor affecting pore volume and surface area. Thus, the observed developed large pores may influence the pore size distribution proportions of shales, leading to higher ratios of macropores. However, the contribution of macropores to the surface area is small. On the other hand, macropores can contribute to the total pore volume, but are limited by their quantity. Namely, their effect on pore volume and surface area is not as prominent as that of the TOC content.

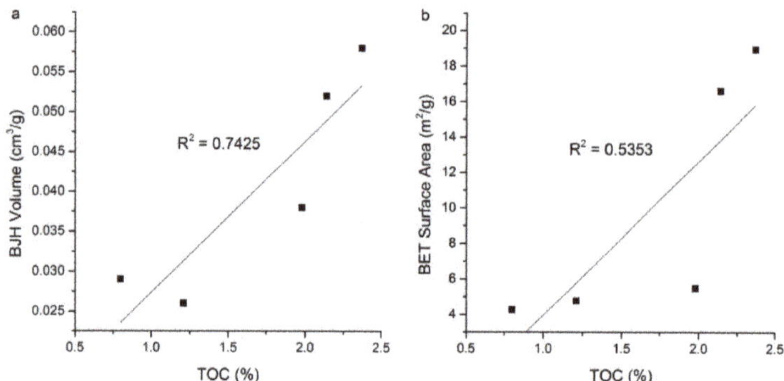

Figure 11. Relationships between pore structure parameters and TOC contents (mass fraction). (**a**) BJH volume; (**b**) BET surface area. R^2 is the correlation coefficient.

Based on the nitrogen adsorption data, the calculated pore volume and surface area versus fractal dimension was drawn as follows (Figure 12). As shown in the diagram, the fractal dimension D_2 calculated from the data of the relative pressure range from 0.5 to 1 is greater than D_1 for each sample. This means that larger pores have rougher inner surfaces and more complex pore structures. This can be verified by the observation that micropores were found to cause irregular surfaces on the walls of larger pores from HIM images (Figure 10e).

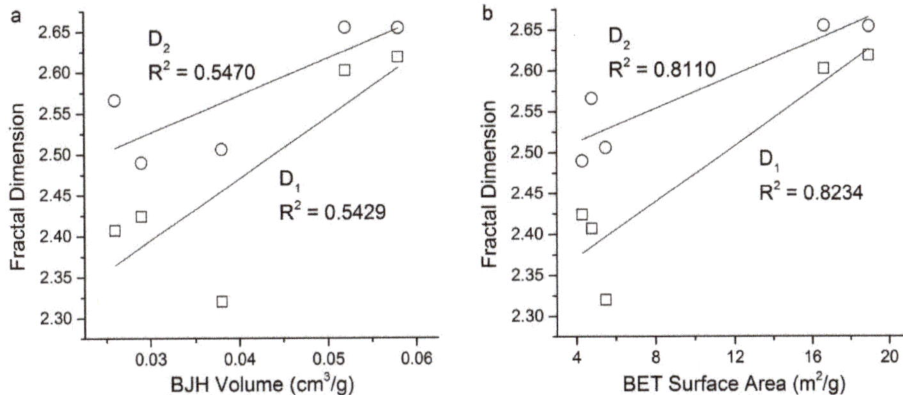

Figure 12. Relationships between pore structure parameters and fractal dimension (mass fraction). (a) BJH volume; (b) BET surface area. The hollow rectangles denote the fractal dimension D_1, and the hollow circles indicate the fractal dimension D_2. R^2 is the correlation coefficient.

In addition, a good positive correlation between pore structure parameters and fractal dimensions was obtained (Figure 12). The positive relationships are consistent with previous studies and suggest that shale samples with higher values of fractal dimension tend to have a greater surface area and larger pore volume [26,63]. In general, the pore size distribution and HIM images show us the well-developed mesopores or even micropores in the samples. Furthermore, the huge pore surface area means enormous numbers of these small pores, which greatly increase the complexity of the pore structure, leading to high fractal dimensions. Above all, this means good hydrocarbon storage capacity and development potential of the shales. Nevertheless, to test this conclusion, more experimental and analytical works are needed because of the limited sample quantity and the strong anisotropism of shale.

Compared with the important shale blocks in China, especially the Fuling shale gas field not far away, the samples of the Laifeng–Xianfeng Block show lower values of pore volume and pore surface area [48,64]. It was found that the pores are not well developed from both the microscopy and nitrogen adsorption experiment results, though both of the pore features maintain an upward trend with the increase of organic matter content.

6. Conclusions

In this paper, the pore-structure characteristics of Lower Silurian Longmaxi Formation shales collected from Well LD1 of the Laifeng–Xianfeng Block in the Upper Yangtze region were investigated using ultra-low N_2 adsorption and helium ion microscope experiments. Colligating the above results and analysis, some conclusions can be drawn as follows:

(1) The pore volume, surface area, and pore size distribution of the samples in Well LD1 were calculated based on N_2 adsorption data. The results were verified by HIM images, which can provide a visual appreciation of pores. These two experiments both indicated that mesopores are the main component of the pore structure, although there are some slight differences in pore size between the samples.

(2) The TOC contents of shales contribute significantly to the pore volume and surface area. This demonstrates that organic matter is the main carrier of pores in shale, in other words, organic pores are the principal parts of the pore structures.

(3) Based on the N_2 adsorption data, two different fractal dimensions, D_1 and D_2, were determined. The value of D_2, calculated from the data of the relative pressure range from 0.5 to 1, is greater than D_1.

This suggests that larger pores have larger fractal dimensions, which represent more complex pore structures and rougher surfaces.

(4) There is a positive correlation between fractal dimension and pore structure parameters, including pore volume and surface area. The increase of these two parameters essentially means a greater number of pores, as well as a more complex pore structure. This eventually leads to an increase of the fractal dimension, and also means stronger adsorption capacity of the shale.

Author Contributions: Conceptualization, Y.J. and C.H.; methodology, C.H.; software, K.Y.; validation, H.Z. and Y.Q.; investigation, C.H. and H.Z.; resources, H.Z., K.Y., and Y.S..; data curation, C.H. and Y.Q.; writing—original draft preparation, C.H.; writing—review and editing, Y.J., H.Z., Y.Q., Y.S. and L.J.; visualization, C.H. and K.Y.; supervision, Y.J.; project administration, Y.J.

Funding: This research was financially supported by the National Natural Science Foundation of China (Grant No. 41530315, 41372213, 41872160), the National Science and Technology Major Project of China (Grant No. 2016ZX05066), the "Climate Change: Carbon Budget and Related Issues" Strategic Priority Research Program of the Chinese Academy of Sciences (Grant No. XDA05030100), and the Sichuan Science and Technology Support Program (Grant No. 2016JZ0037).

Conflicts of Interest: The authors declare no conflict of interest.

References

1. Zou, C.; Zhai, G.; Zhang, G.; Wang, H.; Zhang, G.; Li, J.; Wang, Z.; Wen, Z.; Ma, F.; Liang, Y.; et al. Formation, distribution, potential and prediction of global conventional and unconventional hydrocarbon resources. *Pet. Explor. Dev.* **2015**, *42*, 14–28. [CrossRef]
2. Wang, Q.; Chen, X.; Jha, A.N.; Rogers, H. Natural gas from shale formation–the evolution, evidences and challenges of shale gas revolution in United States. *Renew. Sustain. Energy Rev.* **2014**, *30*, 1–28. [CrossRef]
3. Chalmers, G.R.; Bustin, R.M. The organic matter distribution and methane capacity of the Lower Cretaceous strata of Northeastern British Columbia, Canada. *Int. J. Coal Geol.* **2007**, *70*, 223–239. [CrossRef]
4. Gasparik, M.; Ghanizadeh, A.; Bertier, P.; Gensterblum, Y.; Bouw, S.; Krooss, B.M. High-pressure methane sorption isotherms of black shales from the Netherlands. *Energy Fuels* **2012**, *26*, 4995–5004. [CrossRef]
5. Ross, D.J.K.; Bustin, R.M. The importance of shale composition and pore structure upon gas storage potential of shale gas reservoirs. *Mar. Pet. Geol.* **2009**, *26*, 916–927. [CrossRef]
6. Chalmers, G.R.; Bustin, R.M.; Power, I.M. Characterization of gas shale pore systems by porosimetry, pycnometry, surface area, and field emission scanning electron microscopy/transmission electron microscopy image analyses: Examples from the Barnett, Woodford, Haynesville, Marcellus, and Doig units. *AAPG Bull.* **2012**, *96*, 1099–1119.
7. Curtis, M.E.; Cardott, B.J.; Sondergeld, C.H.; Rai, C.S. Development of organic porosity in the Woodford Shale with increasing thermal maturity. *Int. J. Coal Geol.* **2012**, *103*, 26–31. [CrossRef]
8. Clarkson, C.R.; Solano, N.; Bustin, R.M.; Bustin, A.M.M.; Chalmers, G.R.L.; He, L.; Melnichenko, Y.B.; Radliński, A.P.; Blach, T.P. Pore structure characterization of North American shale gas reservoirs using USANS/SANS, gas adsorption, and mercury intrusion. *Fuel* **2013**, *103*, 606–616. [CrossRef]
9. Wang, G.; Ju, Y.; Yan, Z.; Li, Q. Pore structure characteristics of coal-bearing shale using fluid invasion methods: A case study in the Huainan-Huaibei Coalfield in China. *Mar. Pet. Geol.* **2015**, *62*, 1–13. [CrossRef]
10. Cai, J.; Lin, D.; Singh, H.; Wei, W.; Zhou, S. Shale gas transport model in 3D fractal porous media with variable pore sizes. *Mar. Pet. Geol.* **2018**, *98*, 437–447. [CrossRef]
11. Hao, F.; Zou, H.; Lu, Y. Mechanisms of shale gas storage: Implications for shale gas exploration in ChinaMechanisms of Shale Gas Storage. *AAPG Bull.* **2013**, *97*, 1325–1346. [CrossRef]
12. Zou, C.; Dong, D.; Wang, S.; Li, J.; Li, X.; Wang, Y.; Li, D.; Cheng, K. Geological characteristics and resource potential of shale gas in China. *Pet. Explor. Dev.* **2010**, *37*, 641–653. [CrossRef]
13. Afsharpoor, A.; Javadpour, F.; Wu, J.; Ko, L.T.; Liang, Q. Network modeling of liquid flow in Yanchang shale. *Interpretation* **2017**, *5*, SF99–SF107. [CrossRef]
14. Ju, Y.; Sun, Y.; Tan, J.; Bu, H.; Han, K.; Li, X.; Fang, L. The composition, pore structure characterization and deformation mechanism of coal-bearing shales from tectonically altered coalfields in eastern China. *Fuel* **2018**, *234*, 626–642. [CrossRef]

15. Zhu, H.; Ju, Y.; Qi, Y.; Huang, C.; Zhang, L. Impact of tectonism on pore type and pore structure evolution in organic-rich shale: Implications for gas storage and migration pathways in naturally deformed rocks. *Fuel* **2018**, *228*, 272–289. [CrossRef]

16. Qi, Y.; Ju, Y.; Huang, C.; Zhu, H.; Bao, Y.; Wu, J.; Meng, S.; Chen, W. Influences of organic matter and kaolinite on pore structures of transitional organic-rich mudstone with an emphasis on S2 controlling specific surface area. *Fuel* **2019**, *237*, 860–873. [CrossRef]

17. Tan, J.; Horsfield, B.; Mahlstedt, N.; Zhang, J.; Boreham, C.J.; Hippler, D. Natural gas potential of Neoproterozoic and lower Palaeozoic marine shales in the Upper Yangtze Platform, South China: Geological and organic geochemical characterization. *Int. Geol. Rev.* **2015**, *57*, 305–326. [CrossRef]

18. Chen, S.; Zhu, Y.; Wang, H.; Liu, H.; Wei, W.; Fang, J. Shale gas reservoir characterisation: A typical case in the southern Sichuan Basin of China. *Energy* **2011**, *36*, 6609–6616. [CrossRef]

19. Wang, Y.; Huang, J.; Li, X.; Dong, D.; Wang, S.; Guan, Q. Quantitative characterization of fractures and pores in shale beds of the Lower Silurian, Longmaxi Formation, Sichuan Basin. *Nat. Gas Ind. B* **2015**, *2*, 481–488. [CrossRef]

20. Tang, X.; Jiang, Z.; Jiang, S.; Li, Z. Heterogeneous nanoporosity of the Silurian Longmaxi Formation shale gas reservoir in the Sichuan Basin using the QEMSCAN, FIB-SEM, and nano-CT methods. *Mar. Pet. Geol.* **2016**, *78*, 99–109. [CrossRef]

21. Yang, R.; He, S.; Hu, Q.; Sun, M.; Hu, D.; Yi, J. Applying SANS technique to characterize nano-scale pore structure of Longmaxi shale, Sichuan Basin (China). *Fuel* **2017**, *197*, 91–99. [CrossRef]

22. Hu, J.; Tang, S.; Zhang, S. Investigation of pore structure and fractal characteristics of the Lower Silurian Longmaxi shales in western Hunan and Hubei Provinces in China. *J. Nat. Gas Sci. Eng.* **2016**, *28*, 522–535. [CrossRef]

23. Bernard, S.; Horsfield, B.; Schulz, H.M.; Wirth, R.; Schreiber, A.; Sherwood, N. Geochemical evolution of organic-rich shales with increasing maturity: A STXM and TEM study of the Posidonia Shale (Lower Toarcian, Northern Germany). *Mar. Pet. Geol.* **2012**, *31*, 70–89. [CrossRef]

24. Lewis, R.; Singer, P.; Jiang, T.; Rylander, E.; Sinclair, S.; Mclin, R.H. NMR T2 distributions in the Eagle Ford shale: Reflections on pore size. In Proceedings of the SPE Unconventional Resources Conference-USA, The Woodlands, TX, USA, 10–12 April 2013; Society of Petroleum Engineers: Richardson, TX, USA, 2013.

25. King, H.E., Jr.; Eberle, A.P.R.; Walters, C.C.; Kliewer, C.E.; Ertas, D.; Huynh, C. Pore architecture and connectivity in gas shale. *Energy Fuels* **2015**, *29*, 1375–1390. [CrossRef]

26. Yang, R.; He, S.; Yi, J.; Hu, Q. Nano-scale pore structure and fractal dimension of organic-rich Wufeng-Longmaxi shale from Jiaoshiba area, Sichuan Basin: Investigations using FE-SEM, gas adsorption and helium pycnometry. *Mar. Pet. Geol.* **2016**, *70*, 27–45. [CrossRef]

27. Ju, Y.; Huang, C.; Sun, Y.; Wan, Q.; Lu, X.; Lu, S.; He, H.; Wang, X.; Zou, C.; Wu, J.; et al. Nanogeosciences: Research history, current status, and development trends. *J. Nanosci. Nanotechnol.* **2017**, *17*, 5930–5965. [CrossRef]

28. Gao, F.; Song, Y.; Li, Z.; Xiong, F.; Chen, L.; Zhang, X.; Chen, Z.; Moortgat, J. Quantitative characterization of pore connectivity using NMR and MIP: A case study of the Wangyinpu and Guanyintang shales in the Xiuwu basin, Southern China. *Int. J. Coal Geol.* **2018**, *197*, 53–65. [CrossRef]

29. Yu, K.; Shao, C.; Ju, Y.; Qu, Z. The genesis and controlling factors of micropore volume in transitional coal-bearing shale reservoirs under different sedimentary environments. *Mar. Pet. Geol.* **2019**, *102*, 426–438. [CrossRef]

30. Huang, C.; Ju, Y.; Xu, T.; Sun, Y.; Jia, T.; Neupane, B.; Han, K.; Qi, Y.; Zhu, H.; Cai, J. Full-scale and multi-method combined characterization of micro/nano pores in organic shale. *J. Nanosci. Nanotechnol.* **2017**, *17*, 6634–6644. [CrossRef]

31. Pfeifer, P.; Avnir, D. Chemistry in noninteger dimensions between two and three. I. Fractal theory of heterogeneous surfaces. *J. Chem. Phys.* **1983**, *79*, 3558–3565. [CrossRef]

32. Krohn, C.E. Fractal measurements of sandstones, shales, and carbonates. *J. Geophys. Res. Solid Earth* **1988**, *93*, 3297–3305. [CrossRef]

33. Cai, Y.; Liu, D.; Pan, Z.; Yao, Y.; Li, J.; Qiu, Y. Pore structure and its impact on CH_4 adsorption capacity and flow capability of bituminous and subbituminous coals from Northeast China. *Fuel* **2013**, *103*, 258–268. [CrossRef]

34. Bu, H.; Ju, Y.; Tan, J.; Wang, G.; Li, X. Fractal characteristics of pores in non-marine shales from the Huainan coalfield, eastern China. *J. Nat. Gas Sci. Eng.* **2015**, *24*, 166–177. [CrossRef]

35. Yao, Y.; Liu, D.; Tang, D.; Tang, S.; Huang, W. Fractal characterization of adsorption-pores of coals from North China: An investigation on CH4 adsorption capacity of coals. *Int. J. Coal Geol.* **2008**, *73*, 27–42. [CrossRef]

36. Mahamud, M.M.; Novo, M.F. The use of fractal analysis in the textural characterization of coals. *Fuel* **2008**, *87*, 222–231. [CrossRef]

37. Yang, F.; Ning, Z.; Liu, H. Fractal characteristics of shales from a shale gas reservoir in the Sichuan Basin, China. *Fuel* **2014**, *115*, 378–384. [CrossRef]

38. Pan, J.; Niu, Q.; Wang, K.; Shi, X.; Li, M. The closed pores of tectonically deformed coal studied by small-angle X-ray scattering and liquid nitrogen adsorption. *Microporous Mesoporous Mater.* **2016**, *224*, 245–252. [CrossRef]

39. Wang, P.; Jiang, Z.; Chen, L.; Yin, L.; Li, Z.; Zhang, C.; Tang, X.; Wang, G. Pore structure characterization for the Longmaxi and Niutitang shales in the Upper Yangtze Platform, South China: Evidence from focused ion beam–He ion microscopy, nano-computerized tomography and gas adsorption analysis. *Mar. Pet. Geol.* **2016**, *77*, 1323–1337. [CrossRef]

40. Zhou, L.; Kang, Z. Fractal characterization of pores in shales using NMR: A case study from the Lower Cambrian Niutitang Formation in the Middle Yangtze Platform, Southwest China. *J. Nat. Gas Sci. Eng.* **2016**, *35*, 860–872. [CrossRef]

41. Stevens, S.H.; Moodhe, K.D.; Kuuskraa, V.A. China shale gas and shale oil resource evaluation and technical challenges. In Proceedings of the SPE Asia Pacific Oil and Gas Conference and Exhibition, Jakarta, Indonesia, 22–24 October 2013; Society of Petroleum Engineers: Richardson, TX, USA, 2013.

42. Guo, T.; Zhang, H. Formation and enrichment mode of Jiaoshiba shale gas field, Sichuan Basin. *Pet. Explor. Dev.* **2014**, *41*, 31–40. [CrossRef]

43. Zhu, H.; Ju, Y.; Huang, C.; Han, K.; Qi, Y.; Shi, M.; Yu, K.; Feng, H.; Li, W.; Ju, L.; et al. Pore structure variations across structural deformation of Silurian Longmaxi Shale: An example from the Chuandong Thrust-Fold Belt. *Fuel* **2019**, *241*, 914–932. [CrossRef]

44. Zou, C.; Wei, G.; Xu, C.; Du, J.; Xie, Z.; Wang, Z.; Hou, L.; Yang, C.; Li, J.; Yang, W. Geochemistry of the Sinian–Cambrian gas system in the Sichuan Basin, China. *Org. Geochem.* **2014**, *74*, 13–21. [CrossRef]

45. Huang, J.; Zou, C.; Li, J.; Dong, D.; Wang, S.; Wang, S.; Cheng, K. Shale gas generation and potential of the lower Cambrian Qiongzhusi formation in the southern Sichuan Basin, China. *Pet. Explor. Dev.* **2012**, *39*, 75–81. [CrossRef]

46. Dai, J.; Zou, C.; Liao, S.; Dong, D.; Ni, Y.; Huang, J.; Wu, W.; Gong, D.; Huang, S.; Hu, G. Geochemistry of the extremely high thermal maturity Longmaxi shale gas, southern Sichuan Basin. *Org. Geochem.* **2014**, *74*, 3–12. [CrossRef]

47. Wang, G.; Ju, Y.; Huang, C.; Long, S.; Peng, Y. Longmaxi-Wufeng Shale lithofacies identification and 3-D modeling in the northern Fuling Gas Field, Sichuan Basin. *J. Nat. Gas Sci. Eng.* **2017**, *47*, 59–72. [CrossRef]

48. Wang, R.; Hu, Z.; Sun, C.; Liu, Z.; Zhang, C.; Gao, B.; Du, W.; Zhao, J.; Tang, W. Comparative analysis of shale reservoir characteristics in the Wufeng-Longmaxi (O3 w-S1 l) and Niutitang (Є1 n) Formations: A case study of the Wells JY1 and TX1 in southeastern Sichuan Basin and its neighboring areas, southwestern China. *Interpretation* **2018**, *6*, 1–46. [CrossRef]

49. Li, B.; Wei, G.; Hong, K.; Peng, C.; Hu, X.; Zhu, L. Evaluation and understanding on the shale gas wells in complex tectonic provinces outside Sichuan Basin, South China: A case study from Well Laiye 1 in Laifeng–Xianfeng Block, Hubei. *Nat. Gas Ind.* **2016**, *36*, 29–35. (In Chinese with English Abstract)

50. Jiang, S.; Li, B.; Peng, C.; Peng, C.; Hu, X.; Hong, K.; Zhu, L. Characteristics and gas content of Wufeng-Longmaxi Formation shale in the Well LD2 of the Laifeng-Xianfeng block. *Geol. Explor.* **2018**, *54*, 203–210. (In Chinese with English Abstract)

51. Xu, Z.; Yao, G.; Huang, L.; Dong, Y.; Wang, P.; Yu, G. Risk analysis and play evaluation of marine residual basins in South China. *Pet. Geol. Exp.* **2013**, *35*, 9–16, 23. (In Chinese with English Abstract)

52. Liu, H.; Wang, H. Ultra-low water saturation characteristics and the identification of over-pressured play fairways of marine shales in South China. *Nat. Gas Ind.* **2013**, *33*, 140–144. (In Chinese with English Abstract)

53. Nie, H.; Tang, X.; Bian, R. Controlling factors for shale gas accumulation and prediction of potential development area in shale gas reservoir of South China. *Acta Pet. Sin.* **2009**, *30*, 484–491. (In Chinese with English Abstract)

54. Chen, K.; Zhang, J.; Tang, X. Gas Content Logging Evaluation of Lower Silurian Longmaxi Shale in Western Hunan-Hubei. *Spec. Oil Gas Reserv.* **2016**, *23*, 16–20, 151–152. (In Chinese with English Abstract)

55. Rietveld, H.M. Line profiles of neutron powder-diffraction peaks for structure refinement. *Acta Crystallogr.* **1967**, *22*, 151–152. [CrossRef]

56. Bustin, R.M.; Bustin, A.M.M.; Cui, A.; Ross, D.; Pathi, V.M. Impact of shale properties on pore structure and storage characteristics. In Proceedings of the SPE Shale Gas Production Conference, Fort Worth, TX, USA, 16–18 November 2008; Society of Petroleum Engineers: Richardson, TX, USA, 2008.

57. Wang, G.; Ju, Y. Organic shale micropore and mesopore structure characterization by ultra-low pressure N_2 physisorption: Experimental procedure and interpretation model. *J. Nat. Gas Sci. Eng.* **2015**, *27*, 452–465. [CrossRef]

58. Han, Y.; Kwak, D.; Choi, S.; Shin, C.; Lee, Y.; Kim, H. Pore structure characterization of shale using gas physisorption: Effect of chemical compositions. *Minerals* **2017**, *7*, 66. [CrossRef]

59. Xi, Z.; Tang, S.; Wang, J.; Yi, J.; Guo, Y.; Wang, K. Pore structure and fractal characteristics of Niutitang shale from China. *Minerals* **2018**, *8*, 163. [CrossRef]

60. Loucks, R.G.; Reed, R.M.; Ruppel, S.C.; Jarvie, D.M. Morphology, genesis, and distribution of nanometer-scale pores in siliceous mudstones of the Mississippian Barnett Shale. *J. Sediment. Res.* **2009**, *79*, 848–861. [CrossRef]

61. Löhr, S.C.; Baruch, E.T.; Hall, P.A.; Kennedy, M.J. Is organic pore development in gas shales influenced by the primary porosity and structure of thermally immature organic matter? *Org. Geochem.* **2015**, *87*, 119–132. [CrossRef]

62. Gu, X.; Mildner DF, R.; Cole, D.R.; Rother, G.; Slingerland, R.; Brantley, S.L. Quantification of organic porosity and water accessibility in Marcellus shale using neutron scattering. *Energy Fuels* **2016**, *30*, 4438–4449. [CrossRef]

63. Liu, X.; Xiong, J.; Liang, L. Investigation of pore structure and fractal characteristics of organic-rich Yanchang formation shale in central China by nitrogen adsorption/desorption analysis. *J. Nat. Gas Sci. Eng.* **2015**, *22*, 62–72. [CrossRef]

64. Hu, H.; Hao, F.; Lin, J.; Lu, Y.; Ma, Y.; Li, Q. Organic matter-hosted pore system in the Wufeng-Longmaxi (O_3w-$S_1$1) shale, Jiaoshiba area, eastern Sichuan Basin, China. *Int. J. Coal Geol.* **2017**, *173*, 40–50. [CrossRef]

Article

Tectonic and Thermal Controls on the Nano-Micro Structural Characteristic in a Cambrian Organic-Rich Shale

Hongjian Zhu [1], Yiwen Ju [1,*], Cheng Huang [1], Yu Qi [1], Liting Ju [1], Kun Yu [1], Wuyang Li [1], Xin Su [1,2], Hongye Feng [1] and Peng Qiao [1]

[1] Key Laboratory of Computational Geodynamics, College of Earth and Planetary Sciences, University of Chinese Academy of Sciences, Beijing 100049, China; zhj8641@163.com (H.Z.); huangcheng150@126.com (C.H.); qiuqiuyu911@163.com (Y.Q.); jlt@ucas.ac.cn (L.J.); yukun@cumt.edu.cn (K.Y.); wuyang.li@outlook.com (W.L.); suxin115@mails.ucas.edu.cn (X.S.); fhy0205@163.com (H.F.); qiaopeng18@mails.ucas.edu.cn (P.Q.)

[2] Shanxi Huapu Testing Technology Co., Ltd., Taiyuan 030000, China

* Correspondence: juyw03@163.com

Received: 16 April 2019; Accepted: 27 May 2019; Published: 10 June 2019

Abstract: Until recently, the characteristics of nano-microscale structures in the naturally deformed, overmature, marine shales were poorly known. Thermally overmature Lujiaping shales in the complex tectonic area of the northeast part of the upper Yangtze area, China have experienced strong tectonic deformation and are considered as potentially important strata for shale gas exploration. Naturally deformed samples from the main source rocks are selected from the Lower Cambrian Lujiaping Formation in the Dabashan Thrust-fold Belt to investigate nanometer- to micrometer-sized structures. A combination of scanning electron microscope (SEM), low-pressure nitrogen adsorption (LPNA), and low-field nuclear magnetic resonance (NMR) suggests that the pore types are dominantly fracture-related pores with a lesser abundance of mineral-hosted pores. These two pore types account for the 90% of total pore space. Organic matter (OM)-hosted pores are rare and make up a small part of the pore systems (less than 10%) due to high thermal maturity and intensive tectonic compression. Overall, the Lujiaping deformed, overmature samples have abundant nanometer- to micrometer-sized inorganic pores. High-resolution SEM images provide direct evidence of the formation of nano- and microsized structures such as OM–clay aggregates and silica nanograins. OM–clay aggregates are commonly observed in samples, which also exhibit abundant open microfractures and interparticle pores. Quartz can occur as silica nanograins and botryoids typically 20–100 nm in size, which may influence porosity through the creation or occupying interparticle pore space.

Keywords: nanostructure; OM–clay aggregate; silica nanograin; tectonic and thermal evolution

1. Introduction

Unconventional natural gas production from organic rich shales makes up an ever increasing percentage of total natural gas production all over the world [1,2]. Shale gas exploration and development, initiated successfully in the North America and extended to China, will have application in several other countries in the coming years [1–4]. Organic-rich shale currently produces commercial unconventional hydrocarbons and exhibits a wide variation in nano- and microscale geological attributes and geochemical properties [1–5]. The lower Cambrian Lujiaping Formation marine, organic-rich shales in the Dabashan Thrust-fold Belt of South China are being studied for their potential as shale gas reservoirs [6–10]. Several studies have discussed structural style and conditions of shale gas enrichment [6,10], while other studies have characterized microstructures and associated reservoir quality [7–9]. Such investigations concentrate on nano- to microscale pore systems in the Lujiaping Shale using high-pressure methane sorption analysis, mercury injection capillary pressure (MICP), low-pressure nitrogen adsorption (LPNA), low-field nuclear magnetic resonance (NMR), and focused ion beam-scanning electron microscope imaging (FIB-SEM). These techniques have been used both qualitatively and quantitatively to determine pore sizes and pore types in shale for many years [11–18]. In this paper, the tectonic and thermal controls on the evolution of nanometer- to micrometer-size pore structures and inorganic/organic structures in the Lujiaping shales are discussed.

The contribution of thermogenic, OM-hosted nanopores to overall porosity is such that organic porosity can be as high as 40% of total porosity [19–24]. However, recent studies have suggested that organic porosity in naturally deformed shales are poorly developed, and that fractured-related and mineral-hosted pores are more widely recognized as being the most significant components of porosity in naturally deformed shales [25–27]. The discovery of mineral-hosted and fracture-related pore systems in naturally deformed shales sparked many questions concerning their characteristics, prediction, and reservoir quality: (1) Are such pore networks primary (hosted within initially particles and shale matrix) versus secondary (formed during tectonic deformation)? (2) Is the lack of organic pore development due to thermal evolution, tectonism, or the combination of the both? (3) What is the function of tectonism in the occurrence and thermal evolution of organic pore systems? In addition, the formation of nanosized to microsized structures (clay–organic nanocomposites and nanograins) in the process of tectonic evolution has also attracted much attention from scholars. The answers to such questions are central to making wise decisions on the benefits or detriment of microstructural properties in shale gas reservoirs.

The structural evolution of organic-rich shale in complex tectonic areas of South China has received little attention [7–9,25,26]. In this paper, we use outcrop samples from the fractured and folded shale layers of the Lujiaping Formation to study the influence of the tectonic deformation and thermal maturity on the nanometer- to micrometer-sized structures. The main objectives of this paper are to (1) describe the deformation microstructures and mineralogy of the Lujiaping Shale samples, (2) characterize the common pore types, (3) quantify the pore size distribution (PSD), (4) describe the features of OM–clay aggregates and silica nanograins, and (5) discuss the evolution of reservoir characteristics in Cambrian organic-rich shale during thermal maturation and structural deformation. We hope these results will provide a fundamental starting point for continued research on the reservoir properties in naturally deformed, overmature shale systems.

2. Geological Background and Sample Characterization

The Sichuan Basin, which is a huge superposed basin and the largest shale gas-producing region in China, is geologically complicated due to a complex interaction of multistage tectonic movements and sedimentary history [28–31]. Since the Yanshanian movement (~180–140 Ma), fold belts and detachment belts, as well as thrust belts, have formed in the peripheral areas by intensive lateral compression, which has caused complex folds and faults combinations [29]. The basin is bounded by the Longmenshan fault in the west, the Micangshan uplift in the north, the Dabashan intracontinental orogen in the northeast, the Daloushan in the southeast, and the Emei-Daliangshan fold belt in the southwest [29] (Figure 1a). The Dabashan thrust-fold belt is separated into the South Dabashan and the North Dabashan by the NW-striking Chengkou fault in the transitional zone of the northeast edge of the upper Yangtze block; thus forming a thrusting nappe, ductile shear zone in the front edge of the orogen [6,7,10] (Figure 1b). Shale gas in these deformed rocks is well-preserved and exploration activity has revealed a high gas content [7,8].

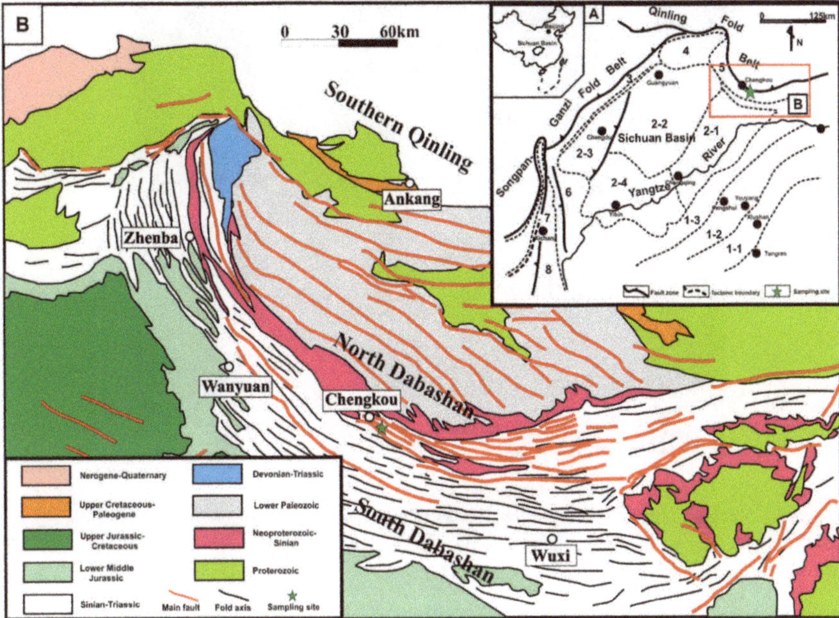

Figure 1. Simplified geological maps and location of the Sichuan Basin (**a**) and Dabashan (**b**) (Modified from Li et al. [6], Ma et al. [7], Kang et al. [10]). 1-1: Xuefengshan fold subzone; 1-2: Western Hubei-Eastern Chongqing-Northern Guizhou fold subzone; 1-3: Qiyueshan-Jinfoushan-Daloushan fold subzone; 2-1: Eastern Sichuan fold belt; 2-2: Central Sichuan fold belt; 2-3: Western Sichuan fold belt; 2-4: Southern Sichuan fold belt; 3: Longmenshan fold belt; 4: Michangshan fold belt; 5: Dabashan fold belt; 6: Emeishan-Daliangshan block-fault belt; 7: Xichang basin; 8: Kangdian tectonic belt.

Representative shale samples were selected from two groups of organic-rich, overmature samples characterized by different deformation structures and from two outcrops (e.g., Longtian and Xiuqi in Chongqing) [9,25,26]. Nanosized to microsized structural characteristics within the Lujiaping Formation are studied in relation to structural deformation and maturation. The sample numbers are abbreviated as L1, L2, L3, and L4 in the fractured zone of Longtian and D1, D2, D3, and D4 in the intensive folds of Xiuqi. Primary sedimentary structures from the Longtian outcrop are difficult to discern due to the presence of brittle deformation structures and abundant calcite veins. Some microfractures developed along grain cleavage or other preexisting weak zones are filled with calcite and partially faulted, showing certain displacement and movement directions and strongly brittle deformation (Figure 2a,b). Four deformed shales taken from the Xiuqi outcrop have been mylonitized almost completely, resulting in distinct anisotropy. We have observed abundant evidence of ductile or brittle deformation such as microfolds, S-C fabric, microfracture, and cataclastic flow in these shales at millimeter and micrometer scales (Figure 2c,d).

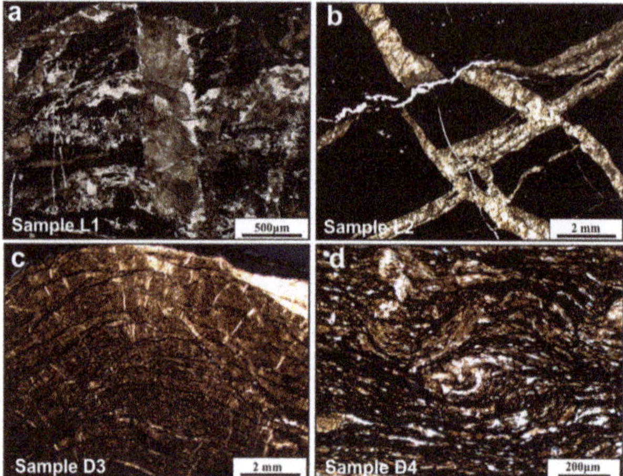

Figure 2. Optical photomicrograph of sample L1 (**a**) and L2 (**b**). Brittle deformation and "X" joints can be found, showing complex cross-cutting relationships. Optical photomicrograph of sample D3 (**c**) and D4 (**d**). Deformation continued after veins formation, resulting in the development of fault-fold microstructures, rotated porphyroclast, and further veining.

3. Methods

3.1. Geochemistry and Petrology

Before experimental analyses of organic geochemistry, each sample was ground to a fine powder (less than 100 mesh) and removed carbonates using hydrochloric acid. The total organic carbon (TOC) was measured by a LECO CS-230 Carbon/Sulfur Analyzer (San Joseph, MI, USA). A 3Y-Leica DMR XP microscopy (Lycra, Frankfurt, Germany) equipped with a microphotometer (MPV-III) was used to measure the R_o of shale samples with vitrinite particles measured at least 30 times. The samples prepared for whole-rock mineralogical analysis were crushed and sieved to 80 mesh size, mixed with ethanol, and then smear mounted on glass slides for random powder X-ray diffraction (XRD) analysis. Scanning range is 5°–80° and speed 4°/min.

3.2. Thin Section Microscopy and SEM Imaging Analyses

The nano-microscale structures were directly observed and described using thin section and SEM. All polished thin sections were analyzed for rock fabrics and microstructures. Standard procedures were used in preparing the thin-section. Thin sections are 30–40 μm thick and 3 cm long, and petrographic studies were done using a Leica DMLP polarizing microscope (Lycra, Frankfurt, Germany) with a Leica DFC450C camera system. The SEM is equipped with low and high secondary electron probes and an X-ray spectrometer. Samples were cut into 5 mm × 5 mm × 5 mm chips and polished to 0.1 mm thick using argon ion-milling. Gold coating is 20 nm thick. Structures can be measured on SEM images using the ImageJ software that is chosen for 2-D and 3-D modeling [32].

3.3. Low-Pressure Nitrogen Adsorption (LPNA)

LPNA analyses have been used to quantify pore structures using N_2 adsorption at −196 °C by an automated pore size analyzer (Autosorb IQ, Quantachrome, Boynton Beach, FL, USA). The pore size distribution analysis was calculated using the density function theory (DFT) calculation model, which can provide a more accurate approach for pore size analysis in the micro- to mesopore-scale [26]. Therefore, in the present study, powered samples were analyzed with N_2 adsorption to obtain information about micropores and mesopores; macropores are also observed by SEM imaging.

3.4. Low-Field Nuclear Magnetic Resonance (NMR)

The samples were also taken for low-field NMR, which has been widely used in porous rocks, such as coal, gas shale, and oil shale [13,19,33–35]. The theory of this method is that T_2 can provide abundant information associated with pore fluids in porous rocks in a typical low-field NMR measurement. In other words, the pore fluid shows a linear relationship with the NMR T_2. All samples were resaturated with water for NMR analyses. Based on this theory, the pore size distribution (PSD) can be reflected in the NMR T_2 spectrum, with a lager pore space having a longer NMR T_2 relaxation time and smaller pores having a shorter relaxation time [19]. A detailed description of such methods can be found in Yao et al. [13].

4. Results

4.1. Organic Geochemistry and Mineralogy

Organic geochemistry and mineralogy of Lujiaping sediments are well-documented as rich source rocks [7–9]. It is important to characterize the TOC, R_o, and mineral composition of samples because each of these parameters has a significant effect on pore structural evolution and pore type [19,36]. Organic geochemistry and mineralogy data of the shale samples are shown in Table 1. All shale samples display high-quality organic carbon (TOC > 2.17%), while R_o ranges from 2.92% to 3.26%. The dominant mineral constituents throughout all samples are quartz, carbonate, and feldspar with an average of 48%, 16%, and 11%, respectively; clay minerals and pyrite with an average of 17% and 6.5%, respectively. These samples have a high content of brittle minerals.

The literature shows that the Lujiaping Formation organic-rich shales have variable TOC content, R_o, and mineral composition [7–9]. Ma et al. [7] and Han et al. [8] measured the TOC values of 0.44% and 6.91%, and vitrinite-like macerals reflectance (R_o) between 3.3% and 4.3% with type II kerogen for the Lower Cambrian Lujiaping shale in the Dabashan arc-like thrust-fold belt, southwestern China. Other studies of the Lujiaping Formation describe similar mineralogy to those presented here [9]. As pointed out by Zhu et al. [9], the Lujiaping Shale samples are all quartz and carbonate minerals rich and all contain greater than 50% relatively brittle minerals (quartz, carbonate, and feldspar). In general, the shale with high levels of brittle minerals can increase the effectiveness of hydraulic fracturing of a shale gas reservoir.

Table 1. Total organic carbon (TOC), R_o and XRD mineral composition characteristics of the shale samples.

Location of Samples	Sample ID	Quartz (%)	Carbonate (%)	Clays (%)	Pyrite (%)	Feldspar (%)	TOC (%)	R_o (%)
Longtian outcrop, Chongqing	L1	38	27	13	6	8	6.55	3.14
	L2	56	23	10	3	6	4.14	3.03
	L3	65	10	13	6	4	5.24	3.01
	L4	57	20	10	3	7	3.57	3.26
Xiuqi outcrop, Chongqing	D1	27	22	17	8	24	2.17	2.92
	D2	42	7	23	11	14	3.49	2.93
	D3	47	9	11	8	19	8.57	2.99
	D4	53	7	35	7	8	2.28	3.02

4.2. Nanometer- to Micrometer-Sized Pore Structures

Gas storage structures of shale form by both depositional and diagenetic processes, affected by thermal evolution and tectonic deformation [9,12,19,23,25,26,36]. Abundant nanometer- to micrometer-sized pores can be observed in shale matrix exhibiting variance in size, shape, and abundance [12,19,20]. Loucks et al. [20] suggested that geoscientists working on shales use pore size terminology whereby nanopores have widths less than 1000 nm. Pore space can also be divided into micropores (pore width <2 nm), mesopores (pore width between 2 and 50 nm), and macropores (pore width >50 nm) as recommended by the International Union of Pure and Applied Chemistry (IUPAC). Loucks et al. [20] and Slatt and O'Brien [37] used SEM to characterize and define abundant types of gas storage space, such as OM pores, interparticle pores, intraparticle pore, microchannels, and microfractures.

A major aim of this study is to document and characterize visible pore types using FIB-SEM and to define major pore networks. Several recent articles have addressed pore types and pore networks in the Lujiaping Shale using FIB-SEM, gas adsorption, high-pressure mercury injection, and low-field NMR [7–9]. However, none of these studies clarified the size distribution of different types of shale pores and compared these results. Several types of pores occur within the samples: (1) OM-hosted pore, (2) mineral-hosted pore (e.g., intraparticle pore and interparticle pore), and (3) fracture-related pore (e.g., microchannel and microfracture). OM-hosted pore is one of the important matrix-related pore types and associated with organic particles; mineral-hosted pore is another matrix-related pore space and associated with mineral grains, and lastly, fracture-related pore is linear nanometer to micrometer size opening whose derivation is not controlled by individual matrix particles [20]. The relative abundance of each pore type quantified by this investigation is plotted on a histogram (Figure 3), following the Slatt and O'Brien [37], Loucks et al. [20], and Ko et al. [38] classification. These pore types were quantified by point counting SEM images using ImageJ software, as discussed in the Section 3.2. Note that there is plenty of evidence of pore networks within OM–clay aggregates of the deformed shales studied in this investigation. Characterizing different pore types and pore networks in the OM–clay aggregates is discussed later.

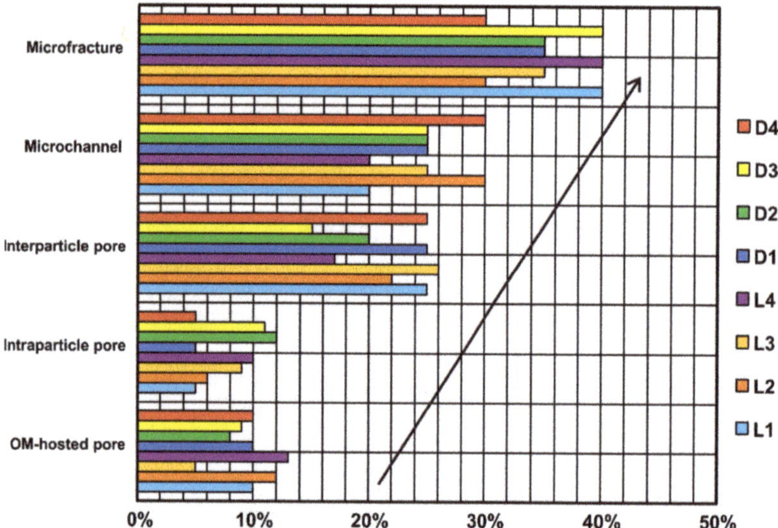

Figure 3. Relative abundance of pore types within all shales. Fracture-related pores (e.g., microchannels and microfractures) account for the 60% of the total pore types, mineral-hosted pores (e.g., interparticle pores and intraparticle pores) account for ~30%, and OM-hosted pores account for ~10%.

4.2.1. Fracture-Related Pores

Microchannels and microfractures are the two most common fracture-related pores in the naturally deformed Lujiaping samples (Figure 4). The relative abundance of microchannels and microfractures appears to vary with tectonism. For example, in deformed shales, microchannels and microfractures are more abundant [9,11,25,39,40] than that in undeformed shales [19,20,37]. A large quantity of open microchannels and microfractures has been found in the samples that we have investigated using SEM-based imaging. These open fracture-related pores are nonmatrix pores, which can have an important effect on hydrocarbon production [11,20,37]. Lujiaping shale gas reservoirs have large amounts of microchannels and microfractures that are not cemented and impermeable. Therefore, the presence of preexisting fractures can have a strong influence on induced fracture propagation. These open pores are linear nanometer to micrometer-sized openings, whereas microchannels have relatively smaller size and a stronger local connection than that of the microfracture [37]. Microchannels occur within the matrix of shale samples have various sizes and shapes (Figure 4a,b), and may provide significant permeability pathways. Such microchannels observed by SEM are generally more than 100 nm in width and 200 nm in length. Microfractures in the Lujiaping deformed shales occur at a variety of widths and lengths (Figure 4c,d). They are more than 300 nm in width and 2000 nm in length. Such pore types are open and may connect to adjacent pores (example Figure 4b) and create even better permeability [20]. Figure 4 shows fracture-related pores within the deformed Lujiaping Shale that are similar to pores described by Zhu et al. [9].

Note that fracture-related pores can be produced during both thermal maturity and tectonic deformation [9,22–25]. Formation of thermally controlled fractures is related to thermal cracking within OM or at the boundary between OM and mineral particles [22–24]. Instead, tectonically controlled fractures are primarily related to the presence of brittle mineral grains [9,25,39,40]. Such fractures appear extend far enough to cut through or around mineral particles.

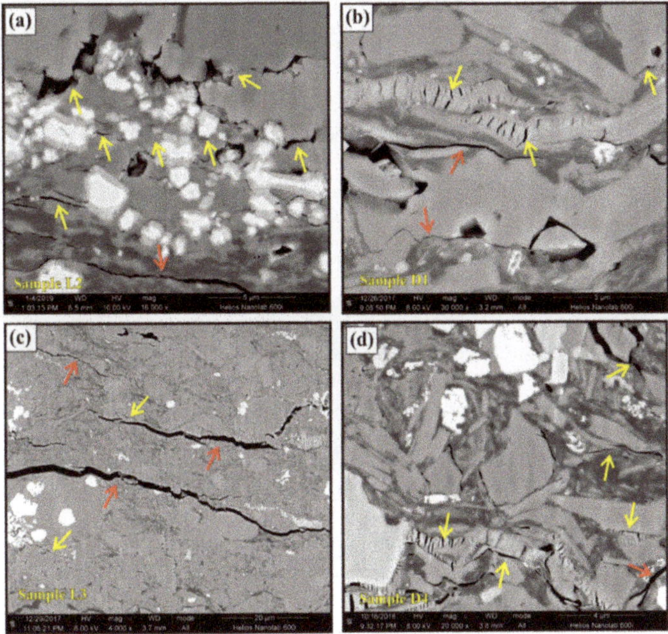

Figure 4. SEM photomicrograph images of fracture-related pores in Lujiaping deformed shales. Natural microchannels are indicated with yellow arrows whereas the microfractures are indicated with red arrows. (**a**) Fracture-related pores within sample L2; (**b**) fracture-related pores within sample D1; (**c**) fracture-related pores within sample L3; (**d**) fracture-related pores within sample D4.

4.2.2. Mineral-Hosted Pores

Mineral-hosted pores described in this study consist of interparticle and intraparticle pores. Interparticle pore is the second most common pore type (Figure 5) and includes interparticle clay pores (Figure 5a) and the open pores between framework brittle mineral grains (Figure 5b–d). Many interparticle pores have been damaged or reformed by tectonic deformation. During this process, pores also develop a preferred orientation (Figure 5a). These tectonism-induced interparticle pores within the Lujiaping Formation have been seen in other studies [7,9], and are most likely well connected and may provide interconnected pathways for shale gas flow. Zhu et al. [9] showed excellent examples of interparticle pores that exist along the edges of OM particles or clays, or occur between rigid grains (their Figure 6) of deformed shales. Such pores range between 100 and 5000 nm within the macropore size range and can significantly contribute to an effective pore network [9].

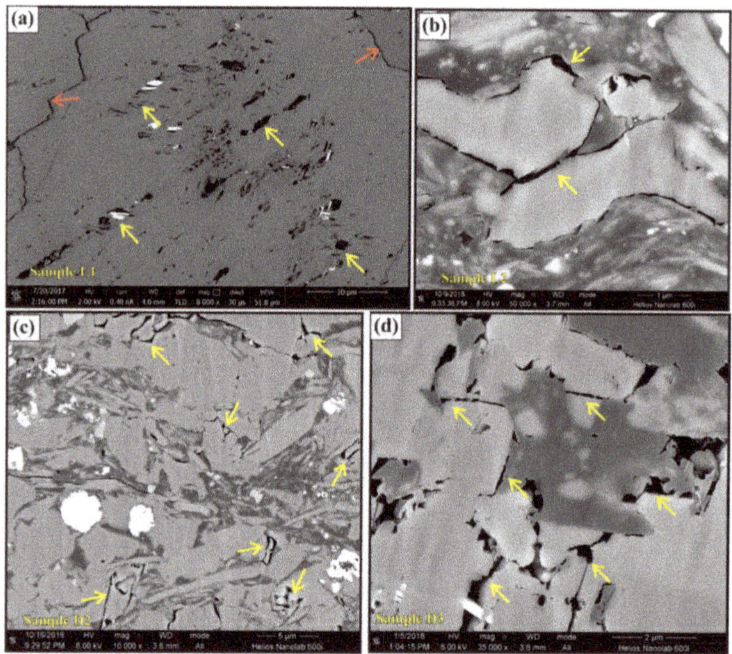

Figure 5. SEM photomicrograph images of interparticle pores in Lujiaping shales. Natural fracture-related pores are indicated with red arrows whereas the interparticle pores are indicated with yellow arrows. (**a**) SEM image showing abundant interparticle clay pores within sample L1; (**b**) SEM image showing abundant interparticle brittle mineral pores within sample L2; (**c**) SEM image showing abundant interparticle brittle mineral pores within sample D2; (**d**) SEM image showing abundant interparticle brittle mineral pores within sample D3.

Intraparticle pore is the third most common pore type (Figure 6) and range in size from 20 nm to 5000 nm with the mesopore to macropore size range. Most intraparticle pores are within calcite and feldspar and form by dissolution in unstable grains. These pores develop in local domains and may not be well connected to the effective pore network. Loucks et al. [20] also recognized intraparticle dissolution pores in undeformed marine shales. According to Loucks et al. [20], this type of intraparticle dissolution pore occurs along crystal rims or within grains and is most likely developed in the subsurface by corrosive fluids. It is generally believed that these pores are formed by the dissolution of unstable minerals by organic acids generated during hydrocarbon generation of organic carbon. Zhu et al. [25] suggested that a large number of dissolution pores developed in deformed shales may have resulted from the open porous network systems produced by tectonism in shale. Such open systems can potentially transmit acidic fluids, resulting in minerals dissolution.

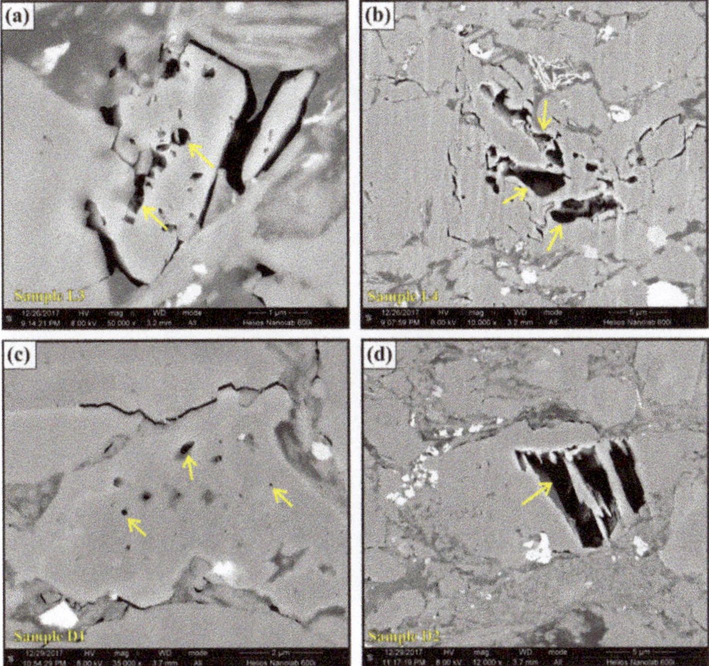

Figure 6. SEM photomicrograph images of intraparticle dissolution pores in Lujiaping deformed shales. The dissolution pores are indicated with yellow arrows. (**a**) SEM image showing abundant intraparticle dissolution pores within sample L3; (**b**) SEM image showing abundant intraparticle dissolution pores within sample L4; (**c**) SEM image showing abundant intraparticle dissolution pores within sample D1; (**d**) SEM image showing abundant intraparticle dissolution pores within sample D2.

4.2.3. OM-Hosted Pores

Recent studies have shown that OM-hosted pores usually display isolated, bubble-like, elliptical cross-sections, and are generally produced during hydrocarbon maturation [19–24]. OM pores are relatively uncommon in all of the deformed samples (Figure 7). OM-hosted pores can only be observed in samples D3 and D4 due to rigid mineral frameworks (Figure 7c,d). There is no evidence of OM pores in the samples L1–L4 (Figure 7a,b). The OM-hosted pores have angular or subangular shapes, with most of the pores being smaller than 50 nm. In addition, the sizes of these OM-hosted pores in our deformed shales appear to be five orders of magnitude smaller than those found in other marine shales (e.g., Barnett Shale and Longmaxi Shale) [19,20,32,36,37]. Similarly, rare OM-hosted pores were observed in deformed shales from previous investigations [7,9,25]. A large number of observations of the Cambrian shales in South China [7–9] have confirmed that organic porosity is not well developed. The reasons for a lack of OM-hosted pores may include (1) the burial depth (more than 7000 m) and strong compaction, (2) high degree of thermal evolution ($R_o > 3\%$), and (3) strongly tectonic deformation. Ma et al. [7] and Zhu et al. [9] stated that few OM pores would develop in Lujiaping shale during the folding and thrusting movements from the Late Triassic time. Furthermore, the angularity of the pores in samples D3 and D4 may be due to one or more of the reasons outlined above [25,26].

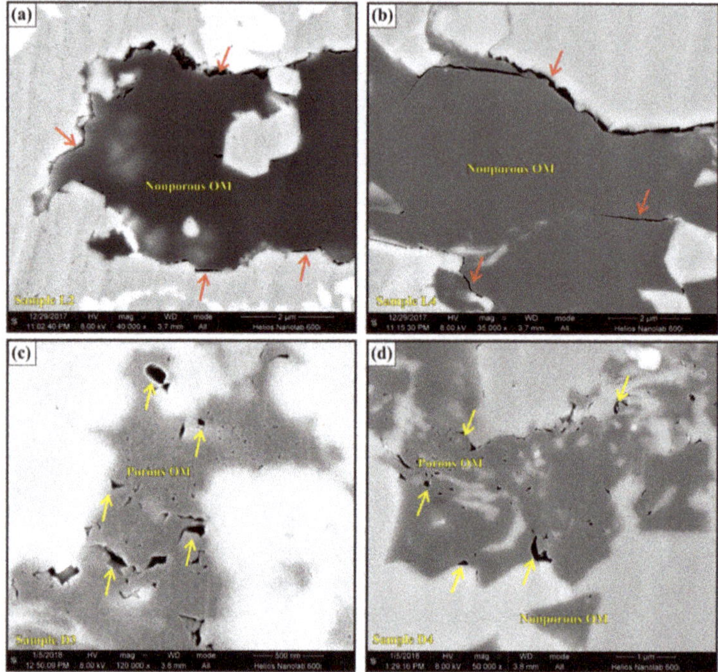

Figure 7. SEM photomicrograph images of OM-hosted pores in Lujiaping deformed shales. Natural thermally controlled fractures are indicated with red arrows whereas the OM pores are indicated with yellow arrows. (**a**) SEM image showing shrinkage crack pores within OM of sample L2; (**b**) SEM image showing shrinkage crack pores within OM of sample L4; (**c**) SEM image showing OM-hosted pores within sample D3; (**d**) SEM image showing OM-hosted pores within sample D4.

Only a few OM pores can be preserved because the organic grains are gradually undergoing significant hydrocarbon generation followed by tectonic compression. Therefore, we use Figure 8 to reveal the structure evolution of OM pores during both thermal maturity and tectonic deformation. With hydrocarbon maturation increase, the OM pores may experience a series of processes, such as development, growth, and mineral-filling. If the shales are influenced by the tectonic compaction, pore shapes tend to appear less rounded, and show more branch-like and line-like edges due to lack of support and protection of the surrounding minerals. These parallel-elongate secondary pores show rectilinear alignment, which is a regularity perhaps related to the structural deformation.

Also, we use the aspect ratio of pores to discuss the evolution of pore morphology during hydrocarbon maturation and tectonism. In general, pores within shale matrix appear to be nonequant. Loucks et al. [19] suggested that average aspect ratios (length divided by width) of pores within OM from Barnett shale have a mean of 2.8:1. Figure 9 illustrates four pores of different aspect ratios and extending directions taken from OM pores of the Lujiaping shales. The detailed features of pore shape in two dimensions and three dimensions are identified by SEM and the software ImageJ Pore shapes can be extracted from gray value, showing a very clear pore shape and surface morphology.

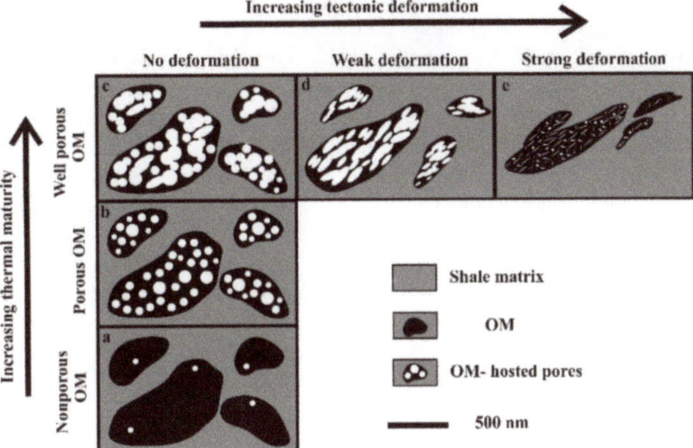

Figure 8. Schematic diagrams showing OM pore evolution of organic-rich shales during both thermal maturity (**a–c**) and tectonic deformation (**c–e**).

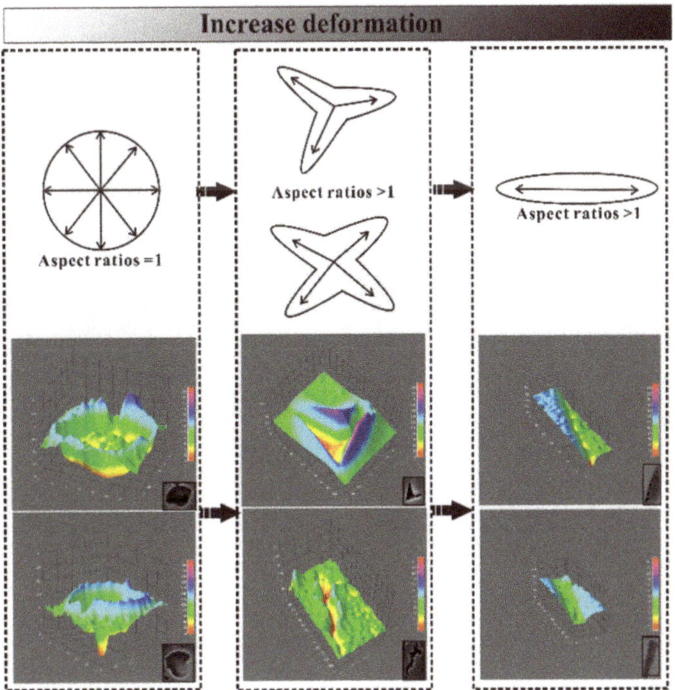

Figure 9. Evolution of the OM pore aspect ratio with increasing tectonic compression.

4.2.4. Pore Size Distribution (PSD)

Pore size distribution (PSD) was documented using a combination of LPNA to measure the micro- to mesopore range and low-field NMR to measure micro- to macropore PSD [9,13,26,33–35]. Both of these two methods indirectly measure nanometer-size pore structures. Nanometer- to micrometer-size pore structures and pore types can be observed qualitatively by SEM imaging [19,20].

The results of the LPNA analysis reveal that there is a distinct bimodal PSD within the Lujiaping Formation (Figure 10a,b). Two subgroups of pore sizes are evident: one ranging from 0.8 to 3 nm and the second ranging from 3 to 10 nm (Figure 10a,b). Figure 10a,b suggests that micropores dominate more of the pore volume than mesopores. The total pore volume of these samples is largely controlled by pore sizes smaller than 10 nm. All of the eight samples have a similar bimodal PSD. Low-field NMR measurement was conducted on eight samples from the Lujiaping Formation. As noted in the methods section, samples were resaturated with water for analyses. T_2 relaxation time curves can provide an estimation of PSD of shale samples. Such curves cannot provide quantitative data on actual pore sizes; however they could indirectly reflect PSD. The shape of the T_2 relaxation time curves of samples, such as L1, L4, and D4, shown in Figure 10c,d, reveals a relative PSD similar to that of the LPNA curves in Figure 10a,b. The T_2 relaxation time curves of most of the other samples show unimodal PSDs, indicating that the samples are dominated by micro- to mesopores. Under the examination of SEM, it is evident that most macropores larger than 50 nm are fracture-related and mineral-hosted pores, whereas pores smaller than 50 nm may be OM-hosted pores and part mineral-hosted pores (Figures 4–7). All samples develop abundant fracture-related and mineral-hosted pores. These fracture-related and mineral-hosted pores are macropore-sized. Most larger pores cannot be measured by LPNA or shown by NMR T_2 relaxation time curves. A combination of SEM, LPNA, and low-field NMR could lend support to our study of nanometer- to micrometer-size pore structures in the Lujiaping samples.

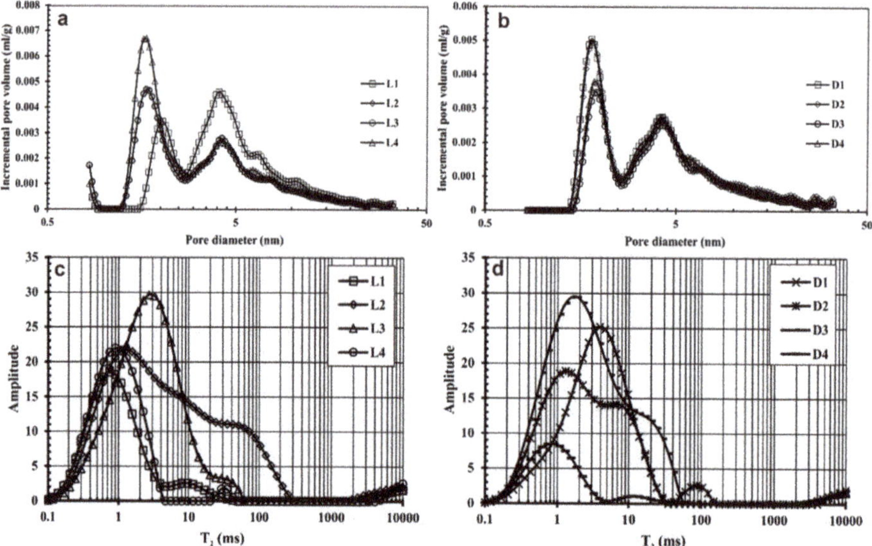

Figure 10. (**a,b**) Plots of pore volume versus pore diameter (nm) showing a distinct bimodal pore size distribution (PSD) within samples obtained from LPNA measurement. (**c,d**) NMR T_2 relaxation time curves of shale samples.

4.3. Nanometer- to Micrometer-Size Material Structures

4.3.1. OM–Clay Aggregates

Lujiaping Shale has been defined as a rock containing abundant OM and clay minerals. OM and clay minerals are frictionally weak relative to other common grains (e.g., quartz, feldspar, pyrite, and calcite). Some important deformation microstructures may be controlled by the content and mechanical properties of organic grains and clay minerals in shale [41–43]. Observations of our naturally deformed samples reveal that OM–clay aggregates within the shale matrix are common (Figure 11). Other studies of the Lujiaping Formation describe similar OM–clay aggregates to those shown in this paper [7,9]. We infer that such OM–clay aggregates can affect the preservation of organic carbon in marine black shale, perhaps especially in deformed shales. A large volume of literature has presented data on the preservation mechanism of organic carbon revealing that large amounts of the OM preserved in most shales are intimately associated with clay minerals [41–45]. Organic grains adsorb onto clay mineral surfaces or concentrate in interlayer or interparticle pores of clays.

Pore structures within OM–clay aggregates were commonly observed and well developed in all deformed shales (Figure 11). The aggregates contain abundant microfractures and interparticle pores. The aggregates are generally open and with good connectivity between aggregates, which is a function of both thermal evolution and tectonic deformation. Other Lujiaping studies have noted similar pore structures in the OM–clay aggregates [7,9]. Zhu et al. [9] report well-developed and high-connectivity pore networks in the Lujiaping Shale. Four types of pores have been identified in OM–clay aggregates (see their Figure 8) [9]. They also showed an evolutionary model of the formation of OM–clay aggregates during tectonic deformation (see their Figure 9), suggesting that these nanometer- to micrometer-sized material structures are naturally formed and are beneficial to the preservation of organic carbon during thermal and tectonic evolution.

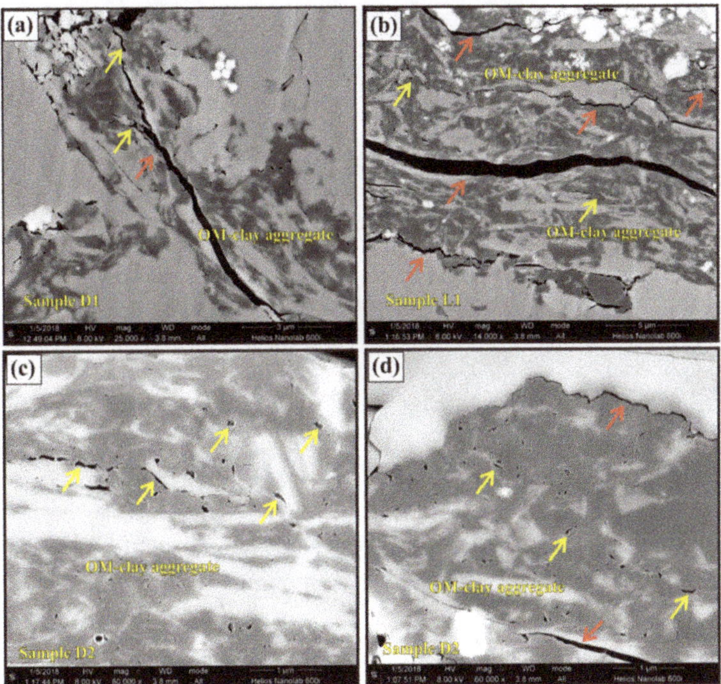

Figure 11. SEM photomicrograph images of OM–clay aggregates and their related pore networks. Natural thermally or tectonically controlled fractures are indicated with red arrows whereas the OM pores are indicated with yellow arrows. (**a**) SEM image showing OM–clay aggregates of sample D1; (**b**) SEM image showing OM–clay aggregates and their related shrinkage crack pores and microfractures of sample L1; (**c**) SEM image showing OM–clay aggregates and their related pore systems within sample D2; (**d**) SEM image showing OM–clay aggregates and their related pore systems within sample D2.

4.3.2. Silica Nanograins

The XRD analyses show that Lujiaping Shale is a very siliceous mudstone with 27 wt %–65 wt % silica (average 48%). A study of shale samples with high-resolution SEM reveals that quartz minerals occur as silica nanograins and botryoids typically 20–100 nm in size and can be only observed under high magnification (Figure 12). Drake et al. [46] also recognized abundant silica nanospheres in the less argillaceous facies of the Upper Devonian Woodford Shale. They used SEM to observe four distinct forms of quartz: (1) randomly distributed angular detrital silica grains (less than 60 μm), (2) siliceous microfossils, (3) tiny euhedral quartz overgrowths and quartz crystals (less than 10 μm), and (4) silica nanospheres (200–500 nm). They believed that such silica nanospheres are ubiquitous in the Woodford Shale, and 2–3 times more common than the other three types of silica grains. Interparticle quartz nanopores between the quartz nanograins are common and apparent. However, the origin of these nanograins is unknown. It is not clear whether they are related to thermal and tectonic evolution in our samples. Drake et al. [46] suggested that such silica nanograins may be a product of microbial precipitation.

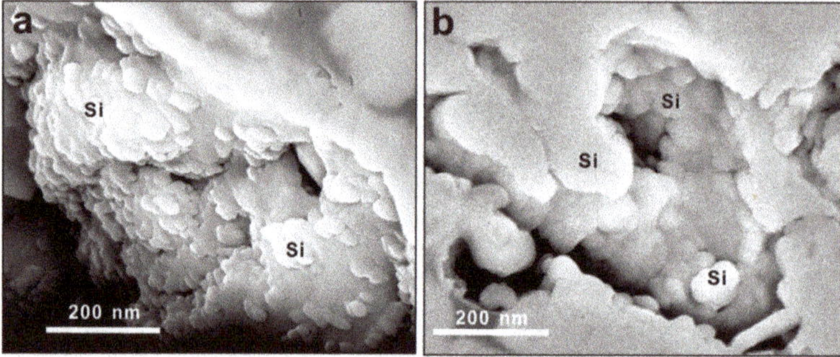

Figure 12. SEM images showing the development of abundant rounded silica nanograins. Silica nanograins have a minimum size of ~20 nm and a maximum size of ~100 nm. Black areas are the open interparticle pore space. (**a**) Features of the rounded silica nanograins within sample L1 and (**b**) features of the rounded silica nanograins within sample D3.

5. Discussion

5.1. Influence of Combined Thermal Evolution and Tectonic Deformation on the Development of the Nanometer- to Micrometer-Sized Structures

Nanometer- to micrometer-sized structures vary due to tectonic deformation and systematically across thermal maturity in organic-rich rocks, but a comprehensive dual study of the effect of tectonism and organic maturity on pore structures and processes in organic-rich shales is lacking. OM-hosted pores are mainly nanometer- to micrometer-sized, and are widely developed in gas shales, such as the Longmaxi Shale, the Yanchang Shale, the Barnett Shale, and the Marcellus Shale [19,20,36,37,47–51]. Recent studies of the Barnett Shale in North America that use SEM imaging suggest that the porosity within OM particles can be higher than 40% of total porosity [22–24]. Some scholars believe that organic nanopores are produced by exsolution of gaseous hydrocarbons during the secondary thermal cracking of liquid hydrocarbons in the gas window [20–22,52], but other geologists disagree with this conclusion, and have observed and found that OM-hosted nanopores can be formed within organic carbon in oil window [53]. The debate is not settled and additional investigations are needed to better understand the thermal control of pores in OM.

On the other hand, several authors have noted that tectonism can significantly affect nanometer- to micrometer-sized sedimentary structures and pore networks along with the proportion of various pore types in organic-rich rocks [7,9,25,54–60]. Based on the results of a structural evolutionary analysis on organic-rich shale, Zhu et al. [9] suggested that clay–organic aggregates and related interparticle pores and open microfractures are formed by tectonic deformation. Also, a conclusive relationship between nanometer- to micrometer-sized pore structure evolution and tectonism was found in other cases, such as the deformed Longmaxi Shale [25]. In particular, Zhu et al. [25] show the development of abundant macropores in the Longmaxi Shale and suggest, similar to the findings of Liang et al. [26], that the proportions of different pore types were significantly changed by structural deformation, and that strongly deformed shales have the least organic pores and the lowest adsorption capacity. The microstructures of tectonically deformed coal (TDC) have also been analyzed in detail by coal geologists [54–62]. These studies have mainly focused on three different aspects: (1) development of TDCs by brittle and ductile deformation of coal seams, (2) macro- and micro-deformation behavior of TDCs, and (3) impact of tectonism and coal rank on the characteristics of pore structures. Brittle deformation in coals has been shown to increase porosity, permeability, and pore size relative to undeformed coals, whereas ductile deformation could result in the development of more nanosized pores, which has a significant effect on the adsorption capacity of coal. Note that gas outburst and strongly ductile deformation are closely related due to their occurrence in coalfield shear zones [54–60]. Such relationships have been documented by different workers [54–60]. For example, Pan et al. [55] summarized the importance of different deformation mechanisms on the evolution of coal pores. Brittle deformation can produce more mesopores and enhance the interconnectivity of the pore systems, whereas ductile deformation strongly affects micropores, methane adsorption capacity, as well as coalbed methane contents. Li et al. [56] described and recognized how tectonic deformation could influence the macromolecular structures and the presence of nanoscale pores smaller than 100 nm.

In addition, other factors, such as deposition, compaction, and magmatic intrusion, are also well known [20]. However, the impact of combined thermal evolution and tectonic deformation in the development of the nanometer- to micrometer-sized pore and material structures has not been well studied and understood. This observation is mainly based on qualitative and quantitative comparisons between shales of varying maturity and structural deformation. The geological controls on organic porosity development are complex. Thermal maturity and the TOC content are the two most important controls on organic nanopores of the gas shales, but the trend is broad, which indicates other secondary factors are present [12,21–24,28,48]. For example, clay doping, clay catalysis, microbial degradation, water film, tectonism, etc. may mask or exacerbate the effects of such thermal controls.

Compared to the previous studies on most undeformed mature or high mature shale reservoirs [19–24,36–38,47], our results demonstrate significant variations in pore structures related to tectonic deformation and thermal maturity. Generally, we expect to find positive relationships between OM-hosted pore volumes and the whole pore network, or between OM-hosted pores and TOC. In our set of samples, however, no relationships were detected between the total pore volumes and OM-hosted pores or between OM-hosted pore volumes and TOC (Figure 7). We suggest that important relationships may be apparent among samples of different maturity or TOC content in undeformed shales, whereas in the case of our deformed shale samples, these relationships are obstructed by strong influences in tectonism. In addition, this lack of correlation between OM-hosted pore volumes and the total pore volume within these shale samples suggest that both fracture-related and mineral-hosted pores are significant contributors to the pore network. Tectonic deformation usually has a positive correlation with the fracture-related pores [9,25]. All of our eight samples have a higher count of microfractures than the undeformed shales. A positive relationship between mineral-hosted pores and tectonism, as well as a generally negative correlation between OM-hosted pores and tectonism also implies that tectonic deformation may significantly control pore types within organic-rich Lujiaping shales. Organic porosity is no longer the prevailing dominant pore type in deformed shales due to both strong tectonic deformation and overmature organic carbon [7,9,25]. Samples that have more fracture-related and mineral-hosted pores are dominated by abundant brittle mineral grains. Such naturally deformed overmature shales show greater connectivity and gas storage space because they have more open fractures and mineral-hosted, microscale pore volumes compared to undeformed mature or highly mature shales.

The authors suggest that OM pores are poorly-developed in deformed shales and further suggest that mineral-hosted porosity, microchannels, and microfractures are significant aspects of the storage and migration of gas. Therefore, our study supports the ideas that (1) thermal maturity is mainly responsible for the formation of organic pore structures, (2) tectonism is largely responsible for mineral-hosted or fracture-related pore structure evolution, (3) total porosity is altered during thermal maturity and structural deformation, (4) a decrease in organic pores may be related to structural deformation, and (5) mineral-hosted pores and microfractures can contribute to gas storage and migration pathways in deformed overmature shales. In the present study, no significant correlations have been observed between abundance of OM and pore volumes measured via SEM, in part due to the limitation of resolution, but organic micropores and mesopores could be indirectly obtained by LPNA and NMR curves. Our overall observations suggest that fracture-related pores together with mineral-hosted pores jointly contribute to the pore network in shales, which is in agreement with other Lujiaping shale studies [7–9].

5.2. Evolution of Reservoir Characteristics in Cambrian Organic-Rich Shale during Thermal Maturation and Structural Deformation

Figure 13 illustrates diagrammatically some hypothetical relationships between arrangement and abundance of the three different pore types and porosity, permeability, connectivity, and gas storage characteristics in deformed, organic-rich shale. If organic pores are dominant in shales, then modest porosity, permeability, and connectivity of OM within shales is expected. Shale gas in this case will mainly be adsorbed gas. If mineral-hosted pores are dominant, then bad to poor porosity, and permeability should be expected. Shale gas storage, where mineral-hosted pores are dominant, will be free gas. Finally, if the microfractures are dominant in shales, these layers may produce relatively good porosity and permeability within shales, and shale gas is also predominantly free gas.

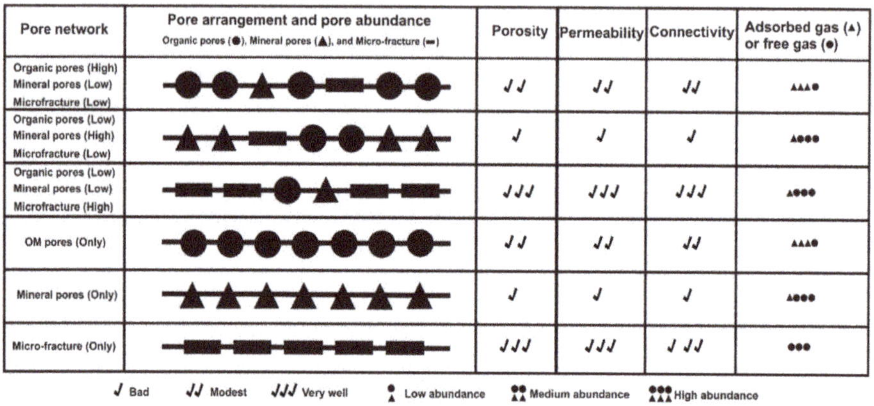

Pore network	Pore arrangement and pore abundance Organic pores (●), Mineral pores (▲), and Micro-fracture (▬)	Porosity	Permeability	Connectivity	Adsorbed gas (▲) or free gas (●)
Organic pores (High) Mineral pores (Low) Microfracture (Low)	●●▲●▬●●	✓✓	✓✓	✓✓	▲▲▲●
Organic pores (Low) Mineral pores (High) Microfracture (Low)	▲▲▬●●▲▲	✓	✓	✓	▲●●●
Organic pores (Low) Mineral pores (Low) Microfracture (High)	▬▬●▲▬▬	✓✓✓	✓✓✓	✓✓✓	▲●●●
OM pores (Only)	●●●●●●●	✓✓	✓✓	✓✓	▲▲▲●
Mineral pores (Only)	▲▲▲▲▲▲	✓	✓	✓	▲●●●
Micro-fracture (Only)	▬▬▬▬▬	✓✓✓	✓✓✓	✓✓✓	●●●

✓ Bad   ✓✓ Modest   ✓✓✓ Very well   ▲ Low abundance   ▲▲ Medium abundance   ▲▲▲ High abundance

Figure 13. Quantitative and qualitative description of hypothetical relationships between arrangement and abundance of three different pore types and porosity, permeability, connectivity, as well as gas storage characteristics.

6. Conclusions

In this study, we investigated the variations and features in the nanometer- to micrometer-sized structures related to thermal maturity and tectonic deformation in the Lower Cambrian Lujiaping overmature shale. The Lujiaping Formation in the Dabashan Thrust-fold Belt has experienced several episodes of intensive tectonic motion after the end of hydrocarbon generation. The deformation of organic-rich shale is clearly evident. Our results reveal the following conclusions:

(1) All shale samples show high-quality OM content (TOC > 2.17%) and the R_o ranges from 2.92% to 3.26%. The dominant mineral constituents are relatively brittle minerals such as quartz, carbonate, and feldspar, with a combined average of 70 wt %.

(2) The Lujiaping Formation pore network is dominated by inorganic porosity, such as fracture-related and mineral-hosted pores. OM-hosted pores observed by SEM are not dominant contributors to the pore network, but may exist as micropores and mesopores.

(3) Our study demonstrates a link between pore structure, and pore type thermal evolution and tectonism. Further research is needed to comprehensively assess the relative importance and abundance of mineral-hosted versus OM-hosted pore types as well as fracture-related pores throughout the Lujiaping Shale. Moreover, we expect our findings to be applicable to other organic-rich, overmature, and strongly deformed shales developed in complex tectonic areas.

(4) We use high-resolution SEM images to observe two main types of nanometer- to micrometer-sized material structures, such as OM–clay aggregates and silica nanograins. Such structures could increase initial gas storage space through the formation of the microfractures and interparticle pore space.

Author Contributions: Conceptualization, H.Z. and Y.J.; Methodology, C.H.; Software, Y.Q.; Validation, L.J., K.Y. and Y.L.; Formal Analysis, X.S.; Investigation, H.F.; Resources, C.H.; Data Curation, P.Q.; Writing—Original Draft Preparation, H.Z.; Writing—Review and Editing, Y.J.; Visualization, H.Z.; Supervision, H.Z.; Project Administration, Y.J. and W.L.; Funding Acquisition, Y.J. and W.L.

Funding: This research was financially supported by the National Natural Science Foundation of China (Grant Nos. 41530315, 41872160, 41372213, 41672201), the National Natural Science Foundation of China for Youth (Grant No. 41804080), the National Science and Technology Major Project of China (Grant Nos. 2016ZX05066003, 2016ZX05066006), the Strategic Priority Research Program of the Chinese Academy of Sciences (Grant No. XDA05030100), and the Sichuan Science and Technology Support Program (Grant No. 2016JZ0037).

Acknowledgments: We thank anonymous reviewers for their critical and constructive reviews. The authors would like to thank Mengyan Shi, Yuzhen Han, Lilong Yan, Libing Wang, Kai Chen, and Le Zhang for sampling work and their constructive comments.

References

1. Gu, Y.; Wan, Q.; Qin, Z.; Luo, T.; Li, S.; Fu, Y.; Yu, Z. Nanoscale Pore Characteristics and Influential Factors of Niutitang Formation Shale Reservoir in Guizhou Province. *J. Nanosci. Nanotechnol.* **2017**, *17*, 6178–6189. [CrossRef]

2. Aguilera, R. Shale gas reservoirs: Theoretical, practical and research issues. *Pet. Res.* **2016**, *1*, 10–26. [CrossRef]

3. Hochella, M.F., Jr. Nanoscience and technology: the next revolution in the Earth sciences. *Earth Planet. Sci. Lett.* **2002**, *203*, 593–605. [CrossRef]

4. Jia, B.; Tsau, J.S.; Barati, R. A review of the current progress of CO_2 injection EOR and carbon storage in shale oil reservoirs. *Fuel* **2019**, *236*, 404–427. [CrossRef]

5. Li, Z.; Liang, Z.; Jiang, Z.; Gao, F.; Zhang, Y.; Yu, H.; Xiao, L.; Yang, Y. The Impacts of Matrix Compositions on Nanopore Structure and Fractal Characteristics of Lacustrine Shales from the Changling Fault Depression, Songliao Basin, China. *Minerals* **2019**, *9*, 127. [CrossRef]

6. Li, Z.; Liu, S.; Luo, Y.; Liu, S.; Xu, G. Structural style and deformation mechanism of the southern Dabashan foreland fold-and-thrust belt, central China. *Front. Earth Sci. China* **2007**, *1*, 181–193. [CrossRef]

7. Ma, Y.; Zhong, N.; Li, D.; Pan, Z.; Cheng, L.; Liu, K. Organic matter/clay mineral intergranular pores in the Lower Cambrian Lujiaping Shale in the north-eastern part of the upper Yangtze area, China: A possible microscopic mechanism for gas preservation. *Int. J. Coal Geol.* **2015**, *137*, 38–54. [CrossRef]

8. Han, H.; Zhong, N.; Ma, Y.; Huang, C.; Wang, Q.; Chen, S.; Lu, J. Gas storage and controlling factors in an over-mature marine shale: A case study of the Lower Cambrian Lujiaping shale in the Dabashan arc-like thrust–fold belt, southwestern China. *J. Nat. Gas Sci. Eng.* **2016**, *33*, 839–853. [CrossRef]

9. Zhu, H.; Ju, Y.; Qi, Y.; Huang, C.; Zhang, L. Impact of tectonism on pore type and pore structure evolution in organic-rich shale: Implications for gas storage and migration pathways in naturally deformed rocks. *Fuel* **2018**, *228*, 272–289. [CrossRef]

10. Kang, J.; Sun, Y.; Men, Y.; Tian, J.; Yu, Q.; Yan, J.; Liu, J. Shale gas enrichment conditions in the frontal margin of Dabashan orogenic belt, south China. *J. Nat. Gas Sci. Eng.* **2018**, *54*, 11–24. [CrossRef]

11. Ougier-Simonin, A.; Renard, F.; Boehm, C.; Vidal-Gilbert, S. Microfracturing and microporosity in shales. *Earth Sci. Rev.* **2016**, *162*, 198–226. [CrossRef]

12. Ross, D.J.; Bustin, R.M. The importance of shale composition and pore structure upon gas storage potential of shale gas reservoirs. *Mar. Pet. Geol.* **2009**, *26*, 916–927. [CrossRef]

13. Yao, Y.; Liu, D.; Che, Y.; Tang, D.; Tang, S.; Huang, W. Petrophysical characterization of coals by low-field nuclear magnetic resonance (NMR). *Fuel* **2010**, *89*, 1371–1380. [CrossRef]

14. Zhu, H.; Ju, Y.; Lu, W.; Han, K.; Qi, Y.; Neupane, B.; Han, Y. The characteristics and evolution of micro-nano scale pores in shales and coals. *J. Nanosci. Nanotechnol.* **2017**, *17*, 6124–6138. [CrossRef]

15. Ju, Y.; Huang, C.; Sun, Y.; Wan, Q.; Lu, X.; Lu, S.; He, H.; Wang, X.; Zou, C.; Wu, J.; et al. Nanogeosciences: Research History, Current Status, and Development Trends. *J. Nanosci. Nanotechnol.* **2017**, *17*, 5930–5965. [CrossRef]

16. Chen, F.; Lu, S.; Ding, X.; Zhao, H.; Ju, Y. Total Porosity Measured for Shale Gas Reservoir Samples: A Case from the Lower Silurian Longmaxi Formation in Southeast Chongqing, China. *Minerals* **2019**, *9*, 5. [CrossRef]

17. Xi, Z.; Tang, S.; Wang, J.; Yi, J.; Guo, Y.; Wang, K. Pore structure and fractal characteristics of Niutitang shale from China. *Minerals* **2018**, *8*, 163. [CrossRef]

18. Curtis, M.E.; Sondergeld, C.H.; Ambrose, R.J.; Rai, C.S. Microstructural investigation of gas shales in two and three dimensions using nanometer-scale resolution imagingMicrostructure of Gas Shales. *AAPG Bull.* **2012**, *96*, 665–677. [CrossRef]

19. Loucks, R.G.; Reed, R.M.; Ruppel, S.C.; Jarvie, D.M. Morphology, genesis, and distribution of nanometer-scale pores in siliceous mudstones of the Mississippian Barnett Shale. *J. Sediment. Res.* **2009**, *79*, 848–861. [CrossRef]

20. Loucks, R.G.; Reed, R.M.; Ruppel, S.C.; Hammes, U. Spectrum of pore types and networks in mudrocks and a descriptive classification for matrix-related mudrock pores. *AAPG Bull.* **2012**, *96*, 1071–1098. [CrossRef]

21. Curtis, M.E.; Cardott, B.J.; Sondergeld, C.H.; Rai, C.S. Development of organic porosity in the Woodford Shale with increasing thermal maturity. *Int. J. Coal Geol.* **2012**, *103*, 26–31. [CrossRef]

22. Milliken, K.L.; Rudnicki, M.; Awwiller, D.N.; Zhang, T. Organic matter–hosted pore system, Marcellus formation (Devonian), Pennsylvania. *AAPG Bull.* **2013**, *97*, 177–200. [CrossRef]

23. Löhr, S.C.; Baruch, E.T.; Hall, P.A.; Kennedy, M.J. Is organic pore development in gas shales influenced by the primary porosity and structure of thermally immature organic matter? *Org. Geochem.* **2015**, *87*, 119–132. [CrossRef]

24. Lu, J.; Ruppel, S.C.; Rowe, H.D. Organic matter pores and oil generation in the Tuscaloosa marine shale. *AAPG Bull.* **2015**, *99*, 333–357. [CrossRef]

25. Zhu, H.; Ju, Y.; Huang, C.; Han, K.; Qi, Y.; Shi, M.; Qian, J. Pore structure variations across structural deformation of Silurian Longmaxi Shale: An example from the Chuandong Thrust-Fold Belt. *Fuel* **2019**, *241*, 914–932. [CrossRef]

26. Liang, M.; Wang, Z.; Gao, L.; Li, C.; Li, H. Evolution of pore structure in gas shale related to structural deformation. *Fuel* **2017**, *197*, 310–319. [CrossRef]

27. Ju, Y.; Sun, Y.; Tan, J.; Bu, H.; Han, K.; Li, X.; Fang, L. The composition, pore structure characterization and deformation mechanism of coal-bearing shales from tectonically altered coalfields in eastern China. *Fuel* **2018**, *234*, 626–642. [CrossRef]

28. Hao, F.; Zou, H.; Lu, Y. Mechanisms of shale gas storage: Implications for shale gas exploration in ChinaMechanisms of Shale Gas Storage. *AAPG Bull.* **2013**, *97*, 1325–1346. [CrossRef]

29. Liu, Y.; Qiu, N.; Xie, Z.; Yao, Q.; Zhu, C. Overpressure compartments in the central paleo-uplift, Sichuan Basin, southwest China. *AAPG Bull.* **2016**, *100*, 867–888. [CrossRef]

30. Feng, W.; Wang, F.; Guan, J.; Zhou, J.; Wei, F.; Dong, W.; Xu, Y. Geologic structure controls on initial productions of lower Silurian Longmaxi shale in south China. *Mar. Pet. Geol.* **2018**, *91*, 163–178. [CrossRef]

31. Ju, Y.; Wang, G.; Bu, H.; Li, Q.; Yan, Z. China organic-rich shale geologic features and special shale gas production issues. *J. Rock Mech. Geotech. Eng.* **2014**, *6*, 196–207. [CrossRef]

32. Tang, X.; Jiang, Z.; Jiang, S.; Li, Z. Heterogeneous nanoporosity of the Silurian Longmaxi Formation shale gas reservoir in the Sichuan Basin using the QEMSCAN, FIB-SEM, and nano-CT methods. *Mar. Pet. Geol.* **2016**, *78*, 99–109. [CrossRef]

33. Zhang, P.; Lu, S.; Li, J.; Chen, C.; Xue, H.; Zhang, J. Petrophysical characterization of oil-bearing shales by low-field nuclear magnetic resonance (NMR). *Mar. Pet. Geol.* **2018**, *89*, 775–785. [CrossRef]

34. Liu, Y.; Yao, Y.; Liu, D.; Zheng, S.; Sun, G.; Chang, Y. Shale pore size classification: An NMR fluid typing method. *Mar. Pet. Geol.* **2018**, *96*, 591–601. [CrossRef]

35. Zhang, S.; Yan, J.; Hu, Q.; Wang, J.; Tian, T.; Chao, J.; Wang, M. Integrated NMR and FE-SEM methods for pore structure characterization of Shahejie shale from the Dongying Depression, Bohai Bay Basin. *Mar. Pet. Geol.* **2019**, *100*, 85–94. [CrossRef]

36. Chalmers, G.R.; Bustin, R.M.; Power, I.M. Characterization of gas shale pore systems by porosimetry, pycnometry, surface area, and field emission scanning electron microscopy/transmission electron microscopy image analyses: Examples from the Barnett, Woodford, Haynesville, Marcellus, and Doig units Characterization of Gas Shale Pore Systems. *AAPG Bull.* **2012**, *96*, 1099–1119. [CrossRef]

37. Slatt, R.M.; O'Brien, N.R. Pore types in the Barnett and Woodford gas shales: Contribution to understanding gas storage and migration pathways in fine-grained rocks. *AAPG Bull.* **2011**, *95*, 2017–2030. [CrossRef]

38. Ko, L.T.; Loucks, R.G.; Zhang, T.; Ruppel, S.C.; Shao, D. Pore and pore network evolution of Upper Cretaceous Boquillas (Eagle Ford–equivalent) mudrocks: Results from gold tube pyrolysis experiments. *AAPG Bull.* **2016**, *100*, 1693–1722. [CrossRef]

39. Ding, W.; Li, C.; Li, C.; Xu, C.; Jiu, K.; Zeng, W.; Wu, L. Fracture development in shale and its relationship to gas accumulation. *Geosci. Front.* **2012**, *3*, 97–105. [CrossRef]

40. Zeng, W.; Zhang, J.; Ding, W.; Zhao, S.; Zhang, Y.; Liu, Z.; Jiu, K. Fracture development in Paleozoic shale of Chongqing area (South China). Part one: Fracture characteristics and comparative analysis of main controlling factors. *J. Asian Earth Sci.* **2013**, *75*, 251–266. [CrossRef]

41. Kennedy, M.J.; Löhr, S.C.; Fraser, S.A.; Baruch, E.T. Direct evidence for organic carbon preservation as clay-organic nanocomposites in a Devonian black shale; from deposition to diagenesis. *Earth Planet. Sci. Lett.* **2014**, *388*, 59–70. [CrossRef]

42. Zhu, X.; Cai, J.; Wang, G.; Song, M. Role of organo-clay composites in hydrocarbon generation of shale. *Int. J. Coal Geol.* **2018**, *192*, 83–90. [CrossRef]

43. Morley, C.K.; von Hagke, C.; Hansberry, R.; Collins, A.; Kanitpanyacharoen, W.; King, R. Review of major shale-dominated detachment and thrust characteristics in the diagenetic zone: Part II, rock mechanics and microscopic scale. *Earth Sci. Rev.* **2018**, *176*, 19–50. [CrossRef]

44. Salmon, V.; Derenne, S.; Lallier-Vergès, E.; Largeau, C.; Beaudoin, B. Protection of organic matter by mineral matrix in a Cenomanian black shale. *Organ. Geochem.* **2000**, *31*, 463–474. [CrossRef]

45. Kennedy, M.J.; Pevear, D.R.; Hill, R.J. Mineral surface control of organic carbon in black shale. *Science* **2002**, *295*, 657–660. [CrossRef]

46. Drake, W.R.; Longman, M.W.; Kostelnik, J. The Role of Silica Nanospheres in Porosity Preservation in the Upper Devonian Woodford Shale on the Central Basin Platform, West Texas. In Proceedings of the RMAG and DWLS Fall Symposium: Geology and Petrophysics of Unconventional Mudrocks, Golden, CO, USA, 27 September 2017.

47. Tan, J.; Horsfield, B.; Fink, R.; Krooss, B.; Schulz, H.M.; Rybacki, E.; Tocher, B.A. Shale gas potential of the major marine shale formations in the Upper Yangtze Platform, south China, Part III: Mineralogical, lithofacial, petrophysical, and rock mechanical properties. *Energy Fuels* **2014**, *28*, 2322–2342. [CrossRef]

48. Tan, J.; Weniger, P.; Krooss, B.; Merkel, A.; Horsfield, B.; Zhang, J.; Tocher, B.A. Shale gas potential of the major marine shale formations in the Upper Yangtze Platform, South China, Part II: Methane sorption capacity. *Fuel* **2014**, *129*, 204–218. [CrossRef]

49. Li, J.; Zhou, S.; Li, Y.; Ma, Y.; Yang, Y.; Li, C. Effect of organic matter on pore structure of mature lacustrine organic-rich shale: A case study of the Triassic Yanchang shale, Ordos Basin, China. *Fuel* **2016**, *185*, 421–431. [CrossRef]

50. Song, L.; Martin, K.; Carr, T.R.; Ghahfarokhi, P.K. Porosity and storage capacity of Middle Devonian shale: A function of thermal maturity, total organic carbon, and clay content. *Fuel* **2019**, *241*, 1036–1044. [CrossRef]

51. Kuila, U.; McCarty, D.K.; Derkowski, A.; Fischer, T.B.; Topór, T.; Prasad, M. Nano-scale texture and porosity of organic matter and clay minerals in organic-rich mudrocks. *Fuel* **2014**, *135*, 359–373. [CrossRef]

52. Bernard, S.; Wirth, R.; Schreiber, A.; Schulz, H.M.; Horsfield, B. Formation of nanoporous pyrobitumen residues during maturation of the Barnett Shale (Fort Worth Basin). *Int. J. Coal Geol.* **2012**, *103*, 3–11. [CrossRef]
53. Reed, R.M.; Loucks, R.G.; Ruppel, S.C. Comment on "Formation of nanoporous pyrobitumen residues during maturation of the Barnett Shale (Fort Worth Basin)" by Bernard et al. (2012). *Int. J. Coal Geol.* **2014**, *127*, 111–113. [CrossRef]
54. Pan, J.; Wang, K.; Hou, Q.; Niu, Q.; Wang, H.; Ji, Z. Micro-pores and fractures of coals analysed by field emission scanning electron microscopy and fractal theory. *Fuel* **2016**, *164*, 277–285. [CrossRef]
55. Pan, J.; Zhu, H.; Hou, Q.; Wang, H.; Wang, S. Macromolecular and pore structures of Chinese tectonically deformed coal studied by atomic force microscopy. *Fuel* **2015**, *139*, 94–101. [CrossRef]
56. Li, X.; Ju, Y.; Hou, Q.; Li, Z.; Li, Q.; Wang, G. Nanopore Structure Analysis of Deformed Coal from Nitrogen Isotherms and Synchrotron Small Angle X-ray Scattering. *J. Nanosci. Nanotechnol.* **2017**, *17*, 6224–6234. [CrossRef]
57. Liu, X.; Song, D.; He, X.; Nie, B.; Wang, L. Insight into the macromolecular structural differences between hard coal and deformed soft coal. *Fuel* **2019**, *245*, 188–197. [CrossRef]
58. Song, Y.; Jiang, B.; Han, Y. Macromolecular response to tectonic deformation in low-rank tectonically deformed coals (TDCs). *Fuel* **2018**, *219*, 279–287. [CrossRef]
59. Zhao, S.; Li, Y.; Wang, Y.; Ma, Z.; Huang, X. Quantitative study on coal and shale pore structure and surface roughness based on atomic force microscopy and image processing. *Fuel* **2019**, *244*, 78–90. [CrossRef]
60. Yin, Y.; Zhao, T.; Zhang, Y.; Tan, Y.; Qiu, Y.; Taheri, A.; Jing, Y. An Innovative Method for Placement of Gangue Backfilling Material in Steep Underground Coal Mines. *Minerals* **2019**, *9*, 107. [CrossRef]
61. Curtis, J.B. Fractured shale-gas systems. *AAPG Bull.* **2002**, *86*, 1921–1938. [CrossRef]
62. Liu, J.; Yao, Y.; Liu, D.; Pan, Z.; Cai, Y. Comparison of three key marine shale reservoirs in the southeastern margin of the Sichuan basin, SW China. *Minerals* **2017**, *7*, 179. [CrossRef]

Article

Fractal Characteristics and Heterogeneity of the Nanopore Structure of Marine Shale in Southern North China

Kun Yu *, Yiwen Ju *, Yu Qi, Peng Qiao, Cheng Huang, Hongjian Zhu and Hongye Feng

Key Laboratory of Computational Geodynamics, College of Earth and Planetary Sciences, University of Chinese Academy of Sciences, Beijing 100049, China; qiuqiuyu911@163.com (Y.Q.); qiaopeng18@mails.ucas.edu.cn (P.Q.); huangcheng150@126.com (C.H.); zhj8641@163.com (H.Z.); fhy0205@163.com (H.F.)
* Correspondence: yukun@cumt.edu.cn (K.Y.); juyw@ucas.ac.cn (Y.J.); Tel.: +86-188-1069-5655 (K.Y.)

Received: 11 March 2019; Accepted: 17 April 2019; Published: 19 April 2019

Abstract: The characteristics of the nanopore structure in shale play a crucial role in methane adsorption and in determining the occurrence and migration of shale gas. In this study, using an integrated approach of X-ray diffraction (XRD), N_2 adsorption, and field emission scanning electron microscopy (FE-SEM), we systematically focused on eight drilling samples of marine Taiyuan shale from well ZK1 in southern North China to study the characteristics and heterogeneity of their nanopore structure. The results indicated that different sedimentary environments may control the precipitation of clay and quartz between transitional shale and marine shale, leading to different organic matter (OM)–clay relationships and different correlations between total organic carbon (TOC) and mineral content. The shale with high TOC content tended to have more heterogeneous micropores, leading to a higher fractal dimension and a more complex nanopore structure. With the increase of TOC content and thermal evolution of OM, the heterogeneity of the pore structure became more significant. Quartz from marine shale possessed abundant macropores, resulting in a decrease of the Brunauere–Emmette–Teller (BET) surface area (SA) and an increase of the average pore size (APS), while clay minerals developed a large number of micropores which worked together with OM to influence the nanopore structure of shale, leading to the increase of the SA and the decrease of the APS. The spatial order of interlayer pores increased with the increase of mixed-layer illite–smectite (MLIS) content, which naturally reduced the fractal dimensions. In contrast, kaolinite, chlorite, and illite have a small number of nanopores, which might enhance the complexity and reduce the connectivity of the nanopore system by mean of pore-blocking. Taiyuan shale with higher heterogeneity is highly fractal, and its fractal dimensions are principally related to the micropores. The fractal dimensions correlate positively with the SA and total pore volume, suggesting that marine shale with higher heterogeneity may possess a larger SA and a higher total pore volume.

Keywords: southern North China; marine shale; nanopore structure; heterogeneity; fractal dimension

1. Introduction

Minerals and organic matter (OM) are the two most basic units of shale, which together determine the structural characteristics and heterogeneity of nanopores [1–3]. Because shale is a complicated porous material which contains a complex and heterogeneous pore structure with both matrix pores and fracture networks [4,5], the nanopore structure parameters of shale are very vital to understand methane adsorption and desorption [2,6,7]. Thus, the evaluation of the nanopore structure of shale is a key issue in shale gas exploration.

Many qualitative and quantitative experimental methods have been applied to investigate shale pores, such as field emission scanning electron microscopy (FE-SEM) [8,9], small-angle neutron scattering [6], mercury intrusion [10,11], gas adsorption [12–16], and nuclear magnetic resonance (NMR) [11,17]. Among them, low-pressure liquid nitrogen physical adsorption analysis is an effective approach to study the nanopore structure of shale [18,19]. On the basis of N_2 adsorption–desorption, most of the pore structure parameters of shale, including pore geometry, pore volume, specific surface area (SA), and average pore size, can be characterized [20–22]. In order to investigate the complexity and heterogeneity of pore structure, the fractal theory, which is a widely used model to be applied on the basis of N_2 physisorption experiments, should be necessarily introduced. There are four common methods to research the fractal characteristics of the nanopore structure of shale, including the mercury porosity method, gas adsorption method, dispersion method, and scattering method [23–25]. In terms of gas adsorption methods, Brunauere–Emmette–Teller (BET) surface area analysis and Frenkele–Halseye–Hill (FHH) theory are usually adopted to calculate fractal dimensions [26,27]. These adsorption theories provide a simple calculation method and an excellent theoretical basis for the investigation of nanopore structure characteristics and heterogeneity of shale. Khalili et al. [28] firstly used the gas–liquid adsorption isotherm to determine the fractal dimension of solid carbon; Fu et al. [29] investigated the fractal dimension and classification of nanopores of coal based on coalbed methane migration; Yao et al. [11] investigated the fractal characterization of coals from North China with different thermal maturity; Zhang et al. [30] studied the pore structure of coals by using the FHH model and suggested the degree of coal metamorphism is a key factor to control the fractal dimension; Bu et al. [31] studied the fractal dimension of nanopores in non-marine shales based on the FHH model; Yang et al. [32] studied the characteristics of the nanopore structure of the transitional shale and indicated significant differences between marine shale and continental shale with respect to the major control factors of heterogeneity; Liu et al. [33] investigated the differences of fractal dimension of nanopores between marine shale and terrestrial shale. Although a lot of related research works have been conducted, very few studies have examined the fractal dimension and heterogeneity of the nanopore structure of marine shale in southern North China. These fractal researches of shales and coals are conducive to the study the fractal dimension of nanopore structure and the investigation of the key issues regarding the heterogeneity of nanopore structure in marine shale.

In this study, we systematically collected eight marine shale drilling samples from the Carboniferous-Permian Taiyuan Formation in Huainan coalfield, southern North China. The major aims of this study were to: (1) characterize the pore structure parameters to estimate whether marine Taiyuan shales are fractal; (2) study the relationships between fractal dimensions and total organic content (TOC), maturity of organic matter, clay minerals, quartz; (3) identify the heterogeneity of the nanopore structure of marine shale.

2. Geological Setting

Adjacent to the northern margin of the Qinling–Dabie orogenic belt, southern North China, especially the Huainan coalfield, is rich in coal and coalbed methane resources (Figure 1a) [31,34]. The Huainan coalfield is an important Carboniferous-Permian coal accumulation basin in southern North China (Figure 1a) [35]. The ZK1 well is situated in Xinji coal mine in the western part of the Huainan coalfield (Figure 1a,b) and was drilled through the Carboniferous-Permian Taiyuan Formation (Figure 1b,c) to collect geological information about the marine Taiyuan shale reservoir and further research the fractal characteristics and heterogeneity of the nanopore structure of marine shale.

Southern North China experienced long-term erosion, resulting in the loss of a large number of sedimentary strata in the Early Ordovician [36,37]. In the Early Carboniferous, the crust of the study area began to subside slowly, resulting in prevailing transgression and the development of marine-continental transitional coal-bearing sediments. In the Late Carboniferous, Taiyuan Formation was deposited in the carbonate tidal flat and lagoon facies, forming thick layers of limestone and shale. The Carboniferous-Permian Taiyuan shale is the main marine shale in southern North China, with a shallow burial depth (1000–1200 m) and a thickness ranging from 60 to 100 m (Figure 1c).

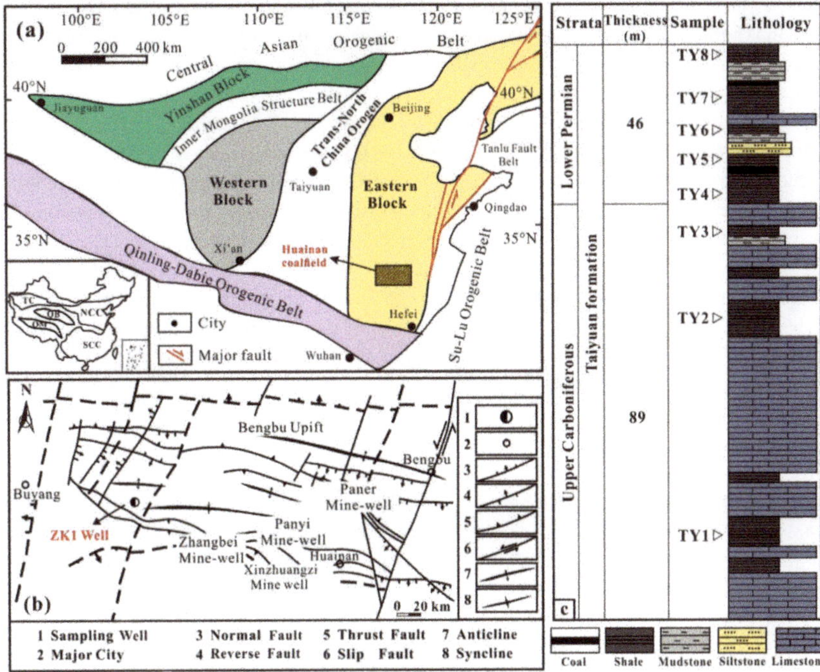

Figure 1. Tectonic map of North China (**a**) [35], sampling well location, detailed structural subdivisions of folds and faults (**b**) [37], and general lithostratigraphic column of the Taiyuan Formation and sampling position of ZK1 well (**c**) [34].

3. Samples and Experiments

3.1. Samples

For the purpose of this study, we selected representative shale samples from well ZK1 that were primely preserved and without weathering. The selected shale samples contained abundant organic matter and fine-dispersed pyrite with extremely thin bedding. In addition, we tried to select samples that were not damaged by drilling. The ZK1 well was drilled at a depth of 1174.3 m in Huainan coalfield, corresponding to the Carboniferous-Permian Taiyuan Formation between 1009 m and 1144 m (Figure 1a,b), and eight shale samples from the marine Taiyuan Formation were obtained from this well (Figure 1c). All shale samples were analyzed for vitrinite reflectance, TOC, composition, and nanopore structure.

3.2. Methods

The shale samples were analyzed on a Leco CS-230 analyzer to obtain the TOC content, following the Chinese Oil and Gas Industry Standard (COGIS) (GB/T4762008) [38]. The vitrinite reflectance experiment was randomly performed on an AXIO Imager Mlm microphotometer produced by ZEISS company, Oberkochen, Germany, following COGIS (SY/T5124-1995) [39]. Quantitative XRD analysis was carried out for the mineralogy of the shale, following COGIS (SY/T5163-2010) [40].

The FE-SEM experiment of the shale samples was carried out in the Analysis and Test Center of Suzhou University, China, using s-4700 cold field emission scanning electron microscopy. Firstly, the sample was broken into blocks of about $10 \times 10 \times 5$ mm^3 in size, and the shale surface was guaranteed to be fresh and pollution-free during the crushing process. In order to increase the sample conductivity and improve imaging quality, it is necessary to spray gold on the sample. This instrument is a secondary electronic imaging system with a super-resolution of 1.2 nm under a high vacuum of 15 KV, which is better than that of the ordinary scanning electron microscope, allowing the observation of nano-scale pores in the shale samples and facilitating the clear observation of nanopore morphology and of the shale samples [9,19].

Low-pressure N_2 adsorption was performed by the Autosorb iQ Station 1 Specific Surface Area Analyzer produced by Quantachrome Ins. Firstly, we crushed the selected shale samples through a 40–60 mesh. In order to remove the residual gas in the sample, the shale sample needed to be vacuum-degassed at 105 °C Celsius for 10 h. The test tubes containing the samples were placed in a Dewar bottle containing liquid nitrogen and connected to the analysis system. Then, high-purity nitrogen was applied for physical adsorption–desorption determination at a test temperature of −195.8 °C. The surface area of the shale samples was obtained by linear regression with the multi-point BET model, and the pore size distribution and pore volume of the shale samples were calculated by using the capillary condensation model Barrett–Joyner–Halenda (BJH) method [41].

4. Results

4.1. Organic Petrology and Mineralogy

The TOC content of the shale samples ranged from 1.24% to 3.38%, with an average value of 2.21% (Table 1). The kerogen of the shale samples was mainly of humic type (Type III), in agreement with the results of a previous works [35]. The mean random reflectance of vitrinite (R_o) of the samples varied from 0.65% to 1.94%, with an average value of 1.33%, indicating a late mature stage of hydrocarbon development. Clay minerals were the main minerals composing the shale samples, whose content ranged from 42.4% to 57.5%, with an average of 47.1%. The content of quartz ranged from 31.9% to 49%, with an average value of 40.3%, and the content of other minerals including carbonate minerals and pyrite were lower (Table 2). Within the clay minerals, the mixed-layer illite–smectite (MLIS) was the main component, with an average value of 28.6%.

Table 1. Partial geological parameters of the Taiyuan marine shale samples. TOC: total organic content.

Sample	Depth (m)	TOC (%)	R_o (%)	Kerogen Type
TY8	1013.1	2.18	0.71	III
TY7	1023.3	1.56	0.65	III
TY6	1034.5	1.91	1.94	III
TY5	1045.7	3.38	1.66	III
TY4	1076.5	2.40	1.7	III
TY3	1083.2	2.32	1.08	III
TY2	1107.3	2.78	1.61	III
TY1	1113.7	1.24	1.31	III

Table 2. Mineral compositions determined by XRD of the Taiyuan marine shale samples. The mineral contents are expressed in weight percentage (wt %). MLIS: mixed-layer illite–smectite.

Sample	Quartz	K-Feldspar	Plagioclase	Calcite	Dolomite	Siderite	Pyrite	Clay	MLIS	Illite	Kaolinite	Chlorite
TY8	39.3	1.1	3.5	0	0	8.2	1.5	45.5	31.7	4.4	8.9	2.5
TY7	41.7	4.4	3.8	0	0	3.3	2.1	44.7	32.5	3.0	9.1	2.4
TY6	40.6	1.6	3.2	0	0	7.5	1.5	45.6	28.5	3.8	8.6	2.3
TY5	49	0	0	0	0	0	8.6	42.4	33.9	4.7	3.0	0.8
TY4	31.9	0.9	1.1	0	0	4.5	4.1	57.5	20.7	7.5	24.7	4.6
TY2	37.1	1.6	0	0	0	8.6	1	51.7	21.7	7.8	18.1	4.1
TY2	41.1	1.2	3.7	0	1.3	0	7.4	45.3	34.4	2.3	6.8	1.8
TY1	36.7	0	0	3.7	2.9	0	7.2	49.5	30.2	6.4	9.9	3.0

4.2. N_2 Adsorption–Desorption Isotherm Characteristics

At present, liquid nitrogen adsorption analysis is an effective way to study the nanopore structure characteristics of materials [42,43]. According to the adsorption isotherm types classified by the

International Union of Pure and Applied Chemistry [44], all of the shale samples here studied correspond to the Type IV isotherm with a hysteresis loop (Figure 2), which is the result of capillary condensation occurring in the mesopore [41], indicating that a large number of mesopores developed in the shale samples. As a whole, the adsorption curve presented an inverse s-shape. At the low-pressure stage of $P/P_0 < 0.4$, the adsorption curve was nearly horizontal, slightly curving upward and rising very slowly, indicating that this stage corresponded to the transition from adsorption monolayer to multilayer. When $0.4 < P/P_0 < 0.8$, the adsorption curve began to rise slowly, and this stage corresponded to the multi-layer adsorption stage. In the $P/P_0 > 0.8$ segment, the adsorption curve rose sharply with the increase of pressure, but when the relative pressure was close to the saturated vapor pressure, the gas adsorption did not reach the saturation state, because capillary condensation occurred in the macropores.

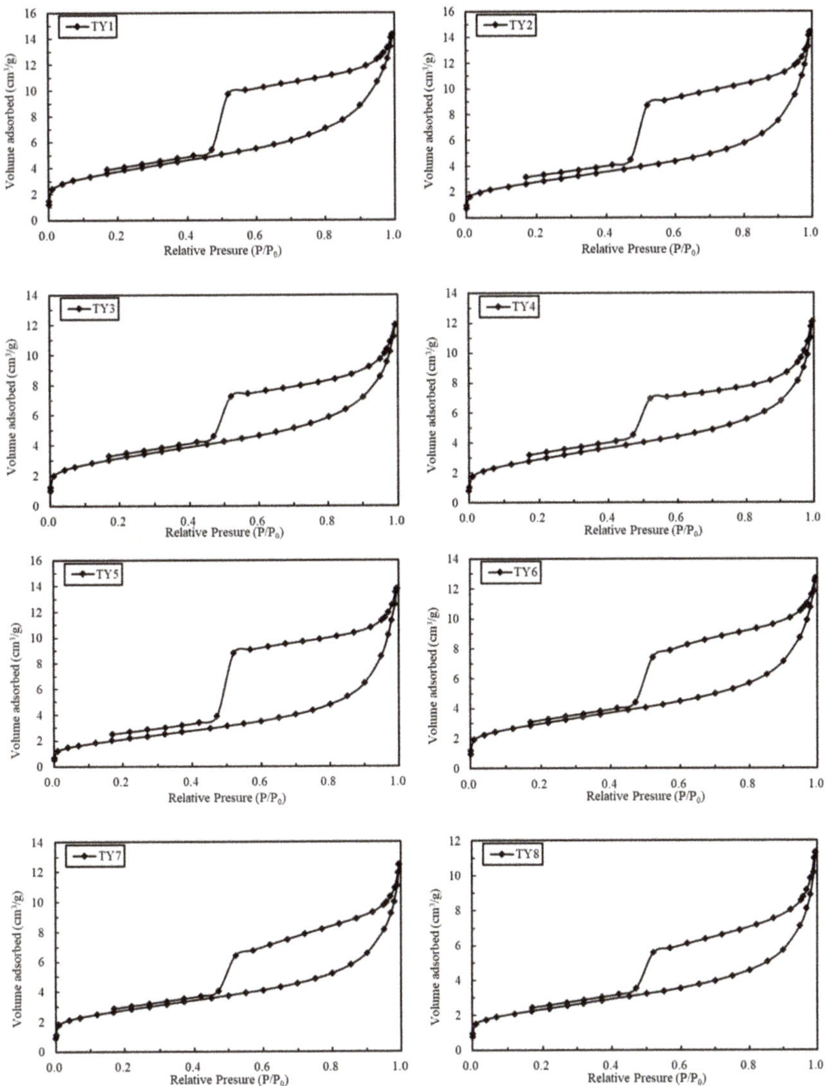

Figure 2. Low-pressure N_2 adsorption–desorption isotherms of the shale samples from Taiyuan Formation.

4.3. Pore Structure Parameter Characteristics

The N_2 adsorption results for the determination of the nanopore structure parameters of the shale samples are presented in Table 3. On the basis of these results, the shale samples exhibited high values of SA, ranging from 8.175 m^2/g to 13.138 m^2/g, with an average value of 10.62 m^2/g. The total pore volumes ranged from 14.341 cm^3/g to 19.508 cm^3/g, with an average value of 16.961cm^3/g.

Table 3. Pore structure parameters of the shale samples from N_2 adsorption isotherms. SA: Brunauere–Emmette–Teller (BET) surface area; APS: average pore size; PV: total pore volume; V_1: micropore volume; V_2: mesopore volume; V_3: macropore volume.

Sample	SA (m^2/g)	APS (nm)	PV (10^{-3} cm^3/g)	V_1 (10^{-3} cm^3/g)	V_2 (10^{-3} cm^3/g)	V_3 (10^{-3} cm^3/g)
TY1	13.138	6.754	19.508	1.551	16.681	1.276
TY2	9.891	9.012	18.994	0.539	16.592	1.863
TY3	11.086	6.696	16.225	1.344	13.461	1.420
TY4	10.331	7.239	15.638	0.883	13.193	1.562
TY5	7.752	11.014	18.166	0.161	16.017	1.988
TY6	10.390	7.372	16.723	1.212	14.271	1.512
TY7	9.728	7.883	15.901	1.266	12.972	1.609
TY8	8.263	8.497	14.298	0.952	11.806	1.603

The pore size distribution (PSD) calculated by the model is shown in Figure 3, showing the pore size of the shale samples ranged from 1 to 100 nm, and the average pore size (APS) was 6.696 nm to 11.014 nm. About the PSD curve (Figure 3a), we observed three different patterns, including micropore, mesopore, and macropore portions. The micropore portion appeared unimodal, with the main peak at 1.4–1.6 nm (Figure 3a). The mesopore portion exhibited multiple peaks, with one major peak in the range of 16–20 nm, and several other peaks (Figure 3a). The macropore portion also appeared unimodal, with a major peak in the range from 60 to 80 nm (Figure 3a). According to the pore SA distribution curve (Figure 3b), there were two peaks in the pore-size range. The micropore and mesopore portions presented a significant peak in the ranges from 1.4 to 1.6 nm and from 2 to 3 nm, respectively.

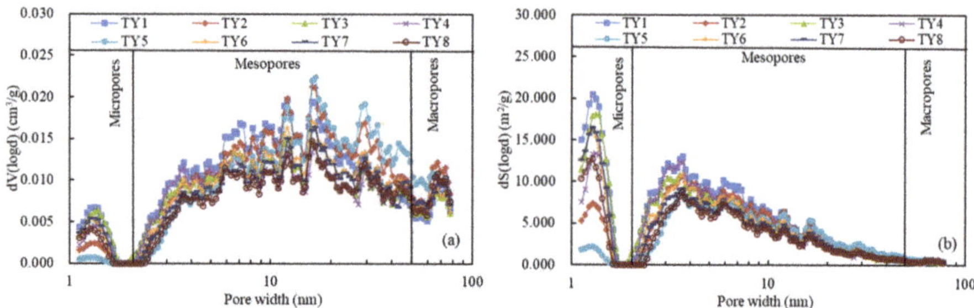

Figure 3. Pore volume (**a**) and pore surface area (**b**) distribution vs pore width.

4.4. Fractal Characteristics of the Pore Structure

Fractal geometry is widely used to describe some fractal systems with no characteristic length scales [45]. Fractal dimension calculated by the gas adsorption method is an effective and reliable petrophysical parameter for describing and quantifying irregular pore structures and complicated surfaces [46]. The calculation method of the FHH model is described as follows: $\ln(V/V_0) = \text{Constant} + K[\ln(\ln(P_0/P))]$. where V is the adsorbed gas volume, V_0 represents the monolayer coverage volume, P_0 refers to the saturation pressure, P is the equilibrium pressure, and K is a constant dependent on the fractal dimension (D) and the adsorption mechanism, which can be obtained by the slope of the plot of $\ln V$ versus $\ln(\ln(P/P_0))$. Then, D can be calculated through the equation used for shales: $D = K + 3$.

The value of D ranges from 2 to 3, and a D value close to 3 indicates that the pore structure is complex and irregular [47].

The plots of lnV versus ln(ln(P/P_0)) and the results of fractal dimensions estimated according to this fractal model are presented in Figure 4 and Table 4. Figure 4a shows that two linear segments (D1 and D2) with different gradients were obvious, which suggested two obvious phases of the N_2 adsorption process, containing monolayer–multilayer adsorption (D2) and pore infilling (D1). The two sections showed a significant linear relationship, with correlation coefficients of more than 0.95, indicating that the marine Taiyuan shales are fractal [47].

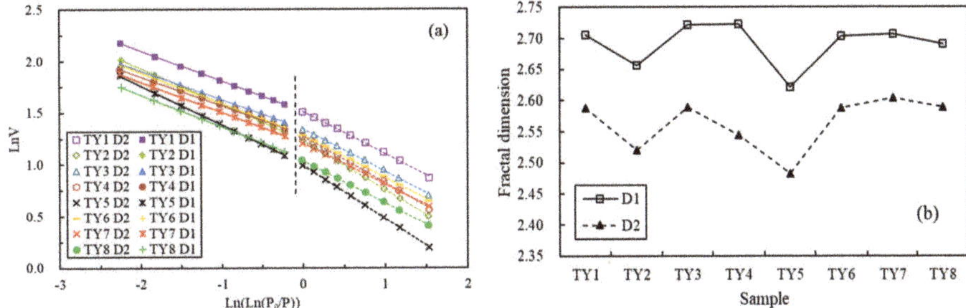

Figure 4. Fractal dimension results calculated from N_2 adsorption isotherms (**a**) and variation trend of fractal dimensions (**b**).

Table 4. Fractal dimensions of marine shale samples calculated from the Frenkele–Halseye–Hill (FHH) model. R^2: correlation coefficient.

Sample	Equation (1)	R^2	K_1	D_1	Equation (2)	R^2	K_2	D_2
TY1	y = −0.2941x + 1.5134	0.9999	−0.2941	2.7059	y = −0.4119x + 1.5149	0.9987	−0.4119	2.5881
TY2	y = −0.3436x + 1.2421	0.9999	−0.3436	2.6564	y = −0.4793x + 1.2333	0.9996	−0.4793	2.5207
TY3	y = −0.2781x + 1.3496	0.9999	−0.2781	2.7219	y = −0.4107x + 1.3456	0.9981	−0.4107	2.5893
TY4	y = −0.2769x + 1.2954	0.9987	−0.2769	2.7231	y = −0.4552x + 1.2775	0.9988	−0.4552	2.5448
TY5	y = −0.3788x + 1.0028	0.9998	−0.3788	2.6212	y = −0.5175x + 0.9936	0.9998	−0.5175	2.4825
TY6	y = −0.2968x + 1.2963	0.9999	−0.2968	2.7032	y = −0.4124x + 1.2883	0.9995	−0.4124	2.5876
TY7	y = −0.2935x + 1.2129	0.9998	−0.2935	2.7065	y = −0.3958x + 1.2069	0.9995	−0.3958	2.6042
TY8	y = −0.3095x + 1.0558	0.9998	−0.3095	2.6905	y = −0.4111x + 1.0416	0.9999	−0.4111	2.5889

The fractal dimension D1 of marine Taiyuan shales varied from 2.6212 to 2.723,1 with an average value of 2.691 (Table 4), which is equal to that of continental shales (2.680, on average), indicating the heterogeneity of nanopores in marine shale is similar to that in continental shale. The fractal dimension D2 from Taiyuan shale showed a wider distribution range from 2.4825 to 2.6042, with an average value of 2.563 (Table 4). In addition, D1 represents the roughness of the pore surface, and D2 refers to the complexity of the pore volume. Figure 4b shows that the variation trend of D1 was similar to that of D2, indicating that the roughness of the pore surface was related to the complexity of the pore volume to some extent. These higher average fractal dimensions suggested that the pore structure of the Taiyuan shale is complex and heterogeneous. Overall, these complex pores make it more difficult for gas to adsorb, diffuse, and flow in marine Taiyuan shale.

5. Discussion

5.1. Relationships between Abundance of Organic Matter and Main Mineral Content

The mineral composition mainly indicates the sedimentary environment of shale in the sedimentary process. The correlations between the abundance of organic matter (TOC content) and quartz and clay content of the Taiyuan marine shale are shown in Figure 5. The results showed that the TOC content was negatively correlated to the clay content (Figure 5a) and positively correlated to the quartz content (Figure 5b), which is consistent with the characteristics of the Wufeng–Longmaxi and Niutitang marine shale in South China and Devonian Horn River shale in Canada [48–53]. However, this result differs from the one obtained for the transitional Permian Shanxi and Xiashihezi shales deposited in North China, which show an opposite correlation [31]. The marine Taiyuan shale was deposited in a carbonate tidal flat facies, a favorable environment for the accumulation of organic matter, which is far away from the land and relatively unfavorable to the enrichment of terrigenous clay minerals, whereas the transitional shale was deposited in a shallow water environment, which is conducive to both sedimentation of terrigenous OM from higher plants and enrichment of terrigenous clay minerals. Therefore, different sedimentary environments result in different OM–clay relationships in transitional shale and marine shale. Basins, depressions, and lagoons are favorable areas for the formation of organic-rich marine shale, because their sedimentary systems are relatively stable. In addition, they are unfavorable for the preservation of terrigenous clay, because the terrigenous material has been transported for a long distance. Shale rich in organic matter and clay minerals possibly develops only in swamps or inland lake basins. In marine sedimentary systems, quartz is mainly derived from the hard bodies of siliceous organisms [13,49,52]. Therefore, there is a positive correlation between TOC content and quartz content in marine shales (Figure 5b). Overall, different sedimentary environments control the precipitation of clay and quartz, leading to different correlations between TOC content and mineral content.

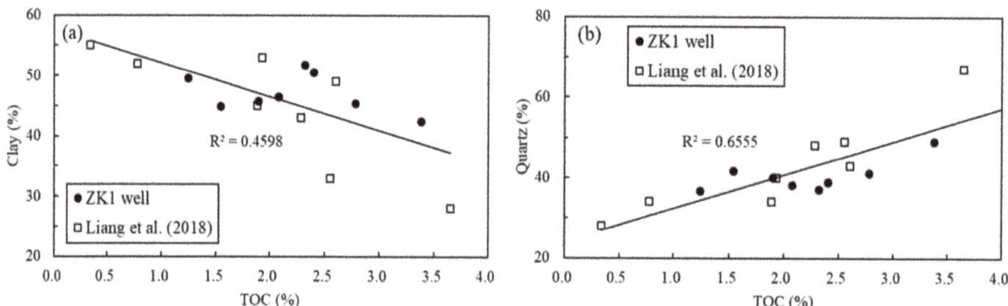

Figure 5. Plots of TOC content versus clay content (**a**) and quartz content (**b**) in the Taiyuan marine shale. The data were obtained from the ZK1 well and from Liang et al. (2018) [53].

5.2. Relationships among Nanopore Structure Parameters

To investigate the relationship among nanopore structure parameters in marine shale, the related plots are given in Figure 6. Significant positive relationships are presented between SA and total pore volume (PV) (Figure 6a–c), which were also demonstrated in the study of over-mature marine shale [51]. This positive relationship was more significant between the SA and the volume of the micropores (R^2 = 0.9924, Figure 6a) and macropores (R^2 = 0.9783, Figure 6c). Furthermore, the micropore SA presented a positive correlation to the total surface area (R^2 = 0.8428, Figure 6d), which indicated that the SA of micropores was dominant among the nanopores of marine shale. Figure 6e exhibits an apparent positive relationship between the mesopore volume and the total pore volume, showing that the mesopore volume dominated the total pore volume in shale, which is consistent with previous studies in marine shale [2,52,54]. The APS showed a negative relationship with the BET SA (R^2 = 0.6632, Figure 6f), indicating that the APS presented a significant negative correlation with SA and PV. Therefore, the negative correlations between APS and SA and PV indicated that shales with smaller APS possess more abundant micropores and mesopores.

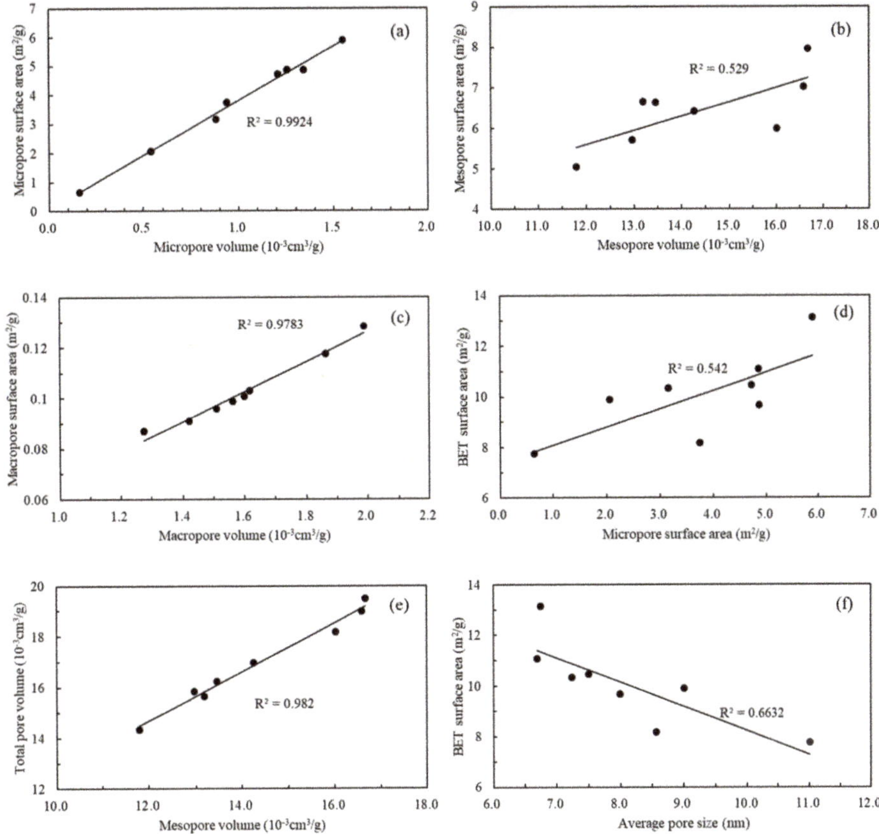

Figure 6. Relationships between SA and PV ((**a**), micropore; (**b**), mesopore; (**c**), macropore), relationship between micropore SA and BET SA (**d**), relationship between mesopore volume and total pore volume (**e**), and relationship between APS and SA (**f**).

5.3. Relationships between Shale Compositions and Nanopore Structure Parameters

The organic-rich shale mainly contains organic matter and minerals, the minerals being clay minerals and quartz. To study the relationships between the main shale composition and nanopore structure parameters in marine shale, the relevant plots are presented in Figure 7. TOC content presented a positive correlation with APS and a negative correlation with SA (Figure 7a), indicating that the studied shale with higher TOC content might possess more macropores and smaller SA and micropore volume. However, the relationships between TOC content and APS and BET SA were not apparent, having correlation coefficients of 0.5584 and 0.4257, respectively, suggesting that the shale with a higher TOC content developed a more complex pore structure, resulting in a stronger heterogeneity of the shale reservoir. According to the results of the FHH fractal dimensions, the studied shale samples are fractal with a high fractal dimensions ($2.6212 < D1 < 2.7231$; $2.4825 < D2 < 2.6042$), which further illustrates that marine shale with high TOC content may develop more heterogeneous micropores, leading to a higher fractal dimension and more complex nanopore structure. Furthermore, the relationships between total pore volume and TOC and Ro were also complex (Figure 7b,c), exhibiting parabolic patterns, which suggests that, with the increase of TOC content and thermal evolution of OM, the variation of the pore structure is more complex. XRD results suggested that quartz and clay were primary minerals in the shale samples (Table 2). We observed moderately positive relationships between BET SA and clay and quartz (Figure 7d), while APS showed opposite correlations with clay and quartz (Figure 7e). Quartz from marine facies possesses a larger number of macropores, resulting in a decrease of the BET SA and an increase of the APS, while clay minerals develop substantial micropores and work together with OM to influence the nanopore system of shale, leading to an increase of the BET SA and a decrease of the APS. However, there was no apparent relationship between total pore volume and clay or quartz (Figure 7f), which indicated that a single mineral cannot determine the total pore volume and may only control the development of a certain type of pore in shale. As for clay minerals, slight relationships are observed in Figure 7g,h, indicating that illite and chlorite had positive relationships with micropore volume and negative relationships with macropore volume. This result suggested that illite and chlorite contain more micropores and fewer macropores, which was also proved by the SEM images presented in Figure 8a,b. Illite is generally leaf-shaped or filamentous, resulting in the development of micro-intergranular pores (Figure 8a); chlorite, originated from biotite diagenetic transformation, usually shows schistose or planar schistose and presents very narrow slit pores among schistose layers (Figure 8b) [55]. Overall, the complicated relationships between pore parameters and OM and the obvious correlation of pore parameters with quart and, clay suggested that OM pores were poorly developed and the complex of OM and minerals might be of great significance in the study of shale pore structure.

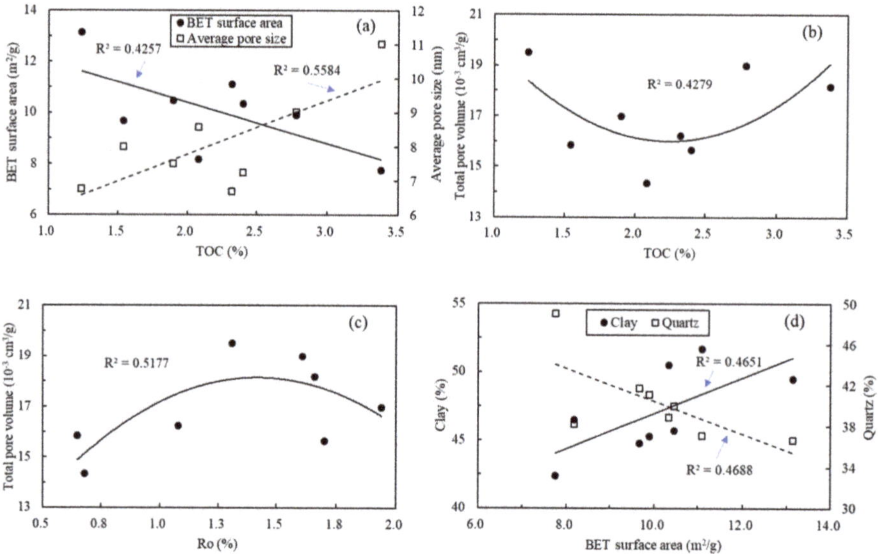

Figure 7. *Cont.*

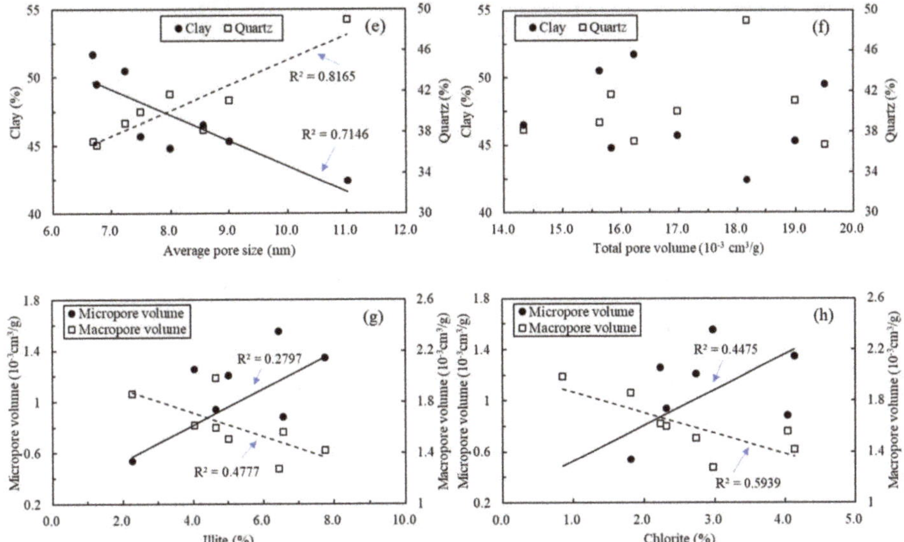

Figure 7. Plots of shale compositions versus nanopore structure parameters in the studied shale. (**a**) BET SA and APS versus TOC content, (**b**) PV versus TOC content, (**c**) PV versus R_o, (**d**) BET SA versus clay and quartz content, (**e**) APS versus clay and quartz content, (**f**) PV versus clay and quartz content, (**g**) micropore volume and macropore volume versus illite content, (**h**) micropore volume and macropore volume versus Chlorite content.

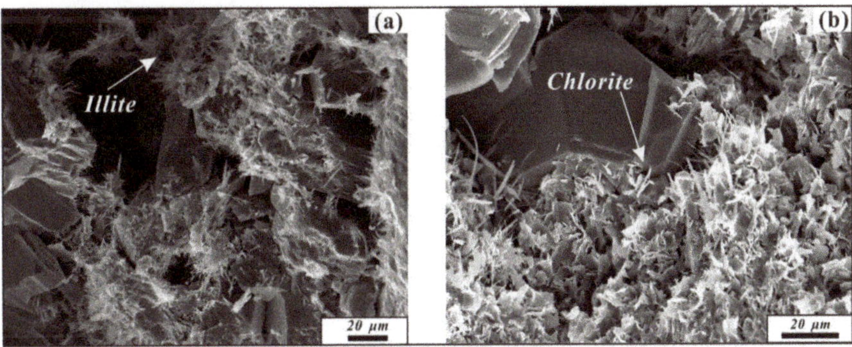

Figure 8. SEM images of illite-associated micro-intergranular pores (**a**) and slit pores among schistose layers of acicular chlorite (**b**).

5.4. Relationships between Mineral Composition and Fractal Dimensions

To study the correlations between minerals and fractal dimensions in marine shale, all linear diagrams between fractal dimensions and quartz and clay minerals are shown in Figure 9. The result showed that the fractal dimensions (D1 and D2) had slight positive correlations with the clay content, whereas a negative relationship existed between the fractal dimensions and the quartz content (Figure 9a,b), which is consistent with the results obtained for the Longmaxi marine shale and Niutitang marine shale [51,54]. These relationships indicated that, with the increase of clay mineral content, the heterogeneity of shale became stronger, and clay minerals possessed a more irregular surface leading to a higher SA compared to quartz (Figure 10). Therefore, the fractal dimension substantially represents the roughness of the surface and is correlated with the SA of minerals. Thus, the greater the value of the fractal dimension, the higher the SA. As for the clay minerals, the fractal dimension showed positive relationships with illite, kaolinite, and chlorite. In contrast, the fractal dimension was negatively correlated to MLIS (Figure 9c,d). For MLIS, the spatial order of interlayer pores increased with the increase of pore content, which naturally reduced the fractal dimension [56]. In contrast, kaolinite, chlorite, and illite have few pores and were more likely to inhibit pore development in the pore system. Therefore, the increase of their content might increase the heterogeneity and complexity of the pore system, thus increasing the value of the fractal dimension of the pore structure, which is consistent with materials with a small number of pores [54,57].

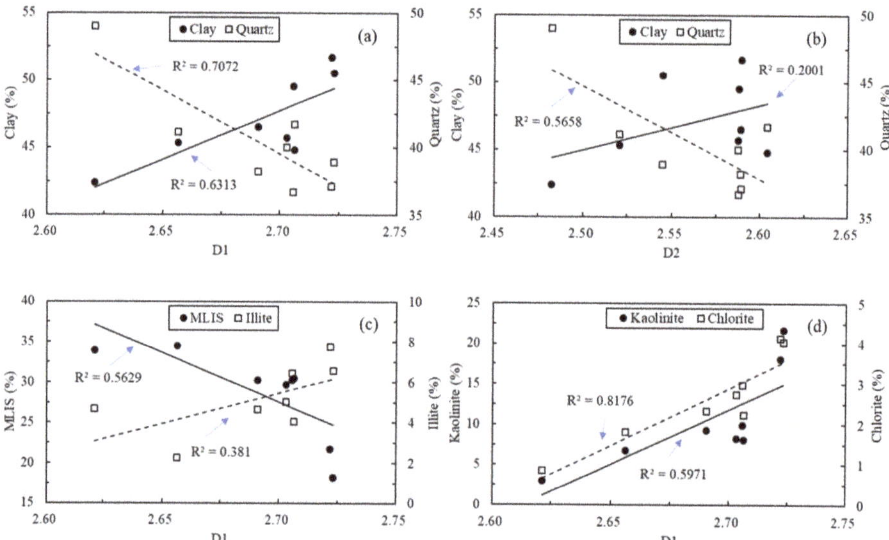

Figure 9. Relationships between shale component and fractal dimensions: (**a**) clay and quartz contents versus D1, (**b**) clay and quartz contents versus D2, (**c**) MLIS and illite contents versus D1, (**d**) kaolinite and chlorite contents versus D1.

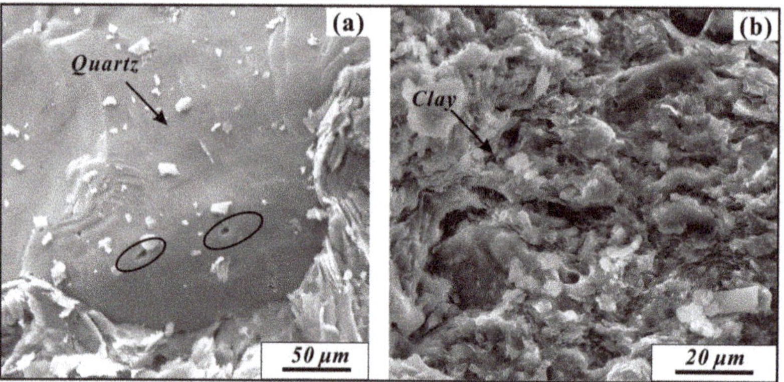

Figure 10. SEM images of quartz and clay minerals in marine Taiyuan shale samples: (**a**) quartz has fewer pores with a smooth surface, (**b**) irregular surfaces and a larger number of pores in clay minerals can be observed.

5.5. Relationships between Fractal Dimension and Nanopore Structure Parameters

As shown in Figure 11, the fractal dimension correlated positively with the BET SA and total pore volume, with correlation coefficients of 0.3724 and 0.5357, respectively (Figure 11a,b), which indicated that marine shale with higher fractal dimensions may have larger BET SA and higher total pore volume. The fractal dimension showed a negative correlation with the APS (Figure 11a), suggesting that shales with complicated pore structures have relatively small APS. Moreover, the fractal dimension exhibited a positive correlation with the micropore surface area and micropore volume (Figure 11c), suggesting that marine shale with higher fractal dimension tend to possess a greater micropore SA and micropore volume. This finding also indicated an irregular and complicated micropore structure in marine shale, while the fractal dimension had no apparent relationship with the mesopore surface area and mesopore volume (Figure 11c), suggesting that the development of micro-fractures of clay minerals (Figure 12a) and organic matter (Figure 12b) might affect or lead to this relationship. In contrast, Figure 11e exhibits a significant negative relationship between the fractal dimension and macropore SA and macropore volume, suggesting that the increase of macropores led to the decrease of heterogeneity of shale. Therefore, the combination of micro-fractures with different dimensions and nanopores with different shapes and sizes led to a complicated pore–fracture system.

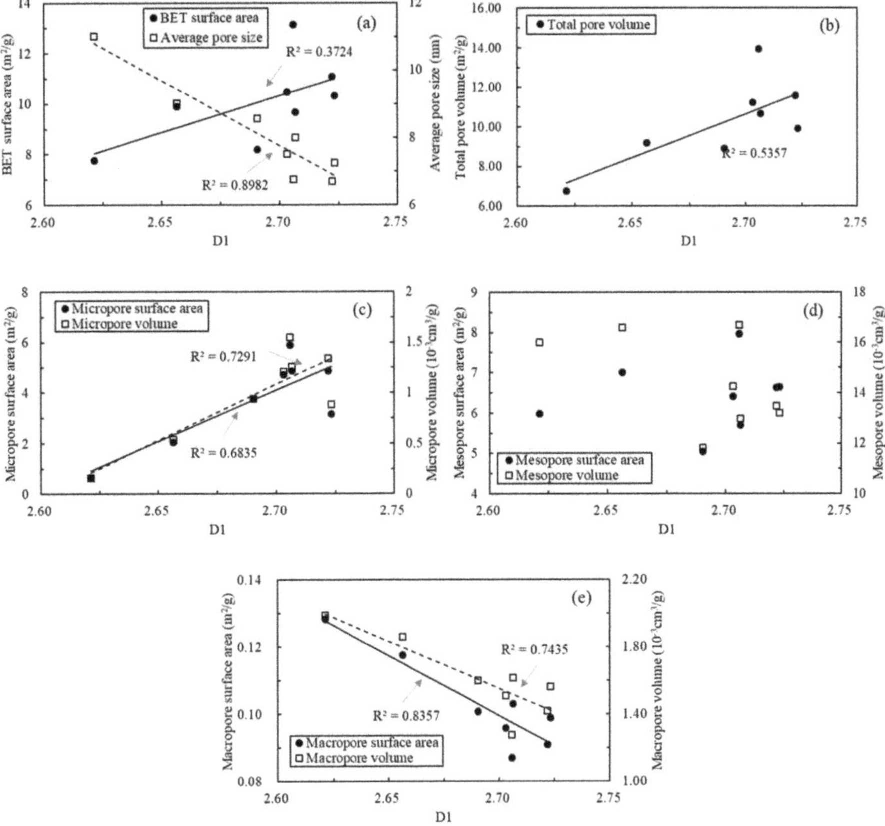

Figure 11. Plots of fractal dimensions versus nanopore structure parameters for the studied shale samples. (**a**) BET SA and APS versus D1, (**b**) PV versus D1, (**c**) micropore surface and volume versus D1, (**d**) mesopore surface and volume versus D1, (**e**) macropore surface and volume versus D1.

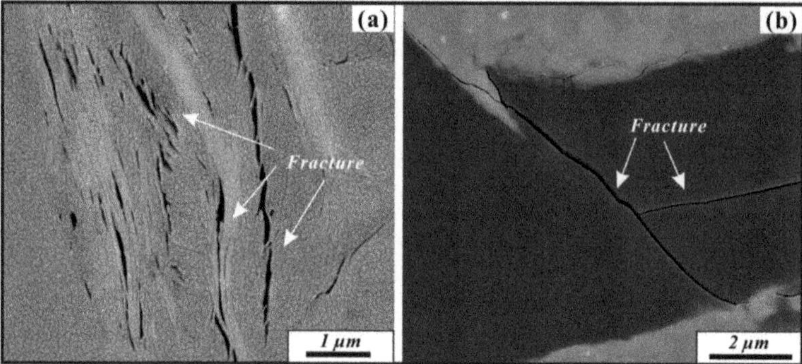

Figure 12. SEM images of micro-fractures in clay minerals (**a**) and organic matter (**b**) of the marine Taiyuan shale.

6. Conclusions

Fractal characteristics and heterogeneity of the nanopore structure of marine shale in southern North China were studied using multiple analysis and testing methods. Compared with continental shale, the marine shale possessed a larger number of SA and PV, with an N_2 adsorption amount of 7.752–13.138 m^2/g and 14.341–19.508 cm^3/g, respectively. The PSD curves of marine shale mainly appeared unimodal, with a major peak corresponding to the pore size ranging from 1.4 nm to 1.6 nm in the microporous range, which is different from what observed for continental shales that possess a wide pore-size interval and pores that are primarily larger than 10 nm. The TOC content showed a negative relationship with the clay content and positively correlated with the quartz content, indicating that different sedimentary facies may control the precipitation of clay and quartz between transitional shale and marine shale, leading to different OM–clay relationships and different correlations between TOC content and mineral content. The shale with high TOC content tended to have more heterogeneous and complex micropores, leading to a higher fractal dimension and more complex nanopore system. With the increase of TOC content and thermal evolution of OM, the heterogeneity of the pore structure became more significant. The quartz from marine facies develops more macropores, resulting in a decrease of the BET SA and an increase of the APS, while clay minerals have abundant micropores and work together with OM to affect the nanopore system of shale, leading to an increase of the BET SA and a decrease of the APS. The spatial order of interlayer pores increased with the increase of MLIS content, which naturally reduced the fractal dimension. In contrast, kaolinite, chlorite, and illite have few pores and are more likely to inhibit pore development in the pore system, significantly increasing the complexity and heterogeneity of the pore–fracture system by mean of pore-blocking. Taiyuan shale with higher heterogeneity is highly fractal, and the fractal dimensions are principally related to micropores. The fractal dimension correlates positively with the BET SA and total pore volume, suggesting that marine shale with higher fractal dimensions may have larger BET SA and higher total pore volume. The combination of micro-fractures with different dimensions and nanopores with different shapes and sizes leads to a complicated pore–fracture system in marine shale.

Author Contributions: K.Y. and Y.J. designed the study and modified the manuscript; K.Y., Y.Q., P.Q. and C.H. contributed to the analysis, data interpretation, diagrams and manuscript preparation; K.Y., H.Z. and H.F. collected and prepared the shale samples; P.Q., Y.J. and K.Y. designed the experiments.

Funding: This research was financially supported by the National Natural Science Foundation of China (Grant Nos. 41530315, 41872160, 41372213), the "Climate Change: Carbon Budget and Related Issues" Strategic Priority Research Program of the Chinese Academy of Sciences (Grant No. XDA05030100), the National Science and Technology Major Project of China (Grant Nos. 2016ZX05066, 2017ZX05064), and Science and Technology Support Program of Sichuan Province, China (Grant No. 2016JZ0037).

Acknowledgments: Our deepest gratitude goes to the anonymous reviewers for their careful work and thoughtful suggestions that have helped improve this paper substantially.

Conflicts of Interest: The authors declare no conflict of interest.

References

1. Curtis, J.B. Fractured shale-gas systems. *AAPG Bull.* **2002**, *86*, 1921–1938.
2. Chen, S.; Zhu, Y.; Wang, H.; Liu, H.; Wei, W.; Fang, J. Shale gas reservoir characterization: A typical case in the southern Sichuan basin of China. *Energy* **2011**, *36*, 6609–6616. [CrossRef]
3. Gasparik, M.; Bertier, P.; Gensterblum, Y.; Ghanizadeh, A.; Krooss, B.M.; Littke, R. Geological controls on the methane storage capacity in organic-rich shales. *Int. J. Coal Geol.* **2014**, *123*, 34–51. [CrossRef]
4. Clarkson, C.R.; Bustin, R.M. The effect of pore structure and gas pressure upon the transport properties of coal: A laboratory and modeling study. 2. adsorption rate modeling. *Fuel* **1999**, *78*, 1345–1362. [CrossRef]
5. Clarkson, C.R.; Bustin, R.M. Binary gas adsorption/desorption isotherms: Effect of moisture and coal composition upon carbon dioxide selectivity over methane. *Int. J. Coal Geol.* **2000**, *42*, 241–271. [CrossRef]
6. Bustin, R.M.; Bustin, A.M.M.; Cui, A.; Ross, D.; Pathi, V.M. Impact of shale properties on pore structure and storage characteristics. In Proceedings of the SPE Paper 119892 Society of Petroleum Engineers Shale Gas Production Conference, Fort Worth, TX, USA, 16–18 November 2008.
7. Ross, D.J.K.; Bustin, R.M. Characterizing the shale gas resource potential of Devonian–Mississippian strata in the Western Canada Sedimentary Basin: Application of an integrated formation evaluation. *AAPG Bull.* **2008**, *92*, 87–125. [CrossRef]
8. Loucks, R.G.; Reed, R.M.; Ruppel, S.C.; Hammes, U. Spectrum of pore types and networks in mudrocks and a descriptive classification for matrixerelatedmudrock pores. *AAPG Bull.* **2012**, *96*, 1071–1098. [CrossRef]
9. Loucks, R.G.; Reed, R.M.; Ruppel, S.C.; Jarvie, D.M. Morphology, genesis, and distribution of nanometerescale pores in siliceous mudstones of the Mississippian Barnett Shale. *J. Sediment. Res.* **2009**, *79*, 848–861. [CrossRef]
10. Lapierre, C.; Leroueil, S.; Locat, J. Mercury intrusion and permeability of louiseville clay. *Can. Geotech. J.* **1990**, *27*, 761–773. [CrossRef]
11. Yao, Y.; Liu, D. Comparison of low-field NMR and mercury intrusion porosimetry in characterizing pore size distributions of coals. *Fuel* **2012**, *95*, 152–158. [CrossRef]
12. Branch, A. Methane and CO_2, sorption and desorption measurements on dry argonnepremium coals: Pure components and mixtures. *Int. J. Coal Geol.* **2003**, *55*, 205–224.
13. Chalmers, G.R.; Bustin, R.M.; Power, I.M. Characterization of gas shale pore systems by porosimetry, pycnometry, surface area, and field emission scanning electron microscopy/transmission electron microscopy image analyses: Examples from the Barnett, Woodford, Haynesville, Marcellus, and Doig units. *AAPG Bull.* **2012**, *96*, 1099–1119.
14. Wang, G.; Ju, Y.; Yan, Z.; Li, Q. Pore structure characteristics of coal-bearing shale using fluid invasion methods: A case study in the Huainan-Huaibei coalfield in China. *Mar. Pet. Geol.* **2015**, *62*, 1–13. [CrossRef]
15. Zhu, H.; Ju, Y.; Qi, Y.; Huang, C.; Zhang, L. Impact of tectonism on pore type and pore structure evolution in organic-rich shale: Implications for gas storage and migration pathways in naturally deformed rocks. *Fuel* **2018**, *228*, 272–289. [CrossRef]
16. Lyu, Q.; Long Xi Ranjith, P.G.; Tan, J.; Kang, Y.; Wang, Z. Experimental investigation on the mechanical properties of a low-clay shale with different adsorption times in sub-/super-critical CO_2. *Energy* **2018**, *147*, 1288–1298. [CrossRef]
17. Clarkson, C.R.; Freeman, M.; He, L.; Agamalian, M.; Melnichenko, Y.B.; Mastalerz, M.; Bustin, R.M.; Radlinski, A.P.; Blach, T.P. Characterization of tight gas reservoir pore structure using USANS/SANS and gas adsorption analysis. *Fuel* **2012**, *95*, 371–385. [CrossRef]
18. Ross, D.J.K.; Bustin, R.M. The importance of shale composition and pore structure upon gas storage potential of shale gas reservoirs. *Mar. Pet. Geol.* **2009**, *26*, 916–927. [CrossRef]
19. Milliken, K.L.; Rudnicki, M.; Awwiller, D.N.; Zhang, T. Organic matter–hosted pore system, Marcellus formation (Devonian), Pennsylvania. *AAPG Bull.* **2013**, *97*, 177–200. [CrossRef]
20. Ge, X.M.; Fan, Y.R.; Li, J.T.; Zhid, M.A. Pore structure characterization and classification using multifractal theory- an application in tight reservoir of Santanghu basin in Western China. *J. Pet. Sci. Eng.* **2015**, *127*, 297–304. [CrossRef]

21. Pan, J.; Peng, C.; Wan, X.; Zheng, D.; Lv, R.; Wang, K. Pore structure characteristics of coal-bearing organic shale in Yuzhou coalfield, China using low pressure N_2 adsorption and FE-SEM methods. *J. Pet. Sci. Eng.* **2017**, *153*, 234–243. [CrossRef]

22. Ju, Y.; Huang, C.; Sun, Y.; Wan, Q.; Lu, X.; Lu, S.; He, H.; Wang, X.; Zou, C.; Wu, J.; et al. Nanogeosciences: Research History, Current Status, and Development Trends. *J. Nanosci. Nanotechnol.* **2017**, *17*, 5930–5965. [CrossRef]

23. Schlueter, E.M.; Zimmerman, R.W.; Witherspoon, P.A.; Cook, N.G.W. The fractal dimension of pores in sedimentary rocks and its influence on permeability. *Eng. Geol.* **1997**, *48*, 199–215. [CrossRef]

24. Qi, Y.; Ju, Y.; Huang, C.; Zhu, H.; Bao, Y.; Wu, J.; Meng, S.; Chen, W. Influences of organic matter and kaolinite on pore structures of transitional organic-rich mudstone with an emphasis on S2 controlling specific surface area. *Fuel* **2019**, *237*, 860–873. [CrossRef]

25. Pérez Bernal, J.L.; Bello, M.A. Fractal geometry and mercury porosimetry: Comparison and application of proposed models on building stones. *Appl. Surf. Sci.* **2001**, *185*, 99–107. [CrossRef]

26. Watt-Smith, M.J.; Edler, K.J.; Rigby, S.P. An experimental study of gas adsorption on fractal surfaces. *Langmuir* **2005**, *21*, 2281–2292. [CrossRef]

27. Ahmad, A.L.; Idrus, N.F.; AbdShukor, S.R. Surface fractal dimension of Perovskite-doped alumina membrane: Influence of calcining temperature. *J. Am. Ceram. Soc.* **2006**, *89*, 1694–1698. [CrossRef]

28. Khalili, N.R.; Pan, M.; Sandí, G. Determination of fractal dimensions of solid carbons from gas and liquid phase adsorption isotherms. *Carbon* **2000**, *38*, 573–588. [CrossRef]

29. Fu, X.; Qin, Y.; Zhang, W.; Wei, Z.; Zhou, R. Pore fractal classification and natural classification based on the coal bed methane migration. *Chin. Sci. Bull.* **2005**, *50*, 51–55. [CrossRef]

30. Zhang, S.; Tang, S.; Tang, D.; Huang, W.; Pan, Z. Determining fractal dimensions of coal pores by FHH model: Problems and effects. *J. Nat. Gas Sci. Eng.* **2014**, *21*, 929–939. [CrossRef]

31. Bu, H.; Ju, Y.; Tan, J.; Wang, G.; Li, X. Fractal characteristics of pores in non-marine shales from the Huainan coalfield, eastern China. *J. Nat. Gas Sci. Eng.* **2015**, *24*, 166–177. [CrossRef]

32. Yang, R.; He, S.; Yi, J.; Hu, Q. Nano-scale pore structure and fractal dimension of organic-rich Wufeng-Longmaxi shale from Jiaoshiba area, Sichuan Basin: Investigations using FE-SEM, gas adsorption and helium pycnometry. *Mar. Pet. Geol.* **2016**, *70*, 27–45. [CrossRef]

33. Liu, J.; Yao, Y.; Liu, D.; Cai, Y.; Cai, J. Comparison of pore fractal characteristics between marine and continental shales. *Fractals* **2018**, *26*, 1840016. [CrossRef]

34. Yu, K.; Shao, C.; Ju, Y.; Qu, Z. The genesis and controlling factors of micropore volume in transitional coal-bearing shale reservoirs under different sedimentary environments. *Mar. Pet. Geol.* **2019**, *102*, 426–438. [CrossRef]

35. Yu, K.; Ju, Y.; Qian, J.; Qu, Z.; Shao, C.; Yu, K.; Shi, Y. Burial and thermal evolution of coal-bearing strata and its mechanisms in the southern North China Basin since the late Paleozoic. *Int. J. Coal Geol.* **20198**, *198*, 110–115. [CrossRef]

36. Zaslow, J.; Orloff, T. Zircon U–Pb age and Lu–Hf isotope constraints on Precambrian evolution of continental crust in the Songshan area, the south-central North China Craton. *Precambrian Res.* **2013**, *226*, 1–20.

37. Ju, Y.; Sun, Y.; Tan, J.; Bu, H.; Han, K.; Li, X. The composition, pore structure characterization and deformation mechanism of coal-bearing shales from tectonically altered coalfields in eastern China. *Fuel* **2018**, *234*, 626–642. [CrossRef]

38. *Determination of Carbon and Hydrogen in Coal*; General Administration of Quality Supervision; Inspection and Quarantine of China: Beijing, China, 2008; No. GB/T4762008.

39. *Method of Determining Microscopically the Reflectance of Vitrinite in Sedimentary*; National Energy Administration: Beijing, China, 1995; No. SY/T5124-1995.

40. *Analysis Method for Clay Minerals and Ordinary Ono-Clay Minerals in Sedimentary Rocks by the X-ray Diffraction*; National Energy Administration: Beijing, China, 2010; No. SY/T5163-2010.

41. Gregg, S.J.; Sing, K.S.W. *Adsorption, Surface Area and Porosity*, 2nd ed.; Academic Press: London, UK, 1982.

42. Lastoskie, C.; Gubbins, K.E.; Quirke, N. Pore size distribution analysis of microporous carbons: A density functional theory approach. *J. Phys. Chem.* **1993**, *97*, 1012–1016. [CrossRef]

43. Seaton, N.A.; Walton JP, R.B.; Quirke, N. A new analysis method for the determination of the pore size distribution of porous carbons from nitrogen adsorption measurements. *Carbon* **1989**, *27*, 853–861. [CrossRef]

44. Rouquerol, J.; Avnir, D.; Fairbridge, C.W.; Everett, D.H.; Haynes, J.M.; Pernicone, N.; Ramsay, J.D.F.; Sing, K.S.W.; Unger, K.K. Recommendations for the characterization of porous solids. *Int. Union Pure Appl. Chem.* **1994**, *66*, 1739–1758. [CrossRef]

45. Mandelbrot, B.B. *The Fractal Geometry of Nature*; Freeman: New York, NY, USA, 1983.

46. Mahamud, M.M.; Novo, M.F. The use of fractal analysis in the textural characterization of coals. *Fuel* **2008**, *87*, 222–231. [CrossRef]

47. Jaroniec, M. Evaluation of the fractal dimension from a single adsorption isotherm. *Langmuir* **1995**, *11*, 2316–2317. [CrossRef]

48. Dong, T.; Harris, N.B.; Ayranci, K.; Twemlow, C.E.; Nassichuk, B.R. Porosity characteristics of the Devonian Horn River shale, Canada: Insights from lithofacies classification and shale composition. *Int. J. Coal Geol.* **2015**, *141*, 74–90. [CrossRef]

49. Tian, H.; Pan, L.; Xiao, X.; Wilkins RW, T.; Meng, Z.; Huang, B. A preliminary study on the pore characterization of lower Silurian black shales in the Chuandong thrust fold belt, southwestern China using low pressure N_2 adsorption and FE-SEM methods. *Mar. Pet. Geol.* **2013**, *48*, 8–19. [CrossRef]

50. Liang, C.; Jiang, Z.; Zhang, C.; Guo, L.; Yang, Y.; Li, J. The shale characteristics and shale gas exploration prospects of the lower Silurian Longmaxi shale, Sichuan basin, south China. *J. Nat. Gas Sci. Eng.* **2014**, *21*, 636–648. [CrossRef]

51. Xi, Z.; Tang, S.; Wang, J.; Yi, J.; Guo, Y.; Wang, K. Pore structure and fractal characteristics of Niutitang shale from China. *Minerals* **2018**, *8*, 163. [CrossRef]

52. Yang, F.; Ning, Z.; Liu, H. Fractal characteristics of shales from a shale gas reservoir in the Sichuan basin, China. *Fuel* **2014**, *115*, 378–384. [CrossRef]

53. Liang, Q.; Zhang, X.; Tian, J.; Sun, X.; Chang, H. Geological and geochemical characteristics of marine-continental transitional shale from the Lower Permian Taiyuan Formation, Taikang Uplift, southern North China Basin. *Mar. Pet. Geol.* **2018**, *98*, 229–242. [CrossRef]

54. Wang, P.; Jiang, Z.; Ji, W.; Zhang, C.; Yuan, Y.; Chen, L. Heterogeneity of intergranular, intraparticle and organic pores in Longmaxi shale in Sichuan basin, south China: Evidence from SEM digital images and fractal and multifractal geometries. *Mar. Pet. Geol.* **2016**, *72*, 122–138. [CrossRef]

55. Chen, S.; Han, Y.; Fu, C.; Zhang, H.; Zhu, Y.; Zuo, Z. Micro and nano-size pores of clay minerals in shale reservoirs: Implication for the accumulation of shale gas. *Sediment. Geol.* **2016**, *342*, 180–190. [CrossRef]

56. Malekani, K. Comparison of techniques for determining the fractal dimensions of clay minerals. *Clays Clay Miner.* **1996**, *44*, 677–685. [CrossRef]

57. Yao, Y.; Liu, D.; Tang, D.; Tang, S.; Huang, W. Fractal characterization of adsorption-pores of coals from North China: An investigation on CH_4 adsorption capacity of coals. *Int. J. Coal Geol.* **2008**, *73*, 27–42. [CrossRef]

Article

The Impacts of Matrix Compositions on Nanopore Structure and Fractal Characteristics of Lacustrine Shales from the Changling Fault Depression, Songliao Basin, China

Zhuo Li [1,2], Zhikai Liang [1,2,*], Zhenxue Jiang [1,2], Fenglin Gao [1,2], Yinghan Zhang [1,2], Hailong Yu [1,2], Lei Xiao [1,2] and Youdong Yang [1,2]

[1] State Key Laboratory of Petroleum Resources and Prospecting, China University of Petroleum (Beijing), Beijing 102249, China; zhuo.li@cup.edu.cn (Z.L.); jiangzhenxue0405@163.com (Z.J.); feilaifeng2011@gmail.com (F.G.); zhangyinghan2021@163.com (Y.Z.); yhl2017210211@163.com (H.Y.); xlcupb@163.com (L.X.); yydzgsydxbj@163.com (Y.Y.)

[2] Unconventional Natural Gas Research Institute, China University of Petroleum (Beijing), Beijing 102249, China

* Correspondence: liangzhikai2020@163.com; Tel.: +86-010-89739051

Received: 13 January 2019; Accepted: 15 February 2019; Published: 22 February 2019

Abstract: The Lower Cretaceous Shahezi shales are the targets for lacustrine shale gas exploration in Changling Fault Depression (CFD), Southern Songliao Basin. In this study, the Shahezi shales were investigated to further understand the impacts of rock compositions, including organic matters and minerals on pore structure and fractal characteristics. An integrated experiment procedure, including total organic carbon (TOC) content, X-ray diffraction (XRD), field emission-scanning electron microscope (FE-SEM), low pressure nitrogen physisorption (LPNP), and mercury intrusion capillary pressure (MICP), was conducted. Seven lithofacies can be identified according to on a mineralogy-based classification scheme for shales. Inorganic mineral hosted pores are the most abundant pore type, while relatively few organic matter (OM) pores are observed in FE-SEM images of the Shahezi shales. Multimodal pore size distribution characteristics were shown in pore width ranges of 0.5–0.9 nm, 3–6 nm, and 10–40 nm. The primary controlling factors for pore structure in Shahezi shales are clay minerals rather than OM. Organic-medium mixed shale (OMMS) has the highest total pore volumes (0.0353 mL/g), followed by organic-rich mixed shale (ORMS) (0.02369 mL/g), while the organic-poor shale (OPS) has the lowest pore volumes of 0.0122 mL/g. Fractal dimensions D_1 and D_2 (at relative pressures of 0–0.5 and 0.5–1 of LPNP isotherms) were obtained using the Frenkel–Halsey–Hill (FHH) method, with D_1 ranging from 2.0336 to 2.5957, and D_2 between 2.5779 and 2.8821. Fractal dimensions are associated with specific lithofacies, because each lithofacies has a distinctive composition. Organic-medium argillaceous shale (OMAS), rich in clay, have comparatively high fractal dimension D_1. In addition, organic-medium argillaceous shale (ORAS), rich in TOC, have comparatively high fractal dimension D_2. OPS shale contains more siliceous and less TOC, with the lowest D_1 and D_2. Factor analysis indicates that clay contents is the most significant factor controlling the fractal dimensions of the lacustrine Shahezi shale.

Keywords: Changling Fault Depression; Shahezi Formation; fractal dimensions; pore structure; shale lithofacies; lacustrine shales

---

1. Introduction

Organic shales commonly contain complex matrix compositions (organic matters and minerals) and nanopore pore networks, various pore types, and multiscale pore width [1–4]. Scanning electron microscopy, small angle scattering, nuclear magnetic resonance, low pressure gas physisorption, and high-pressure mercury intrusion methods can be used to characterize pore structure in organic shales [1,2,5–10]. Among these methods, low pressure nitrogen physisorption (LPNP) is proposed to be an effective method for characterizing pore structure and fractal dimensions in organic shales [2,11–16]. Fractal theory was proposed [17] and used to study the properties of porous materials and pores with irregular surfaces and shapes [18–20]. Fractal dimensions were applied to study pore geometries and pore size distribution in coals and shales [2,9,20–23]. Lithofacies may refers to the homogeneity of specific geochemical, geological, mineralogical, and petrophysical characteristics of rocks [24]. Shale lithofacies may represent the spatial variations in organic matter richness, and shale properties [25]. Lithofacies classification is also an effective technique to identify favorable shale gas targets [26]. Previous studies about the shale lithofacies, pore structure and fractal characteristics were mainly focused on marine shales [4,13,15,25–29]. Therefore, shale lithofacies and their impacts on pore structure and fractal dimensions of lacustrine shales need further investigation. Lithofacies classification, shale reservoir characteristics, and main controlling factors of pore structure of the lacustrine shales from the Lower Cretaceous Shahezi Formation in the Changling Fault Depression (CFD) of the Songliao Basin have been studied [30]. However, the relationships of pore structure and fractal dimensions with shale lithofacies of the Shahezi shales were still poorly correlated.

In the present work, the Shahezi Shales in CFD were investigated to reveal the impacts of shale compositions on nanopore structure and fractal characteristics (Figure 1). The CFD experienced three successive evolution stages including Early Cretaceous rifting (extensional) stage, Late Cretaceous depression stage, and Latest Cretaceous to Quaternary uplifting stage can be confirmed [31–33]. The Shahezi (K_1sh) shales are consist of deep lake and semi-deep lake sediments, which are the main source rock of the gas accumulated within the gas fields (Ro > 2%) in the study area [34]. The predominant macerals type is vitrinite (82.1%–96.7%) and the main organic type of kerogen is type III of the Shahezi shales [30,34].

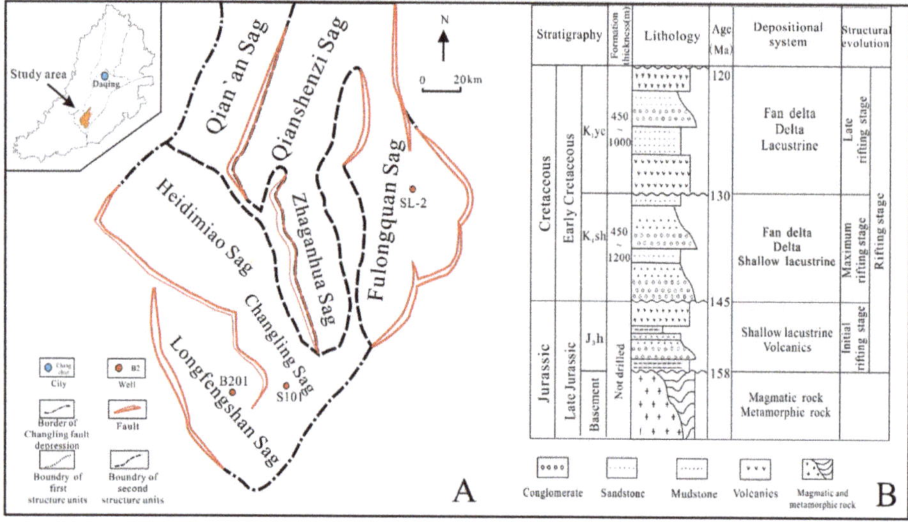

Figure 1. (**A**) Maps showing the location and the structure distributions in the Changling Fault Depression (CFD); (**B**) stratigraphy column showing petrology and depositional facies of the Shahezi Formation in CFD (modified after Gao et al., 2018 [30] and Cai et al., 2017 [31]).

The objectives of this paper are to: (1) identify shale lithofacies based on a mineralogy-based classification scheme for shales; (2) characterize pore structures of each shale lithofacies; (3) calculate fractal dimensions obtain from LPNP isotherms using FHH method; (4) investigate of the impacts of lithofacies on pore structure and fractal dimensions of lacustrine shales in CFD, Songliao Basin.

2. Materials and Methods

A total of twenty-two selected shale samples were obtained from the lower Cretaceous Shahezi Formation in Wells S-101, SL-2 and B-201. The depths are in the ranges of 2392–2577 m and 3430–3943 m, respectively (Table 1). An integrated experiment procedure, including total organic carbon (TOC), X-ray diffraction (XRD), field emission scanning electron microscope (FE-ESEM), and low pressure nitrogen physisorption (LPNP), was conducted.

The TOC content, Rock-Eval and bulk XRD composition measurements were conducted at China University of Petroleum (Beijing). TOC contents were determined by a Leco CS230 carbon/sulfur analyzer (LECO Corporation, St Joseph, MI, USA). Shale samples were treated with hydrochloric acid for two hours. Then the samples were washed out using distilled water and dried before the TOC content analysis. Rock-Eval analysis was carried out using OGE-II rock pyrolyzer (RIPED, Beijing, China) under programmed heating processes. The vitrinite reflectance (Ro) values were calculated by Rock-Eval pyrolysis parameters based on the formula of Jarvie et al. (2001) [35]. The kerogen type was determined by cross-plots of hydrogen index (HI) against maximum cracking temperature (T_{max}) and residual hydrocarbon (S_2) versus TOC [36]. Bulk mineral compositions were determined using a Bruker D8 DISCOVER XRD diffractometer (BRUKER AXS Corporation, Karlsruhe, Germany) using Co Ka-radiation at 45 kV. Quantitative analysis was performed by Rietveld refinement, with customized Ufer models [37].

Ar-ion milling surface of shale samples was prepared on a Hitachi IM4000 apparatus for 2 h before FE-SEM imaging. FE-SEM observation and imaging were conducted using a Zeiss SUPRA 55 Sapphire FE-SEM (Carl Zeiss, Heidenheim, Germany) equipped with secondary electron (SE), backscattered electron (BSE) detectors and an energy dispersive spectrometer at Institute of Geology and Geophysics, Chinese Academy of Sciences. Image-Pro Plus software (Image-Pro® Plus, Media Cybernetics, Rockville, MD, USA) was used for image analysis on shale samples. FE-SEM images were initially transformed to an eight-bit bitmap, then, OM and mineral grains were filtered from transformed SEM images via a color threshold. Finally, the pore binary images were extracted and quantitatively analyzed for pore characteristics.

LPNP experiments were conducted at the Beijing Center for Physical and Chemical Analysis with a Quantachrome NOVA 4200e (Quantachrome, Boynton Beach, FL, USA) following Chinese National Standard GB/T31483-2015. The samples were dried at 110 °C for 12 h before the experiment to remove moisture. The parameters were set at −196.15 °C, with relative pressure range from 0.001 to 0.998. The specific surface area was calculated by the multipoint BET (Brunauer–Emmett–Teller) method and the pore size distribution and pore volume were calculated using the Barrette–Joyner–Halenda (BJH) theory. the FHH model [18] for fractal dimensions calculation can be expressed as:

$$\ln V = (D - 3) \times \ln(\ln P_0 / P) + C \tag{1}$$

where P is the equilibrium pressure, V is the volume corresponding to the equilibrium pressure P; D is the fractal dimension, C is the parameter, and P_0 is the saturation pressure.

The mercury intrusion capillary pressure (MICP) measurements were performed on a Micromeritics Autopore IV 9500 equipment (Micromeritics, Atlanta, GA, USA) at China University of Mining and Technology, with an operating pressure up to 30,000 Psi and the measurable pores ranging from 5.4 nm to 200 μm [30].

Table 1. Mineral and geochemical compositions of the Shahezi shale in CFD.

No.	Sample ID	Well ID	Depth (m)	Lithofacies	Mineral Composition (%)						Geochemical Composition (%)					
					Quartz	Feldspar	Calcite	Pyrite	Clay	TOC (%)	Ro (%)	Tmax (°C)	S_1 (mg/g)	S_2 (mg/g)	HI (mg/g)
1	CL-1	S-101	2392.66	OPS	37.6	23.4			39	0.9704	1.68	492	0.56	2.12	218
2	CL-2	S-101	2397.93	OPS	33	15.0			52	0.8516	1.75	495	0.2	2.11	247
3	CL-3	S-101	2450.64	OPS	56.9		1.7		41.4	0.9204	1.96	507	1.26	2.24	243
4	CL-4	S-101	2463.65	OMSS	31	22.0			47	1.378	2.44	534	0.61	2.08	150
5	CL-5	S-101	2465.24	OMMS	0	48.0	19		30	1.516	1.97	507	0.02	0.04	3
6	CL-6	SL-2	3077.12	OMSS	41	19.0			40	1.122	1.95	506	0.22	0.54	48
7	CL-7	SL-2	3077.77	OPS	41.5	4.8			53.7	0.9264	1.99	508	0.14	0.44	47
8	CL-8	SL-2	3430.14	ORAS	34.4	10.1			50.5	3.569	1.98	508	0.11	0.66	18
9	CL-9	SL-2	3432.44	OMAS	30.6				69.4	1.516	1.35	473	0.16	0.26	17
10	CL-10	SL-2	3433.48	OMAS	36	0.5	0.9		62.6	1.395	1.43	478	0.1	0.35	25
11	CL-11	SL-2	3433.88	ORAS	37.4	6.6	0.8		55.2	2.519	1.64	489	0.12	0.36	14
12	CL-12	SL-2	3435.15	OMAS	35.5	6.2		2.8	55.5	1.186	1.68	491	0.34	0.31	26
13	CL-13	SL-2	3434.58	OMAS	42.4	6.1			51.5	1.791	1.67	491	0.09	0.27	23
14	CL-14	B-201	3742.5	OPS	52.2	16.7			31.1	0.8402	1.88	502	0.19	0.25	13
15	CL-15	B-201	3748.55	ORAS	36.8	6.4		1.6	55.2	3.533	1.90	503	0.87	1.29	154
16	CL-16	B-201	3893.93	OMAS	29.4	4.2	1.9		64.5	1.041	2.06	512	0.13	0.42	12
17	CL-17	B-201	3896.08	OMAS	30.1	2.3			67.6	1.084	2.09	514	0.35	0.51	47
18	CL-18	B-201	3897.67	OMAS	27.1	2.5	1.7		68.7	1.505	1.99	509	0.33	0.73	49
19	CL-19	B-201	3937.69	OMMS	40.4	4.9	6.3		46.1	1.597	2.06	512	0.41	0.78	49
20	CL-20	B-201	3938.2	ORSS	49.7	4.2			46.1	2.159	2.06	513	0.61	0.99	46
21	CL-21	B-201	3942.08	OPS	20.9			25.2	53.9	0.9213	2.08	513	0.07	0.1	11
22	CL-22	B-201	3943.27	ORMS	32.7	12	7.8		42.6	2.852	2.04	511	0.15	0.45	16

OPS: organic-poor shale; OMSS: organic-medium siliceous shale; OMMS: organic-medium mixed shale; ORAS: organic-medium argillaceous shale; OMAS: organic-medium argillaceous shale.

3. Results

3.1. Organic Geochemistry and Mineralogy

The organic geochemistry and mineralogy composition of the Shahezi shale samples are presented in Table 1. The TOC contents are in the range of 0.8402–3.569 wt.%, with an average value of 1.599 wt.%. The Rock-Eval pyrolysis parameter S_1 values are ranging from 0.02 to 1.26 mg of hydrocarbon (HC)/g of rock, with an average of 0.32 mg of HC/g of rock. The S_2 values are in the range of 0.04–2.24 mg of HC/g of rock, and S_2 for most samples is less than 1 mg of HC/g of rock. T_{max} values of the samples are larger than 473 °C, with the highest value up to 534 °C. The average hydrogen index (HI) is 67.1 mg of HC/g of TOC (3–247 mg/g). The dominant kerogen type is type III according to the cross-plots of T_{max} versus HI and S_2 versus TOC (Figure 2). The vitrinite reflectance (Ro) values converted from the T_{max} values of the shale samples range between 1.34% and 2.4%, suggesting the maturity of dry gas stage.

Clay minerals are dominant in the Shahezi shales, with an average content over 50 wt.% (27.6–69 wt.%). Quartz contents vary from 20.9 to 56.9 wt.%, carbonate minerals vary from 2.3 to 15 wt.%, and feldspar are between 2.3 wt.% and 15 wt.% in the Shahezi shales. Trace amount of pyrite were detected, no more than 2.8 wt.% (Table 1).

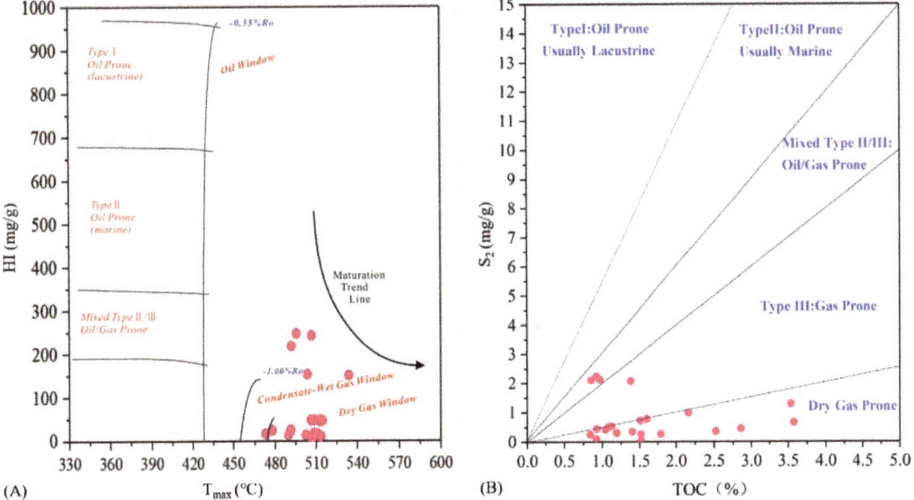

Figure 2. Cross-plots of T_{max} versus hydrogen index (**A**) and S_2 versus total organic carbon (**B**) for identifying the kerogen type of the Shahezi shale in CFD.

3.2. Lithofacies Classification

Shale lithofacies of the lacustrine Shahezi shales were sorted based on TOC and mineral compositions. Based on the classification scheme proposed by Gao et al. (2018) [30], seven shale lithofacies can be identified for the lacustrine shales in the CFD (Figure 3), including the organic-rich argillaceous shale (ORAS), organic-rich siliceous shale (ORSS), organic-rich mixed shale (ORMS), organic-medium argillaceous shale (OMAS), organic-medium siliceous shale (OMSS), organic-medium mixed shale (OMMS), and organic-poor shale (OPS). Since shales with TOC content less than 1 wt.% were not considered as target layers [14,27], they are summed together as organic poor shale (OPS).

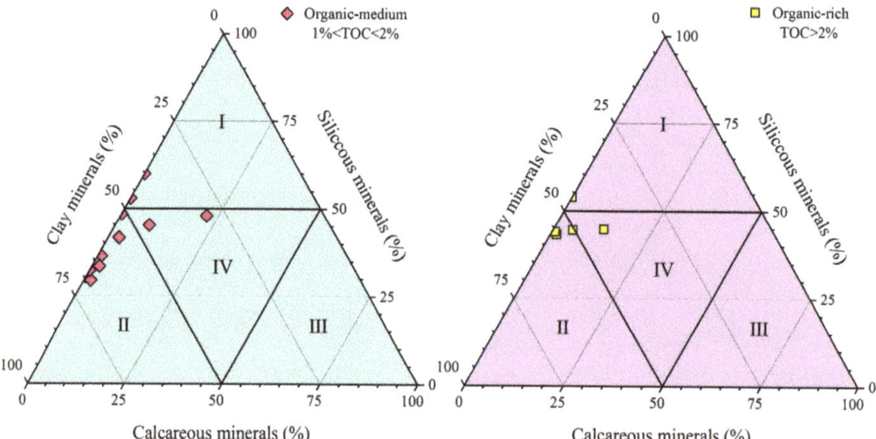

Figure 3. Lithofacies classification of the Shahezi shales in CFD (I: siliceous shale; II: argillaceous shale; III: calcareous shale; IV: mixed shale).

3.3. Field Emission-Scanning Electron Microscope (FE-SEM) Imaging

3.3.1. Organic Matter Pores

As seen from the FE-SEM images, few organic matter (OM) pores were observed in the Shahezi shale samples (Figure 4). Pyrite framboids filled by OM were commonly seen (Figure 4A) with small oval pores (around 30 nm) (Figure 4B). Heterogeneous distribution of OM pores with larger pores in the center and no pores in the rim of the same OM particle were observed (Figure 4C). OM coexist with clay and siliceous minerals host complex pore networks. Sponge-like shaped pores with pore size over 300 nm are generated in OM coexisted with quartz (Figure 4D,E). The OM may also be mixed with clay minerals forming organic-clay composites, which have relatively abundant OM pores (Figure 4F). Larger pores contain smaller sponge-like pores, forming complex pore networks in three dimensions, which may have better connectivity within the OM (Figure 4G–I). The sponge-like pores were also observed where OM coexists with pyrite framboids (Figure 4J,K). Same samples have completely different pore characteristics (Figure 4C,L). The same observations were reported in Woodford shale samples with a measured Ro value of 1.4%, which may be due to organic macerals heterogeneity or OM difference (non-porous kerogen/porous pyrobitumen) [38].

3.3.2. Inorganic Minerals Pores

Inorganic pores, mainly hosted in clay minerals, feldspar and quartz, are abundant in the Shahezi shales in CFD (Figure 5). Frequently observed pores are slit-like intraparticle pores (intraP pores) and interparticle pores (interP pores) within clay mineral composites (Figure 5A–F). Many Linear pores between clay platelets were found in the pressure shadow of the rigid minerals (Figure 5A,B). Intraparticle pores were also observed within the mixed illite-smectite and chlorite aggregates (Figure 5C,D). Many interP pores within clay minerals are filled with pyrite particles (Figure 5E). Micro-fractures are easily formed in the clay aggregates (Figure 5F). Oval intraP pores in siliceous and calcareous minerals are observed (Figure 5G–L). InterP pores of quartz and calcite are also observed with a shape of triangular or polygonal (Figure 5K,L). These pores are probably originated from organic acid dissolution [1].

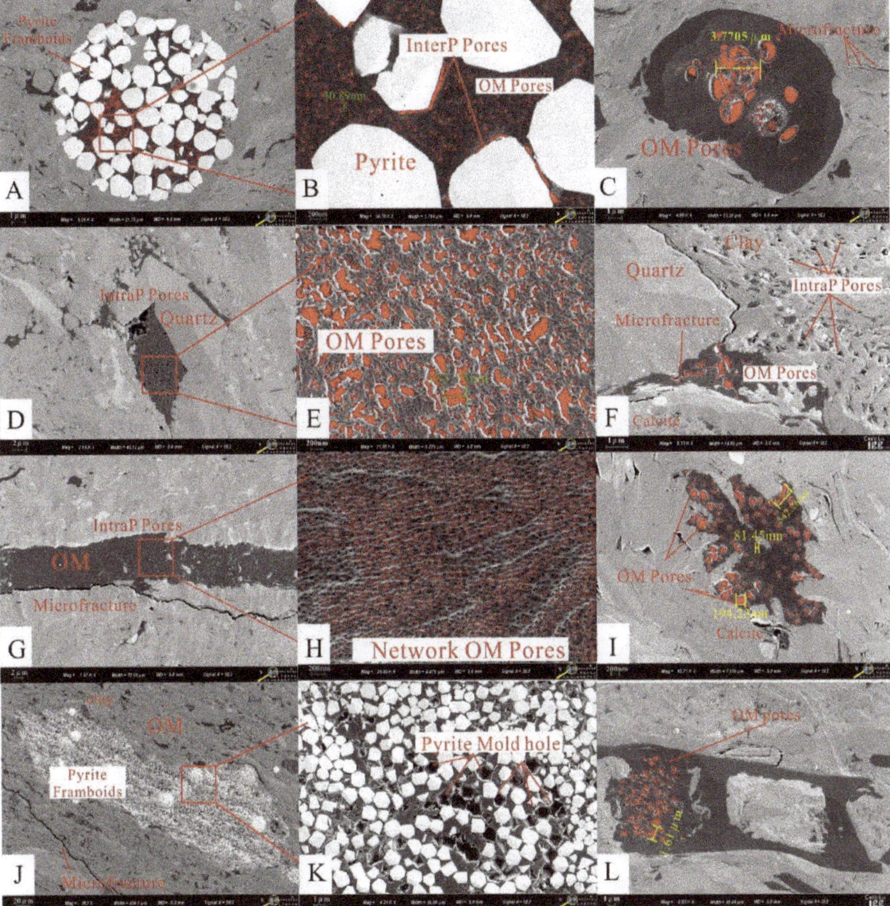

Figure 4. FE-SEM images of OM pores in the Shahezi shales in CFD. (**A**) OM filled in pyrite framboids grains contain small oval OM pores with pore diameters of about 30 nm. (**B**) enlarged square area from image A. (**C**) Heterogeneous characteristics of organic matter (OM) pores. (**D,E**) Sponge-like OM pores hosted in OM coexist with quartz. (**F**) cracks around quartz grains and pores in OM-clay composites. (**G,H**) Complex OM pore networks with small width. (**I**) Heterogeneous OM pore characteristics in the same OM particle with wide pore size. (**J,K**) mold holes and OM pores associated with pyrite framboids. (**L**) Different pore characteristics in same samples.

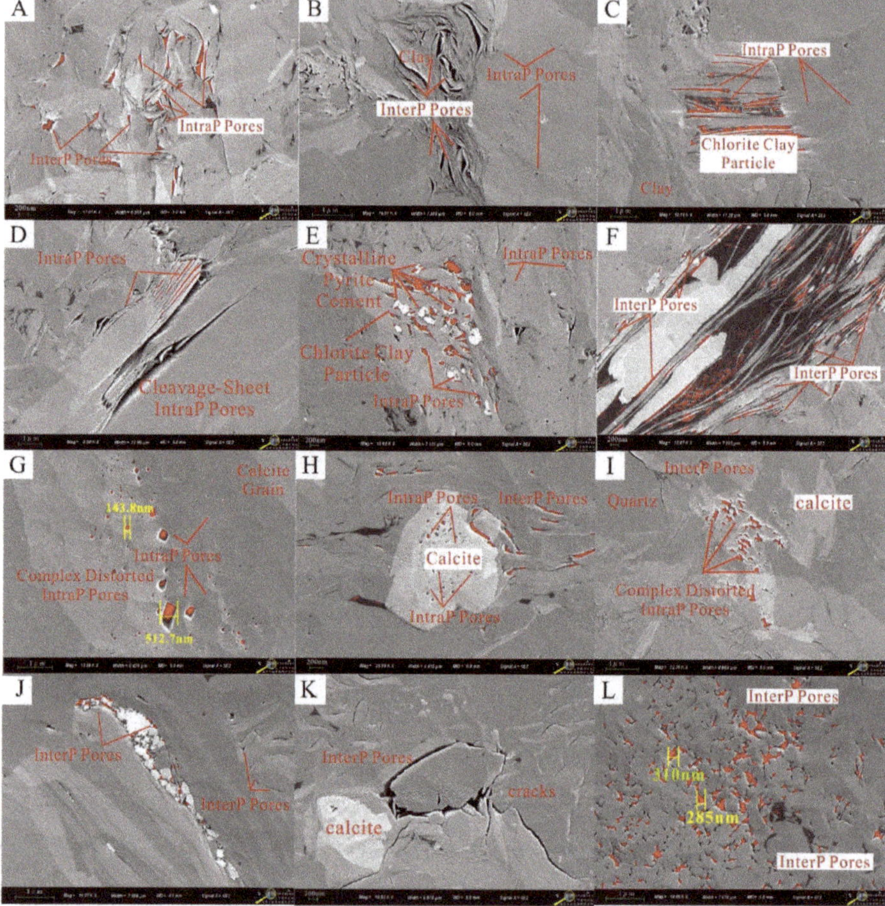

Figure 5. FE-SEM images showing inorganic mineral pores in the Shahezi shales in CFD. (**A–F**) Slit-like interP pores and intraP pores in clay composites. (**G–I**) IntraP pores hosted in siliceous and calcareous minerals. (**J**) InterP pores are filled with pyrite framboids. (**K**) Micro-cracks developed at the edge of inorganic mineral grains. (**L**) Complex intergranular pores between quartz grains.

3.3.3. Image Processing Analyses

Shales display strong heterogeneity at the micro scale and becomes weaker heterogeneity in a larger field of view, therefore, representative elementary area (REA) were taken at a magnification of ×20,000 [39]. REA images of the Shahezi shales were quantitatively studied using Image-Pro Plus software [40]. The parameters of pores image processing, including pore width, length, perimeter, area and fractal dimension, are listed in Table 2. Totally 3669 pores, including OM pores, interP pores and intraP pores, were extracted from the REA images of the Shahezi shales. InterP pores contribute the highest percentage in the total pore systems (58.8%). However, intraP pores contribute the lowest percentage (5.6%). The mean pore sizes of the OM pores, interP pores, and intraP pores are 35.6, 56.2, and 27.9 nm. IntraP pores have the lowest fractal dimension (1.06), while the interP pores have the highest value (1.44) among the three pore types. These fractal dimension data reveal that the intraP pores have the most regularly shaped pores, while the interP pores are the most irregularly shape in the Shahezi shales in CFD.

Table 2. Image processing results of representative elementary area (REA) for the Shahezi shale in CFD.

Pore Type	Number of Pores	Percentage (%)	Pore Size (nm)			Pore Area (nm^2)	Fractal Dimension
			Min Value	Max Value	Mean		
OM pore	1306	35.6	12.6	355.9	35.6	4996.2	1.18
InterP pore	2157	58.8	18.8	890.6	56.2	8762.3	1.44
IntraP pore	206	5.6	25.9	240.4	27.9	2237.8	1.06

3.4. LPNP Isotherms and FHH Fractal Dimensions

The LPNP isotherms of the selected lacustrine shale samples are plotted in terms of shale lithofacies (Figure 6). The hysteresis loops show H3 and H4 shapes according to the classification of IUPAC [41,42], indicating slit-like and ink-bottle pores exist in the Shahezi shales. The organic-rich shales (TOC > 2 wt.%) have relatively higher adsorption volumes than organic lean shales (TOC < 2 wt.%). Among the shale lithofacies, the decreasing order of adsorption capacity is: OMMS > OMAS > ORAS > ORMS > ORSS > OMSS > OPS.

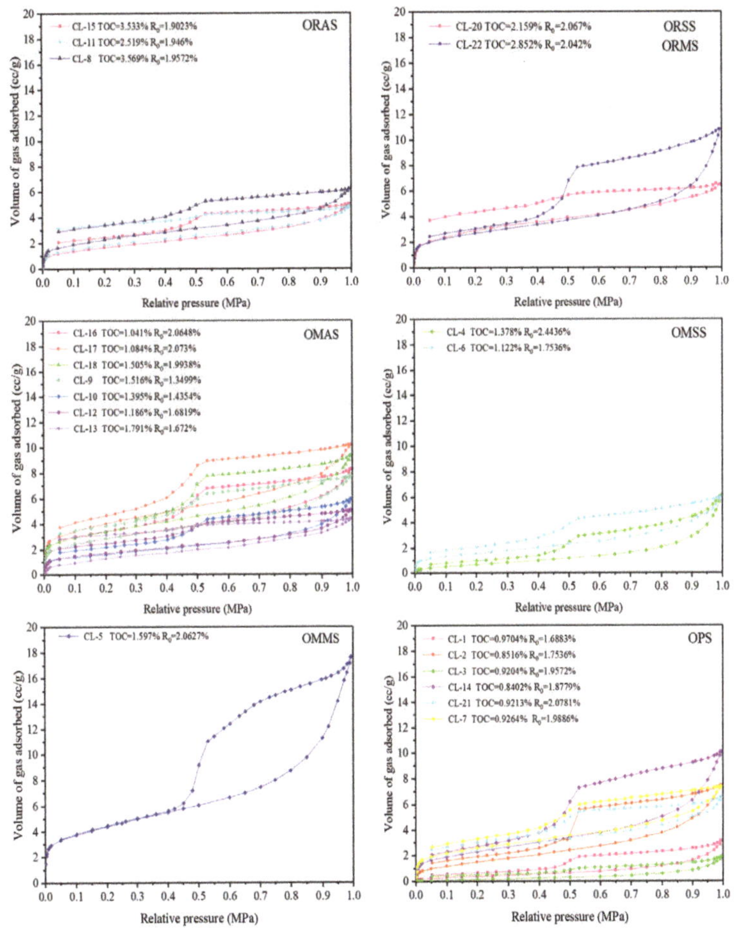

Figure 6. Low pressure nitrogen physisorption (LPNP) isotherms of the Shahezi shales with different lithofacies in CFD.

The FHH plots of the Shahezi shales are shown in Figure 7. Two linear segments in the relative pressure (P/P_0) ranges of 0–0.5 and 0.5–1 were identified. The fitting equations, correlation coefficients and calculated fractal dimensions $(D_1$ and $D_2)$ are summarized in Table 3. The fractal dimension D_1 values are relatively low, ranging from 2.0336 to 2.5957, with a mean value of 2.4385. The fractal dimension D_2 values are in the range of 2.5779–2.8821, with an average of 2.7703, indicating high pore structure complexity in larger pores [9,15,23].

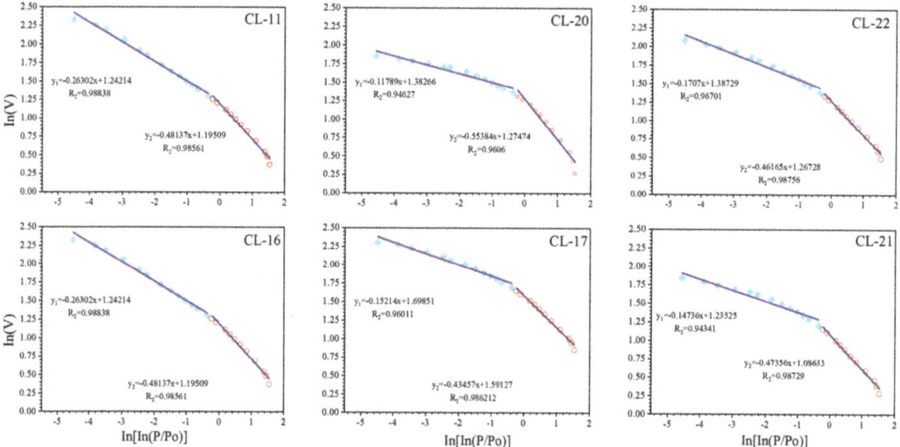

Figure 7. Plots of ln$V$ versus ln(lnP_0/P) obtained from the adsorption branch of LPNP isotherms of the Shahezi shales with different lithofacies in CFD.

3.5. MICP Analysis

The results of MICP are summarized in terms of lithofacies (Figure 8). The cumulative mercury intrusion volumes for each sample are similar. The curves show a rapid increase of mercury intrusion volume to a certain value at low pressures, suggesting cracks may generate during sample preparation process [43]. In the pressure range of 10–10,000 Psi, the cumulative intrusion volumes increase slowly (Figure 8). When intrusion pressure exceeds 10,000 Psi, the intrusion volumes increases rapidly to the maximum values. In the stage of pressure decreases, no obvious variation can be identified in the pressure range of 100–30,000 Psi and extrusion volumes decrease in the pressure range of 1–20 Psi. The variation trends of the MICP curves suggest shales have abundant nanoscale pores with bottle-necked shapes and poor pore connectivity [10,43,44].

Table 3. Fractal dimensions obtained from the Frenkel–Halsey–Hill (FHH) model using LPNP isotherms.

Sample ID	Lithofacies	Fitting Equation	k_1	D_1	Coefficient (R^2)	Fitting Equation	k_2	D_2	Coefficient (R^2)
CL-14	OPS	$y = -0.5988x + 1.0935$	-0.5988	2.4012	0.9713	$y = -0.2698x + 1.1944$	-0.3	2.7302	0.9751
CL-15	ORAS	$y = -0.5406x + 0.7522$	-0.5406	2.4594	0.9817	$y = -0.1774x + 0.8864$	-0.2	2.8226	0.9599
CL-16	OMAS	$y = -0.4616x + 1.2673$	-0.4616	2.5384	0.9888	$y = -0.1707x + 1.3873$	-0.2	2.8293	0.967
CL-17	OMAS	$y = -0.4346x + 1.5913$	-0.4346	2.5654	0.9876	$y = -0.1521x + 1.6985$	-0.2	2.8479	0.9634
CL-18	OMAS	$y = -0.4043x + 1.4172$	-0.4043	2.5957	0.996	$y = -0.1671x + 1.5363$	-0.2	2.8329	0.9618
CL-19	OMMS	$y = -0.4346x + 1.5913$	-0.4346	2.5654	0.9876	$y = -0.1521x + 1.6985$	-0.2	2.8479	0.9634
CL-20	ORSS	$y = -0.5538x + 1.2747$	-0.5538	2.4462	0.9607	$y = -0.1179x + 1.3827$	-0.1	2.8821	0.9363
CL-21	OPS	$y = -0.4736x + 1.0863$	-0.4736	2.5264	0.9886	$y = -0.1474x + 1.2352$	-0.1	2.8526	0.9434
CL-22	ORMS	$y = -0.4814x + 1.1951$	-0.4814	2.5186	0.987	$y = -0.263x + 1.2421$	-0.3	2.737	0.9896
CL-6	OMSS	$y = -0.5332x + 0.711$	-0.5332	2.4668	0.9881	$y = -0.2223x + 0.8492$	-0.2	2.7777	0.9653
CL-7	OPS	$y = -0.4814x + 1.1951$	-0.4814	2.5186	0.987	$y = -0.263x + 1.2421$	-0.3	2.737	0.9894
CL-8	ORAS	$y = -0.4648x + 1.0287$	-0.4648	2.5352	0.993	$y = -0.1601x + 1.1526$	-0.2	2.8399	0.9703
CL-9	OMAS	$y = -0.4508x + 1.2572$	-0.4508	2.5492	0.985	$y = -0.1715x + 1.3519$	-0.2	2.8285	0.9571
CL-10	OMAS	$y = -0.4393x + 0.7203$	-0.4393	2.5607	0.9966	$y = -0.2436x + 0.8057$	-0.2	2.7564	0.9869
CL-11	ORAS	$y = -0.4814x + 1.1951$	-0.4814	2.5186	0.987	$y = -0.263x + 1.2421$	-0.3	2.737	0.9894
CL-12	OMAS	$y = -0.5027x + 0.7836$	-0.5027	2.4973	0.9576	$y = -0.1785x + 0.867$	-0.2	2.8215	0.9763
CL-13	OMAS	$y = -0.842x + 0.5915$	-0.842	2.158	0.9415	$y = -0.1831x + 0.7362$	-0.2	2.8169	0.9388
CL-1	OPS	$y = -0.7984x - 0.5986$	-0.7984	2.2016	0.9905	$y = -0.3638x - 0.4258$	-0.4	2.6362	0.9701
CL-3	OPS	$y = -0.9664x - 1.4317$	-0.9664	2.0336	0.9967	$y = -0.4221x - 1.183$	-0.4	2.5779	0.9696
CL-4	OMSS	$y = -0.7577x - 0.05$	-0.757	2.243	0.9984	$y = -0.3871x + 0.1337$	-0.4	2.6129	0.9824
CL-5	OMMS	$y = -0.4417x + 1.6823$	-0.4417	2.5583	0.9965	$y = -0.2737x + 1.7491$	-0.3	2.7263	0.9698
CL-2	OPS	$y = -0.6771x + 0.7099$	-0.6771	2.3229	0.9921	$y = -0.4417x + 1.6824$	-0.3	2.7264	0.9595

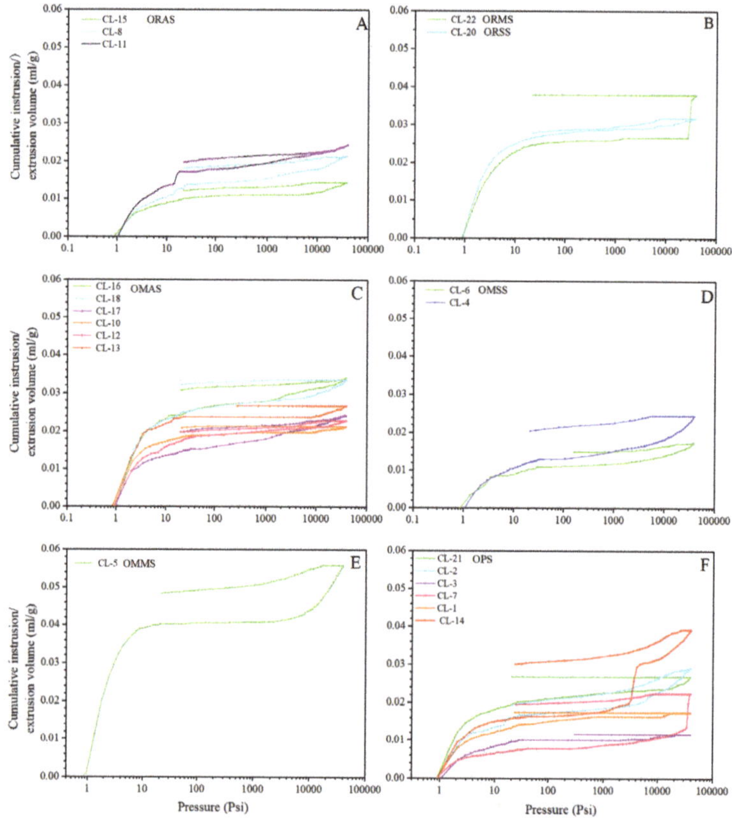

Figure 8. Mercury intrusion capillary pressure (MICP) intrusion and extrusion volumes for the Shahezi shales with different lithofacies in CFD.

3.6. Full-Size Pore Characterization

The full-size pore size distribution (PSD) characteristics of the selected shale samples are obtained by combining LPNP and MICP data (Figure 9). The method was proposed and effectively used for characterization of pore characteristics in organic shales [10,45,46]. The results show that multimodal characteristics were shown in the PSD spectra in 0.5–0.9 nm, 3–6 nm and 10–40 nm (Figure 9). The PSD characteristics vary in different lithofacies. The OMMS has the highest total pore volumes of 0.0328 mL/g, followed by the ORMS. The proportions of micropores, mesopores and macropores for the Shahezi shale samples are shown in Figure 10. The characteristics of nanopore systems show greater variation with the increasing buried depth. The mesopores primarily contribute to the pore volumes of the Shahezi shales. Mesopores and macropores account for 66.0%–88.3% of the total pore volume (Table 4).

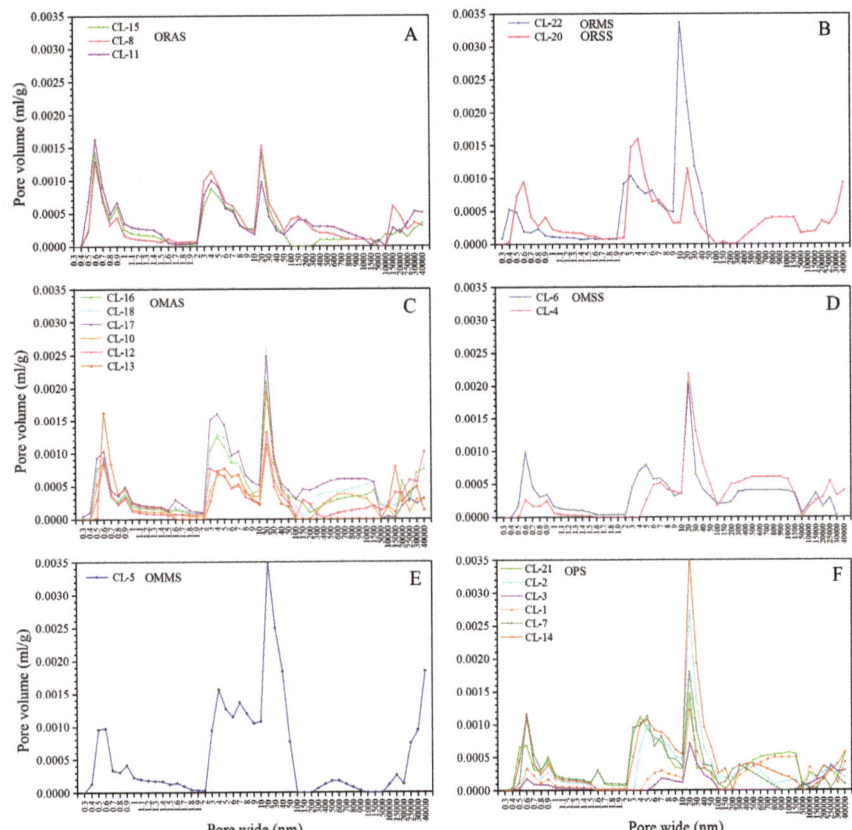

Figure 9. Full-size pore size distribution (PSD) characteristics of pore volumes of the Shahezi shales with different lithofacies in CFD.

Table 4. Pore structure parameters of lacustrine shale samples with different lithofacies in CFD.

Sample ID	Lithofacies	Pore Volume (cm³/g)				Percentage (%)		
		Micropore	Mesopore	Macropore	Total	Micropore	Mesopore	Macropore
CL-15		0.0045	0.0055	0.0022	0.0122	36.80	45.13	18.07
CL-8	ORAS	0.0036	0.0058	0.0067	0.0162	22.47	36.07	41.46
CL-11		0.0044	0.0026	0.0079	0.0149	29.70	17.37	52.93
Mean		0.0042	0.0046	0.0056	0.0144	29.00	32.17	38.84
CL-20	ORSS	0.0039	0.0043	0.0042	0.0123	31.28	34.68	34.04
Mean		0.0039	0.0043	0.0042	0.0123	31.28	34.68	34.04
CL-22	ORMS	0.0023	0.0175	0.0065	0.0263	8.80	66.46	24.75
Mean		0.0023	0.0175	0.0065	0.0263	8.80	66.46	24.75
CL-16		0.0035	0.0103	0.0071	0.0209	16.79	49.22	33.99
CL-17		0.0046	0.0122	0.0067	0.0235	19.63	51.88	28.48
CL-9		0.0041	0.0083	0.0068	0.0193	21.48	43.22	35.30
CL-10	OMAS	0.0030	0.0078	0.0024	0.0132	22.77	59.03	18.20
CL-12		0.0027	0.0053	0.0036	0.0116	23.37	45.55	31.07
CL-13		0.0044	0.0023	0.0019	0.0086	51.38	26.59	22.03
Mean		0.0037	0.0083	0.0049	0.0169	22.06	48.96	28.98

Table 4. *Cont.*

Sample ID	Lithofacies	Pore Volume (cm³/g)				Percentage (%)		
		Micropore	Mesopore	Macropore	Total	Micropore	Mesopore	Macropore
CL-6	OMSS	0.0030	0.0080	0.0039	0.0149	19.93	53.88	26.20
Mean		0.0030	0.0080	0.0039	0.0149	19.93	53.88	26.20
CL-5		0.0042	0.0186	0.0125	0.0353	11.92	52.69	35.41
CL-19	OMMS	0.0036	0.0116	0.0057	0.0209	17.31	55.46	27.23
Mean		0.0039	0.0151	0.0091	0.0281	14.615	54.075	31.32
CL-1		0.0012	0.0048	0.0042	0.0102	11.72	47.10	41.18
CL-3		0.0007	0.0025	0.0020	0.0052	13.62	47.65	38.72
CL-7	OPS	0.0036	0.0087	0.0032	0.0155	23.35	56.01	20.64
CL-21		0.0036	0.0073	0.0050	0.0159	22.48	46.12	31.39
Mean		0.0026	0.0062	0.0034	0.0122	21.60	50.53	27.87

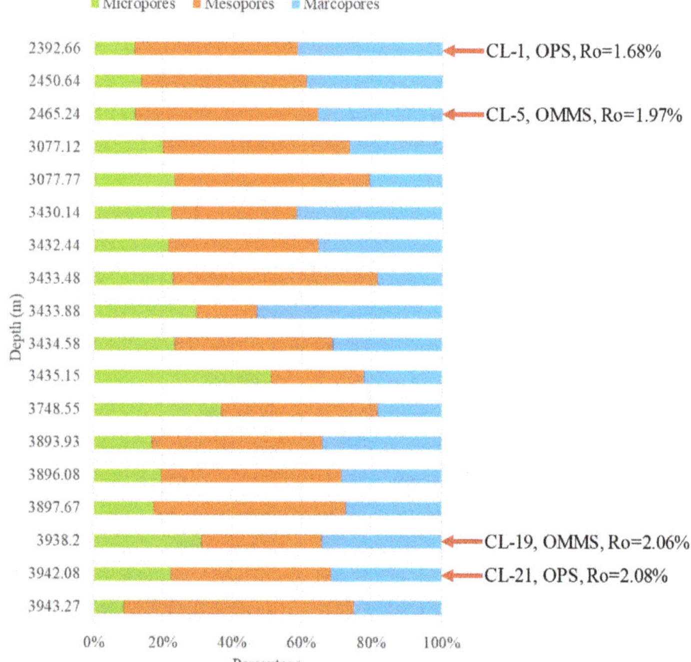

Figure 10. Percentage of pore volumes of the Shahezi shales along with depth in CFD.

4. Discussion

The development of pore structure is the combined effect of multiple factors, including TOC, thermal evolution (burial depth), and mineral composition [1,47,48]. Seven shale lithofacies can be identified for the lacustrine shales in CFD (Figure 3). These lithofacies display variable proportions of quartz, clay, and carbonate. Therefore, the impacts of lithofacies on pore structure and fractal dimensions are discussed in terms of OM richness and mineral compositions.

4.1. The Impacts of Lithofacies on Pore Structure

The ORSS and ORMS with similar TOC, clay mineral and quartz content, vary significantly in pore volumes, which may be caused by different calcite contents with dissolved pores ORMS (Figure 4G–I). Compared with ORAS, the OMAS has larger pore volumes, suggesting the contribution of clay minerals to pore volumes is greater than that of organic matter (Figure 4). The pore volumes of the OPS samples show a large variation. Due to the low TOC content, the mineral composition has a great influence on the pore structure, such as CL-2 and CL-7 clay content are more than 50%.

No obvious correlation between the burial depth and pore percentage were observed in Figure 10. Therefore, burial depth may not independently control the pore size distribution and porosity of shales [41]. By compare the pore volumes of shale samples with similar lithofacies and different burial depth, the possible impacts of burial depth on macro-, meso-, and micropore volumes were discussed. The decline in macropore and mesopore volumes in deeply buried shales samples (CL-19 and CL-21) is probably the results of deepen burial (Figure 10). Larger pores from shales buried over 3500 m may be compacted and greatly decrease the pore size and total pore volumes [41].

The correlations between micropore, mesopore and macropore volumes and TOC are plotted in Figure 11A–C. The micropore and mesopore volumes show slightly positive correlations with TOC contents, suggests the contribution of OM pores to the total pore networks in the Shahezi shale may not be dominant. The TOC contents of in the Shahezi shale is much lower (0.92%–3.57%, averaging 1.69%) than that of the Barnett (3%–13%, averaging 4.5%) and Longmaxi shales (0.87%–8.01%, averaging 3.44%) [12,49–51], which may be insufficient to provide significant pores [48,51]. The heterogeneity of organic pore distribution may be another reason for the weak correlation between pore volumes and TOC contents, which is related to the maceral types of OM [38,50] (Figure 4C,L). Negative correlations of micropores, mesopores, and macropores with quartz are presented in Figure 11D–F. The Shahezi shales are deeply buried (Table 1), therefore, under the effect of compaction, the primary pores between brittle grains gradually shrink and even disappear [52], indicating the siliceous minerals generally have a limited effect on the pore development. With the increase of clay mineral content, the pore volumes of micropores, mesopores and macropores all increase, indicating the various development of clay related pores in different pore size ranges (Figure 11G–I and Table 4). Clay minerals are dominant in the Shahezi samples, which are often associated with OM and quartz (Figure 5). In addition, previous study from study area reveal that clay minerals commonly host abundant clay related pores in the Shahezi shales, which have a much larger pore volume than other minerals [34]. During the transformation of montmorillonite to illite or chlorite, an increase trend in the number of micropores and mesopores appears [34,53,54]. Therefore, clay minerals are the essential controlling factor of pore development in the Shahezi shales in CFD. No obvious correlation was observed between multiscale pore volumes and carbonate components (Figure 11J–L). IntraP pores can be formed due to the solubility of calcite, consequently, abundant calcite contents may play significant roles in pore development in organic shales [7,47,55]. Relatively small amount of calcite and dolomite with dissolved pores are identified in the Shahezi shales, which may be the cause of the scatter correlations in the plots of pore volumes against carbonate contents.

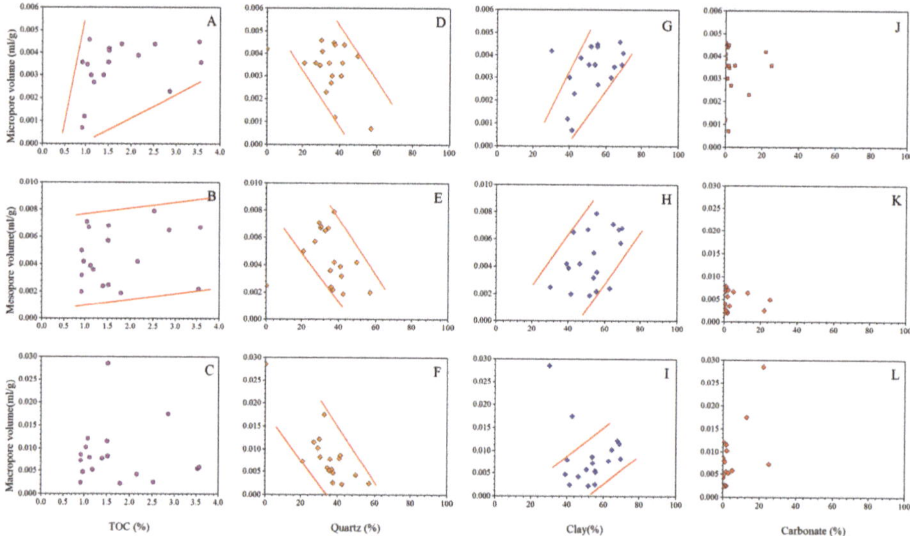

Figure 11. Relationships between pore structure parameters and TOC, mineral compositions: (**A**) Micropore volume versus TOC content; (**B**) Mesopore volume versus TOC content; (**C**) Macropore volume versus TOC content; (**D**) Micropore volume versus quartz content; (**E**) Mesopore volume versus quartz content; (**F**) Macropore volume versus quartz content; (**G**) Micropore volume versus clay mineral content; (**H**) Mesopore volume versus clay mineral content; (**I**) Macropore volume versus clay mineral content; (**J**) Micropore volume versus carbonate mineral content; (**K**) Mesopore volume versus carbonate mineral content; (**L**) Macropore volume versus carbonate mineral content.

4.2. The Impacts of Lithofacies on Pore Fractal Dimensions

The impacts of mineralogy-based lithofacies on nanopore fractal dimensions were discussed in aspects of matrix composition, including TOC, quartz, clay, and calcareous mineral contents. The different proportions of mineral components in the seven lithofacies strongly affect the fractal dimensions in the Shahezi shales. The OMAS has the highest D_1, probably because of the highest clay content in these shales (Table 3). OPS have the lowest D_1 and D_2, which may be caused by the much lower TOC and clay contents than the other lithofacies (Table 3).

The relationships between the fractal dimensions (D_1, D_2) and rock compositions including TOC, clay minerals, quartz, and carbonates are presented in Figure 12. Fractal dimensions show slightly positive correlations with TOC contents (Figure 11A,B). This result is probably due to the relatively low organic richness and insufficient OM pores in the Shahezi shales, and is inconsistent with the fractal dimensions of the over-mature marine shales [9,15,22]. OM pores with smaller pore width may result in more complex pore networks in organic shales [1,8,12], consequently, increase the fractal dimensions D_1 and D_2. In FE-SEM images, few OM pores were observed in the selected lacustrine shales (Figure 4), which may explain the weak correlations of fractal dimensions with TOC content [22].

With the increase of clay contents, both the D_1 and D_2 values increase (Figure 12E,F). Clay minerals are the main components in lacustrine shales of the Shahezi Formation, with an average content over 50% (Table 1). In addition, OM are commonly associated with clay minerals forming complex OM-clay composites (Figure 5A–F). As clay minerals host abundant complex interP and intraP pores, which may enhance the heterogeneity of pore volumes, result in larger fractal dimensions [10,15,22]. The relationships between fractal dimensions and quartz contents display negative correlations (Figure 12C,D) and no obvious correlations of fractal dimensions with calcareous minerals were observed (Figure 12G,H). The slight negative linear relationship between fractal dimension D_1 and brittle minerals (quartz and carbonate) contents, indicating the brittle minerals have a little effect on fractal dimension D_1. This is possibly because the more homogenous nature of pore volumes than organic matter and clay minerals. Fractal dimension D_2 decreases with increasing brittle minerals. This result is probably due to the relatively small amount of calcite and dolomite with dissolved pores shown in the FE-SEM images of the Shahezi shales (Figure 5H,I,L) and the better protection of complex organic pore network from numerous brittle mineral grains (Figure 4J, L), which results in a greater fractal dimension D_2 [22].

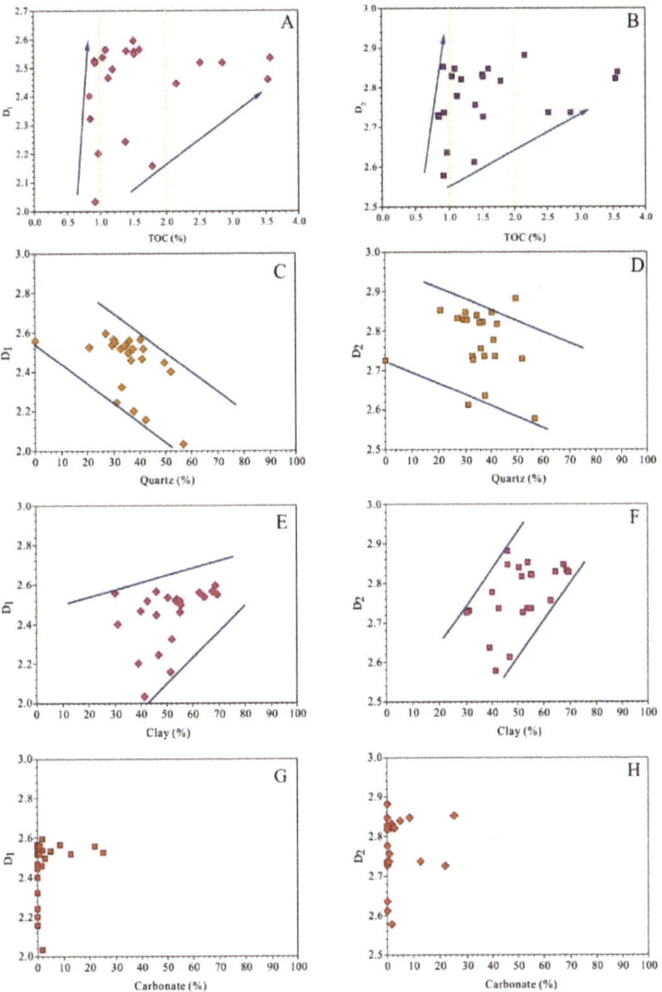

Figure 12. Relationship between fractal dimensions (D1, D2) and OM and inorganic mineral compositions: (**A**) D1 versus TOC content; (**B**) D2 versus TOC content; (**C**) D1 versus quartz content; (**D**) D2 versus quartz content; (**E**) D1 versus clay mineral content; (**F**) D2 versus clay mineral content; (**G**) D1 versus carbonate mineral content; (**H**) D2 versus carbonate mineral content.

5. Conclusions

Based on our studies, the following conclusions can be drawn.

(1) Seven lithofacies, including the organic-rich argillaceous shale (ORAS), organic-rich siliceous shale (ORSS), organic-rich mixed shale (ORMS), organic-medium argillaceous shale (OMAS), organic-medium siliceous shale (OMSS), organic-medium mixed shale (OMMS), and organic-poor shale (OPS), are identified and sorted based on a mineralogy-based classification scheme for the lacustrine shales in the CFD.

(2) Two fractal dimensions, D_1 and D_2, were obtained at relative pressures of 0–0.5 and 0.5–1 using the FHH model. The D_1 and D_2 values are in range of 2.0336–2.5957 and 2.5779–2.8821, respectively. D_2 values are slightly greater than D_1, indicating small pores may form more complex pore networks in the Shahezi shales.

(3) Fractal dimensions of the selected lacustrine shales are affected by shale mineral compositions and pore structure parameters. OMAS shale, rich in clay, have comparatively high fractal dimension D_1. ORAS shale, rich in TOC and clay, have comparatively high fractal dimension D_2. OPS shale, rich in siliceous and lack of TOC, have the lowest D_1 and D_2.

(4) Pore structure and fractal dimensions, which are a combined function of organic and inorganic composition, varies among the shale lithofacies. Samples with higher clay content have larger pore width, whereas samples with low clay content have smaller pore width. clay contents are the most significant factor controlling the pore structure and fractal dimensions of the lacustrine Shahezi shale in CFD. Observations of few organic matter pores and abundant inorganic pores hosted in the Shahezi shales may contribute to these correlations.

Author Contributions: Z.L. (Zhuo Li) and Z.J. designed and supervised the project; F.G., Y.Z. and Z.L. (Zhikai Liang) performed the experiments and analyzed the data; Z.L. (Zhikai Liang) wrote this paper and Z.L. (Zhuo Li) corrected it; H.Y., L.X., and Y.Y. modified the formats.

Funding: This research was funded by National Science and Technology Major Project, grant number 2016ZX05034-001 and the National Natural Science Foundation of China, grant number 41502123 and the APC was funded by National Science and Technology Major Project, grant number 2016ZX05034-001.

Acknowledgments: The northeast Branch of SINOPEC, China is thanked for providing shale samples and geological background references. We are grateful to editors and reviewers for their constructive comments and suggestions.

Conflicts of Interest: The authors declare no conflict of interest.

References

1. Loucks, R.G.; Reed, R.M.; Ruppel, S.C.; Hammes, U. Spectrum of pore types and networks in mudrocks and a descriptive classification for matrix-related mudrock pores. *AAPG Bull.* **2012**, *96*, 1071–1098. [CrossRef]
2. Clarkson, C.R.; Solano, N.; Bustin, R.M.; Bustin, A.M.M.; Chalmers, G.R.L.; He, L.; Melnichenko, Y.B.; Radlinski, A.P.; Blach, T.P. Pore structure characterization of North American shale gas reservoirs using USANS/SANS, gas adsorption, and mercury intrusion. *Fuel* **2013**, *103*, 606–616. [CrossRef]
3. Tang, X.L.; Jiang, Z.X.; Li, Z.; Gao, Z.Y.; Bai, Y.Q.; Zhao, S.; Feng, J. The effect of the variation in material composition on the heterogeneous pore structure of high-maturity shale of the Silurian Longmaxi formation in the southeastern Sichuan Basin, China. *J. Nat. Gas Sci. Eng.* **2015**, *23*, 464–473. [CrossRef]
4. Wang, P.F.; Jiang, Z.X.; Yin, L.S.; Chen, L.; Li, Z.; Zhang, C.; Li, T.W.; Huang, P. Lithofacies classification and its effect on pore structure of the Cambrian marine shale in the Upper Yangtze Platform, South China: Evidence from FE-SEM and gas adsorption analysis. *J. Pet. Sci Eng.* **2017**, *156*, 307–321. [CrossRef]
5. Loucks, R.G.; Reed, R.M.; Ruppel, S.C.; Jarvie, D.M. Morphology, Genesis, and Distribution of Nanometer-Scale Pores in Siliceous Mudstones of the Mississippian Barnett Shale. *J. Sediment. Res.* **2009**, *79*, 848–861. [CrossRef]
6. Chalmers, G.R.L.; Ross, D.J.K.; Bustin, R.M. Geological controls on matrix permeability of Devonian Gas Shales in the Horn River and Liard basins, northeastern British Columbia, Canada. *Int. J. Coal Geol.* **2012**, *103*, 120–131. [CrossRef]

7.   Li, A.; Ding, W.L.; Wang, R.Y.; He, J.H.; Wang, X.H.; Sun, Y.X.; Gu, Y.; Jiao, N.L. Petrophysical characterization of shale reservoir based on nuclear magnetic resonance (NMR) experiment: A case study of Lower Cambrian Qiongzhusi Formation in eastern Yunnan Province, South China. *J. Nat. Gas Sci. Eng.* **2017**, *37*, 29–38. [CrossRef]

8.   Ross, D.J.K.; Bustin, R.M. The importance of shale composition and pore structure upon gas storage potential of shale gas reservoirs. *Mar. Pet. Geol.* **2009**, *26*, 916–927. [CrossRef]

9.   Yang, F.; Ning, Z.F.; Liu, H.Q. Fractal characteristics of shales from a shale gas reservoir in the Sichuan Basin, China. *Fuel* **2014**, *115*, 378–384. [CrossRef]

10. Wang, G.C.; Ju, Y.W.; Yan, Z.F.; Li, Q.G. Pore structure characteristics of coal-bearing shale using fluid invasion methods: A case study in the Huainan-Huaibei Coalfield in China. *Mar. Pet. Geol.* **2015**, *62*, 1–13. [CrossRef]

11. Yao, Y.B.; Liu, D.M.; Tang, D.Z.; Tang, S.H.; Huang, W.H.; Liu, Z.H.; Che, Y. Fractal characterization of seepage-pores of coals from China: An investigation on permeability of coals. *Comput. Geosci.* **2009**, *35*, 1159–1166. [CrossRef]

12. Tian, H.; Pan, L.; Xiao, X.M.; Wilkins, R.W.T.; Meng, Z.P.; Huang, B.J. A preliminary study on the pore characterization of Lower Silurian black shales in the Chuandong Thrust Fold Belt, southwestern China using low pressure N-2 adsorption and FE-SEM methods. *Mar. Pet. Geol.* **2013**, *48*, 8–19. [CrossRef]

13. Liang, L.X.; Xiong, J.; Liu, X.J. An investigation of the fractal characteristics of the Upper Ordovician Wufeng Formation shale using nitrogen adsorption analysis. *J. Nat. Gas Sci. Eng.* **2015**, *27*, 402–409. [CrossRef]

14. Yang, C.; Zhang, J.C.; Wang, X.Z.; Tang, X.; Chen, Y.C.; Jiang, L.L.; Gong, X. Nanoscale pore structure and fractal characteristics of a marine-continental transitional shale: A case study from the lower Permian Shanxi Shale in the southeastern Ordos Basin, China. *Mar. Pet. Geol.* **2017**, *88*, 54–68. [CrossRef]

15. Shao, X.H.; Pang, X.Q.; Li, Q.W.; Wang, P.W.; Chen, D.; Shen, W.B.; Zhao, Z.F. Pore structure and fractal characteristics of organic-rich shales: A case study of the lower Silurian Longmaxi shales in the Sichuan Basin, SW China. *Mar. Pet. Geol.* **2017**, *80*, 192–202. [CrossRef]

16. Bu, H.L.; Ju, Y.W.; Tan, J.Q.; Wang, G.C.; Li, X.S. Fractal characteristics of pores in non-marine shales from the Huainan coalfield, eastern China. *J. Nat. Gas Sci. Eng.* **2015**, *24*, 166–177. [CrossRef]

17. Mandelbrot, B.B. Stochastic models for the Earth's relief, the shape and the fractal dimension of the coastlines, and the number-area rule for islands. *Proc. Natl. Acad. Sci. USA* **1975**, *72*, 3825–3828. [CrossRef]

18. Pfeifer, P. Chemistry in noninteger dimensions between two and three. I. Fractal theory of heterogeneous surfaces. *J. Chem. Phys.* **1984**, *80*, 4573. [CrossRef]

19. Schlueter, E.M.; Zimmerman, R.W. The fractal dimension of pores in sedimentary rocks and its influence on permeability. *Eng. Geol.* **1997**, *48*, 199–215. [CrossRef]

20. Yao, Y.B.; Liu, D.M.; Tang, D.Z.; Tang, S.H.; Huang, W.H. Fractal characterization of adsorption-pores of coals from North China: An investigation on CH_4 adsorption capacity of coals. *Int. J. Coal Geol.* **2008**, *73*, 27–42. [CrossRef]

21. Javadpour, F. Nanopores and Apparent Permeability of Gas Flow in Mudrocks (Shales and Siltstone). *J. Can. Pet. Technol.* **2009**, *48*, 16–21. [CrossRef]

22. Liu, J.Z.; Zhu, J.F.; Cheng, J.; Zhou, J.H.; Cen, K.F. Pore structure and fractal analysis of Ximeng lignite under microwave irradiation. *Fuel* **2015**, *146*, 41–50. [CrossRef]

23. Ji, W.M.; Song, Y.; Jiang, Z.X.; Meng, M.M.; Liu, Q.X.; Chen, L.; Wang, P.F.; Gao, F.L.; Huang, H.X. Fractal characteristics of nano-pores in the Lower Silurian Longmaxi shales from the Upper Yangtze Platform, south China. *Mar. Pet. Geol.* **2016**, *78*, 88–98. [CrossRef]

24. Slatt, R.M.; Rodriguez, N.D. Comparative sequence stratigraphy and organic geochemistry of gas shales: Commonality or coincidence? *J. Nat. Gas Sci. Eng.* **2012**, *8*, 68–84. [CrossRef]

25. Wang, G.C.; Carr, T.R. Methodology of organic-rich shale lithofacies identification and prediction: A case study from Marcellus Shale in the Appalachian basin. *Comput. Geosci.* **2012**, *49*, 151–163. [CrossRef]

26. Loucks, R.G.; Ruppel, S.C. Mississippian Barnett Shale: Lithofacies and depositional setting of a deep-water shale-gas succession in the Fort Worth Basin, Texas. *AAPG Bull.* **2007**, *91*, 579–601. [CrossRef]

27. Tan, J.Q.; Weniger, P.; Krooss, B.; Merkel, A.; Horsfield, B.; Zhang, J.C.; Boreham, C.J.; van Graas, G.; Tocher, B.A. Shale gas potential of the major marine shale formations in the Upper Yangtze Platform, South China, Part II: Methane sorption capacity. *Fuel* **2014**, *129*, 204–218. [CrossRef]

28. Tang, X.L.; Jiang, Z.X.; Huang, H.X.; Jiang, S.; Yang, L.; Xiong, F.Y.; Chen, L.; Feng, J. Lithofacies characteristics and its effect on gas storage of the Silurian Longmaxi marine shale in the southeast Sichuan Basin, China. *J. Nat. Gas Sci. Eng.* **2016**, *28*, 338–346. [CrossRef]

29. Chen, L.; Lu, Y.C.; Jiang, S.; Li, J.Q.; Guo, T.L.; Luo, C. Heterogeneity of the Lower Silurian Longmaxi marine shale in the southeast Sichuan Basin of China. *Mar. Petrol. Geol.* **2015**, *65*, 232–246. [CrossRef]

30. Gao, F.L.; Song, Y.; Li, Z.; Xiong, F.Y.; Chen, L.; Zhang, Y.H.; Liang, Z.K.; Zhang, X.X.; Chen, Z.Y.; Joachim, M. Lithofacies and reservoir characteristics of the Lower Cretaceous continental Shahezi Shale in the Changling Fault Depression of Songliao Basin, NE China. *Mar. Petrol. Geol.* **2018**, *98*, 401–421. [CrossRef]

31. Cai, Q.S.; Hu, M.Y.; Ngia, N.R.; Hu, Z.G. Sequence stratigraphy, sedimentary systems and implications for hydrocarbon exploration in the northern Xujiaweizi Fault Depression, Songliao Basin, NE China. *J. Pet. Sci. Eng.* **2017**, *152*, 471–494. [CrossRef]

32. Huang, W.B.; Lu, S.F.; Osman, S.H. Quality grading system for tight sandstone reservoirs in the Quantou 4 Member, southern Songliao Basin, Northeast China. *Interpret. J. Sub.* **2017**, *5*, T503–T522. [CrossRef]

33. Ding, X.Q.; Hersi, O.S.; Hu, X.; Zhu, Y.; Zhang, S.N.; Miao, C.S. Diagenesis of volcanic-rich tight sandstones and conglomerates: A case study from Cretaceous Yingcheng Formation, Changling Sag, Songliao Basin, China. *Arab. J. Geosci.* **2018**, *11*, 287. [CrossRef]

34. Lin, J.H.; Jiang, T.; Song, L.B.; Cao, Y.; Xia, D.; Wang, Y. The origin and gas vertical distribution of the Harjin mixed-gas reservoir. *Acta Pet. Sin.* **2010**, *34*, 927–932.

35. Jarvie, D.M.; Claxton, B.L.; Henk, F.B.; Breyer, J.T. Oil and Shale Gas from the Barnett Shale, Ft. Worth Basin, Texas, AAPG National Convention, June 3–6, 2001, Denver, CO. *AAPG Bull.* **2001**, *85*, 100.

36. Jarvie, D.M.; Hill, R.J.; Pollastro, R.M. Assessment of the gas potential and yields from shales: The Barnett Shale model. *Oklahoma Geol. Surv. Circ.* **2005**, *110*, 37–50.

37. Ufer, K.; Stanjek, H.; Roth, G.; Dohrmann, R.; Kleeberg, R.; Kaufhold, S. Quantitative phase analysis of bentonites by the Rietveld method. *Clays Clay Miner.* **2008**, *56*, 272–282. [CrossRef]

38. Curtis, M.E.; Cardott, B.J.; Sondergeld, C.H.; Rai, C.S. Development of organic porosity in the Woodford Shale with increasing thermal maturity. *Int. J. Coal Geol.* **2012**, *103*, 26–31. [CrossRef]

39. Tang, X.L.; Jiang, Z.X.; Jiang, S.; Li, Z. Heterogeneous nanoporosity of the Silurian Longmaxi Formation shale gas reservoir in the Sichuan Basin using the QEMSCAN, FIB-SEM, and nano-CT methods. *Mar. Petrol. Geol.* **2016**, *78*, 99–109. [CrossRef]

40. Jiao, K.; Ye, Y.H.; Liu, S.G.; Ran, B.; Deng, B.; Li, Z.W.; Li, J.X.; Yong, Z.Q.; Sun, W. Characterization and Evolution of Nanoporosity in Superdeeply Buried Shales: A Case Study of the Longmaxi and Qiongzhusi Shales from MS Well #1, North Sichuan Basin, China. *Energy Fuels* **2018**, *32*, 191–203.

41. Guo, T.L.; Zhang, H.R. Formation and enrichment mode of Jiaoshiba shale gas field, Sichuan Basin. *Pet. Explor. Dev.* **2014**, *41*, 31–40. [CrossRef]

42. Sing, K.S. Reporting physisorption data for gas/solid systems with special reference to the determination of surface area and porosity (Recommendations 1984). *Pure Appl. Chem.* **1985**, *57*, 603–619. [CrossRef]

43. Gao, Z.Y.; Hu, Q.H. Wettability of Mississippian Barnett Shale samples at different depths: Investigations from directional spontaneous imbibition. *AAPG Bull.* **2016**, *100*, 101–114. [CrossRef]

44. Chen, L.; Jiang, Z.X.; Liu, K.Y.; Tan, J.Q.; Gao, F.L.; Wang, P.F. Pore structure characterization for organic-rich Lower Silurian shale in the Upper Yangtze Platform, South China: A possible mechanism for pore development. *J. Nat. Gas Sci. Eng.* **2017**, *46*, 1–15. [CrossRef]

45. Wei, Z.F.; Wang, Y.L.; Wang, G.; Sun, Z.P.; Xu, L. Pore characterization of organic-rich Late Permian Da-long Formation shale in the Sichuan Basin, southwestern China. *Fuel* **2018**, *211*, 507–516. [CrossRef]

46. Wang, Q.T.; Lu, H.; Wang, T.L.; Liu, D.Y.; Peng, P.A.; Zhan, X.; Li, X.Q. Pore characterization of Lower Silurian shale gas reservoirs in the Middle Yangtze region, central China. *Mar. Pet. Geol.* **2018**, *89*, 14–26. [CrossRef]

47. Fu, H.J.; Wang, X.Z.; Zhang, L.X.; Gao, R.M.; Li, Z.T.; Xu, T.; Zhu, X.L.; Xu, W.; Li, Q. Investigation of the factors that control the development of pore structure in lacustrine shale: A case study of block X in the Ordos Basin, China. *J. Nat. Gas Sci. Eng.* **2015**, *26*, 1422–1432. [CrossRef]

48. Jiang, F.J.; Chen, D.; Wang, Z.F.; Xu, Z.Y.; Chen, J.; Liu, L.; Huyan, Y.Y.; Liu, Y. Pore characteristic analysis of a lacustrine shale: A case study in the Ordos Basin, NW China. *Mar. Pet. Geol.* **2016**, *73*, 554–571. [CrossRef]

49. Curtis, M.E.; Sondergeld, C.H.; Ambrose, R.J.; Rai, C.S. Microstructural investigation of gas shales in two and three dimensions using nanometer-scale resolution imaging. *AAPG Bull.* **2012**, *96*, 665–677. [CrossRef]

50. Milliken, K.L.; Rudnicki, M.; Awwiller, D.N.; Zhang, T. Organic matter–hosted pore system, Marcellus Formation (Devonian), Pennsylvania. *AAPG Bull.* **2013**, *97*, 177–200. [CrossRef]
51. Yang, R.; He, S.; Yi, J.Z.; Hu, Q.H. Nano-scale pore structure and fractal dimension of organic-rich Wufeng-Longmaxi shale from Jiaoshiba area, Sichuan Basin: Investigations using FE-SEM, gas adsorption and helium pycnometry. *Mar. Pet. Geol.* **2016**, *70*, 27–45. [CrossRef]
52. Wu, S.T.; Zhu, R.K.; Cui, J.G.; Cui, J.W.; Bai, B.; Zhang, X.X.; Jin, X.; Zhu, D.S.; You, J.C.; Li, X.H. Characteristics of lacustrine shale porosity evolution, Triassic Chang 7 Member, Ordos Basin, NW China. *Pet. Explor. Dev.* **2015**, *42*, 185–195. [CrossRef]
53. Lu, J.M.; Ruppel, S.C.; Rowe, H.D. Organic matter pores and oil generation in the Tuscaloosa marine shale. *AAPG Bull.* **2015**, *99*, 333–357. [CrossRef]
54. Jarvie, D.M.; Hill, R.J.; Ruble, T.E.; Pollastro, R.M. Unconventional shale-gas systems: The Mississippian Barnett Shale of north-central Texas as one model for thermogenic shale-gas assessment. *AAPG Bull.* **2007**, *91*, 475–499. [CrossRef]
55. Li, T.W.; Jiang, Z.X.; Li, Z.; Wang, P.F.; Xu, C.L.; Liu, G.H.; Su, S.Y.; Ning, C.X. Continental shale pore structure characteristics and their controlling factors: A case study from the lower third member of the Shahejie Formation, Zhanhua Sag, Eastern China. *J. Nat. Gas. Sci. Eng.* **2017**, *45*, 670–692. [CrossRef]

Article

Pore Connectivity Characterization of Lacustrine Shales in Changling Fault Depression, Songliao Basin, China: Insights into the Effects of Mineral Compositions on Connected Pores

Zhuo Li [1,2,*], **Zhikai Liang** [1,2], **Zhenxue Jiang** [1,2], **Hailong Yu** [1,2], **Youdong Yang** [1,2] **and Lei Xiao** [1,2]

1   State Key Laboratory of Petroleum Resources and Prospecting, China University of Petroleum, Beijing 102249, China; Liangzhikai2020@163.com (Z.L.); jiangzhenxue0405@163.com (Z.J.); yhl2017210211@163.com (H.Y.); yydzgsydxbj@163.com (Y.Y.); xlcupb@163.com (L.X.)
2   Unconventional Natural Gas Research Institute, China University of Petroleum, Beijing 102249, China
*   Correspondence: zhuo.li@cup.edu.cn; Tel.: +86-188-1080-3526

Received: 22 February 2019; Accepted: 22 March 2019; Published: 26 March 2019

Abstract: Pore connectivity of lacustrine shales was inadequately documented in previous papers. In this work, lacustrine shales from the lower Cretaceous Shahezi Formation in the Changling Fault Depression (CFD) were investigated using field emission scanning electron microscopy (FE-SEM), mercury intrusion capillary pressure (MICP), low pressure gas (CO_2 and N_2) sorption (LPGA) and spontaneous fluid imbibition (SFI) experiments. The results show that pores observed from FE-SEM images are primarily interparticle (interP) pores in clay minerals and organic matter (OM) pores. The dominant pore width obtained from LPGA and MICP data is in the range of 0.3–0.7 nm and 3–20 nm. The slopes of n-decane and deionized (DI) water SFI are in the range of 0.34–0.55 and 0.22–0.38, respectively, suggesting a mixed wetting nature and better-connected hydrophobic pores than hydrophilic pores in the Shahezi shales. Low pore connectivity is identified by the dominant nano-size pore widths (0.3–20 nm), low DI water SFI slopes (around 0.25), high geometric tortuosity (4.75–8.89) and effective tortuosity (1212–6122). Pore connectivity follows the order of calcareous shale > argillaceous shale > siliceous shale. The connected pores of Shahezi shales is mainly affected by the high abundance and coexistence of OM pores and clay, carbonate minerals host pores.

Keywords: lacustrine shales; pore networks; pore connectivity; spontaneous fluid imbibition; Shahezi shales; Changling Fault Depression

---

1. Introduction

Organic-shales and coals are proposed to contain complex pore systems with various pore types, shapes and wide pore size distribution (PSD) ranges [1–7]. Understanding pore characteristics is essential for assessing shale gas storage capacity and gas transport mechanisms in organic shales [8–13]. Pore structure characterization is also the key issue of shale gas resource and coalbed methane assessment [14–21]. The pore characteristics (types, shapes, geometries, PSDs and evolution mechanisms) of gas shales were extensively studied around the world in previous studies by several researchers [1,3,13,14,21–28]. For example, field emission scanning electron microscopy (FE-SEM) [1,7,29], small angle neutron scattering (SANS) [2,24], low pressure gas adsorption (LPGA) and mercury intrusion capillary pressure (MICP) [2,3,9] were used to characterize the pore structure in shales. However, less attention was paid to the connectivity of pore networks in organic shales. Pore connectivity significantly affects the gas flow distance in the shale matrix, and is critical to shale gas development [30]. High pore connectivity may improve the gas flow through the matrix, while low pore connectivity may be the main cause for the rapid decline of the initial shale gas production [24,31,32]. Pore connectivity in shales can be evaluated by various methods [33–39]. Three dimensional pore system reconstruction by micro- and nano-scale X-ray computed tomography (micro-CT and nano-CT) and focused ion beam-scanning electron microscopy (FIB-SEM) techniques [29,40], the hysteresis loops of LPGA and MICP isotherms [41–44], the spontaneous fluid imbibition (SFI) features [34,45,46] and the divergence of MICP and nuclear magnetic resonance (NMR) results [47,48], can be used to evaluate pore connectivity in organic shales. Among these methods, SFI is a simple and effective technique, because the SFI process is much faster than diffusion, and only requires monitoring mass change over time [32,34,36,49,50]. The pore connectivity of marine shales in North America and southern China were assessed through SFI experiments [33,45,46]. However, only a few studies documented the pore connectivity of lacustrine shales in Sichuan Basin and Ordos Basin of China [24,48]. Therefore, previous studies on the pore connectivity of lacustrine shale are still inadequate.

Lacustrine shales gas accounts for about 32% of the total recoverable shale gas resources in China [51–53]. The lower Cretaceous Shahezi shales were studied as major source rocks for conventional gas resources in Songliao Basin [54,55]. Recently, these shales in the Changling Fault Depression (CFD) have become shale gas exploration targets [48]. Lacustrine shales are significantly distinct from marine shales in terms of the rapid sedimentary environment changes, discontinuous and complex lithofacies association and high concentration of clay minerals [56–58]. Recently, shale lithofacies and their impacts on the pore characteristics and pore fractal dimensions of the lacustrine shales in the Shahezi Formation were investigated [58,59]. In this work, the pore connectivity of these shales was evaluated, and the control of the matrix compositions on pore connectivity was further discussed. Our findings are of significance to an understanding of gas transport mechanisms and the impacts of lacustrine shale gas production.

The objectives of this work are to (1) determine the pore structure characteristics (pore type, pore size distribution and pore volume); (2) assess the pore connectivity; (3) and investigate the controls of OM and mineral compositions on the pore connectivity of Shahezi shales. An integrated experiment procedure including total organic carbon contents (TOC), X-ray diffraction (XRD), FE-SEM, MICP, LPGA and SFI (n-decane and deionized water) experiments was conducted in this study.

2. Samples and Experiments

2.1. Samples

In total, six lacustrine shale core samples (about 9.5 cm in diameter and 2–5 cm in length) were collected from the lower Cretaceous Shahezi Formation in Well TS-6, which was drilled in the CFD of Songliao Basin (seen in Figure 1a,b). The CFD can be subdivided into the west step-fault belt, central depression belt, east slope belt and east structural belt based on structural features [60]. The rift stage for the Songliao Basin occurs during the Jurassic-early Cretaceous episode [61]. From bottom to top, the lower Cretaceous strata consist of the Huoshiling Formation (K_1h), the Shahezi Formation (K_1sh), the Yingcheng Formation (K_1y) and the Denglouku Formation (K_1d) (Figure 1c,d). Black mudstones and shales in the K_1sh and K_1y Formations deposited in semi-deep to deep lacustrine environments are the main source rocks of natural gas (with Ro values above 2%) in the study area [58,59,62]. The sample information, including depth and lithofacies, are listed in Table 1. The lithofacies of the selected samples was determined by core logging data of the Jilin oil field Branch, Petrochina.

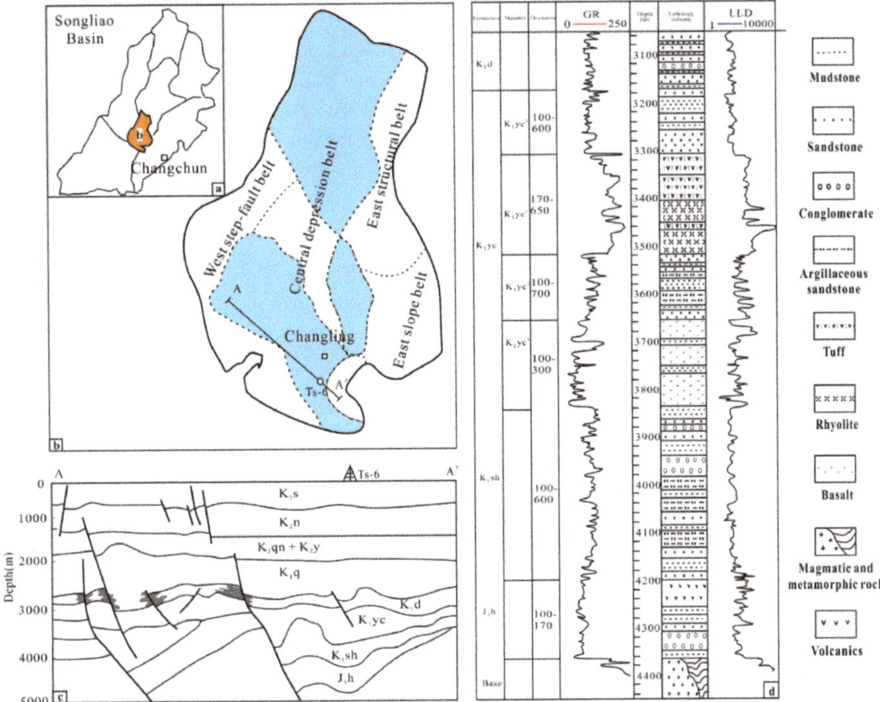

Figure 1. (**a**) Location map of the Changling Fault Depression (CFD). (**b**) Location map of the study area and the sub-tectonic units of the CFD in southern Songliao Basin. (**c**) Cross section across the CFD showing the structural pattern and stratigraphic intervals. (**d**) Stratigraphic column showing the lithology of the strata and lithology in the CFD.

2.2. Experiments

Each shale core sample was subsampled and prepared for the total organic carbon content (TOC), XRD, FE-SEM, LPGA, MICP and SFI experiments. The subsamples were taken as homogeneous as possible in order to correlate the results.

2.2.1. TOC and XRD

The TOC content was determined by a Leco CS230 carbon/sulfur analyzer (LECO Corporation, St. Joseph, MI, USA). In detail, shale samples were crushed to 100 mesh, treated with hydrochloric acid to remove carbonates and washed with distilled water. After being dried at 70 °C, the TOC values were measured. The bulk XRD mineral compositions were measured on powder samples, which were mixed with ethanol on glass slides for an XRD analysis. The XRD experiments were performed on an X-ray diffractometer (Bruker D8 DISCOVER, Bruker AXS Corporation, Karlsruhe, Germany) using Co Ka-radiation of 45 kV and 35 mA. The quantitative analysis of the bulk mineral compositions and clay minerals was performed on the Bruker customized Topas software.

2.2.2. FE-SEM

FE-SEM imaging was performed on an Argon-ion polished surface of shale samples on a Helios NanoLab 650 FEI SEM (Thermo Fisher Scientific, Waltham, MA, USA). The brick shaped samples (about 1 cm $\times$ 1 cm $\times$ 0.5 cm) were ion polished on a Hitachi ion milling IM-4000 System (Hitachi, Ltd., Tokyo, Japan) and imaged using an FE-SEM (Zeiss Helios NanoLab 650, FEI, Carl Zeiss, Heidenheim, Germany) equipped with secondary electron (SE) and backscattered electron (BSE) detectors. The SE mode detects the topographic variation, and the BSE mode shows the compositional variation [1]. The accelerating voltages were 1–10 kV. The working distances from detector to samples were 3–7 mm in this FE-SEM system.

2.2.3. MICP

The MICP experiments were performed on shale core plugs (about 2.54 cm in diameter) on a mercury porosimeter (Micromeritics AutoPore 9510, Micromeritics, Atlanta, GA, USA) with the maximum intrusion pressure set to 400 MPa. Before the MICP test, each shale sample was oven-dried at 80 °C for 24 h to remove moisture. With increasing pressure, the mercury volumes are recorded; conversely, the intruded mercury will extrude from the sample when the pressure drops. The pore throat is obtained based on the Washburn equation, assuming a specific pore cylindrical shape. The inputting physical constants of mercury include the surface tension γ = 485 mN/m and the contact angle θ = 130°.

2.2.4. LPGA

The LPGA (N_2 and CO_2) experiments were conducted on a Micromeritics ASAP-2460 System (Micromeritics, Atlanta, GA, USA). The samples were crushed to 100 mesh and degassed at 110 °C for 12 h. Nitrogen and carbon dioxide physisorption isotherms were obtained at 77.3 K and 273.1 K, respectively. The micro-pore volumes and surface areas were determined from low pressure CO_2 adsorption branches using a DFT (Density Functional Theory) model. The BJH (Barrett-Joyner-Halenda) pore volumes, BET (Brunauer-Emmett-Teller) surfaces areas and average pore sizes were determined by low pressure N_2 adsorption data.

2.2.5. SFI

The SFI experiments were performed following the procedure proposed by Gao and Hu [63]. Briefly, 1 cm-sized cubic samples were cut from the shale cores at each depth intervals provided in Table 1. Two subsamples for each shale sample were used for the DI-water and n-decane imbibition measurements. The SFI direction was parallel to the laminae of the shales. Except for the bottom and top sides, the other four sides were coated with epoxy. Following this, the epoxied cubes were oven-dried at 80 °C for 12 h and cooled to room temperature (23 $\pm$ 0.5 °C). For the DI water and n-decane SFI experiments, the bottom surface was merged at about 1 mm in the fluids. The weight change was automatically recorded by a high precision electric balance. In the homogeneous porous materials with the capillary as the main forces for the SFI, the slopes of the log total imbibition versus log imbibition time curves can be used to assess the pore connectivity [63]. The DI water and n-decane SFI can be used to investigate the connectivity of hydrophilic and hydrophobic pores, respectively [61,64,65].

3. Results

3.1. Mineral Compositions and TOC Content

The detailed sample information, including lithofacies, TOC contents and mineral composition, was listed in Table 1. Quartz and clay mineral contents are dominant in these samples, accounting for over 70 wt % in the Shahezi shale samples. The clay minerals are the most significant mineral components. The clay minerals mainly consist of mixed illite-smectite (66 wt %–79 wt %), illite (13 wt %–23 wt %) and chlorite (9 wt %–13 wt %) in our sample set. The dolomite, pyrite and siderite contents are relatively low. Three lithofacies, including argillaceous shale, siliceous shale and calcareous shale, are classified on the basis of the core logging of the Jilin oil field Branch and verified by bulk XRD results. The TOC contents range from 1.09 wt % to 4.75 wt %, with an average of 2.841 wt % in the samples.

3.2. Pore characterization

3.2.1. FE-SEM Imaging and Image Processing Analysis

The FE-SEM images show pore characteristics in the Shahezi shale samples with different lithofacies (TS-6-4, TS-6-5 and TS-6-7). According to the classification scheme proposed by Loucks et al. [1], organic matter (OM) pores and interparticle pores (interP pores) between clay flakes are dominant in the Shahezi shales. OM pores are observed in siliceous shale samples and these pores are elliptical, bubble-like, irregular (Figure 2a–c) in shape. Many OM pores display better connectivity in 3D dimensions (Figure 2a). OM surrounded by rigid mineral grains could be protected from compaction and well preserved (Figure 2b).

InterP pores are commonly observed between illite flakes and chlorite in Shahezi shale samples (Figure 2d,e). A few larger-sized interP pores between quartz grains were observed as well (Figure 2f). InterP pores connected to OM pores could serve as a significant migration pathway for shale gas [1]. Shale gas is expected to flow along the most conductive pathway associated with the connected pores [64].

A few intraparticle pores (intraP pores) can be observed in the Shahezi shales (Figure 2g–i), which are primary formed due to the dissolution of chemically unstable minerals, such as carbonate and feldspar [1].

An image processing analysis were performed on the FE-SEM images of the Shahezi shales (Figure 2) using Image-Pro Plus software [66]. The results of the image processing include the pore number, pore width and fractal dimension, all listed in Table 2. In total, 869 pores, including OM pores, interP pores and intraP pores, were extracted from the FE-SEM images (Figure 2).

For the siliceous shale sample (TS-6-4), interP pores contribute the highest percentage in the total pore systems (53.21%). However, intraP pores contribute the lowest percentage (16.06%). The mean pore sizes of the OM pores, interP pores, and intraP pores are 113.6, 90.1, and 54.2 nm. IntraP pores have the lowest fractal dimension (1.08), while OM pores have the highest value (1.68). For the calcareous shale sample (TS-6-5), no OM pores were extracted from the FE-SEM image (Figure 2g), while 44 interP pores and 25 intraP pores were extracted. The mean pore sizes of the interP pores and intraP pores are 181.7 and 193.1, respectively. Both InterP pores and intraP pores have close fractal dimensions of 1.17 and 1.21. For the argillaceous shale sample (TS-6-7), 582 pores in total were extracted from the images (Figure 2c–e,h,i). InterP pores have the highest percentage of pore numbers (57.73%), while intraP pores contribute the lowest percentage (19.59%). The mean pore sizes of the OM pores, interP pores, and intraP pores are 56.3, 224.6, and 89.3 nm.

Overall, interP pores provide the highest proportion of pore numbers and a relatively larger mean pore size in the studied shale samples. The pores in the studied Shahezi shales are mostly irregular in shape as displayed by the fractal dimension data, which is consistent with previous image processing work [58].

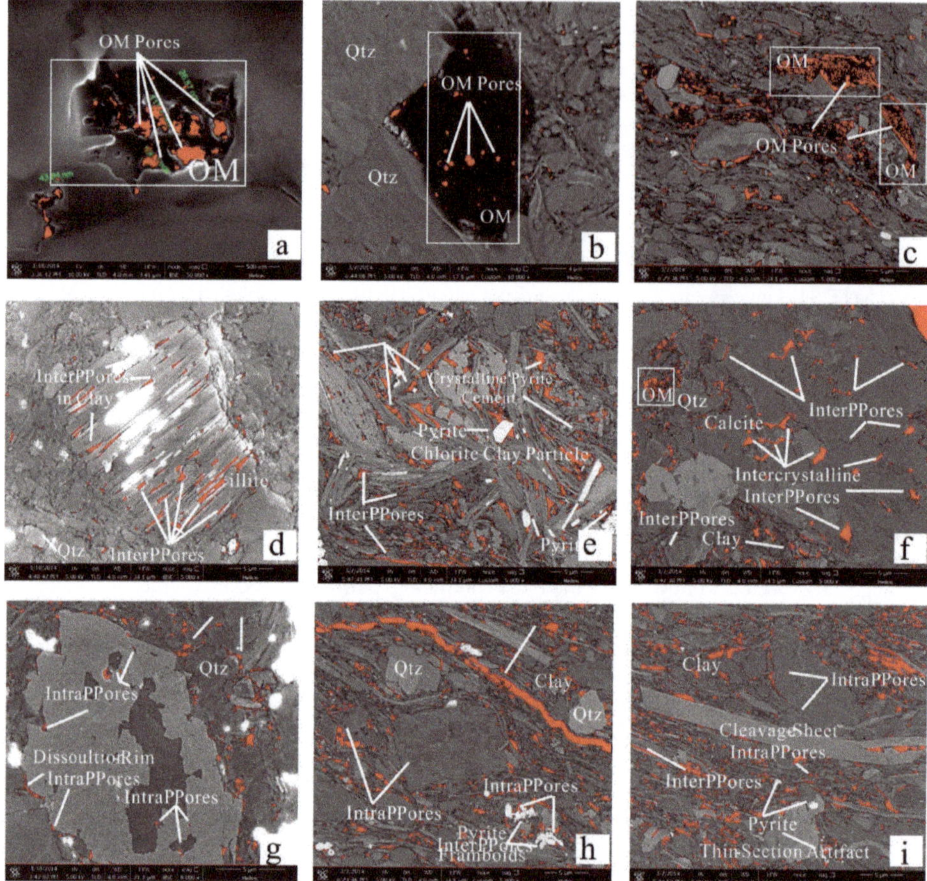

Figure 2. Field emission-scanning electron microcopy (FE-SEM) images of pores in the Shahezi shale samples. (**a**,**b**) Organic particle with relatively large organic matter (OM) pores showing heterogeneous distribution and high connectivity (TS-6-4, siliceous shale); (**c**) organic particle distributed as lumps with poorly developed and isolated OM pores (TS-6-7, argillaceous shale); (**d**,**e**) interP pores in illite grains containing a cleavage-sheet (TS-6-7, argillaceous shale); (**f**) shale sample contains coexisted intercrystal line interP pores, intraP pores and OM pores (TS-6-4, siliceous shale); (**g**) dissolution-rim and intraP pores in calcite crystals (TS-6-5, calcareous shale); (**h**,**i**) linear interP pore along clay grains (TS-6-7, argillaceous shale).

Table 1. Basic properties of lacustrine shale samples from the lower Cretaceous Shahezi Formation in the Changling Fault Depression (CFD).

Sample ID	Depth Beneath to Surface (m)	Lithofacies	Total Organic Carbon (wt %)	Mineral Composition (wt %)									
				Quartz	Feldspar	Calcite	Dolomite	Pyrite	Siderite	Clay	Mixed Illite-Smectite	Illite	Chlorite
TS-6-1	3696.00–3696.03	Argillaceous Shale	1.09	21.1	8.9	1.2	1.6	2.3	0.8	63.2	72	19	9
TS-6-3	3734.30–3743.32	Siliceous Shale	2.27	36.1	13.2	3.8	0.5	1.8	-	45.2	66	23	11
TS-6-4	3738.00–3738.05	Siliceous Shale	4.64	38.5	7.2	2.7	2.1	2.2	0.8	45	74	16	10
TS-6-5	3759.00–3759.02	Calcareous Shale	2.12	30.1	5.2	30.5	-	-	3.1	30.1	67	20	13
TS-6-7	3765.00–3765.03	Argillaceous Shale	4.75	29.6	8.3	0.9	2.4	1.8	0.3	50.9	79	13	8
TS-6-8	3787.30–3787.33	Argillaceous Shale	2.18	30.5	9.6	2.9	1.3	2.6	-	51.3	75	16	9

Table 2. Image processing results of the Shahezi shales with various lithofacies in the CFD.

Sample ID	Lithofacies	Pore Type	Number of Pores	Percentage (%)	Pore Size (nm)			Fractal Dimension
					Min Value	Max Value	Mean	
TS-6-4	Siliceous Shale	OM pore	67	30.73	30.7	890.5	113.6	1.68
		InterP pore	116	53.21	57.8	458.7	90.1	1.24
		IntraP pore	35	16.06	45.9	210.8	54.2	1.08
TS-6-5	Calcareous Shale	OM pore	0	0	-	-	-	-
		InterP pore	44	63.77	192.9	336.5	181.7	1.17
		IntraP pore	25	36.23	78.5	423.9	193.1	1.21
TS-6-7	Argillaceous Shale	OM pore	132	22.68	42.6	155.9	56.3	1.52
		InterP pore	336	57.73	78.4	801.7	224.6	1.72
		IntraP pore	114	19.59	55.9	231.6	89.3	1.38

3.2.2. Full-Size Pore Size Distribution

Figure 3 illustrates the full-size PSD by combining the low pressure gas adsorption (CO_2 and N_2) and MICP data. Micropores (<2 nm), mesopores (2–50 nm), and macropores (>50 nm) are well developed in all shale samples, according to the International Union of Pure and Applied Chemistry (IUPAC) classification [67]. Multimodal PSD characteristics of the lacustrine Shahezi shale samples can be determined in the ranges of 0.3–0.7 nm and, 3–20 nm (Figure 3). The peaks at 10–30 μm determined by MICP are considered fake peaks, which may be due to artificial fractures formed during the sample preparation [11].

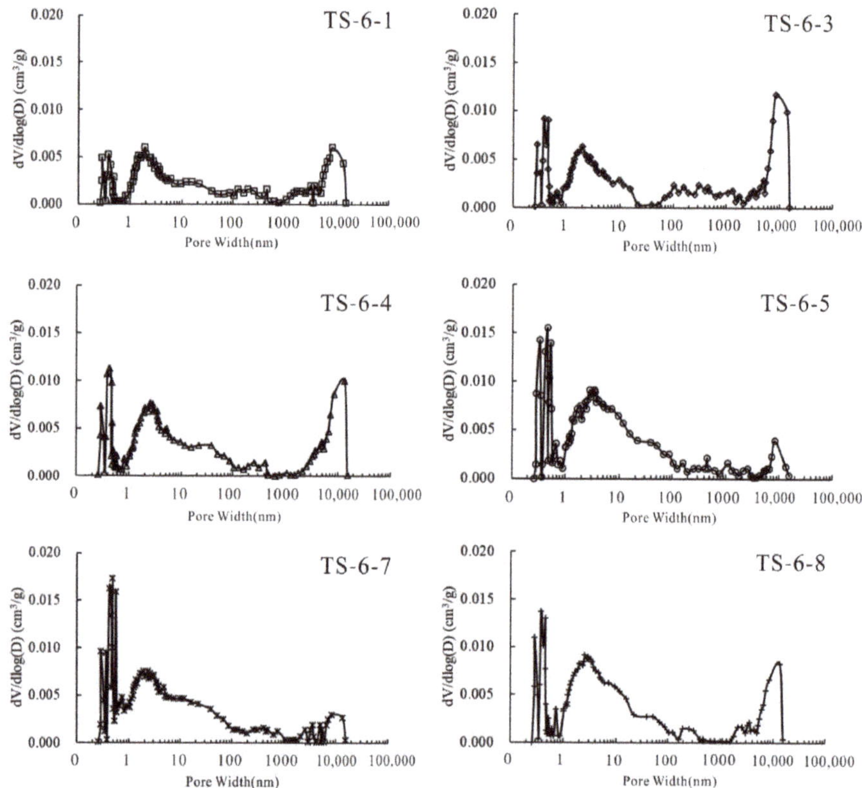

Figure 3. Pore size distribution (PSD) of the selected lacustrine shale samples from CFD. The micropores are obtained from low pressure CO_2 adsorption data, the mesopores from low pressure N_2 adsorption data, and the macropores from mercury intrusion capillary pressure (MICP) experiments.

In this paper, the micropore (pore width <2 nm) volumes of the Shahezi shale samples were determined by low pressure CO_2 adsorption, the mesopore (2 nm < pore width < 50 nm) volumes were determined by low pressure N_2 adsorption, and the macropore (pore width >50 nm) volumes were determined by MICP experiments. The total volume of each sample is the sum of the micropore, mesopore and macropore volumes. The total pore volumes are in the range of 0.79 and 1.38 cm³/100 g for the Shahezi shales (Table 3). The micropore volumes with an average of 0.281 cm³/100 g (0.12–0.44 cm³/100 g) account for approximately 25% of the total pore volume, mesopore volumes with a mean value of 0.503 cm³/100 g (0.38–0.79 cm³/100 g) account for approximately 45% of the total pore volume, and macropore volumes with an average of 0.335 cm³/100 g (0.25–0.61 cm³/100 g) account for approximately 30% of the total pore volume (Table 3, Figure 4a). Therefore, mesopores provide the largest contribution to the total pore volume, followed by macropores. Micropores contribute the least proportion to the total pore volume. Micropores have the highest proportion to the total specific surface area (over 75%), mesopores only account for a small portion (about 24%), and the surface area provided by macropores is negligible (Table 3, Figure 4b).

3.2.3. Tortuosity

Tortuosity refers to the ratio of the actual distance of the fluid transport to the apparent straight-line length through the medium [45], which is calculated from Equation (1):

$$\tau = \frac{D_0}{D_e} = \frac{1}{\Phi}\left(\frac{L_e}{L}\right)^2,\tag{1}$$

where D_0 is the aqueous diffusion coefficient in certain fluids (m^2/s), D_e is the effective diffusion coefficient in porous media (m^2/s), L_e is the actual distance (m) travelled by a molecule, and L is the length a molecule moves between two points in a porous medium (m) [45].

The effect tortuosity values can be calculated from Equation (2) by the MICP results [32]:

$$\tau = \sqrt{\frac{\rho}{24k(1+\rho V_{tot})}\int_{\eta=r_{c,\min}}^{\eta=r_{c,\max}}\eta^2 f_v(\eta)d\eta},\tag{2}$$

where τ is effective tortuosity, ρ is the mercury density (g/cm^3), V_{tot} is the total pore volume (mL/g), $\int_{\eta=r_{c,\min}}^{\eta=r_{c,\max}}\eta^2 f_v(\eta)d\eta$ is the pore throat volume probability density function, $r_{c,}$max is the maximum pore radius, $r_{c,}$min is the minimum pore radius, $f_v(\eta)$ is the density function distributed over v points, η is the pore throat density, and k is the absolute permeability.

The geometrical tortuosity (L_e/L) can be calculated based on Equation (1) by inputting the effective tortuosity (τ) and MICP derived porosity (Φ).

As can be seen in Figures 3 and 4, most pore sizes in the Shahezi shale are smaller than 100 nm, with an average pore throat diameter of 6.13–13.68 nm (Table 3). In addition, the maximum intrusion pressure of MICP in the present study is 400 MPa, which results in a minimum detected pore throat of about 4.5 nm. Therefore, the effective tortuosity values of this study represent the mercury flow characteristics in a predominant pore throat width of 4.5–13.68 nm. Low pore connectivity commonly leads to high tortuosity [32].

The effective tortuosity and geometrical tortuosity (L_e/L) values calculated from the MICP data of the Shahezi shale samples are in the range of 1212–6122 and 4.75–8.89, respectively, which is consistent with the previous reported values of Barnett shales (1772–5372 and 2.13–11.6), Longmaxi shales (123–1716 and 11.1–41.4), Niutitang shales (6284–6807), Dongyuemiao (284 and 11.2) and Dalong shales (104 and 14) [24,32,68]. Tortuosity can be used to evaluate pore connectivity in a porous medium [30]. According to the tortuosity values of this study and previously reported tortuosity values of typical gas shales [24,32,68], low pore connectivity can be identified in the Shahezi shales in the CFD, Songliao Basin.

Table 3. Pore structure parameters obtained from low pressure gas physisorption and mercury intrusion capillary pressure (MICP) of lacustrine shale samples from the Shahezi Formation in the CFD.

Sample ID	CO$_2$ Physisorption		N$_2$ Physisorption			Mercury Injection Capillary Pressure					Total Pore Surfaces Areas (m^2/g)	Total Pore Volumes (cm^3/100 g)
	DFT Surface Area (m^2/g)	DFT Pore Volume (cm^3/100 g)	BET Surface Area (m^2/g)	BJH Pore Volume (cm^3/100 g)	Average Pore Size (nm)	Surface Area (m^2/g)	Pore Volume (cm^3/100 g)	Porosity (%)	Tortuosity	L_e/L		
TS-6-1	6.478	0.12	2.413	0.38	9.85	0.010	0.29	1.29	5922	8.74	8.901	0.79
TS-6-3	8.223	0.18	2.815	0.40	8.85	0.009	0.61	2.42	2495	7.77	11.047	1.19
TS-6-4	14.843	0.26	3.523	0.49	7.99	0.007	0.39	3.01	1879	7.52	18.373	1.14
TS-6-5	6.385	0.44	3.871	0.75	6.13	0.005	0.19	1.86	2112	6.27	10.261	1.38
TS-6-7	8.532	0.36	2.389	0.51	7.04	0.009	0.28	2.08	2458	7.15	10.93	1.15
TS-6-8	13.752	0.33	2.546	0.49	13.68	0.007	0.25	3.22	2160	8.34	16.305	1.07

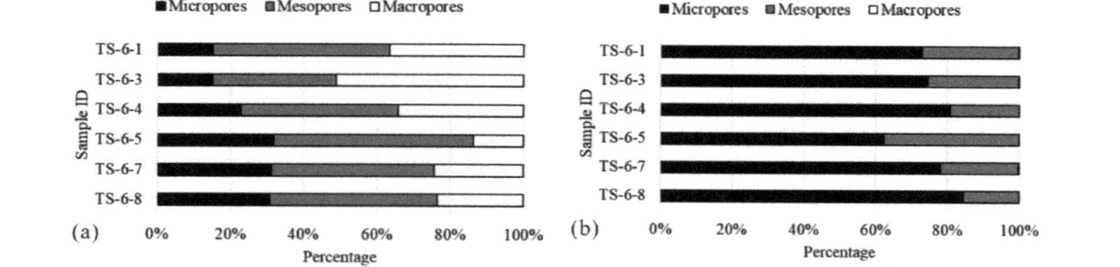

Figure 4. (**a**) The percentage of pore volume and (**b**) specific surface area of micropores, mesopores, macropores of lacustrine shales from the CFD.

3.3. Spontaneous Fluid Imbibition

Log cumulative imbibition (mm)-log time (min) plots of n-decane and DI water SFI of the Shahezi shales are shown in Figures 5 and 6. The DI water SFI curves can be segmented to an initial stage (0–120 s), an increasing stage (120 s–23 h) and a final stage (23–24 h) [38]. The rapid increases and fluctuations of imbibition in the initial stage may result from boundary effects, which is due to the initial contact of samples with fluids and to the migrations of fluid up the outside of the sample [24]. Following this, one or two linear segments with different slopes can be observed. Larger SFI slopes suggest more imbibed fluids and higher imbibition rates. The variations of the SFI slopes may reflect the complex connectivity and spatial wettability of these organic shales [30–34]. For instance, a decrease in the SFI slopes suggests that the imbibed fluids transport from larger pores to smaller pores in the connected pore networks within the shale matrix [36]. Based on the percolation theory, the stable slopes in the log-log plots were used to quantitatively assess the pore connectivity, as high pore connectivity had slopes of about 0.5 (classical Fickian behavior), medium pore connectivity had slopes of 0.26–0.5, and low pore connectivity had slopes of about 0.25 [30,36,46,69,70]. The imbibition increases very little in the final stage (23–24 h), suggesting that the samples are saturated with fluids.

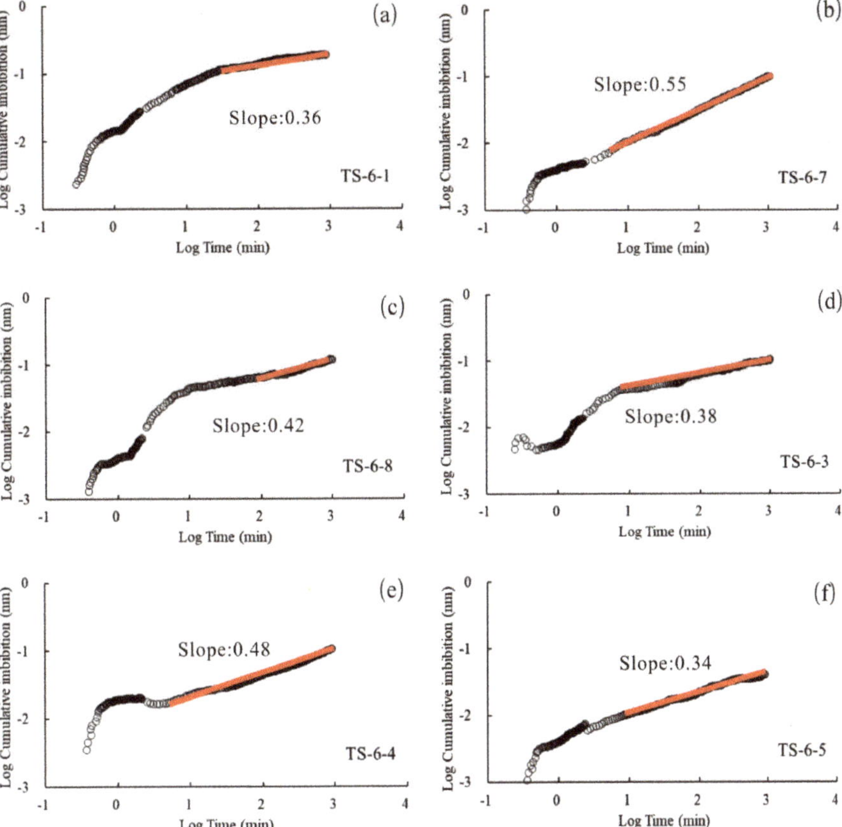

Figure 5. Plots of log cumulative n-decane spontaneous fluid imbibition (SFI) versus log imbibition time for the lacustrine Shahezi shales from the CFD. (**a**) Argillaceous shale (TS-6-1) with slope of 0.36; (**b**) argillaceous shale (TS-6-7) with slope of 0.55; (**c**) argillaceous shale (TS-6-8) with slope of 0.42; (**d**) siliceous shale (TS-6-3) with slope of 0.38; (**e**) siliceous shale (TS-6-4) with slope of 0.48; (**f**) calcareous shale (TS-6-5) with slope of 0.34.

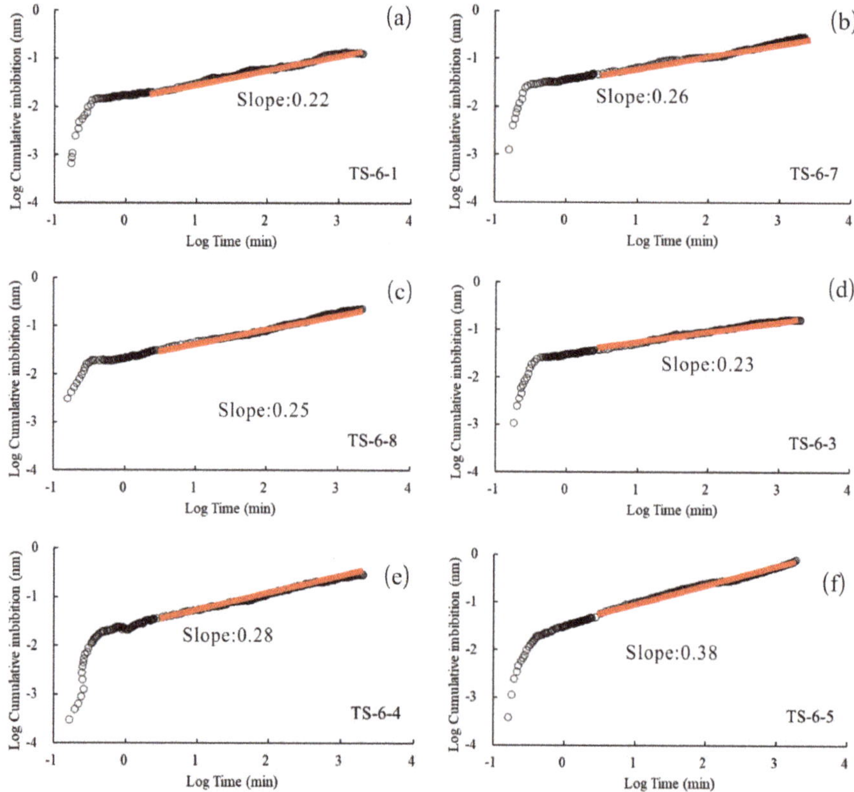

Figure 6. Plots of log cumulative DI water SFI versus log imbibition time for the lacustrine Shahezi shales from the CFD. (**a**) Argillaceous shale (TS-6-1) with slope of 0.22; (**b**) argillaceous shale (TS-6-7) with slope of 0.26; (**c**) argillaceous shale (TS-6-8) with slope of 0.25; (**d**) siliceous shale (TS-6-3) with slope of 0.23; (**e**) siliceous shale (TS-6-4) with slope of 0.28; (**f**) calcareous shale (TS-6-5) with slope of 0.38.

The n-decane SFI slopes of the Shahezi shales are in the range of 0.34–0.55 (Figure 5a–f). The samples with TOC contents over 4 wt % have higher SFI slopes (0.48–0.55) than other samples (Figure 5b,e). The n-decane SFI slopes are in the range of 0.36–0.55 for the argillaceous shales (Figure 5a–c), 0.34 for the calcareous shales (Figure 5d), and 0.38–0.48 for the siliceous shales (Figure 5e,f). The DI water SFI slopes of the Shahezi shales are in the range of 0.22–0.38 (Figure 6a–f). The DI water SFI slopes are in the range of 0.22–0.26 and 0.23–0.28 for argillaceous shales (Figure 6a–c) and siliceous shales (Figure 6e,f), respectively. The calcareous shale sample has the highest SFI slope, at 0.38 (Figure 6d).

4. Discussion

4.1. Pore Connectivity Obtained from SFI Slopes and Tortuosity Values

SFI is an effective way to assess pore connectivity in organic shales [30–34,36,63]. The SFI slopes of <0.25, 0.25–0.5 and >0.5 refer to poor-connected, moderate-connected, and well-connected pore networks for specific SFI fluids [30,36,46,71,72]. Specifically, DI water may preferentially be imbibed into the hydrophilic pores, while n-decane may be primarily imbibed into the hydrophobic pores [24]. The SFI process is affected by the wettability of both pores and fluids [63,73]. Rocks can have a good pore connectivity for a wetting fluid, but a poor connection for a nonwetting fluid [30]. Previous studies suggest that OM pores are commonly oil wetting and inorganic pores are water wetting [68,69]. A mixture of oil-wetting and water-wetting pores will result in the mixed wettability nature of organic shales [38]. The wettability of shale can be divided into more-water-wet, more-oil-wet, and mixed wet [36]. Different slopes of DI water and n-decane SFI can be applied to study the wettability of the shale matrix [24,30–34]. The ratio of the n-decane imbibition slope to DI water imbibition slope may provide useful information about the wettability. The ratios that are lower than one mean that the pore surface is more hydrophilic. Compared with DI water, the n-decane SFI of the studied Shahezi shales shows larger slopes (Figures 5 and 6), suggesting that the hydrophobic pore networks are better connected than the hydrophilic pores in the Shahezi shales and that the studied shale samples are all more oil-wet. The well-connected OM pores and inorganic pores contribute to the mixed-wetting characteristics of the lacustrine Shahezi shales. This result can be supported by OM pores observed in the FE-SEM images (Figure 2a–c).

The pore connectivity of the Shahezi shales vary in different shale lithofacies (Figure 7). The calcareous shale has the highest DI water SFI slope, at 0.38, followed by the siliceous shale and argillaceous shale samples, with slopes ranging from 0.22 to 0.28. These results suggest that hydrophilic pore networks are moderate-connected in calcareous shale and poor-connected in siliceous and argillaceous shales. In addition, the hydrophobic pore connectivity evaluated by the n-decane SFI of argillaceous shales (0.42–0.55) is better than for siliceous shales (0.38–0.48) and calcareous shales (0.34). The TS-6-7 sample has the largest n-decane SFI slope, suggesting well-connected hydrophobic pores in the argillaceous shale sample. Well-connected OM pores were observed in sample TS-6-4 (Figure 2a,b). Considering the similar TOC contents and thermal maturity of samples TS-6-7 and TS-6-4, it can be inferred that OM pores with a high connectivity formed in sample TS-6-7. According to the indication of the SFI slope of 0.25 [46], a high pore connectivity for n-decane and low connectivity for DI water can be determined [36,71–74].

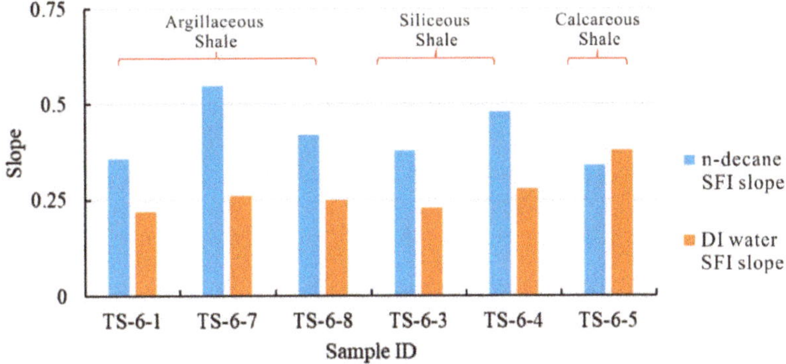

Figure 7. Slopes of the n-decane SFI and DI water SFI of different lithofacies of the lacustrine Shahezi shales from the CFD.

Tortuosity is another parameter for the pore connectivity evaluation [24,32]. The effective tortuosity obtained from the non-wetting MICP data is correlated with the pore connectivity but unrelated to the pore wettability [24,68]. The tortuosity represents both connected OM pores and inorganic pores in shales [24,30–33]. We can observe negative correlations of the effective tortuosity (Figure 8a) and geometrical tortuosity (Figure 8b) with the DI water SFI slopes, suggesting that poor-connected pores in the Shahezi shale samples lead to high tortuosity [32]. The geometrical tortuosity (L_e/L) values also correlated well with the DI water SFI slopes and indicate the pore connectivity of hydrophilic pore networks. Specifically, the maximum L_e/L value is 8.89, suggesting that non-wetting mercury has to travel 8.89 cm to move a linear distance of 1 cm in the shale matrix. Therefore, the low pore connectivity of the Shahezi shales results in this high tortuosity [24]. The shale sample TS-6-5, which has the highest water SFI slope, has the shortest diffusion distance (lowest geometrical tortuosity), while sample TS-6-1, with the lowest water imbibition slope, has the farthest diffusion distance (highest geometrical tortuosity) (Table 3). Shales with low pore connectivity have significant proportions of unconnected pores, which will slow the gas diffusion in the shale matrix and increase the effective tortuosity values [24,32]. Overall, moderate-connected oil-wetting and poor-connected water-wetting pore networks were identified in the Shahezi shales, which is consistent with gas shales in the US with limited connected water-wet pore systems [31,32].

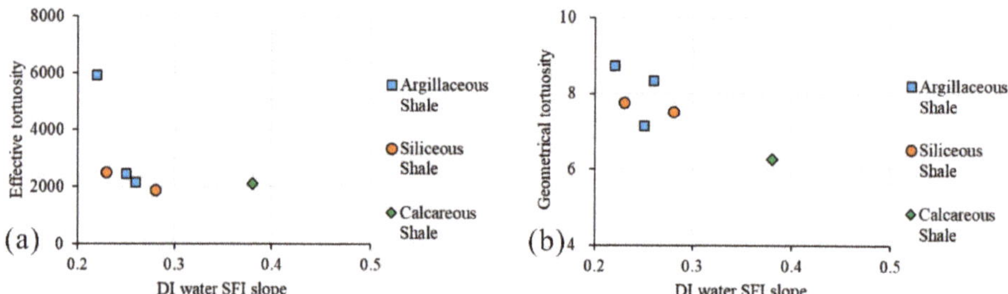

Figure 8. Correlations of (**a**) the effective tortuosity and (**b**) geometrical tortuosity with the DI water SFI slopes of different lithofacies of the lacustrine Shahezi shales from the CFD.

Shales are characterized by an extremally low matrix permeability on the nanodarcy scale [75–79]. Therefore, a diffusive mechanism is proposed to be the mass transfer process within complex matrix pore systems [75,80,81]. Low permeability and limited pore connectivity will result in the extremely slow fluid transport within the shale matrix, which may explain the steep decline of hydrocarbon production within the initial years [32]. The Shahezi shale samples are predominantly in the nanometer size range (Figure 3 and Table 3), which is expected to have an extremely low permeability and slow fluid transport. The low connectivity of pore networks will also slow the diffusive transport.

4.2. Effects of Matrix Compositions on Pore Connectivity

The effects of shale matrix compositions on the pore connectivity of the studied shale samples were discussed with respect to lithofacies (Figure 7). The argillaceous shale contains the most abundant clay minerals. The siliceous shale and the calcareous shale are rich in quartz and calcite, respectively (Table 1). Therefore, OM and mineral compositions in Shahezi shales exert different impacts on pore connectivity.

The clay mineral compositions of the studied sample are mainly mixed-layer illite/smectite (I/S), illite and chlorite (Table 1). I/S minerals have a larger water-swelling potential [82], which may alter the pore structure of shales and control the gas flow [83]. In addition, clay mineral swelling may be the reason for the lower imbibition capacity of the samples [84]. Consequently, the pore structure of the Shahezi shale samples could be altered due to the swelling of clay minerals. Gao and Hu [36] performed triplicate water imbition experiments, and the results show poor reproducibility, which indicates a strong interaction between the imbibed DI water and clay minerals. Therefore, the water or n-decane imbibition slope obtained in the first SFI experiment is used to evaluate the initial connectivity and wettability of our samples [35–37].

Pore connectivity varies in different shale lithofacies. The n-decane SFI slopes are all larger than 0.25 (0.36–0.55) for argillaceous shales rich in clay minerals, especially for sample TS-6-7 (TOC = 4.75 wt %), which displays the highest slope: 0.55 (Figure 7). The DI water SFI slopes of argillaceous shales are close to 0.25 (0.22–0.26) (Figure 5a–c). In addition, sample TS-6-1 (TOC = 1.09 wt %) has the highest effective tortuosity and geometrical tortuosity at 6122 and 8.89, indicating low pore connectivity. Well-connected OM pores may lead to higher n-decane SFI slopes, while insufficient OM pores may result in low n-decane SFI slopes [24,32]. The FE-SEM images show pores within OM-clay composites in the argillaceous shales (Figure 2d,e). InterP pores formed when coming in contact with OM-clay are often connected to OM pores to increase the pore connectivity of the argillaceous shales (Figure 2d,e).

The siliceous shale samples also have higher SFI slopes for n-decane as well as a lower effective tortuosity and geometrical tortuosity than DI water, also suggesting a higher hydrophobic pore connectivity (Figure 5d,e and Figure 6d,e). The siliceous shale samples show a good pore connectivity for n-decane, but a moderate pore connectivity for the DI water fluid. The n-decane SFI slope of 0.48 in sample TS-6-4 with a TOC content of 4.64 wt % is consistent with the well-connected OM pores shown in the FE-SEM images (Figure 2a,b) [30,32,46].

The calcareous shale sample (TS-6-5) has similar SFI slopes for both DI water (0.38) and n-decane (0.34) (Figures 5f and 6f), suggesting a mixed wettability and moderate-connected pores in this sample [30]. This result is also supported by the fact that it exhibits the lowest geometrical tortuosity, at 6.27 (Figure 8b). Well-developed inorganic pores and poor-developed OM pores in calcareous shales are more accessible to water due to the good connection of hydrophobic pores (Figure 2g).

The pore connectivity of various lithofacies with TOC values of about 2 wt % is following the order in terms of SFI slopes and geometrical tortuosity (L_e/L), from high to low: calcareous shale (TS-6-5), argillaceous shale (TS-6-1 and TS-6-7 and TS-6-8) and siliceous shale (TS-6-3 and TS-6-4) (Table 3). The correlations of geometrical tortuosity (L_e/L) with matrix compositions, including TOC, clay minerals, quartz, and carbonate mineral content, are presented in Figure 9. The geometrical tortuosity (L_e/L) values show no obvious correlations with the TOC and quartz contents (Figure 9a,b). This result is probably due to the insufficient OM pores and quartz host pores in the Shahezi shales. The geometrical tortuosity (L_e/L) values are negatively correlated with carbonate mineral contents and positively correlated with the clay minerals (Figure 9c,d). Abundant complex interP and intraP pores are formed in the OM-clay composites, which may enhance the pore connectivity.

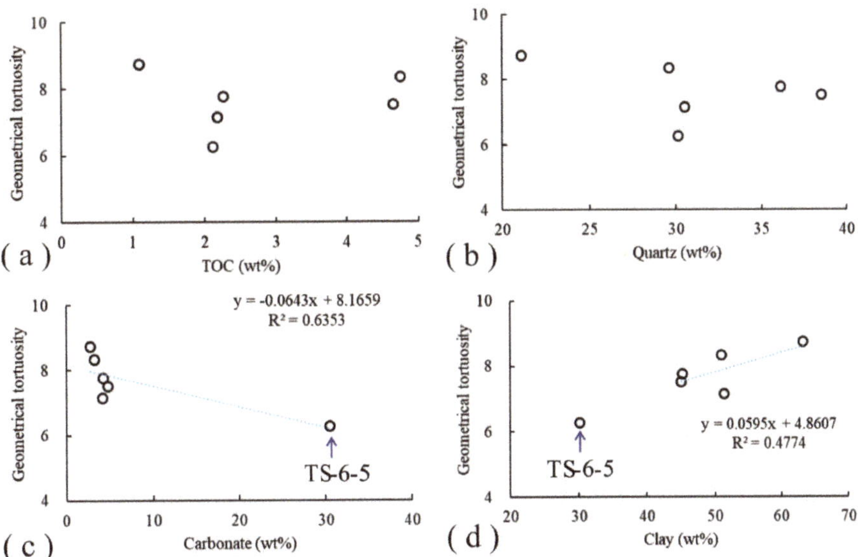

Figure 9. Correlations of the geometrical tortuosity (L_e/L) with (**a**) the total organic carbon (TOC) contents, (**b**) quartz contents; (**c**) carbonate contents and (**d**) clay mineral contents of the lacustrine Shahezi shales from the CFD.

A small amount of calcite and dolomite with dissolved pores is shown in the FE-SEM images of the Shahezi shales (Figure 2g). High contents of calcite should play an important role in pore development, because of the solubility of calcite and the large number of dissolved pores within the calcite. However, the samples with a calcite content <5 wt % show a negative correlation (Figure 9c). Therefore, we speculate that only a high concentration of calcite (over 20 wt %) plays significant roles in pore development and in the enhancement of pore connectivity [59,85].

5. Conclusions

Based on our studies, the following conclusions can be drawn.

(1) FE-SEM images show that the most observed pores in the Shahezi shales are clay minerals related to interP pores and OM pores. The primary pore width calculated by combining LPGA and MICP data is in the range of 0.3–0.7 nm and 3–20 nm.

(2) The n-decane and DI water SFI slopes (0.34–0.55 and 0.22–0.38) indicate a mixed wetting nature and relatively better-connected hydrophobic pores than hydrophilic pores in the Shahezi shales.

(3) The limited pore connectivity of the Shahezi shales is identified by the dominant pore widths (0.3–20 nm), low DI water SFI slopes (around 0.25), high geometric tortuosity (4.75–8.89) and effective tortuosity (1212–6122).

(4) The pore connectivity, affected by both the OM and inorganic compositions, varies among the shale lithofacies and follows the connectivity order of calcareous shale > argillaceous shale > siliceous shale. The high concentration of clay and calcite (over 20 wt %) significantly controls the pore connectively of the Shahezi shale in the CFD.

Author Contributions: Z.L. (Zhuo Li) and Z.J. designed and supervised the project; H.Y. and Z.L. (Zhikai Liang) performed the experiments and analyzed the data; Z.L. (Zhuo Li) wrote this paper and Z.J. corrected it; L.X. and Y.Y. modified the formats.

Funding: This study is supported by the National Major Project of China (2016ZX05034-001) and National Natural Science Foundation of China (41502123).

Acknowledgments: The Jilin oil company Branch of PetroChina is thanked for providing shale samples and geological background references. We are grateful to editors and reviewers for their constructive comments and suggestions.

References

1. Loucks, R.; Reed, R.; Ruppel, S.; Hammes, U. Spectrum of pore types and networks in mudrocks and a descriptive classification for matrix-related mudrock pores. *AAPG Bull.* **2012**, *96*, 1071–1098. [CrossRef]

2. Clarkson, C.R.; Solano, N.; Bustin, R.M.; Bustin, A.M.M.; Chalmers, G.R.L.; He, L.; Melnichenko, Y.B.; Radlinski, A.P.; Blach, T.P. Pore structure characterization of North American shale gas reservoirs: Using USANS/SANS, gas adsorption, and mercury intrusion. *Fuel* **2013**, *103*, 606–616. [CrossRef]

3. Wang, G.; Ju, Y.; Yan, Z.; Li, Q. Pore structure characteristics of coal-bearing shale using fluid invasion methods: A case study in the Huainan–Huaibei Coalfield in China. *Mar. Pet. Geol.* **2015**, *62*, 1–13. [CrossRef]

4. Loucks, R.G.; Reed, R.M.; Ruppel, S.C.; Jarvie, D.M. Morphology, genesis and distribution of nanometer-scale pores in siliceous mudstones of the Mississippian Barnett Shale. *J. Sediment. Res.* **2009**, *79*, 848–861. [CrossRef]

5. Mastalerz, M.; Drobniak, A.; Strapoc, D.; Acosta, W.S.; Rupp, J. Variations in pore characteristics in high volatile bituminous coals; implications for coalbed gas content. *Int. J. Coal Geol.* **2008**, *76*, 205–216. [CrossRef]

6. Chalmers, G.R.L.; Bustin, R.M.; Power, I.M. Characterization of gas shale pore systems by porosimetry, pycnometry, surface area, and field emission scanning electron/transmission electron microscopy image analyses: Examples from the Barnett Woodford, Haynesville, Marcellus and Doig units. *AAPG Bull.* **2012**, *96*, 1099–1119. [CrossRef]

7. Karayiğit, A.İ.; Mastalerz, M.; Oskay, R.G.; Buzkan, İ. Bituminous coal seams from underground mines in the Zonguldak Basin (NW Turkey): Insights from mineralogy, coal petrography, Rock-Eval pyrolysis, and meso- and microporosity. *Int. J. Coal Geol.* **2018**, *199*, 91–122. [CrossRef]

8. Ross, D.J.K.; Bustin, R.M. Characterizing the shale gas resource potential of Devonian-Mississippian strata in the Western Canada sedimentary basin: Application of an integrated formation evaluation. *AAPG Bull.* **2015**, *92*, 87–125. [CrossRef]

9. Ross, D.J.K.; Bustin, R.M. The importance of shale composition and pore structure upon gas storage potential of shale gas reservoirs. *Mar. Pet. Geol.* **2009**, *26*, 916–927. [CrossRef]

10. Bowker, K.A. Barnett shale gas production, Fort Worth Basin: Issues and discussion. *AAPG Bull.* **2007**, *91*, 523–533. [CrossRef]

11. Thommes, M.; Kaneko, K.; Neimark, A.V.; Olivier, J.P.; Rodriguez-Reinoso, F.; Rouquerol, J.; Sing, K.S. Physisorption of gases, with special reference to the evaluation of surface area and pore size distribution (IUPAC Technical Report). *Pure Appl. Chem.* **2015**, *87*, 1051–1069. [CrossRef]

12. Mastalerz, M.; He, L.; Melnichenko, B.Y.; Rupp, J.A. Porosity of coal and shale: Insights from gas adsorption and SANS/USANS techniques. *Energy Fuels* **2012**, *26*, 5109–5120. [CrossRef]

13. Mastalerz, M.; Schimmelmann, A.; Drobniak, A.; Chen, Y. Porosity of Devonian and Mississippian New Albany Shale across a maturation gradient: Insights from organic petrology, gas adsorption, and mercury intrusion. *AAPG Bull.* **2013**, *97*, 1621–1643. [CrossRef]

14. Milliken, K.L.; Rudnicki, M.; Awwiller, D.N.; Zhang, T. Organic matter-hosted pore system, Marcellus Formation (Devonian), Pennsylvania. *AAPG Bull.* **2013**, *97*, 177–200. [CrossRef]

15. Furmann, A.; Mastalerz, M.; Bish, D.; Schimmelmann, A.; Pedersen, P.K. Porosity and pore size distribution in mudrocks from the Belle Fourche and Second White Specks formations in Alberta, Canada. *AAPG Bull.* **2016**, *100*, 1265–1288. [CrossRef]

16. Wang, P.; Jiang, Z.; Ji, W.; Zhang, C.; Yuan, Y.; Chen, L.; Yin, L. Heterogeneity of intergranular, intraparticle and organic pores in Longmaxi shale in Sichuan Basin, South China: Evidence from SEM digital images and fractal and multifractal geometries. *Mar. Pet. Geol.* **2016**, *72*, 122–138. [CrossRef]

17. Cao, T.T.; Song, Z.G.; Wang, S.B.; Cao, X.X.; Li, Y.; Xia, J. Characterizing the pore structure in the Silurian and Permian shales of the Sichuan Basin, China. *Mar. Pet. Geol.* **2015**, *61*, 140–150. [CrossRef]

18. Chen, L.; Jiang, Z.X.; Liu, K.Y.; Tan, J.Q.; Gao, F.L.; Wang, P.F. Pore structure characterization for organic-rich lower Silurian shale in the upper Yangtze platform, South China: A ssible mechanism for pore development. *J. Nat. Gas Sci. Eng.* **2017**, *46*, 1–15. [CrossRef]

19. Chen, S.; Tang, D.; Tao, S.; Xu, H.; Li, S.; Zhao, J.; Ren, P.; Fu, H. In-situ stress measurements and stress distribution characteristics of coal reservoirs in major coalfields in China: Implication for coalbed methane (CBM) development. *Int. J. Coal Geol.* **2017**, *182*, 66–84. [CrossRef]

20. Lee, M.C.; Richard, J.H.; Hywel, J.S.; Thomas, R. Dual porosity modelling of the coupled mechanical response of coal to gas flow and adsorption. *Int. J. Coal Geol.* **2019**, *205*, 115–125.

21. İnan, S.; Al Badairy, H.; İnan, T.; Al Zahrani, A. Formation and occurrence of organic matter-hosted porosity in shales. *Int. J. Coal Geol.* **2018**, *199*, 39–51. [CrossRef]

22. Jiang, F.J.; Chen, D.; Wang, Z.F.; Xu, Z.Y.; Chen, J.; Liu, L.; Huyan, Y.Y.; Liu, Y. Pore characteristic analysis of a lacustrine shale: A case study in the Ordos Basin. NW China. *Mar. Pet. Geol.* **2016**, *73*, 554–571. [CrossRef]

23. Chen, L.; Jiang, Z.X.; Liu, K.Y.; Wang, P.F.; Ji, W.M.; Gao, F.L.; Li, P.; Hu, T.; Zhang, B.; Huang, H.X. Effect of lithofacies on gas storage capacity of marine and continental shales in the Sichuan Basin, China. *J. Nat. Gas Sci. Eng.* **2016**, *36*, 773–785. [CrossRef]

24. Sun, M.D.; Yu, B.S.; Hu, Q.H.; Yang, R.; Zhang, Y.F.; Li, B. Pore connectivity and tracer migration of typical shales in south China. *Fuel* **2017**, *203*, 32–46. [CrossRef]

25. Curtis, M.E.; Sondergeld, C.H.; Ambrose, R.J.; Rai, C.S. Microstructural investigation of gas shales in two and three dimensions using nanometer-scale resolution imaging. *AAPG Bull.* **2012**, *96*, 665–677. [CrossRef]

26. Xiong, J.; Liu, X.; Liang, L. Experimental study on the pore structure characteristics of the Upper Ordovician Wufeng Formation shale in the southwest portion of the Sichuan Basin, China. *J. Nat. Gas. Sci. Eng.* **2015**, *22*, 530–539. [CrossRef]

27. Yang, F.; Ning, Z.; Liu, H. Fractal characteristics of shales from a shale gas reservoir in the Sichuan Basin, China. *Fuel* **2015**, *115*, 378–384. [CrossRef]

28. Yang, C.; Zhang, J.C.; Wang, X.Z.; Tang, X.; Chen, Y.C.; Jiang, L.L.; Gong, X. Nanoscale pore structure and fractal characteristics of a marine-continental transitional shale: A case study from the lower Permian Shanxi Shale in the southeastern Ordos Basin, China. *Mar. Pet. Geol.* **2017**, *88*, 54–68. [CrossRef]

29. Tang, X.L.; Jiang, Z.X.; Jiang, S.; Li, Z. Heterogeneous nanoporosity of the Silurian Longmaxi Formation shale gas reservoir in the Sichuan Basin using the QEMSCAN, FIB-SEM, and nano-CT methods. *Mar. Pet. Geol.* **2016**, *78*, 99–109. [CrossRef]

30. Hu, Q.H.; Liu, X.G.; Gao, Z.Y.; Liu, S.G.; Zhou, W.; Hu, W.X. Pore structure and tracer migration behavior of typical American and Chinese shales. *Pet. Sci.* **2015**, *12*, 651–663. [CrossRef]

31. Gao, Z.Y.; Hu, Q.H. Estimating permeability using median pore-throat radius obtained from mercury intrusion porosimetry. *J. Geophys. Eng.* **2013**, *10*, 025014. [CrossRef]

32. Hu, Q.H.; Ewing, R.P.; Rowe, H.D. Low nanopore connectivity limits gas production in Barnett formation. *J. Geophys. Res.-Solid Earth* **2015**, *120*, 8073–8087. [CrossRef]

33. Hu, Q.H.; Gao, X.; Gao, Z.; Ewing, R.; Dultz, S.; Kaufmann, J. Pore accessibility and connectivity of mineral and kerogen phases for shales. In Proceedings of the Unconventional Resources Technology Conference, Denver, CO, USA, 25–27 August 2014; pp. 1188–1204.

34. Lan, Q.; Dehghanpour, H.; Wood, J.; Sanei, H. Wettability of the Montney Tight Gas Formation. *SPE Reserv. Eval. Eng.* **2015**, *18*, 417–431. [CrossRef]

35. Gao, Z.; Yang, S.; Jiang, Z.; Zhang, K.; Chen, L. Investigating the spontaneous imbibition characteristics of continental Jurassic Ziliujing Formation shale from the northeastern Sichuan Basin and correlations to pore structure and composition. *Mar. Pet. Geol.* **2018**, *98*, 697–705. [CrossRef]

36. Gao, Z.Y.; Hu, Q.H. Wettability of Mississippian Barnett Shale samples at different depths: Investigations from directional spontaneous imbibition. *AAPG Bull.* **2016**, *100*, 101–114. [CrossRef]

37. Gao, Z.; Hu, Q. Pore structure and spontaneous imbibition characteristics of marine and continental shales in China. *AAPG Bull.* **2018**, *102*, 1941–1961. [CrossRef]

38. Yang, R.; Hao, F.; He, S.; He, C.C.; Guo, X.S.; Yi, J.Z.; Hu, H.Y.; Zhang, S.W.; Hu, Q.H. Experimental investigations on the geometry and connectivity of pore space in organic-rich Wufeng and Longmaxi shales. *Mar. Pet. Geol.* **2017**, *84*, 225–242. [CrossRef]

39. Wang, Y.; Pu, J.; Wang, L.H.; Wang, J.Q.; Jiang, Z.; Song, Y.F.; Wang, C.C.; Wang, Y.F.; Jin, C. Characterization of typical 3D pore networks of Jiulaodong formation shale using nanotransmission X-ray microscopy. *Fuel* **2016**, *170*, 84–91. [CrossRef]

40. Saif, T.; Lin, Q.Y.; Bijeljic, B.; Martin, J.B. Microstructural imaging and characterization of oil shale before and after pyrolysis. *Fuel* **2017**, *197*, 562–574. [CrossRef]

41. Liu, H.L.; Seaton, N.A. Determination of the connectivity of porous solids from nitrogen sorption measurements III. Solids containing large mesopores. *Chem. Eng. Sci.* **1994**, *49*, 1869–1878. [CrossRef]

42. Liu, H.L.; Zhang, L.; Seaton, N.A. Determination of the connectivity of porous solids from nitrogen sorption measurements II. Generalization. *Chem. Eng. Sci.* **1992**, *47*, 4393–4404. [CrossRef]

43. Liu, H.L.; Lai, B.T.; Chen, J.H. Unconventional spontaneous imbibition into shale matrix: Theory and a methodology to determine relevant parameters. *Transp. Porous Media* **2016**, *111*, 41–57. [CrossRef]

44. Seaton, N.A. Determination of the connectivity of porous solids from nitrogen sorption measurements. *Chem. Eng. Sci.* **1991**, *46*, 1895–1909. [CrossRef]

45. Armatas, G.S.; Salmas, C.E.; Louloudi, M.; Androutsopoulos, G.P.; Pomonis, P.J. Relationships among pore size, connectivity, dimensionality of capillary condensation, and pore structure tortuosity of functionalized mesoporous silica. *Langmuir* **2003**, *19*, 3128–3136. [CrossRef]

46. Hu, Q.; Ewing, R.P.; Dultz, S. Low pore connectivity in natural rock. *J. Contam. Hydrol.* **2012**, *133*, 76–83. [CrossRef] [PubMed]

47. Gao, F.; Song, Y.; Li, Z.; Xiong, F.; Chen, L.; Zhang, X.; Chen, Z.; Moortgat, J. Quantitative characterization of pore connectivity using NMR and MIP: A case study of the Wangyinpu and Guanyintang shales in the Xiuwu basin, Southern China. *Int. J. Coal Geol.* **2018**, *197*, 53–65. [CrossRef]

48. Ning, C.; Jiang, Z.; Gao, Z.; Su, S.; Li, T.; Wang, G.; Chen, L. Characteristics and controlling factors of reservoir space of mudstone and shale in Es3x in the Zhanhua Sag. *Mar. Pet. Geol.* **2017**, *88*, 214–224. [CrossRef]

49. Ghanbari, E.; Dehghanpour, H. The fate of fracturing water: A field and simulation study. *Fuel* **2016**, *163*, 282–294. [CrossRef]

50. Dehghanpour, H.; Zubair, H.A.; Chhabra, A.; Ullah, A. Liquid Intake of Organic Shales. *Energy Fuels* **2012**, *26*, 5750–5758. [CrossRef]

51. Jiang, S.; Tang, X.L.; Cai, D.S.; Xue, G.; He, Z.L.; Long, S.X.; Peng, Y.M.; Gao, B.; Xu, Z.Y.; Dahdah, N. Comparison of marine, transitional, and lacustrine shales: A case study from the Sichuan Basin in China. *J. Pet. Sci. Eng.* **2017**, *150*, 334–347. [CrossRef]

52. Zhang, D.; Li, Y.; Zhang, J.; Qiao, D.; Jiang, W.; Zhang, J. *National Survey and Assessment of Shale Gas Resource Potential in China*; Geologic Publishing House: Beijing, China, 2012; pp. 8–92. (In Chinese)

53. Jiang, S.; Xu, Z.; Feng, Y.; Zhang, J.; Cai, D.; Chen, L.; Wu, Y.; Zhou, D.S.; Bao, S.J.; Long, S.X. Geologic characteristics of hydrocarbon-bearing marine, transitional and lacustrine shales in china. *J. Asian Earth Sci.* **2015**, *115*, 404–418. [CrossRef]

54. Lu, R.J. The Evaluation of Deep Source Rocks in Changling Fault Depression of Songliao Southern Basin. Ph.D. Thesis, Northeast Petroleum University, Daqing, China, 2011. (In Chinese)

55. Sun, D. The Prediction of Source Rocks Distribution and the Research of Hydrocarbon Generation Potential of Changling Fault Depression. Ph.D. Thesis, Northeast Petroleum University, Daqing, China, 2014. (In Chinese)

56. Tang, Y.; Yang, R.Z.; Zhu, J.; Yin, S.; Fan, T.; Dong, Y.; Hou, Y. Analysis of continental shale gas accumulation conditions in a rifted basin: A case study of Lower Cretaceous shale in the southern Songliao Basin, northeastern China. *Mar. Pet. Geol.* **2019**, *101*, 389–409. [CrossRef]

57. Hao, F.; Zou, H.Y.; Lu, Y.C. Mechanisms of shale gas storage: Implications for shale gas exploration in China. *AAPG Bull.* **2013**, *97*, 1325–1346. [CrossRef]

58. Li, Z.; Liang, Z.K.; Jiang, Z.X.; Gao, F.L.; Zhang, Y.H.; Yu, H.L.; Xiao, L.; Yang, Y.D. The Impacts of Matrix Compositions on Nanopore Structure and Fractal Characteristics of Lacustrine Shales from the Changling Fault Depression, Songliao Basin, China. *Minerals* **2019**, *9*, 127. [CrossRef]

59. Gao, F.L.; Song, Y.; Li, Z.; Xiong, F.Y.; Chen, L.; Zhang, Y.H.; Liang, Z.K.; Zhang, X.X.; Chen, Z.Y.; Joachim, M. Lithofacies and reservoir characteristics of the Lower Cretaceous continental Shahezi Shale in the Changling Fault Depression of Songliao Basin, NE China. *Mar. Pet. Geol.* **2018**, *98*, 401–421. [CrossRef]

60. Yang, G.; Zhao, Z.Y.; Shao, M.L. Formation of carbon dioxide and hydrocarbon gas reservoirs in the Changling fault depression, Songliao Basin. *Pet. Explor. Dev.* **2011**, *38*, 52–58.

61. Wu, H.Y.; Liang, X.D.; Xiang, C.F.; Wang, Y.W. Characteristics of petroleum accumulation in syncline of the Songliao basin and discussion on its accumulation mechanism. *Sci. China* **2007**, *50*, 702–709. [CrossRef]

62. Lin, J.H.; Jiang, T.; Song, L.B.; Cao, Y.; Xia, D.; Wang, Y. The origin and gas vertical distribution of the Harjin mixed-gas reservoir. *Acta Pet. Sin.* **2010**, *34*, 927–932.

63. Gao, Z.Y.; Hu, Q.H. Initial water saturation and imbibition fluid affect spontaneous imbibition into Barnett shale samples. *J. Nat. Gas Sci. Eng.* **2016**, *34*, 541–551. [CrossRef]

64. Graue, A.; Fernø, M.A. Water mixing during spontaneous imbibition at different boundary and wettability conditions. *J Pet. Sci. Eng.* **2011**, *78*, 586–595. [CrossRef]

65. Mason, G.; Fischer, H.; Morrow, N.R.; Ruth, D.W.; Wo, S. Effect of sample shape on counter-current spontaneous imbibition production vs time curves. *J. Petrol. Sci. Eng.* **2009**, *66*, 83–97. [CrossRef]

66. Jiao, K.; Ye, Y.H.; Liu, S.G.; Ran, B.; Deng, B.; Li, Z.W.; Li, J.X.; Yong, Z.Q.; Sun, W. Characterization and Evolution of Nanoporosity in Superdeeply Buried Shales: A Case Study of the Longmaxi and Qiongzhusi Shales from MS Well #1, North Sichuan Basin, China. *Energy Fuels* **2018**, *32*, 191–203.

67. Sing, K.S.W.; Everett, D.H.; Haul, R.A.W.; Mescou, L.; Pierotti, R.A.; Rouquerol, J.; Siemieniewska, T. Reporting physisorption data for gas/solid systems with special reference to the determination of surface area and porosity. *Pure Appl. Chem.* **1985**, *57*, 603–619. [CrossRef]

68. Yang, R.; Hu, Q.H.; He, S.; Hao, F.; Guo, X.S.; Yi, J.Z.; He, X.P. Pore structure, wettability and tracer migration in four leading shale formations in the Middle Yangtze Platform, China. *Mar. Pet. Geol.* **2018**, *89*, 415–427. [CrossRef]

69. Gao, Z.Y.; Hu, Q.H. Investigating the Effect of Median Pore-Throat Diameter on Spontaneous Imbibition. *J. Porous Media* **2015**, *18*, 1231–1238. [CrossRef]

70. Dehghanpour, H.; Lan, Q.; Saeed, Y.; Fei, H.; Qi, Z. Spontaneous Imbibition of Brine and Oil in Gas Shales: Effect of Water Adsorption and Resulting Microfractures. *Energy Fuels* **2013**, *27*, 3039–3049. [CrossRef]

71. Hu, M.Q.; Persoff, P.; Wang, J.S.Y. Laboratory measurement of water imbibition into low-permeability welded tuff. *J. Hydrol.* **2001**, *242*, 64–78. [CrossRef]

72. Ma, S.; Morrow, N.R.; Zhang, X. Generalized scaling of spontaneous imbibition data for strongly water-wet systems. *J. Petrol. Sci. Eng.* **1997**, *18*, 165–178.

73. Guo, C.; Xu, J.; Wu, K.; Wei, M.; Liu, S. Study on gas flow through nano pores of shale gas reservoirs. *Fuel* **2015**, *143*, 107–117. [CrossRef]

74. Ghanizadeh, A.; Clarkson, C.R.; Aquino, S.; Ardakani, O.H.; Sanei, H. Petrophysical and geomechanical characteristics of Canadian tight oil and liquid-rich gas reservoirs: I. Pore network and permeability characterization. *Fuel* **2015**, *153*, 664–681. [CrossRef]

75. Anovitz, L.M.; Cole, D.R. Characterization and analysis of porosity and pore structures. *Rev. Mineral. Geochem.* **2015**, *80*, 61–164. [CrossRef]

76. Pommer, M.; Milliken, K. Pore types and pore-size distributions across thermal maturity, Eagle Ford Formation, southern Texas Pores across Thermal Maturity, Eagle Ford. *AAPG Bull.* **2015**, *99*, 1713–1744. [CrossRef]

77. Hsieh, P.A.; Tracy, J.V.; Neuzil, C.E.; Bredehoeft, J.D.; Silliman, S.E. A transient laboratory method for determining the hydraulic properties of 'tight'rocks—I. Theory. In *International Journal of Rock Mechanics and Mining Sciences & Geomechanics Abstracts*; Pergamon: Oxford, UK, 1981; pp. 245–252.

78. Cui, X.; Bustin, A.M.M.; Bustin, R.M. Measurements of gas permeability and diffusivity of tight reservoir rocks: Different approaches and their applications. *Geofluids* **2009**, *9*, 208–223. [CrossRef]

79. Jia, B.; Feng, R.; Tsau, J.S.; Barati, R. Multiphysical Flow Behavior in Shale and Permeability Measurement by Pulse-Decay Method. In *Petrophysical Characterization and Fluids Transport in Unconventional Reservoirs*; Elsevier: Amsterdam, The Netherlands, 2019; pp. 301–324.

80. Javadpour, F.; Fisher, D.; Unsworth, M. Nanoscale gas flow in shale gas sediments. *J. Can. Pet. Technol.* **2007**, *46*, 55–61. [CrossRef]

81. Hughes, J.D. Energy: A reality check on the shale revolution. *Nature* **2013**, *494*, 307–308. [CrossRef]

82. Hensen, E.J.; Smit, B. Why clays swell. *J. Phys. Chem. B.* **2002**, *106*, 12664–12667. [CrossRef]

83. Odusina, E.; Sondergeld, C.; Rai, C. An NMR study on shale wettability. In Proceedings of the Canadian Unconventional Resources Conference, Calgary, AL, Canada, 15–17 November 2011; pp. 1–15.

84. Shen, Y.; Ge, H.; Li, C.; Yang, X.; Ren, K.; Yang, Z.; Su, S. Water imbibition of shale and its potential influence on shale gas recovery—A comparative study of marine and continental shale formations. *J. Nat. Gas Sci. Eng.* **2016**, *35*, 1121–1128. [CrossRef]

85. Li, T.; Jiang, Z.; Li, Z.; Wang, P.; Xu, C.; Liu, G.; Su, S.; Ning, C. Continental shale pore structure characteristics and their controlling factors: A case study from the lower third member of the Shahejie Formation, Zhanhua Sag, Eastern China. *J. Nat. Gas Sci. Eng.* **2017**, *45*, 670–692. [CrossRef]

Article

Two Hydrothermal Events at the Shuiyindong Carlin-Type Gold Deposit in Southwestern China: Insight from Sm–Nd Dating of Fluorite and Calcite

Qinping Tan [1,*], Yong Xia [1,*], Zhuojun Xie [1], Zepeng Wang [2], Dongtian Wei [3], Yimeng Zhao [1,4], Jun Yan [1,2,5] and Songtao Li [1,2,3]

[1] State Key Laboratory of Ore Deposit Geochemistry, Institute of Geochemistry, Chinese Academy of Sciences, Guiyang 550081, China; xiezhuojun@vip.gyig.ac.cn (Z.X.); zhaoyimeng@mail.gyig.ac.cn (Y.Z.); xiao127721@sina.com (J.Y.); lisongtaozgh@163.com (S.L.)

[2] No.105 Geological Team, Guizhou Bureau of Geology and Mineral Exploration & Development, Guiyang 550018, China; wangzepengyu@126.com

[3] Guangxi Key Laboratory of Exploration for Hidden Metallic Ore Deposits, College of Earth Sciences, Guilin University of Technology, Guilin 541006, China; weidongtian2016@163.com

[4] University of Chinese Academy of Sciences, Beijing 100039, China

[5] College of Resources and Environmental Engineering, Guizhou Institute of Technology, Guiyang 550003, China

* Correspondence: tanqinping@vip.gyig.ac.cn (Q.T.); xiayong@vip.gyig.ac.cn (Y.X.)

Received: 8 March 2019; Accepted: 4 April 2019; Published: 12 April 2019

Abstract: The Shuiyindong Gold Mine hosts one of the largest and highest-grade, strata-bound Carlin-type gold deposits discovered to date in Southwestern China. The outcrop stratigraphy and drill core data of the deposit reveal Middle–Upper Permian and Lower Triassic formations. The ore is mainly hosted in Upper Permian bioclastic limestone near the axis of an anticline. The gold is mainly hosted in arsenian pyrite and arsenopyrite, mainly existing in the form of crystal lattice gold, submicroscopic particles and nanoparticles. Fluorite commonly occurs at the vicinity of an unconformity between the Middle–Upper Permian formations, which is proposed to be the structural conduit that fed the ore fluids. Calcite commonly fills fractures at the periphery of decarbonated rocks, which contain high grade orebodies. This study aimed to verify the occurrence of two distinct hydrothermal events at the Shuiyindong, based on Sm–Nd isotope dating of the fluorite and calcite. For this purpose, rare-earth element (REE) concentrations, Sm/Nd isotope ratios, and Sm–Nd isochron ages of the fluorite and calcite were determined. The fluorite and calcite contain relatively high total concentrations of REE (12.3–25.6 µg/g and 5.71–31.7 µg/g, respectively), exhibit variable Sm/Nd ratios (0.52–1.03 and 0.57–1.71, respectively), and yield Sm–Nd isochron ages of 200.1 ± 8.6 Ma and 150.2 ± 2.2 Ma, with slightly different initial $\varepsilon_{Nd}(t)$ values of −4.4 and −1.1, respectively. These two groups of Sm–Nd isochron ages suggest two episodes of hydrothermal events at the Shuiyindong gold deposit. The age of the calcite probably represents the late stage of the gold mineralization period. The initial $\varepsilon_{Nd}(t)$ values of the fluorite and calcite indicate that the Nd was probably derived from mixtures of basaltic volcanic tuff and bioclastic limestone from the Permian formations.

Keywords: Sm–Nd dating; Shuiyindong; Carlin-type gold deposit; Southwestern China

---

1. Introduction

The Dian-Qian-Gui "Golden Triangle", located at the junction of Yunnan, Guizhou, and Guangxi Provinces in Southwestern China (Figure 1), is famous for hosting clusters of Carlin-type gold deposits [1–4]. More than 200 Carlin-type gold deposits and occurrences have been identified in the

"Golden Triangle", with a total proven gold reserve of more than 800 tons [3,5]. It is the second-largest Carlin-type gold mineralized area in the world, after the largest in Nevada, USA [3,5].

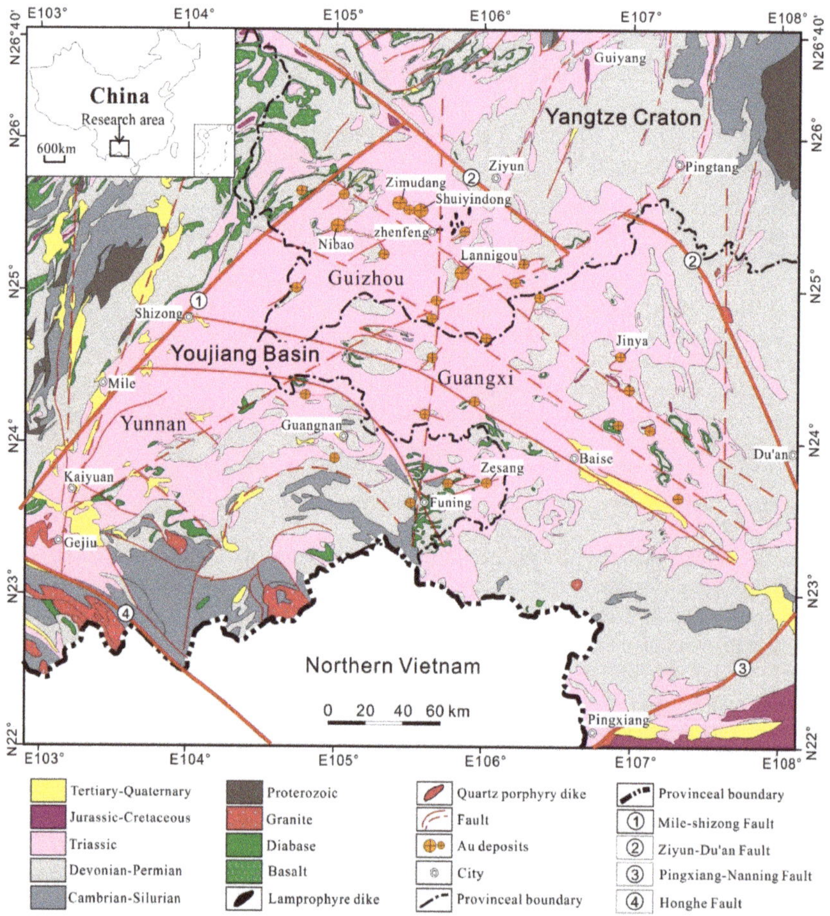

Figure 1. Geologic map of the Dian-Qian-Gui "Golden Triangle" region in Southwestern China (modified after Su et al. [5] and 1:2,500,000 Chinese geological map [6]).

Direct dating of hydrothermal deposits is critical for properly evaluating their relationships to tectonic, magmatic, and metamorphic events. However, despite previous detailed investigations, the ages of Carlin-type gold deposits in Southwestern China are poorly constrained. Chen et al. [7] obtained a ^{40}Ar–^{39}Ar plateau age of 194.6 ± 2 Ma, using sericite in quartz veins from the Lannigou gold deposit. Chen et al. [8] reported ages of 204 ± 19 Ma, 206 ± 22 Ma, and 235 ± 33 Ma based on Re–Os isotopes of arsenopyrite from the Lannigou, Jinya, and Shuiyindong deposits, respectively. Pi et al. [9] dated hydrothermal rutile and sericite from the Zhesang gold deposit and obtained an in situ U–Pb isochron age of 213.6 ± 4.6 Ma for rutile and a ^{40}Ar–^{39}Ar plateau age of 215.3 ± 1.9 Ma for sericite. In addition, Sm–Nd isochron ages of 134 ± 3 Ma and 136 ± 3 Ma have been reported for calcite from the Shuiyindong gold deposit [10]. Wang [11] reported another Sm–Nd isochron age of 148.4 ± 4.8 Ma of calcite from the Zimudang gold deposit. Chen et al. [12] obtained weighted-mean secondary ion mass spectrometry (SIMS) Th–Pb age of 141 ± 3 Ma for apatite from the Nibao gold deposit. These ages can be mainly classified into two groups—ca. 130–150 Ma and 200–230 Ma. Nevertheless, further research is required on whether there were two episodes of hydrothermal or metallogenesis events during the formation of the Carlin-type gold deposits in the Dian-Qian-Gui "Golden Triangle" area in Southwestern China.

Sm and Nd have similar chemical characteristics. Therefore, the daughter ^{143}Nd decayed from the parent ^{147}Sm is often preserved in the mineral lattice, i.e., the Sm–Nd isotope system has high likelihood of being closed and capable of resisting weathering and alteration to some degree [13]. Therefore, Sm–Nd isotope dating is an effective method for precisely determining the time of hydrothermal events [14–16], even for relatively young mineralization [17]. It has been successfully used for dating hydrothermal Ca-bearing minerals, such as calcite [18–20], scheelite [15,21,22], fluorite [23–25], and tourmaline [15,21].

The ideal method for determining the age of gold deposits is to analyze minerals that are known to have formed coevally with the gold. Fluorite and calcite are common gangue minerals in Carlin-type gold deposits in Southwestern China. Previous research indicated that the rare-earth element (REE) patterns of fluorite and calcite are unique, characterized by middle rare-earth element (MREE) enrichment and relatively variable Sm/Nd ratios [10,26], which is favorable for Sm–Nd isotope dating. The Shuiyindong gold deposit is the largest gold deposit reported thus far in Southwestern China, with a total gold reserve of over 260 tons [27,28]. In this study, we selected the Shuiyindong gold deposit as a case study, and explore whether there were two distinct hydrothermal events based on Sm–Nd isotope dating of fluorite and calcite.

2. Geological Setting

Carlin-type gold deposits in the Dian–Qian–Gui area are restricted to the Youjiang Basin (Figure 1). The basin is bound to the northwest and northeast by the Mile–Shizong fault and Ziyun–Du'an fault, respectively, separating the basin from the Yangtze Craton [5,8]. The southwestern and southeastern margins of the basin are separated from the Simao and Cathaysia blocks by the Honghe and Pingxiang–Nanning faults, respectively [8,29].

The Shuiyindong deposit, located in the northern part of the Youjiang Basin, lies approximately 20 km northwest of Zhenfeng in Guizhou Province (Figure 1). The mining area has been divided into four ore blocks, Shuiyindong, Xionghuangyan, Bojitian, and Nayang (Figure 2). Detailed geological descriptions of the deposit are available in Su et al. [30] and Tan et al. [28,31].

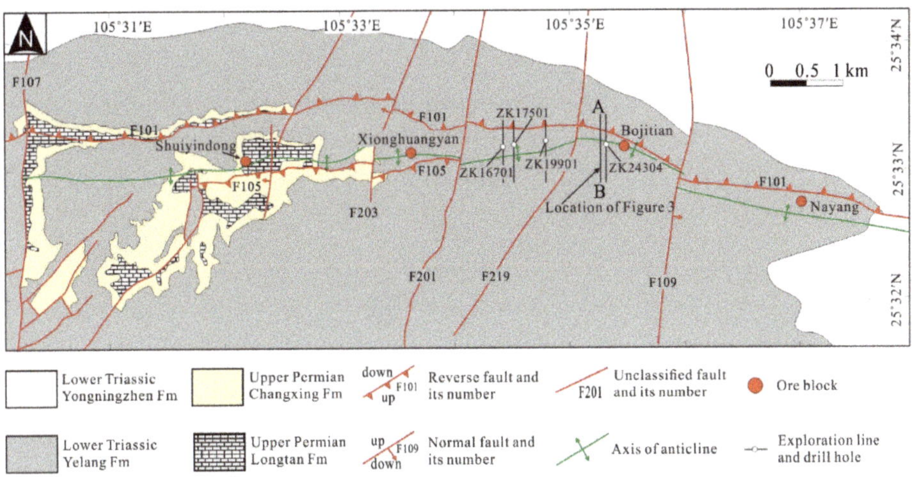

Figure 2. Geological map of the Shuiyindong gold deposit. The Shuiyindong mining area has been divided into four ore blocks: Shuiyindong, Xionghuangyan, Bojitian, and Nayang.

The outcrop stratigraphy and drill core data in the Shuiyindong mining area reveal Permian and Triassic formations that consist of muddy limestone, bioclastic limestone, siltstone, and argillite (Figures 2 and 3). The Middle Permian Maokou formation, a massive bioclastic limestone, is conformably overlain by the Upper Permian Longtan, Changxing, and Dalong and Lower Triassic Yelang formations (Figure 3). These strata were deformed into a nearly east–west-trending anticline with north and south limbs cut by reverse faults F101 and F105, respectively (Figure 2; [32]). The Maokou and Longtan formations are separated by an unconformity (SBT), consisting of silicified, brecciated argillite and limestone. SBT has been proposed as the structural conduit that fed ore fluids into the anticline core [28].

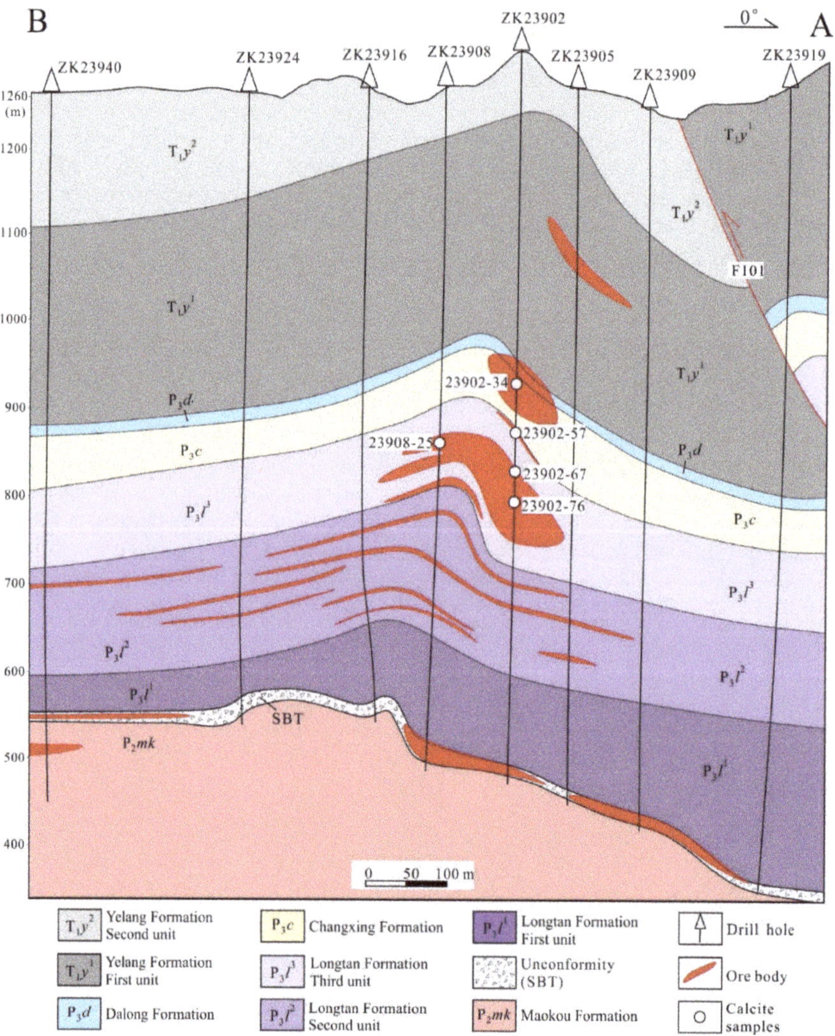

Figure 3. Geological cross section A–B through the Shuiyindong mining area (looking west) showing its major structures and stratigraphic units.

In this deposit, gold mineralization occurs mainly at the vicinity of the anticlinal core, and is preferentially disseminated in bioclastic limestone and calcareous siltstone of the Longtan formation at depths of 100–1400 m below the surface (Figure 3). In addition, SBT hosts low-grade orebodies (Figure 3). The mineralization is closely associated with decarbonatization (carbonate dissolution), silicification, sulfidation, and dolomitization, similar to Carlin-type gold deposits in Nevada, USA [33]. The gold is mainly hosted in arsenian pyrite and arsenopyrite as invisible forms including submicroscopic particles, nanoparticles, and crystal lattice gold, which suggests that sulfidation took place during the main mineralization stage [34]. Sulfides formed from sulfidation consist mainly of arsenian pyrite, arsenopyrite, marcasite, and lesser orpiment, realgar (Figure 4a,b), and stibnite. Gangue minerals consist of quartz, dolomite, calcite (Figure 4a,b), fluorite (Figure 4c,d), and clay minerals (e.g., kaolinite and illite). Fluorite generally occurs at the vicinity of SBT, and calcite commonly fills fractures at the periphery of highly porous decarbonated rocks that host high-grade orebodies.

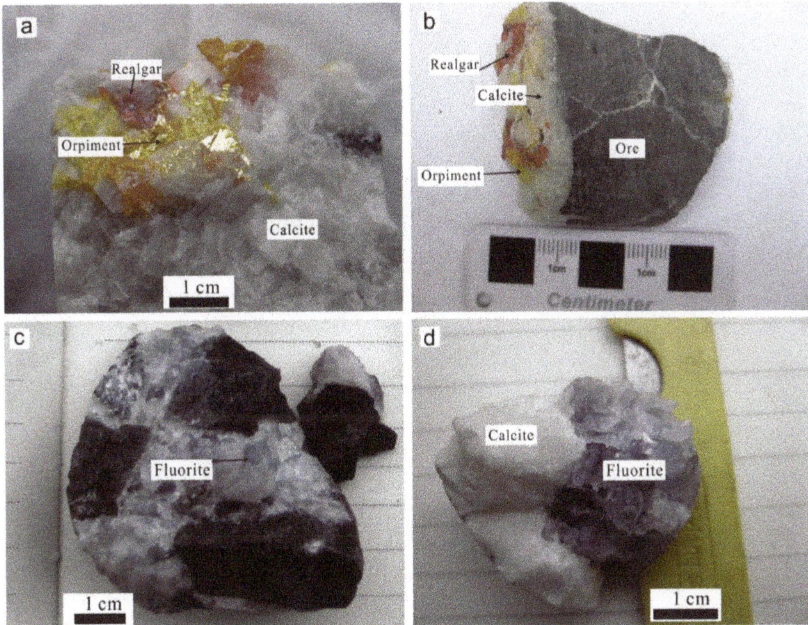

Figure 4. Photographs of calcite and fluorite samples from Shuiyindong. (**a,b**) Calcite samples containing intergrown realgar and orpiment from drill cores at the periphery of decarbonated rocks; (**c,d**) fluorite samples cementing argillite breccia or intergrowth with calcite at the vicinity of SBT.

3. Sampling and Analytical Methods

Six fluorite samples were collected from drill holes ZK16701, ZK17501, ZK19901, and ZK24304 of SBT at depths of 600–747 m below the surface (Figure 2). Five calcite samples, containing intergrown realgar and orpiment, were collected from drill holes ZK23908 and ZK23902 at depths of 376–522 m below the surface (Figure 3). Pure fluorite and calcite separates were hand-picked under a binocular microscope and crushed to 200 mesh in an agate mortar.

Prior to isotopic analysis, concentrations of REE in subsamples from the separated fluorite and calcite were determined by a Perkin-Elmer Sciex ELAN 6000 inductively coupled plasma quadrupole mass spectrometer at the Institute of Geochemistry at the Chinese Academy of Sciences. Sm and Nd concentrations and isotope ratio measurements were performed using a MAT-261 thermal ionization mass spectrometer at the Tianjin Institute of Geology and Mineral Resources at the Chinese Academy of Geological Sciences. Detailed analytical procedures are available in Peng et al. [13], Su et al. [10], and Zhang et al. [35]. Nd ratios were normalized to a $^{146}Nd/^{144}Nd$ ratio of 0.7219, using a power law fractionation correction. The reproducibility of isotopic ratios is better than 0.005% (2σ); the precision for Sm and Nd concentrations is less than 0.5% of the quoted values (2σ). The average element concentrations and isotopic ratios of the standard BCR-1 determined during this study were 6.57 μg/g for Sm, 28.75 μg/g for Nd, and 0.512644 ± 0.000005 (2σ, *n* = 6) for ^{143}Nd/^{144}Nd, which are consistent with the values of 6.58 μg/g for Sm and 28.8 μg/g for Nd in the literature [22]. Replicate analyses of the Johnson and Mattey® Nd standard (JMC) provided an average ^{143}Nd/^{144}Nd ratio of 0.511132 ± 0.000005 (2σ, *n* = 6). Average blanks were 0.03 ng for Sm and 0.05 ng for Nd. The decay constant of $λ^{147}$Sm = 6.54 × 10^{-12}/year was used in the age calculation. Sm–Nd isochron ages were calculated using the computer program ISOPLOT 2.9 [36].

4. Results

REE concentrations and Sm–Nd isotopic compositions of fluorite and calcite are shown in Figure 5 and listed in Tables 1 and 2, respectively. All the samples contained considerable ΣREE concentrations (12.3–25.6 μg/g for fluorites and 5.71–31.7 μg/g for calcites) and variable Sm/Nd ratios (0.52–1.03 for fluorites and 0.57–1.71 for calcites). Chondrite-normalized REE patterns of fluorite and calcite all showed MREE enrichment characteristics (Figure 5a,b). However, fluorite and calcite exhibited certain apparent differences between their REE patterns. Fluorite was characterized by negative Eu anomalies (δEu = 0.67–0.78; Table 1) and a parabolic shape, with the peak between Gd and Ho (Figure 5a), whereas calcite was characterized by positive Eu anomalies (δEu = 1.01–1.54; Table 1) and a hump shape with the peak on Eu (Figure 5b).

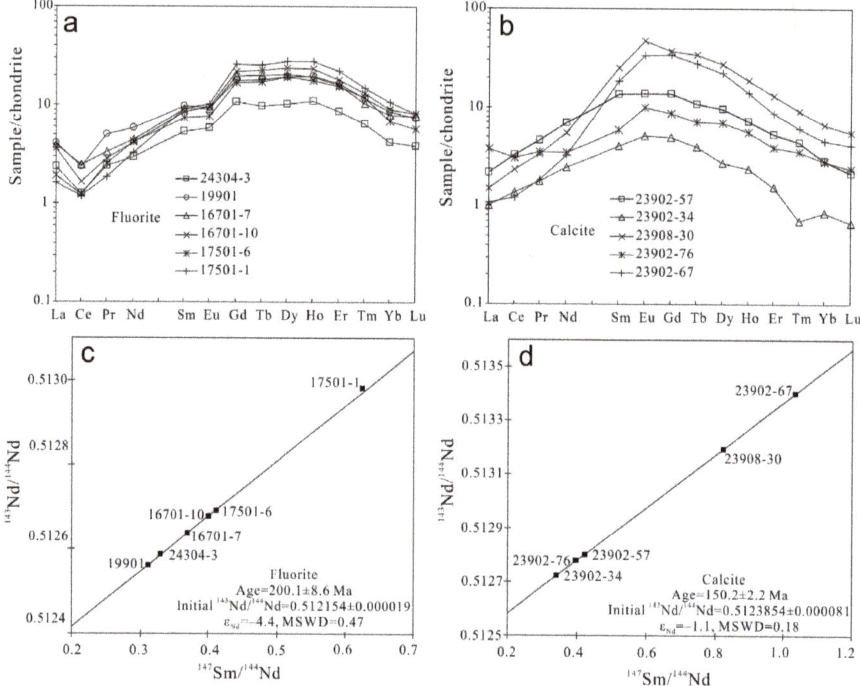

Figure 5. Chondrite-normalized REE patterns (**a,b**) and corresponding Sm–Nd isochron ages (**c,d**) for fluorite and calcite from the Shuiyindong deposit.

Table 1. REE data (µg/g) for fluorite and calcite from the Shuiyindong deposit.

Sample Number	Minerals	Locality	La	Ce	Pr	Nd	Sm	Eu	Gd	Tb	Dy	Ho	Er	Tm	Yb	Lu	ΣREE	δEu
16701-7	Fluorite	ZK16701, 615 m	0.938	1.48	0.317	2.06	1.40	0.523	4.02	0.745	5.17	1.10	2.64	0.266	1.34	0.196	22.2	0.67
16701-10	Fluorite	ZK16701, 627 m	0.850	0.999	0.269	1.94	1.29	0.552	4.43	0.841	6.06	1.32	2.97	0.335	1.54	0.209	23.6	0.71
17501-1	Fluorite	ZK17501, 600 m	0.386	0.723	0.175	1.52	1.40	0.593	5.26	0.946	6.95	1.56	3.65	0.380	1.84	0.210	25.6	0.67
17501-6	Fluorite	ZK17501, 605 m	0.451	0.750	0.239	1.97	1.13	0.442	3.48	0.635	4.87	0.99	2.57	0.306	1.18	0.146	19.2	0.68
19901	Fluorite	ZK19901, 675 m	0.959	1.43	0.476	2.71	1.48	0.543	3.64	0.671	4.87	1.11	2.74	0.298	1.47	0.187	22.6	0.71
24304-3	Fluorite	ZK24204, 747 m	0.559	0.768	0.226	1.38	0.815	0.340	2.20	0.365	2.62	0.62	1.43	0.166	0.714	0.099	12.3	0.78
23902-34	Calcite	ZK23902, 376 m	0.239	0.838	0.169	1.14	0.609	0.296	1.01	0.148	0.69	0.13	0.25	0.018	0.144	0.017	5.71	1.15
23902-57	Calcite	ZK23902, 441 m	0.511	2.000	0.439	3.24	2.06	0.794	2.81	0.401	2.44	0.40	0.88	0.112	0.484	0.054	16.6	1.01
23902-67	Calcite	ZK23902, 487 m	0.254	0.745	0.174	1.50	2.83	1.94	7.02	1.03	5.66	0.79	1.43	0.153	0.773	0.104	24.4	1.33
23902-76	Calcite	ZK23902, 522 m	0.892	1.88	0.330	1.60	0.894	0.572	1.77	0.263	1.76	0.31	0.64	0.090	0.480	0.060	11.5	1.39
23908-30	Calcite	ZK23908, 427 m	0.354	1.43	0.316	2.56	3.85	2.72	7.59	1.26	6.88	1.05	2.16	0.234	1.12	0.138	31.7	1.54

Note: $\delta Eu = Eu_N / \sqrt{Sm_N \times Gd_N}$.

Table 2. Sm and Nd isotope compositions of fluorite and calcite from the Shuiyindong deposit.

Sample Number	Occurrence	Sm (µg/g)	Nd (µg/g)	Sm/Nd	$^{147}Sm/^{144}Nd$	$^{143}Nd/^{144}Nd$ <2σ>	ε_{Nd}
16701-7	Fl	1.4365	2.3481	0.61	0.3699	0.512637 ± 0.000006	−4.4
16701-10	Fl	1.1948	1.7974	0.66	0.4019	0.512678 ± 0.000013	−4.4
17501-1	Fl + Cal	1.3179	1.2764	1.03	0.6242	0.512981 ± 0.000023	−4.2
17501-6	Fl + Cal	1.2285	1.8001	0.68	0.4126	0.512692 ± 0.000009	−4.4
19901	Fl + Cal + Qz	1.6196	3.1428	0.52	0.3115	0.512561 ± 0.000005	−4.4
24304-3	Fl	0.8801	1.6124	0.55	0.3300	0.512587 ± 0.000003	−4.4
23902-34	Cal	0.6734	1.1878	0.57	0.3427	0.512721 ± 0.000010	−1.1
23902-57	Cal + Rlg + Orp	2.1858	3.1230	0.70	0.4231	0.512798 ± 0.000011	−1.2
23902-67	Cal + Rlg + Orp	2.7132	1.5875	1.71	1.0332	0.513401 ± 0.000008	−1.1
23902-76	Cal + Rlg + Orp	1.0964	1.6630	0.66	0.3986	0.512778 ± 0.000004	−1.1
23908-30	Cal + Rlg + Orp	3.9756	2.9210	1.36	0.8228	0.513194 ± 0.000017	−1.1

Abbreviations: Cal—calcite, Rlg—realgar, Orp—orpiment, Fl—fluorite, Qz—quartz.

The six fluorite samples collected from SBT showed $^{147}Sm/^{144}Nd$ and $^{143}Nd/^{144}Nd$ values ranging from 0.3115 to 0.6242 and from 0.512561 to 0.512981, respectively, and yielded a Sm–Nd isochron age of 200.1 ± 8.6 Ma (Figure 5c), with a low mean square of weighted deviates (MSWD) of 0.47 and an initial $^{143}Nd/^{144}Nd$ ratio of 0.512154 ± 0.000019 (initial $\varepsilon_{Nd}(t)$ = −4.4). The five calcite samples collected from orebodies showed $^{147}Sm/^{144}Nd$ and $^{143}Nd/^{144}Nd$ values ranging from 0.3427 to 1.0332 and from 0.512721 to 0.513401, respectively, and yielded a different Sm–Nd isochron age of 150.2 ± 2.2 Ma (Figure 5d), with a low MSWD of 0.18 and an initial $^{143}Nd/^{144}Nd$ ratio of 0.5123854 ± 0.000081 (initial $\varepsilon_{Nd}(t)$ = −1.1). The low MSWD values reflect the excellent fit of the data to a straight line.

5. Discussion

The linear relationships shown in Figure 5c,d represent isochrons or mixed lines with two end members having quite different $^{143}Nd/^{144}Nd$ and $^{147}Sm/^{144}Nd$ ratios. In the former, the slopes of the straight lines determine the ages of fluorite and calcite; in the latter, the slopes have no meaning but simply reflect the isotopic compositions and Sm/Nd ratios of the two end members. The variation in Sm/Nd ratios is not attributable to the mixing of the two end members because this would result in variable initial $\varepsilon_{Nd}(t)$ values. The $\varepsilon_{Nd}(t)$ values of fluorite and calcite range from −4.2 to −4.4 and from −1.1 to −1.2, both showing slight variations. In addition, no linear relationships could be determined from the 1/Nd vs. $^{143}Nd/^{144}Nd$ diagrams for fluorite and calcite (Figure 6). Therefore, the possibility of a mixing line can be rejected.

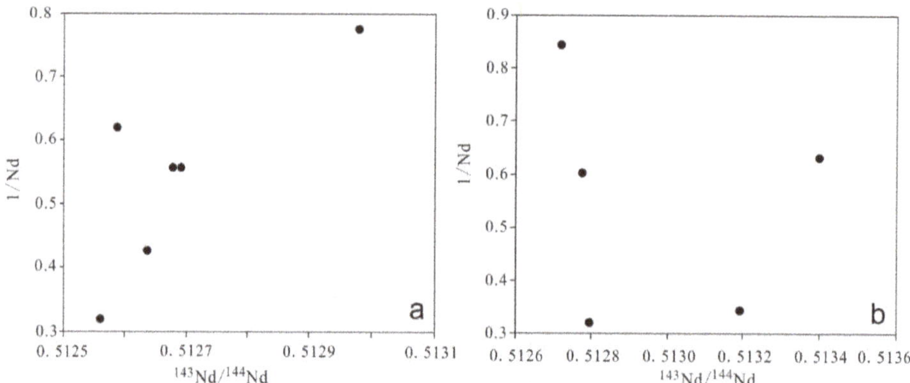

Figure 6. Diagrams of 1/Nd vs. $^{143}Nd/^{144}Nd$ of (**a**) fluorite and (**b**) calcite from the Shuiyindong deposit.

The slopes of the straight lines in Figure 5c,d determine the ages of fluorite (200.1 ± 8.6 Ma) and calcite (150.2 ± 2.2 Ma), respectively. The difference in ages suggests the occurrence of two episodes of hydrothermal events at the Shuiyindong deposit. Some important progress has been made for dating hydrothermal minerals from Carlin-type gold deposits in Southwestern China. Some hydrothermal minerals (e.g., calcite and apatite) were deposited during the late Yanshanian Period (ca. 130–150 Ma), based on the Sm–Nd isochron age (calcite; [10,11]) and the SIMS Th–Pb age (apatite; [12]). However, other hydrothermal minerals (e.g., sericite, arsenopyrite, and rutile) were precipitated during the Indosinian Period (ca. 200–230 Ma), based on the $^{40}Ar-^{39}Ar$ plateau age (sericite; [7]), the Re–Os age (arsenopyrite; [8]), and the in situ U–Pb age (rutile; [9]). The two age groups may also imply two distinct low-temperature hydrothermal events in Southwestern China.

To date, there is no consensus on the age of the Shuiyindong gold deposit owing to the lack of minerals clearly related to gold mineralization. The Re–Os isotope age of arsenopyrite (235 ± 33 Ma; [8]), interpreted as the age of gold mineralization, may be a mixed age. The arsenopyrite is very difficult to separate from pre-ore pyrite, as it is commonly intergrown with zoned pyrite containing pre-ore Au-poor pyrite cores and Au-rich pyrite overgrowth rims (Figure 7a,b; [2,5]).

Hydrothermal calcite around orebodies is believed to be a product of decarbonatization of the host rocks that served as the main source of Fe for sulfidation during gold deposition [37]. The gold often occurs in invisible forms, including submicroscopic particles, nanoparticles, and crystal lattice gold. Gold submicroscopic particles and nanoparticles are mostly distributed in arsenian pyrite or quartz veinlets, while lattice gold always occurs as crystal lattice in the overgrowth rims of arsenian pyrite [30,34]. Decarbonatization and sulfidation are responsible for providing the Fe and S to form hydrothermal arsenian pyrite, which is the most important mineral coprecipitated with gold. For Carlin-type gold deposits, numerous studies have shown that gold deposition is closely associated with decarbonatization of host strata [28,33,37]. After decarbonatization, host rocks commonly develop high porosity (Figure 7c), and calcite veins fill fractures at the periphery of decarbonated rocks. It is noteworthy that decarbonatization did not occur in the wall rocks, which are nonporous (Figure 7d). Therefore, the age of the calcite (150.2 ± 2.2 Ma) formed from decarbonatization potentially reflects the age of decarbonatization and gold deposition during the late stages of the Yanshanian Period.

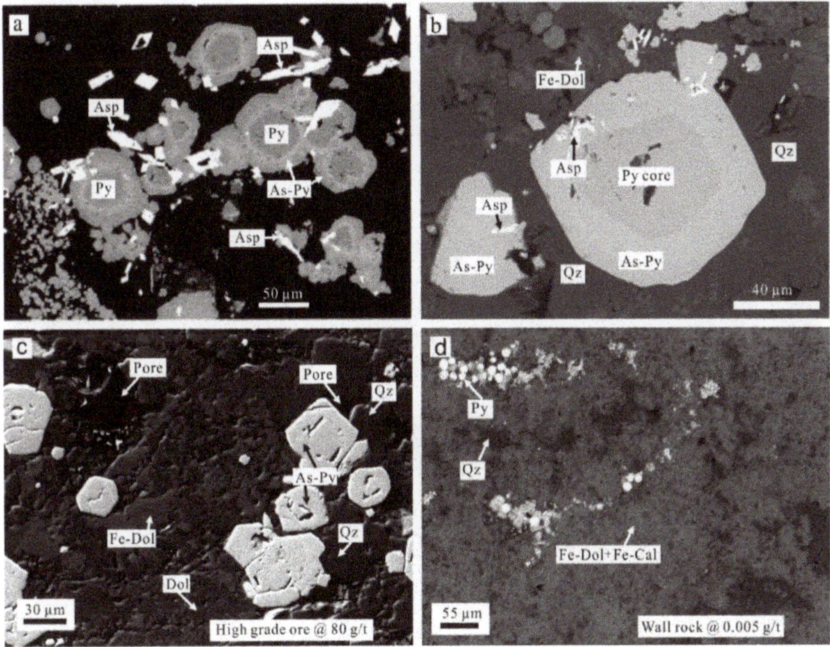

Figure 7. Backscattered electron (BSE) image of selected ore as well as wall rock from Shuiyindong. (**a,b**) BSE image of ores containing zoned arsenian pyrite (showing As-rich rims on As-poor pyrite cores) and arsenopyrite (after Su et al. [5,38]). (**c**) BSE image of high-grade ore showing decarbonatization and high porosity. (**d**) BSE image of wall rock showing no decarbonatization and nonporous. Abbreviations: Asp—arsenopyrite, As-py—arsenian pyrite, Dol—dolomite, Fe-Cal—ferroan calcite, Fe-Dol—ferroan dolomite, Py—pyrite, Qz—quartz.

Fluorite veins commonly cement argillite breccia or are intergrown with calcite veins at the vicinity of SBT at the Shuiyindong deposit (Figure 4c,d). However, no fluorite appears at decarbonated high-gold-grade rocks or around other types of ores. Until now, the relationship between the precipitation of fluorite and gold mineralization remains uncertain based on geological evidence (e.g., mineral assemblage and paragenetic sequence). The age of the hydrothermal fluorite (200.1 ± 8.6 Ma) likely records an episode of low-temperature hydrothermal events in the Indosinian Period in Southwestern China.

Isotope data can also be used to shed some light on the source of the Nd in the fluorite and calcite. The Maokou formation limestone and Longtan formation shale from the Permian period in Southeastern China have low $\varepsilon_{Nd}(t)$ values of −6.3 and −12.5 at 150 Ma, as calculated using the data of Peng et al. [39] and Chen and Jahn [40], respectively. The basaltic volcanic tuff from the Permian period has an $\varepsilon_{Nd}(t)$ value of +1.5, based on the data in Chung and Jahn [41] at 150 Ma. The initial $\varepsilon_{Nd}(t)$ values of fluorite and calcite are −4.4 and −1.1, respectively, and are in the range of values for Permian limestone (−6.3) and basaltic volcanic tuff (+1.5). This indicates that the Nd in the fluorite and calcite probably originated from mixtures of basaltic volcanic tuff and bioclastic limestone of the Permian formations, such as the Longtan formation.

6. Conclusions

Fluorite and calcite from the Shuiyindong gold deposit contain considerable concentrations of REE, and exhibit variable Sm/Nd ratios, facilitating the direct dating of associated hydrothermal events. The fluorite and calcite yielded Sm–Nd isochron ages of 200.1 ± 8.6 (initial $\varepsilon_{Nd}(t)$ = −4.4; MSWD = 0.47) and 150.2 ± 2.2 Ma (initial $\varepsilon_{Nd}(t)$ = −1.1; MSWD = 0.18), respectively. These two groups of Sm–Nd isochron ages suggest two episodes of hydrothermal events in Shuiyindong. The age of the calcite probably represents the late stage of the gold mineralization period. Initial Nd isotopic compositions indicate that the Nd in the fluorite and calcite was likely derived from mixtures of basaltic volcanic tuff and bioclastic limestone of the Permian formations.

Author Contributions: Y.X. designed the study concept and revised the manuscript; Q.T. contributed to the analysis, data interpretation and manuscript preparation; Q.T., Z.X., Z.W., and D.W. collected the calcite and fluorite samples; Y.Z., J.Y., and S.L. drawn the geologic maps.

Funding: This work was financially supported by the National Key R&D Program of Deep penetrating Geochemistry (2016YFC0600607) and the National Natural Science Foundation of China (41803046).

Acknowledgments: We wish to thank the geological workers from the Guizhou Bureau of Geology and Mineral Resource Guizhou and the Guizhou Zijin Gold Mines for their help during our field investigation.

Conflicts of Interest: The authors declare no conflict of interest.

References

1. Yan, J.; Hu, R.Z.; Liu, S.; Lin, Y.T.; Zhang, J.C.; Fu, S.L. NanoSIMS element mapping and sulfur isotope analysis of Au-bearing pyrite from Lannigou Carlin-type Au deposit in SW China: New insights into the origin and evolution of Au-bearing fluids. *Ore Geol. Rev.* **2018**, *92*, 29–41. [CrossRef]

2. Xie, Z.J.; Xia, Y.; Cline, J.S.; Koenig, A.; Wei, D.T.; Tan, Q.P.; Wang, Z.P. Are There Carlin-Type Gold Deposits in China? A Comparison of the Guizhou, China, Deposits with Nevada, USA, Deposits. In *Diversity of Carlin-Style Gold Deposits, Reviews in Economic Geology*; Muntean, J.L., Ed.; Society of Economic Geologists: Littleton, CO, USA, 2018; Volume 20, pp. 187–233.

3. Hu, R.Z.; Fu, S.L.; Huang, Y.; Zhou, M.F.; Fu, S.H.; Zhao, C.H.; Wang, Y.J.; Bi, X.W.; Xiao, J.F. The giant South China Mesozoic low-temperature metallogenic domain: Reviews and a new geodynamic model. *J. Asian Earth Sci.* **2017**, *137*, 9–34. [CrossRef]

4. Xie, Z.J.; Xia, Y.; Cline, J.S.; Pribil, M.J.; Koenig, A.; Tan, Q.P.; Wei, D.T.; Wang, Z.P.; Yan, J. Magmatic Origin for Sediment-Hosted Au Deposits, Guizhou Province, China: In Situ Chemistry and Sulfur Isotope Composition of Pyrites, Shuiyindong and Jinfeng Deposits. *Econ. Geol.* **2018**, *113*, 1627–1652.

5. Su, W.C.; Dong, W.D.; Zhang, X.C.; Shen, N.P.; Hu, R.Z.; Hofstra, A.H.; Cheng, L.Z.; Xia, Y.; Yang, K.Y. Carlin-Type Gold Deposits in the Dian-Qian-Gui "Golden Triangle" of Southwest China. In *Diversity of Carlin-Style Gold Deposits, Reviews in Economic Geology*; Muntean, J.L., Ed.; Society of Economic Geologists: Littleton, CO, USA, 2018; Volume 20, pp. 157–185.

6. 1:2,500,000 Chinese Geological Map, The National Geological Archives of China. Available online: http://geodata.ngac.cn/Document/Map.aspx?MapId=EC7E1A7A7CF81954E0430100007F182E (accessed on 12 April 2019).

7.   Chen, M.H.; Huang, Q.W.; Hu, Y.; Chen, Z.Y.; Zhang, W. Genetic type of Phyllosilicate (Micas) and its Ar-Ar dating in Lannigou gold deposit, Guizhou Province, China. *Acta Mineral. Sin.* **2009**, *29*, 353–362. (In Chinese with English Abstract)

8.   Chen, M.H.; Mao, J.W.; Li, C.; Zhang, Z.Q.; Dang, Y. Re–Os isochron ages for arsenopyrite from Carlin-like gold deposits in the Yunnan–Guizhou–Guangxi "golden triangle", southwestern China. *Ore Geol. Rev.* **2015**, *64*, 316–327.

9.   Pi, Q.H.; Hu, R.Z.; Xiong, B.; Li, Q.L.; Zhong, R. In situ SIMS U-Pb dating of hydrothermal rutile: Reliable age for the Zhesang Carlin-type gold deposit in the golden triangle region, SW China. *Miner. Depos.* **2017**, *52*, 1179–1190. [CrossRef]

10.   Su, W.C.; Hu, R.Z.; Xia, B.; Xia, Y.; Liu, Y.P. Calcite Sm-Nd isochron age of the Shuiyindong Carlin-type gold deposit, Guizhou, China. *Chem. Geol.* **2009**, *258*, 269–274. [CrossRef]

11.   Wang, Z.P. Genesis and Dynamic Mechanism of the Epithermal Ore Deposits, SW Guizhou, China—A Case Study of Gold and Antimony Deposits. Ph.D. Thesis, University of Chinese Academy of Sciences, Beijing, China, 2013. (In Chinese with English Abstract)

12.   Chen, M.H.; Bagas, L.; Liao, X.; Zhang, Z.Q.; Li, Q.L. Hydrothermal apatite SIMS ThPb dating: Constraints on the timing of low-temperature hydrothermal Au deposits in Nibao, SW China. *Lithos* **2019**, *324–325*, 418–428.

13.   Peng, J.T.; Hu, R.Z.; Burnard, P.G. Samarium-neodymium isotope systematics of hydrothermal calcites from the Xikuangshan antimony deposit (Hunan, China): The potential of calcite as a geochronometer. *Chem. Geol.* **2003**, *200*, 129–136. [CrossRef]

14.   Jiang, S.Y.; Slack, J.F.; Palmer, M.R. Sm-Nd dating of the giant Sullivan Pb-Zn-Ag deposit, British Columbia. *Geology* **2000**, *28*, 751–754.

15.   Anglin, C.D.; Jonasson, I.R.; Franklin, J.M. Sm-Nd dating of scheelite and tourmaline; implications for the genesis of Archean gold deposits, Val d'Or, Canada. *Econ. Geol.* **1996**, *91*, 1372–1382. [CrossRef]

16.   Chesley, J.T.; Halliday, A.N.; Kyser, T.K.; Spry, P.G. Direct dating of mississippi valley-type mineralization; use of Sm-Nd in fluorite. *Econ. Geol.* **1994**, *89*, 1192–1199. [CrossRef]

17.   Chesley, J.T.; Halliday, A.N.; Scrivener, R.C. Samarium-neodymium direct dating of fluorite mineralization. *Science* **1991**, *252*, 949–951. [CrossRef] [PubMed]

18.   Zou, Z.C.; Hu, R.Z.; Bi, X.W.; Wu, L.Y.; Feng, C.X.; Tang, Y.Y. Absolute and relative dating of Cu and Pb-Zn mineralization in the Baiyangping area, Yunnan Province, SW China: Sm-Nd geochronology of calcite. *Geochem. J.* **2015**, *49*, 103–112. [CrossRef]

19.   Wang, J.S.; Wen, H.J.; Fan, H.F.; Zhu, J.J. Sm-Nd geochronology, REE geochemistry and C and O isotope characteristics of calcites and stibnites from the Banian antimony deposit, Guizhou Province, China. *Geochem. J.* **2012**, *46*, 393–407. [CrossRef]

20.   Uysal, I.T.; Zhao, J.X.; Golding, S.D.; Lawrence, M.G.; Glikson, M.; Collerson, K.D. Sm-Nd dating and rare-earth element tracing of calcite: Implications for fluid-flow events in the Bowen Basin, Australia. *Chem. Geol.* **2007**, *238*, 63–71. [CrossRef]

21.   Roberts, S.; Palmer, M.R.; Waller, L. Sm-Nd and REE characteristics of tourmaline and scheelite from the Bjorkdal gold deposit, northern Sweden: Evidence of an intrusion-related gold deposit? *Econ. Geol.* **2006**, *101*, 1415–1425. [CrossRef]

22.   Bell, K.; Anglin, C.; Franklin, J. Sm-Nd and Rb-Sr isotope systematics of scheelites: Possible implications for the age and genesis of vein-hosted gold deposits. *Geology* **1989**, *17*, 500–504. [CrossRef]

23.   Chernyshev, I.V.; Golubev, V.N.; Aleshin, A.P.; Larionova, Y.O.; Gol'tsman, Y.V. Fluorite as an Sm-Nd geochronometer of hydrothermal processes: Dating of mineralization hosted in the Strel'tsovka uranium ore field, eastern Baikal region. *Geol. Ore Depos.* **2016**, *58*, 447–455. [CrossRef]

24.   Xu, W.G.; Fan, H.R.; Hu, F.F.; Santosh, M.; Yang, K.F.; Lan, T.G.; Wen, B.J. Geochronology of the Guilaizhuang gold deposit, Luxi Block, eastern North China Craton: Constraints from zircon U-Pb and fluorite-calcite Sm-Nd dating. *Ore Geol. Rev.* **2015**, *65*, 390–399. [CrossRef]

25.   Munoz, M.; Premo, W.; Courjault-Rade, P. Sm-Nd dating of fluorite from the worldclass Montroc fluorite deposit, southern Massif Central, France. *Miner. Depos.* **2005**, *39*, 970–975. [CrossRef]

26.   Tan, Q.P.; Xia, Y.; Wang, X.Q.; Xie, Z.J.; Wei, D.T. Carbon-oxygen isotopes and rare earth elements as an exploration vector for Carlin-type gold deposits: A case study of the Shuiyindong gold deposit, Guizhou Province, SW China. *J. Asian Earth Sci.* **2017**, *148*, 1–12. [CrossRef]

27. Liu, J.Z.; Yang, C.F.; Wang, Z.P.; Wang, D.F.; Qi, L.S.; Li, J.H.; Hu, C.W.; Xu, L.Y. Geological research of Shuiyindong gold deposit in Zhenfeng County, Guizhou Province. *Geol. Surv. China* **2017**, *4*, 32–41. (In Chinese with English Abstract)

28. Tan, Q.P.; Xia, Y.; Xie, Z.J.; Yan, J. Migration paths and precipitation mechanisms of ore-forming fluids at the Shuiyindong Carlin-type gold deposit, Guizhou, China. *Ore Geol. Rev.* **2015**, *69*, 140–156. [CrossRef]

29. Zhu, J.J.; Hu, R.Z.; Richards, J.P.; Bi, X.-W.; Stern, R.; Lu, G. No genetic link between Late Cretaceous felsic dikes and Carlin-type Au deposits in the Youjiang basin, Southwest China. *Ore Geol. Rev.* **2017**, *84*, 328–337. [CrossRef]

30. Su, W.C.; Xia, B.; Zhang, H.T.; Zhang, X.C.; Hu, R.Z. Visible gold in arsenian pyrite at the Shuiyindong Carlin-type gold deposit, Guizhou, China: Implications for the environment and processes of ore formation. *Ore Geol. Rev.* **2008**, *33*, 667–679. [CrossRef]

31. Tan, Q.P.; Xia, Y.; Xie, Z.J.; Yan, J.; Wei, D.T. S, C, O, H, and Pb isotopic studies for the Shuiyindong Carlin-type gold deposit, Southwest Guizhou, China: Constraints for ore genesis. *Chin. J. Geochem.* **2015**, *34*, 525–539. [CrossRef]

32. Tan, Q.P.; Xia, Y.; Wang, X.Q.; Xie, Z.J.; Wei, D.T. Tectonic model and tectonic-geochemistry characteristics of the Huijiabao gold orefield, SW Guizhou Province. *Geotecton. Metallog.* **2017**, *41*, 291–304. (In Chinese with English Abstract)

33. Cline, J.S. Nevada's Carlin-Type Gold Deposits: What We've Learned During the Past 10 to 15 Years. In *Diversity of Carlin-Style Gold Deposits, Reviews in Economic Geology*; Muntean, J.L., Ed.; Society of Economic Geologists: Littleton, CO, USA, 2018; Volume 20, pp. 7–37.

34. Li, X.X.; Zhu, X.Q.; Gu, Y.T.; Ling, K.Y.; Sheng, X.Y.; He, L. Mineralogy and Geochemistry Characteristics and Genetic Implications for Stratabound Carlin-Type Gold Deposits in Southwest Guizhou, China. *J. Nanosci. Nanotechnol.* **2017**, *17*, 6307–6317. [CrossRef]

35. Zhang, J.R.; Wen, H.J.; Qiu, Y.Z.; Zhang, Y.X.; Li, C. Ages of sediment-hosted Himalayan Pb-Zn-Cu-Ag polymetallic deposits in the Lanping basin, China: Re-Os geochronology of molybdenite and Sm-Nd dating of calcite. *J. Asian Earth Sci.* **2013**, *73*, 284–295. [CrossRef]

36. Ludwig, R.K. ISOPLOT: A plotting and regression program for radiogenic isotope data (Version 2.9). *U.S. Geol. Surv. Open-File Rep.* **1996**, *91*, 445–447.

37. Su, W.C.; Heinrich, C.A.; Pettke, T.; Zhang, X.C.; Hu, R.Z.; Xia, B. Sediment-hosted gold deposits in Guizhou, China: Products of wall-rock sulfidation by deep crustal fluids. *Econ. Geol.* **2009**, *104*, 73–93. [CrossRef]

38. Su, W.C.; Zhang, H.T.; Hu, R.Z.; Ge, X.; Xia, B.; Chen, Y.Y.; Zhu, C. Mineralogy and geochemistry of gold-bearing arsenian pyrite from the Shuiyindong Carlin-type gold deposit, Guizhou, China: Implications for gold depositional processes. *Miner. Depos.* **2012**, *47*, 653–662. [CrossRef]

39. Peng, J.T.; Hu, R.Z.; Jiang, G.H. Samarium-Neodymium isotope system of fluorites from the Qinglong antimony deposit, Guizhou Province: Constrains on the mineralizing age and ore-forming materials' sources. *Acta Petrol. Sin.* **2003**, *19*, 785–791. (In Chinese with English Abstract)

40. Chen, J.F.; Jahn, B.M. Crustal evolution of southeastern China: Nd and Sr isotopic evidence. *Tectonophysics* **1998**, *284*, 101–133. [CrossRef]

41. Chung, S.-L.; Jahn, B.-M. Plume-lithosphere interaction in generation of the Emeishan flood basalts at the Permian-Triassic boundary. *Geology* **1995**, *23*, 889–892. [CrossRef]

Article

Green Preparation of Nanoporous Pyrrhotite by Thermal Treatment of Pyrite as an Effective Hg(II) Adsorbent: Performance and Mechanism

Ping Lu, Tianhu Chen *, Haibo Liu, Ping Li, Shuchuan Peng and Yan Yang

Laboratory for Nano-minerals and Environmental Materials, School of Resources and Environmental Engineering, Hefei University of Technology, Hefei 230009, China; luping9305@163.com (P.L.); liuhaibosky116@hfut.edu.cn (H.L.); liping@hfut.edu.cn (P.L.); Pengshuchuan@hfut.edu.cn (S.P.); yan.yang@ucd.ie (Y.Y.)
* Correspondence: chentianhu@hfut.edu.cn; Tel.: +86-0551-62903990

Received: 10 November 2018; Accepted: 22 January 2019; Published: 27 January 2019

Abstract: The removal of Hg(II) from aqueous solutions by pyrrhotite derived from the thermal activation of natural pyrite was explored by batch experiments. The adsorption isotherms demonstrated that the sorption of Hg(II) by modified pyrite (MPy) can be fitted well by the Langmuir model. The removal capacity of Hg(II) on MPy derived from the Langmuir model was determined to 166.67 mg/g. The adsorption process of Hg(II) on MPy was well fitted by a pseudo-second-order model. The sorption of Hg(II) on MPy was a spontaneous and endothermic process. The removal of Hg(II) by MPy was mainly attributed to a chemical reaction resulting in cinnabar formation and the electrostatic attraction between the negative charges in MPy and positive charges of Hg(II). The results of our work suggest that the thermal activation of natural pyrite is greatly important for the effective utilization of ore resources for the removal of Hg(II).

Keywords: pyrite; thermal treatment; pyrrhotite; Hg(II) removal; mechanism

---

1. Introduction

With the rapidly increasing industrialization, heavy metal pollution has received increasingly more attention by the populace. Heavy metal ions are highly toxic even at low concentrations, and when released into the environment, they can cause devastating public health hazards [1,2]. As one of the most toxic metals ever discovered, Hg(II) is carcinogenic and stable with high cellular toxicity. Research has shown that Hg(II) can cause considerable damage to human health by causing toxicity to the central nervous system, kidneys, lung tissues, and reproductive system, resulting in health problems including paralysis, dysfunction of the central nervous system, intestinal and urinary complications, and even death in extreme cases [3,4]. Furthermore, trace amounts of Hg(II) in water are puzzling due to its complexation and mobility features at low concentrations, bioaccumulation during metabolic processes, wide distribution, and control difficulties. The United States Environment Protection Agency set a mandatory discharge limit of 10 μg/L for the total mercury content in wastewater, and the limit in drinking water is 1 μg/L [5]. Hence, it is very important to remove Hg(II) effectively from wastewater [6,7].

Various techniques have been put forward for aqueous Hg(II) removal such as electrolysis, precipitation, coagulation co-precipitation, membrane filtration, ionic exchange, and adsorption methods [8]. Among them, adsorption methods have been widely studied because they are cost-effective, environmentally friendly, and easy to conduct. Numerous adsorbents have been extensively studied, such as activated carbon [9,10], lichens [11], amine-modified attapulgite [8], poly(2-aminothiazole) [12], and mesoporous silica [5]. However, those adsorbents are either too costly or have low adsorption capacity. Recently, many researchers have committed to the application of natural mineral materials as adsorbents due to their low cost and physicochemical properties. Hg(II) sorption to sulfides is a potentially important Hg(II) sequestration mechanism due to the strong chemical bond between Hg(II) and sulfur [13].

The application of natural pyrite is limited by the low adsorption capacity because of its low specific surface area and the strong S–S bond in the crystal structure. Natural pyrrhotite ($Fe_{1-x}S$, $0 < X < 0.125$) as a reactive iron sulfide mineral has a nanostructure but a small specific surface area [14]. Therefore, in order to gain highly reactive products, our group developed a cost-effective method by thermally activating pyrite [15]. Thermally activated pyrrhotite as a promising adsorbent has been studied for the removal of various contaminants, such as Cu(II) [16], Pb(II) [17,18], U(VI) [19], and Hg^0 [20]. They all exhibit excellent adsorption characteristics. Therefore, it is valuable to explore the remove of Hg(II) from wastewater using modified pyrite (MPy) in detail [21]. The widespread application of the modified pyrite as a filtering media or a recyclable magnetic sorbent for the remediation of Hg(II)-polluted environments is of great importance.

In this work, batch experiments are conducted to study the feasibility of MPy as a medium for adsorbing Hg(II). The objectives of this work are to fabricate nanoporous and magnetically recycled MPy to explore the properties and mechanisms of Hg(II) removal with MPy and to explore the application of pyrite as an effective adsorbent in environmental cleanup. It is believed that the experimental data are helpful for the application of MPy for Hg(II) removal.

2. Experimental Procedure

2.1. Sample Preparation

MPys were prepared by calcinating the natural pyrite in an N_2 atmosphere at different temperatures (550, 600, 700, 800 °C) for 0.5 h. Natural pyrite was collected from the Lujiang Mine, Anhui Province, China. First, the pyrite was fractured and ground to 74 μm. Then, the sample was soaked for 2 h with 5% HCl to remove the oxidation film and washed by Milli-Q water 4 times. The obtained samples were dried using a lyophilizer and then kept in a drier. Thermally-activated samples at different temperatures were obtained by setting different heating times. The as-prepared samples were labeled MPy-*T* (*T* denotes the annealing temperature). For instance, MPy-550 means that the pyrite was thermally treated at 550 °C.

2.2. Characterization

The phase composition of MPy was observed using XRD (Dandonghaoyuan 2700, D/max-rB, Rigaku, Tokyo, Japan, a voltage of 40 kV, an electric current of 30 mA, and ascan rate of 4°/min). The morphology and size distribution of MPy were characterized through gold-sputtering, field emission scanning electron microscopy (SEM, SU8020, Hitachi, Tokyo, Japan), and transmission electron microscopy (TEM, JEM-2100F, JEOL, Tokyo, Japan). The zeta potentials of MPy were recorded by a Zetasizer Nano ZS (Nano-ZS90, Malvern Panalytical Ltd, Malvern, UK). Thermogravimetric analysis and differential thermal analysis (TG-DTA) were performed on a Thermogravimetric Analyzer (TG/DTA7300, NSK, Tokyo, Japan). The hysteresis loop of MPy was demonstrated using MPMS (MPMS XL-7, Quantum Design, Inc., San Diego, CA, USA). X-ray photoelectron spectra (XPS) of Fe2p, S2p, O1s, and Hg4f spectral regions were characterized using X-ray photoelectron spectroscopy (Thermo, ESCALAB 250Xi, Thermo Fisher Scientific, Waltham, MA, USA). IR spectra of MPy were examined using a Fourier transform infrared spectrometer (FTIR, Vertex-70, KBr, Bruker, Ettlingen, Germany). The spectra were recorded using a Raman spectrometer (Horiba Jobin Yvon, HR Evolution, HORIBA Scientific, Kyoto, Japan) with a laser at 532 nm in these experiments.

2.3. Batch Experiments

In this research, batch experiments were conducted to evaluate the performance of MPy on Hg(II) adsorption. The solution of Hg(II) was obtained by dissolving $HgNO_3$ (analytically pure) in pure water, and the pH was adjusted to 3 for preservation. The effect of the activation temperature was evaluated by 1.0 g/L MPy prepared at different temperatures in different Hg(II)concentrations in the presence of 0.01 mol/L $NaNO_3$ for 24 h. The performance of suspension pH and ionic strengths was explored with 1.0 g/L MPy-600 at the range of 2.0–7.0 by using trace amounts of 0.01–1.0 mol/L HNO_3 and NaOH solution in 10 mg/L of Hg(II) for 210 min. The effect of the adsorbent dose on Hg(II) adsorption was evaluated by adding a different concentration (0.2–1.0 g/L) of MPy-600 in 20 mg/L of Hg(II) in the presence of 0.01 mol/L $NaNO_3$ at pH 6 for 24 h. The kinetic analysis was explored with 0.4 g/L MPy-600 in 10 mg/L of Hg(II) in the presence of 0.01 mol/L $NaNO_3$. The effect of the reaction temperature was investigated at different temperatures (298, 308 and 318 K). After the reaction, the solid-liquid phases were separated with the 0.22 μm membrane. The content of residual Hg(II) in the solution was measured using a direct mercury analyzer (DMA-80, Milestone Systems, Borneo Municipality, Denmark). The removal (adsorption, %) and sorption capacity (Qs, mg/g) were obtained by Equations (1) and (2), respectively:

$$Adsorption(\%) = (C_0 - C_e)/C_0 \times 100\% \tag{1}$$

$$Q_s = V \times (C_0 - C_e)/m \tag{2}$$

where C_0 (mg/L) is the starting concentration, C_e (mg/L) is the concentration after adsorption, V (L) is the suspension volume, and m (g) is the mass of adsorbent.

3. Results and Discussion

3.1. Characterization of Naturally Derived Pyrrhotite

Figure 1A shows the XRD spectrum of natural pyrite and MPy calcined at different temperatures. The appearance of a weak peak of MPy-550 at $2\theta = 44.08°$ indicates the formation of monoclinic pyrrhotite. When the calcination temperature is 600 °C, the disappearance of the reflections of pyrite demonstrates that pyrite is completely decomposed into monocline pyrrhotite. The enhancement in the reflections of MPy-700 suggests a higher crystallinity. As the temperature rises to 800 °C, monoclinic pyrrhotite translates into hexagonal pyrrhotite [22].

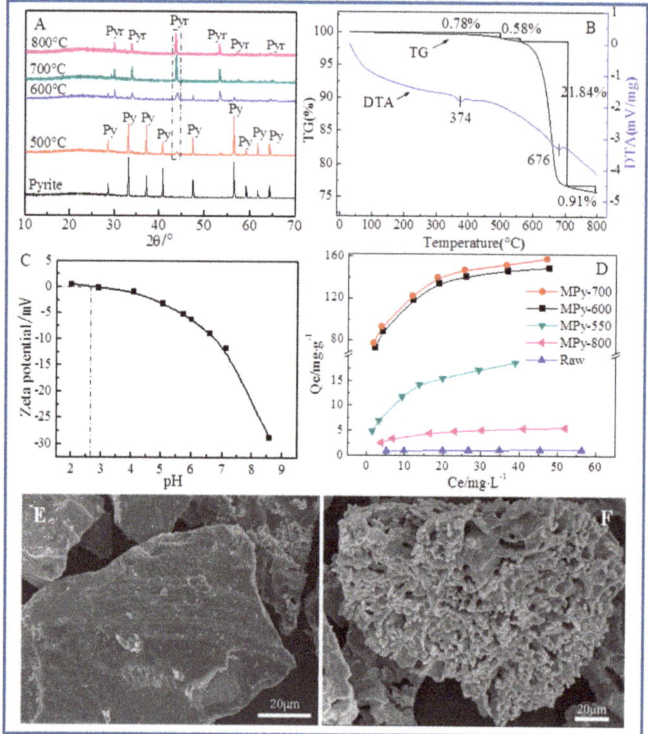

Figure 1. (**A**) XRD spectrum of natural pyrite and MPy. Py—pyrite; Pyr—pyrrhotite. (**B**) TG-DTA spectrum of natural pyrite in a N_2 atmosphere. (**C**) The zeta potential of MPy-600. (**D**) Removal isotherms of Hg(II) by pyrite and MPy. (**E,F**) SEM images of natural pyrite and MPy-600, respectively.

Apparently, from the TG curve in Figure 1B, four weight loss stages are found in the thermal degradation of pyrite. However, the DTA curves reveal two endothermic peaks that correspond well to the weight losses. The first region (under 493 °C) and the endothermic peak at 374 °C due to the dehydration and dehydroxylation. The second region (493–531 °C) is associated with the process of ferrous sulfate decomposed to hematite and sulfur dioxide gas [23]. The third region (531–697 °C) and the endothermic peak at 676 °C corresponds to the process that the pyrite transformed to pyrrhotite by desulfuring as follows in Equation (3) [18]. Lastly, the fourth region (under 800 °C) is assigned to the conversion of monoclinic pyrrhotite to hexagonal pyrrhotite [24].

$$FeS_2 \rightarrow Fe_{1-x}S + S \tag{3}$$

The zeta potential of MPy-600 is shown in Figure 1C. The pH_{ZPC} (zero-point charge) of MPy-600 is observed at pH 2.6. The findings indicate that the negative charge of the MPy-600 surface is obtained by releasing protons at pH > 2.6. The MPy-600 surface possesses positive charges by protonating amphoteric ions at pH < 2.6 [25]. SEM images of natural pyrite and MPy-600 are shown in Figure 1E,F. The crystal size of natural pyrite is large with no porous texture. However, the MPy-600 surface has nanometer-sized structures with abundant inhomogeneous pores that provide ample active sites and high activity.

3.2. Adsorption Isotherms

The sorption capacities of Hg(II) on pyrite and MPy calcined at different temperatures are presented in Figure 1D. The removal of Hg(II) by MPy and pyrite obviously increases with the initial concentration increase. The sorption capacities of MPy-600 and MPy-700 are higher than those of the other samples because of the nanoporous structure and high specific surface area (27.62 m^2/g) [15]. According to the research results, MPy-600 is an ideal material for the study of Hg(II) removal with respect to the consumed energy for grinding. The saturated sorption capacity is approximately 148.65 mg/g. The sorption process of Hg(II) by MPy-600 can be matched by the Langmuir model and the Freundlich model. The Langmuir type describes monolayer sorption which has identical and equal-energy sorption sites, while the Freundlich type is used for heterogeneous adsorption.

Langmuir isotherm:

$$\frac{C_e}{Q_e} = \frac{1}{Q_m K_L} + \frac{C_e}{Q_m} \tag{4}$$

Freundlich isotherm:

$$lnQ_e = lnK_F + \frac{1}{n}lnC_e \tag{5}$$

where C_e is the equilibrium concentration (mg/L), Q_e is the equilibrium sorption capacity (mg/g), Q_m is the saturated sorption capacity (mg/g), K_L is the Langmuir sorption constant (L/mg), K_F is the equilibrium coefficient ((mg/g)/(mg L)$^{-n}$), and $1/n$ is the sorption exponent associated with the heterogeneity of sorption sites. The D-R (Dubinin-Radushkevich) isotherm is more applicable than the Langmuir, and it assumes neither a uniform sorption potential nor a homogeneous surface. The D-R model in a linear formation can be expressed as:

$$q_e = q_m exp\left(-\beta\varepsilon^2\right) \tag{6}$$

where β is the activity coefficient of the average sorption energy (mol^2/kJ^2) and ε is the Polanyi potential, which is equivalent to:

$$\varepsilon = RTln\left(1 + \frac{1}{C_e}\right) \tag{7}$$

where R (8.314 J/(mol·K)) is an ideal gas constant, and T (K) is the absolute temperature in Kelvin (K). E (kJ/mol) represents the free energy change. The value of E can be calculated using the following expression:

$$E = \frac{1}{\sqrt{2\beta}} \tag{8}$$

The magnitude of E can provide an idea regarding the type of adsorption process, that is, whether the process is physical or chemical. When E is below 8 kJ/mol, the adsorbent process is considered to be physical. When E ranges from 8 to 16 kJ/mol, the adsorbent process is triggered by ion exchange. If the value of $E > 16$ kJ/mol, then the adsorbent process is of chemical properties.

The relative parameters of the above models are presented in Table 1. It can be inferred that the Langmuir type ($R^2 = 0.9991$) fit the experimental data better regarding the sorption of Hg(II) onto MPy-600 compared with Freundlich model ($R^2 = 0.9724$). This fact indicates that the adsorption of Hg(II) onto MPy-600 is monolayer sorption. The Q_m calculated from the Langmuir model is 166.67 mg/g. The maximum sorption capacity q_m worked out from the D-R model is lower than the Langmuir model, which may be ascribed to the different hypotheses of the sorption models [26,27]. A contrast of the Q_m in Table 2 indicates that MPy-600 presents a high adsorption ability of Hg(II).

Table 1. Isotherm parameters of the Langmuir, Freundlich, and D-R models.

Models	Parameters			
Langmuir	Q_m (mg/g) 166.67		K_L (L/mg) 0.30	R^2 0.9991
Freundlich	k_F ((mg/g)/(mg L)$^{-n}$) 63.06		$1/n$ 0.22	R^2 0.9724
D-R	β (mol^2/kJ^2) 6.06×10^{-6}	q_m (mg/g) 149.66	E (kJ/mol) 287	R^2 0.9879

Table 2. Comparison of the maximum adsorption capacities (Q_{max}) of Hg(II) on various adsorbents.

Material	Experimental Conditions	Q_{max} (mg/g)	References
Lichens	pH 6.0, 293 K	82.5	[11]
Activated carbon made from sago waste	pH 5.0, 303 K	55.6	[9]
MPy-600	pH 6.0, 303 K	166.7	this work
Activated carbon derived from (AEC)	pH 5.0, 303 K	28.4	[10]
Amine-modified attapulgite	pH 6.0, 298 K	93.2	[8]
Synthesis of poly (2-aminothiazole)	pH 6.5, 288 K	291.5	[12]
Al2O3-supported nanoscale FeS	pH 6.0, 303 K	142.7	[28]

3.3. The Effects of pH and Ionic Strength

The impact of pH and ionic strength on the Hg(II) sorption by MPy-600 is presented in Figure 2. The pH of the solution plays a large role in the removal of Hg(II) onto MPy-600. The removal quantity of Hg(II) rises from 0.8 to 4.3 mg/g as the pH varies from 2–6. However, with a further increase of pH from 6.0–7.0, the adsorption capacity exhibited a negligible change. According to the pH$_{PZC}$ of MPy-600, when pH < 2.6, the surface charge is positive. However, for pH > 2.6, the MPy-600 surface presents an abundant negative surface charge. Figure 2Bshows the Hg(II) species distribution in the solution. Hg(II) mainly exists as Hg^{2+} species at a pH < 3.1. But at the pH of 3–3.5, 40% of Hg(II) exists as Hg^{2+} species, 40% of Hg(II) exists in the form of Hg(OH)$_2$, and 20% of Hg(II) exists in the form of Hg(OH)$^+$.Furthermore, 50% of Hg(II) exists in the form of Hg(OH)$_2$, and 50% of Hg(II) exists in the hydrolyzed mononuclear and multinuclear species (i.e., Hg^{2+}, Hg$_2$(OH)$^{3+}$, Hg$_3$(OH)$_3$$^{3+}$, and Hg(OH)$^+$) at a pH range of 3.5–4.0. Hg(OH)$_2$ is the main variety of the Hg(II) aqueous solution at a pH > 4.0. Consequently, the slightly increased adsorption of Hg(II) onto MPy-600 at pH < 3 probably ascribes to the electrostatic repulsion between Hg^{2+} and the positive surface, and also the competition adsorption between the Hg(II) and H$^+$ of the binding sites on the MPy-600. The increased adsorption at the pH range from 3.0–6.0 can be ascribed to the electrostatic attraction which occurs between the positive charges of Hg(II) (i.e., Hg$_2$(OH)$^{3+}$, Hg$_3$(OH)$_3$$^{3+}$, Hg(OH)$^+$) and the negative MPy-600 surface. When the pH > 6.0, the adsorption capacity exhibits a neglectable change that is caused by the formation or precipitation of the hydroxyl complexes by metal ions. Given all these considerations, an initial pH of 6.0 is suitable as the optimal value for Hg(II) adsorption [3,11].

The impact of the ionic strength on the adsorption of Hg(II) onto MPy-600 is observed in Figure 2A. As presented in Figure 2A, with the increase of ion strength, the sorption of Hg(II) by MPy-600 increases. The electrical conductivity of the solution increases, which strengthens the electrostatic attraction of Hg(II) to the surface of MPy-600. It is certified that the inner surface complexation is irrelevant to the ionic strength, otherwise, the outer surface complexation is more sensitive to it. Consequently, this infers that the outer surface complexation mainly dominated the sorption process of Hg(II) onto MPy-600 [1,29]. It can be speculated that Hg(II) adsorbed on the MPy-600is due to the newly formed ferric hydroxide.

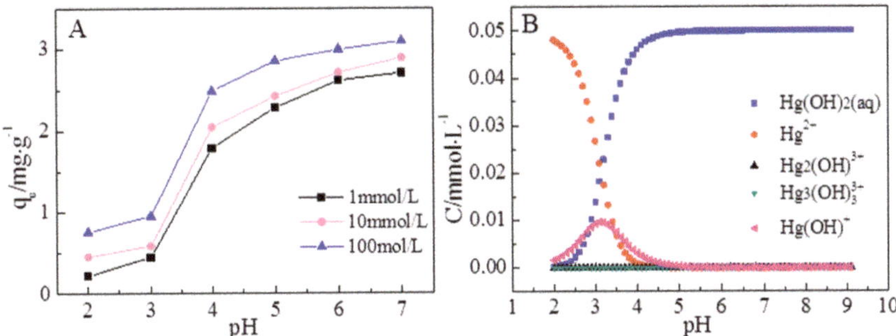

Figure 2. (**A**) The effect of pH and ionic strength. (**B**) Distribution of Hg(II) in water solutions (C_0 = 10 mg/L, I = 0.01 mol/L, T = 303 K).

3.4. The Effect of the AdsorbentDose

The impact of the MPy-600 concentration (0.2–1.0 g/L) on Hg(II) adsorption is presented in Figure 3. As presented in Figure 3, the removal rate of Hg(II) by MPy-600 increases with increasing adsorbent dose, whereas the removal quantity correspondingly decreases. With the increase of the adsorbent dose, there are more active sites available for sorption of Hg(II), which is more likely to facilitate penetration of Hg(II) to the adsorption points. The removal is turned up to be 80% at an MPy-600 concentration of 0.4 g/L. A further increase in MPy-600 concentration over 0.4 g/L does not lead to an obvious improvement in the removal but a reduction in the removal quantity of Hg(II) because of the surplus adsorption sites related to the quantity of the Hg(II) solution. Hence, the optimalMPy-600 concentration is 0.4 g/L for further experiments.

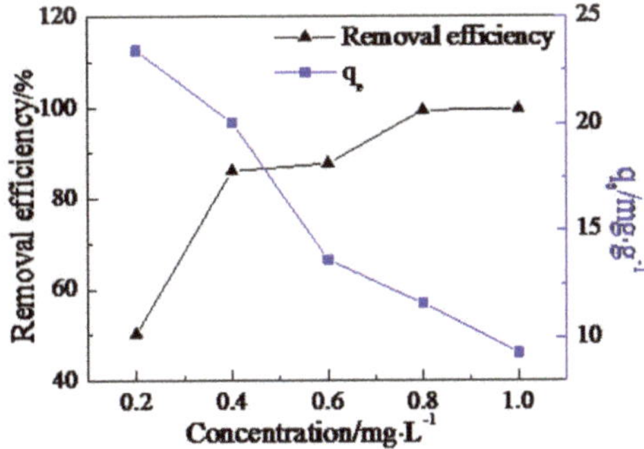

Figure 3. The effect of MPy-600 concentration on the adsorption of Hg(II).

3.5. Adsorption Kinetics

Adsorption kinetics measurements are conducted to estimate the performance of Hg(II) on MPy-600andto develop an understanding of the underlying mechanisms of Hg(II) adsorption on this sorbent and the potential rate-controlling steps through valuable data. As given in Figure 4A, the sorption capacity of Hg(II) onto MPy-600 distinctly increases with the increasing reaction time from 0–240 min, and the adsorption capacity rises slowly until equilibrium is attained during the experimental time period. To gain more information about the mechanisms, the data of the adsorption process are fitted with three model equations: The pseudo-first-order model, pseudo-second-order model, and Weber-Morris intraparticle diffusion model.

The pseudo-first-order and pseudo-second-order models are rate controlled, which is the strength of the adsorption capacity other than the solution concentration. Their linear forms of those models are given in Equations (9) and (10):

$$ln(q_e - q_t) = lnq_e - \frac{k_1}{2.303}t \tag{9}$$

$$\frac{t}{q_t} = \frac{1}{k_2 q_e^2} + \frac{t}{q_e} \tag{10}$$

where q_e (mg/g) and q_t (mg/g) represent the quantity of metal adsorbed at equilibrium and at a certain time, respectively, t (min) represents time, k_1 is the adsorption rate constants (g/(mg·min)) of the pseudo-first-order kinetic model, and k_2 is the adsorption rate constants (g/(mg·min)) of the pseudo-second-order kinetic model.

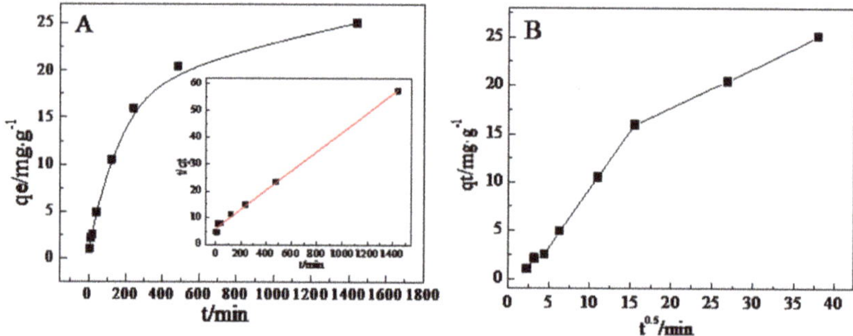

Figure 4. The adsorption kinetics of MPy-600 on Hg(II): (**A**) The pseudo-second-order model and (**B**) the Web-Morris model.

The fitting parameters of the pseudo-first-order and pseudo-second-order kinetic models are shown in Table 3. As presented in Figure 4, the results are fitted more accurately by the pseudo-second-order kinetic model (R^2 = 0.995) than the pseudo-first-order kinetic model (R^2 = 0.989) of Hg(II) on MPy-600, which follows that primary mechanism is chemical adsorption rather than physical adsorption.

Table 3. The fitted parameters of the pseudo-first-order model and pseudo-second-order kinetic model of Hg(II) sorption on MPy-600.

C_0 (mg/L)	Pseudo-First-Order			Pseudo-Second-Order		
	q_e (mg/g)	k_1	R^2	q_e (mg/g)	k_2	R^2
10	34.411	0.0038	0.952	27.925	0.0002	0.996

The Weber-Morris plot is used to describe the process of intraparticle diffusion. The Weber-Morris model can be expressed as

$$q_t = k \times t^{1/2} + c \tag{11}$$

where q_t (mg/g) represents the adsorbed amount at time t, k (mg/(g·min$^{1/2}$)) represents the rate constant, and c represents a constant. As shown in Figure 4B, the multilinear plots indicate that more than one mechanism might be responsible for Hg(II) adsorption onto MPy-600 [5].

The adsorption process can consist of two stages. The first stage from 0–20 min is for the surface adsorption or instantaneous adsorption on the most effective adsorbing sites of the MPy-600 surface. The second linear stage covering up to 20–40 min is for intraparticle diffusion. The primary effect is intraparticle diffusion, so the sorption rate slows down with increasing diffusion resistance until the diffusion process reaches equilibrium. During the adsorption process, many other steps may be involved, but these steps may be indistinguishable from the two main stages [30,31]. The plot of q_t vs. $t^{1/2}$ of the second regression stage must be linear and pass through the origin, which proves that the mechanism of the intra-particle diffusion is controlled by the rate. In Figure 4B, the plot does not go across the origin, which may be ascribed to the distinction of the mass transfer rate of the starting and final adsorption stops. These results suggest that mainly controlling the mechanism for the sorption of Hg(II) onto MPy-600 is not intra-particle diffusion [27,32,33]. The relevant parameters are listed in Table 4.

Table 4. Kinetic Parameters of the Weber-Morris Model.

Sample	Parameters		R^2
MPy-600	$K_1 = 1.78$	$C_1 = -4.20$	0.9997
	$K_2 = 0.60$	$C_2 = 14.12$	0.9997

3.6. Thermodynamics

The thermodynamic parameters reveal the reaction mechanism of Hg(II) sorption onto MPy-600. The fitting for the thermodynamic model of Hg(II) onto MPy-600 is shown in Figure 5. The related parameters, including the standard free energy change (ΔG), standard enthalpy change (ΔH), and the standard entropy change (ΔS) for adsorption process can be calculated as follows (Equations (12)–(14)):

$$\Delta G = -RTlnK_c \tag{12}$$

$$lnK_c = \frac{\Delta S}{R} - \frac{\Delta H}{RT} \tag{13}$$

$$lnK_d = (C_0 - C_e)/C_e \times (V/m) \tag{14}$$

where R (8.314 J/(mol·K)) represents the ideal gas constant, and T (K) represents the temperature in Kelvin. The parameter K_d (L/g) is the distribution coefficient. The constant lnK_c (L/g) is the adsorption equilibrium constant obtained by plotting lnK_d vs. C_e and then extrapolating C_e to zero.

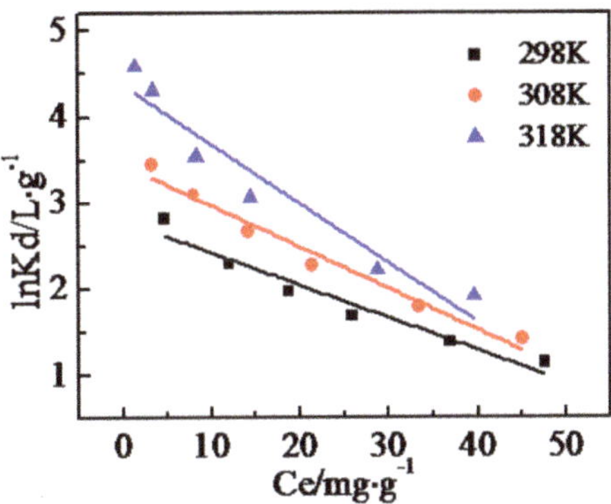

Figure 5. The linear plots of lnK_d vs. C_e of Hg(II) onto MPy-600.

The line graphs of lnK_d vs. C_e of Hg(II) adsorption onto MPy-600 at different temperatures are presented in Figure 5. The corresponding parameters for the sorption of Hg(II) onto MPy-600 are shown in Table 5. The negative ΔG values testify that the sorption of Hg(II) onto MPy-600 is a spontaneous process, and ΔG decreases as temperature is increasing, revealing that the sorption is more beneficial at higher temperatures. The positive ΔH value suggests this reaction process is endothermic. The positive ΔS implies that the molecular arrangement becomes more chaotic during the reaction process, which leads to an increasing disorder in the reaction system, and the Hg(II) adsorption process is driven by enthalpy. The value of ΔS is between 0 and 22 J/(mol·K), which indicates that both physical and chemical adsorption processes exist during the adsorption.

Table 5. Thermodynamic parameters for the sorption of Hg(II) on MPy-600.

T (K)	ΔG (kJ/mol)	ΔH (kJ/mol)	ΔS (J/(mol·K))
298	−0.6930		
308	−0.8843	62.0557	0.2310
318	−1.1569		

3.7. Magnetization

As shown in Figure 6, the magnetic property of natural pyrite and MPy-600 are measured with the applied field of −45,000–45,000 Oe at room temperature. The pyrite shows a negligible magnetization, and the saturation magnetization of it is just 0.1 emu/g. However, the excellent magnetization of MPy-600 makes it possible to separate the sample from the solution by magnetic separation, which has a high saturation magnetization of 13.3 emu/g. The coercivity of MPy-600is 61.75, and Oe indicates an obvious magnetization hysteresis. MPy-600 does not show superparamagnetism, and the permanent magnetization is as low as 4.278 emu/g. The saturation magnetization of MPy-600 after adsorption decreases to 12 emu/g [20,34]. Thus, the magnetic property of MPy-600 guarantees the convenient magnetic separation from the aqueous solution in adsorption applications.

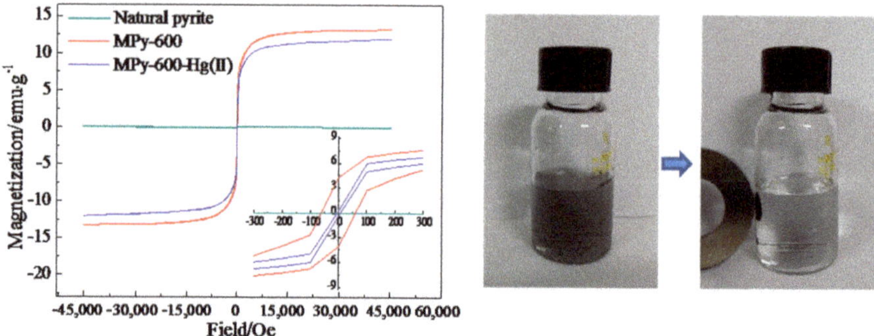

Figure 6. Magnetization characteristics of natural pyrite, MPy600, and MPy-600 after adsorption.

3.8. Adsorption Mechanism

3.8.1. XRD Analyses

Figure 7 shows the XRD pattern of MPy-600 after Hg(II) adsorption compared to the PDF standard card of pyrrhotite and HgS. From the figure, the reflections at 30.1°, 34.0°, 43.9°, and 53.2° correspond to the pyrrhotite [22]. Compared to the standard sample peaks of cinnabar, the peaks at $2\theta = 26.5°$ of the reacted MPy-600 material are evident, which indicates that HgS is formed. The intensity of the peaks is relatively low for HgS compared to that of MPy-600, which reveals a lower content. Furthermore, the XRD pattern of the used MPy-600 preliminarily illustrates that the removal of Hg(II) onto MPy-600 is owing to the form of the HgS from the chemical reaction.

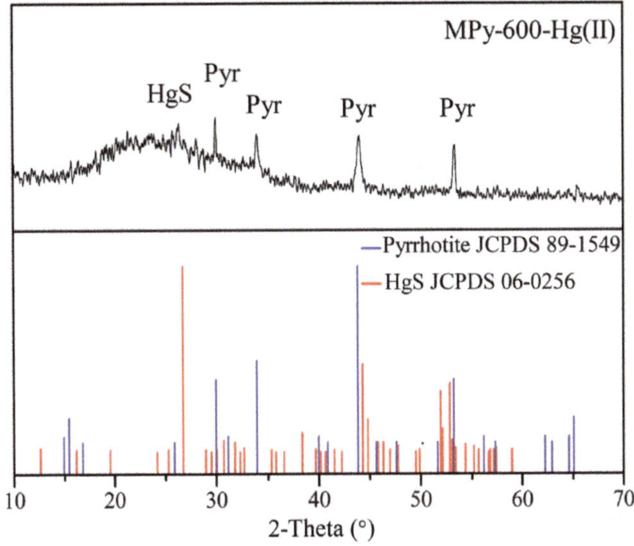

Figure 7. The XRD pattern of the reacted MPy-600 material. Pyr—pyrrhotite.

3.8.2. The FTIR and Raman Spectra

The FTIR spectra of MPy-600 and Hg(II)-adsorbed MPy-600 are exhibited in Figure 8A. The characteristic absorbance lines at 1076 cm^{-1} and 483 cm^{-1} are assigned to pyrrhotite. The peak at 3649 cm^{-1} refers to the $-$OH vibrations [12,35]. The characteristic absorbance peak at 3649 cm^{-1} of MPy-600-Hg(II) is higher, which can be ascribed to the formation of ferric hydroxide from the dissolution of iron, and more Hg(OH)$_2$ is formed by attracting more $-$OH molecules to Hg(II) [36]. The relative intensities of the MPy-600-Hg(II) peaks at 483 cm^{-1} and 1076 cm^{-1} are lower, which probably suggests that MPy-600 reacted with Hg(II) [37]. This result is consistent with XRD results, and the equations are as follows (Equations (15)–(20)):

$$Fe_{1-x}S_{(s)} \leftrightarrow (1-3x)Fe^{2+} + S^{2-} + 2xFe^{3+} \tag{15}$$

$$Hg^{2+} + S^{2-} \rightarrow HgS \tag{16}$$

$$Hg^{2+} + 2OH^- \rightarrow Hg(OH)_2 \tag{17}$$

$$Fe^{3+} + 3H_2O \rightarrow Fe(OH)_3 + 3H^+ \tag{18}$$

$$K_{sp} \ of \ FeS = \left[Fe^{2+}\right]\left[S^{2-}\right] = 1.59 \times 10^{-19} \tag{19}$$

$$K_{sp} \ of \ HgS = \left[Hg^{2+}\right]\left[S^{2-}\right] = 6.44 \times 10^{-53} \tag{20}$$

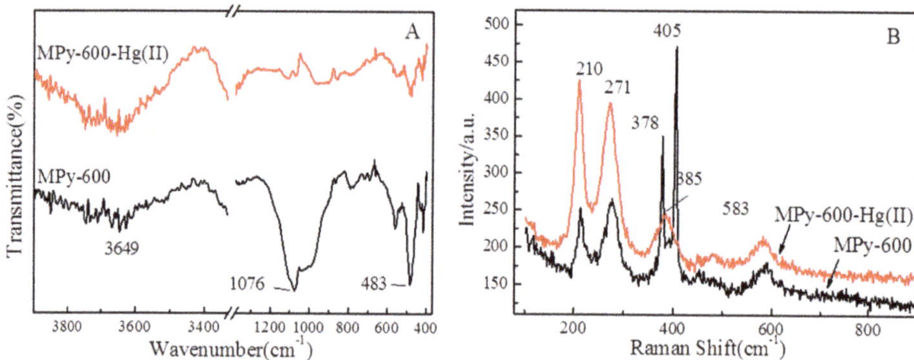

Figure 8. (**A**) The FTIR and (**B**) Raman spectra of MPy-600 and MPy-600 after Hg(II) adsorption.

The Raman spectra of MPy-600 and MPy-600 after Hg(II) sorption are presented in Figure 8B. Some research studies have shown that the frequency window of pyrrhotite vibrations is approximately 300–450 cm^{-1}. As exhibited in Figure 8B, the distinct vibration peaks at 378cm^{-1} and 405 cm^{-1} are consistent with vibrations of the Fe–S band to pyrrhotite [38,39]. The stretching vibrational frequency of Hg–S is approximately at 180–400 cm^{-1}. Therefore, the bands observed at 210 cm^{-1} and 271 cm^{-1} are ascribed to the S–Hg–S oscillations [35]. The weak and broad peak at 385 cm^{-1} is ascribed to a reduction in MPy-600, which is involved in the response [40]. The formation of a band at 583 cm^{-1} belongs to FeOOH (RRUFF). The appearance of the Hg-S band confirms the removal capacity of MPy-600 on Hg(II) [35].

3.8.3. XPS Analyses

The XPS is applied to characterize the elemental states to reveal the mechanism of Hg(II) adsorption onto MPy-600 [41]. The XPS spectrogram for Hg 4f, Fe 2p, S 2p, and O 1s of MPy-600 before and after the reaction is presented in Figure 9. As presented in Figure 9a,d, the Fe 2p peaks correspond to the fresh and used MPy-600, respectively. The fresh MPy-600 has four obvious peaks of Fe 2p on its surface. The binding energies centered at approximately 711.5 eV, 715.5 eV, and 723.8 eV may be assigned to Fe^{2+} bonded with S^{2-}, Fe^{3+} bonded with −OH, and Fe^{2+} bonded with SO_4^{2-}, respectively. Among them, the peaks at 711.5 eV and 724.9 eV assign to the Fe 2p3/2 and Fe 2p1/2 peaks (split both by 13.1eV), respectively. The used MPy-600 has four obvious Fe 2p peaks. These peaks are positioned at an offset to certain peak values. The binding energies centered at approximately 710.6 eV, 713.2 eV, and 719 eV may be ascribed to Fe^{2+} bonded with S^{2-}, Fe^{3+} bonded with –OH, and Fe^{2+} bonded with SO_4^{2-}, respectively. As shown in Figure 9a,d, the percentage increase in Fe^{2+} can be ascribed to $Fe_{1-x}S$ dissolved to Fe^{2+} and S^{2-} and iron diffused to the surface from the interior [42]. The increasing Fe-oxide peaks reveal that MPy-600 is oxidized with increasing reaction time.

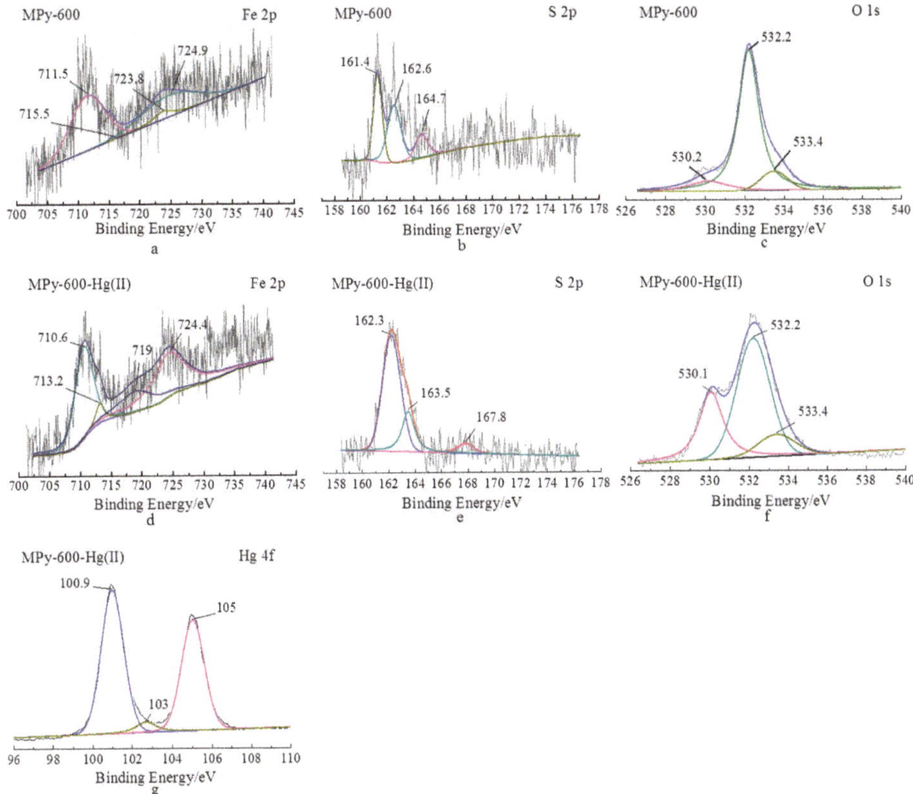

Figure 9. XPS spectra of MPy-600 (**a–c**) and MPy-600 after Hg(II) adsorption(**d–g**) in the Fe 2p, O 1s, S 2p, and Hg 4f spectral regions.

As observed in Figure 9b,e, the S 2p peak of MPy-600 centered at 161.4 eV is ascribed to S^{2-}, and the values at 162.6 eV and 164.7 eV represent S_2^{2-} and SO_4^{2-}, respectively [43]. In addition, the peak positions of MPy-600 after the removal of Hg(II) shifted to high energies because the surface electrons need plenty of energy for removal. The S 2p spectra of MPy-600 after Hg(II) adsorption exhibited peaks at 162.3 eV, 163.5 eV, and 167.8 eV that are ascribed to S^{2-}, S_2^{2-}, and SO_4^{2-}, respectively [34]. The amount of S^{2-} significantly increases after adsorption, which is ascribed to the dissolution of $Fe_{1-x}S$ and the formation of HgS. The amount of S_2^{2-} decreases because the S_2^{2-} is oxidized to disulfide and sulfate [44]. The oxidation of adsorbed Hg(II) mainly involved S_2^{2-} on MPy-600, from which it can be deduced that the percent of S_2^{2-} on MPy-600 has declined dramatically after Hg(II) adsorption [45].

As observed in Figure 9c,f, the O 1s at 530.2 eV ascribed to O^{2-} sufficiently exists on MPy-600. The binding energy appeared at 532.2 eV, which is owing to the O in the SO_4^{2-}. In addition, the binding energy of the O 1s spectrum at 533.4 eV is attributed to the O in the SiO_2. The higher spectral peaks are ascribed to iron diffusion, and when combined with minute amounts of oxygen, hydroxide, and water, ferric-hydroxides are formed [20]. Figure 9g exhibits the Hg 4f spectrum of MPy-600 after Hg(II) removal. The bond energies of Hg 4f7/2 at 100.9 eV and Hg 4f5/2 at 105 eV correspond to Hg^{2+} bonded with S^{2-}. Therefore, the above XPS conclusions confirmed that the HgS successfully formed on MPy-600. The other peak at 103 eV appeared, which could be due to the Si 2p of the SiO_2 [46]. Therefore, it suggested that the adsorption of Hg(II) by MPy-600 is mainly ascribed to the formation of HgS as the result of a chemical reaction.

3.8.4. SEM and TEM Analyses

The major composition and surface morphology of Hg(II) on the used MPy-600are characterized by SEM. Figure 10a shows the formation of a large number of new particles on the surface after Hg(II) adsorption, which has a nanometer-sized and plate-like shape. Compared to MPy-600 before adsorption, it can be derived that HgS is formed on the MPy-600. The content of Hg(II) in the EDS confirms the above discourse [16]. Figure 10b displays the SEM images with an EDX mapping of the MPy-600 after the reaction. In the Hg-EDX map, the distribution of the points which are characterized by the concentration of Hg(II) suggests the discrete Hg phase has formed. In the S-EDX map, the distribution of the points suggests the removal of Hg(II) is mainly ascribed to the form of HgS. SEM analysis of the used MPy-600 indicates that a large amount of HgS is produced of the material surface. Hence, it can be inferred that the reaction is driven by the solubility products of the sulfides and pyrrhotite, and the form of HgS is the principal factor of Hg(II) removal.

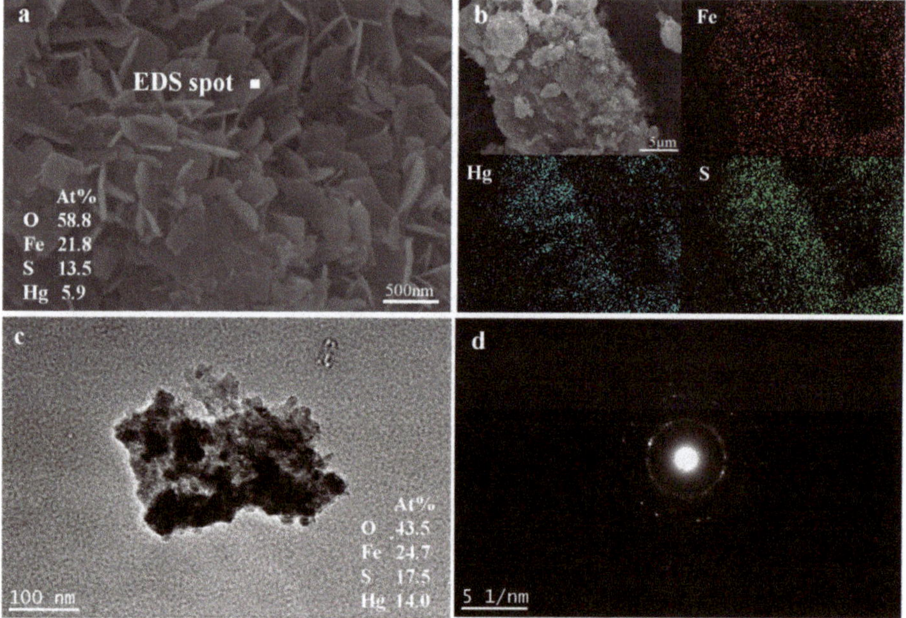

Figure 10. (**a**) SEM-EDS characterization. (**b**) SEM image with EDX mapping for different elements (Fe, S, and Hg). (**c**) TEM image, and (**d**) electron diffraction spot pattern of the used MPy-600.

As observed in Figure 10c, the diameter of the used MPy-600 particles is on a nanoscale. In addition, the EDS indicates that 14% Hg is detected. As shown in Figure 10d, the diffraction ring pattern of the specimen indicates that the material is polycrystalline [47]. The diffraction ring diameters are 3.3 Å, 2.45 Å, and 2.07 Å, which are indexed to the (101), (103), and (110) planes of cinnabar, respectively. In conclusion, HgS formed on theMPy-600 surface after the Hg(II) sorption. These results are in accordance with the proposed analysis. Therefore, it can be inferred that the primary mechanism on Hg(II) removal by MPy-600 is via a chemical reaction.

3.8.5. Product Analyses

The proportion of mercury precipitation in the products is evaluated by soaking the MPy-600 after reaction with 0.5% HCl to remove the Hg(OH)$_2$ and then measuring the content of residual Hg(II)in the leachate. According to the result, the content of Hg(OH)$_2$ accounts for 13% of the total Hg(II) adsorption capacity. Therefore, it demonstrates that the Hg(II) sorption was mainly ascribed by the form of the HgS.

4. Conclusions

In this study, batch experiments were conducted to explore the property of MPy derived from the decomposition of pyrite with the adsorption of Hg(II). The experiments showed that the removal of Hg(II) onto MPy-600 can be effectively simulated by the Langmuir model with the maximum adsorptivity of 166.67 mg/g. The sorption of Hg(II) was raised as the pH increased from 2–6. The sorption of Hg(II) onto MPy-600 increased as the ionic strength increased, reflecting that the outer surface complexation was mainly a controlling factor in the reaction. The sorption of Hg(II) ontoMPy-600 can be fitted effectively by a pseudo-second-order kinetic model ($R^2 = 0.995$). The fitting of the thermodynamic model of Hg(II) on MPy-600 indicated that the reaction process is endothermic, spontaneous, and driven by entropy. Hg(II)-loaded MPy-600 can be easily removed from solution using magnetic separation through its magnetic property. From all the studies, the removal of Hg(II) by MPy-600 is ascribed to the chemical reactions and electrostatic attraction between negative charges in MPy-600 and the positive charges of Hg(II). This research proves that MPy is a highly effective sorbent for Hg(II) sorption. Therefore, the high property of MPy-600 for Hg(II) removal indicates that MPy can play a crucial role as a suitable substance for the removal for Hg(II) on iron sulfide in environmental purification territory.

Author Contributions: Conceptualization, P.L. (Ping Lu) and T.C.; methodology, H.L.; validation, P.L. (Ping Li), S.P. and Y.Y.; formal analysis, P.L. (Ping Lu); investigation, P.L. (Ping Lu); resources, T.C.; data curation, H.L.; writing—original draft preparation, P.L. (Ping Lu); writing—review and editing, P.L. (Ping Lu); visualization, P.L. (Ping Lu); supervision, T.C.; project administration, H.L.; funding acquisition, T.C.

Funding: This work was funded by the National Natural Science Foundation of China (41772035, 41702043, 41402029).

Conflicts of Interest: The authors declare no conflict of interest.

References

1. Huang, J.; Wu, Z.; Chen, L.; Sun, Y. The sorption of Cd(II) and U(VI) on sepiolite: A combined experimental and modeling studies. *J. Mol. Liq.* **2015**, *209*, 706–712. [CrossRef]

2. Liu, T.; Xue, L.; Guo, X.; Liu, J.; Huang, Y.; Zheng, C. Mechanisms of Elemental Mercury Transformation on alpha-Fe$_2$O$_3$(001) Surface from Experimental and Theoretical Study: Influences of HCl, O$_2$, and SO$_2$. *Environ. Sci. Technol.* **2016**, *50*, 13585–13591. [CrossRef] [PubMed]

3. Huang, S.; Ma, C.; Liao, Y.; Min, C.; Du, P.; Jiang, Y. Removal of Mercury(II) from Aqueous Solutions by Adsorption on Poly(1-amino-5-chloroanthraquinone) Nanofibrils: Equilibrium, Kinetics, and Mechanism Studies. *J. Nanomater.* **2016**, *2016*, 1–11. [CrossRef]

4. Wajima, T.; Sugawara, K. Adsorption behaviors of mercury from aqueous solution using sulfur-impregnated adsorbent developed from coal. *Fuel Process. Technol.* **2011**, *92*, 1322–1327. [CrossRef]

5. Dash, S.; Chaudhuri, H.; Gupta, R.; Nair, U.G.; Sarkar, A. Fabrication and Application of Low-Cost Thiol Functionalized Coal Fly Ash for Selective Adsorption of Heavy Toxic Metal Ions from Water. *Ind. Eng. Chem. Res.* **2017**, *56*, 1461–1470. [CrossRef]

6. Wang, J.; Deng, B.; Chen, H.; Wang, X.; Zheng, J. Removal of Aqueous Hg(II) by Polyaniline: Sorption Characteristics and Mechanisms. *Environ. Sci. Technol.* **2009**, *43*, 5223–5228. [CrossRef]

7. Zhang, F.S.; Nriagu, J.O.; Itoh, H. Mercury removal from water using activated carbons derived from organic sewage sludge. *Water Res.* **2005**, *39*, 389. [CrossRef]

8. Cui, H.; Qian, Y.; Li, Q.; Wei, Z.; Zhai, J. Fast removal of Hg(II) ions from aqueous solution by amine-modified attapulgite. *Appl. Clay Sci.* **2013**, *72*, 84–90. [CrossRef]

9. Kadirvelu, K.; Kavipriya, M.; Karthika, C.; Vennilamani, N.; Pattabhi, S. Mercury (II) adsorption by activated carbon made from sago waste. *Carbon* **2004**, *42*, 745–752. [CrossRef]

10. Kadirvelu, K.; Kanmani, P.; Senthilkumar, P.; Subburam, V. Separation of Mercury(II) from Aqueous Solution by Adsorption onto an Activated Carbon Prepared from Eichhornia crassipes. *Adsorpt. Sci. Technol.* **2004**, *22*, 207–222. [CrossRef]

11. Tuzen, M.; Sari, A.; Mendil, D.; Soylak, M. Biosorptive removal of mercury(II) from aqueous solution using lichen (Xanthoparmelia conspersa) biomass: Kinetic and equilibrium studies. *J. Hazard. Mater.* **2009**, *169*, 263–270. [CrossRef] [PubMed]

12. Wang, X.; Lv, P.; Zou, H.; Li, Y.; Li, X.; Liao, Y. Synthesis of Poly(2-aminothiazole) for Selective Removal of Hg(II) in Aqueous Solutions. *Ind. Eng. Chem. Res.* **2016**, *55*, 4911–4918. [CrossRef]

13. Couture, R.M.; Rose, J.; Kumar, N.; Mitchell, K.; Wallschlager, D.; van Cappellen, P. Sorption of arsenite, arsenate, and thioarsenates to iron oxides and iron sulfides: A kinetic and spectroscopic investigation. *Environ. Sci. Technol.* **2013**, *47*, 5652–5659. [CrossRef] [PubMed]

14. Ma, B.; Kang, M.; Zheng, Z.; Chen, F.; Xie, J.; Charlet, L.; Liu, C. The reductive immobilization of aqueous Se(IV) by natural pyrrhotite. *J. Hazard. Mater.* **2014**, *276*, 422–432. [CrossRef] [PubMed]

15. Chen, T.; Yang, Y.; Li, P.; Liu, H.; Xie, J.; Xie, Q.; Zhan, X. Performance and characterization of calcined colloidal pyrite used for copper removal from aqueous solutions in a fixed bed column. *Int. J. Miner. Process.* **2014**, *130*, 82–87. [CrossRef]

16. Yang, Y.; Chen, T.; Li, P.; Liu, H.; Xie, J.; Xie, Q.; Zhan, X. Removal and Recovery of Cu and Pb from Single-Metal and Cu–Pb–Cd–Zn Multimetal Solutions by Modified Pyrite: Fixed-Bed Columns. *Ind. Eng. Chem. Res.* **2014**, *53*, 18180–18188. [CrossRef]

17. Chen, T.; Shi, Y.; Liu, H.; Chen, D.; Li, P.; Yang, Y.; Zhu, X. A novel way to prepare pyrrhotite and its performance on removal of phosphate from aqueous solution. *Desalin. Water Treat.* **2016**, *57*, 23864–23872. [CrossRef]

18. Chen, T.H.; Wang, J.Z.; Wang, J.; Xie, J.J.; Zhu, C.Z.; Zhan, X.M. Phosphorus removal from aqueous solutions containing low concentration of phosphate using pyrite calcinate sorbent. *Int. J. Environ. Sci. Technol.* **2014**, *12*, 885–892. [CrossRef]

19. Liu, H.; Zhu, Y.; Xu, B.; Li, P.; Sun, Y.; Chen, T. Mechanical investigation of U(VI) on pyrrhotite by batch, EXAFS and modeling techniques. *J. Hazard. Mater.* **2017**, *322*, 488–498. [CrossRef]

20. Liao, Y.; Chen, D.; Zou, S.; Xiong, S.; Xiao, X.; Dang, H.; Chen, T.; Yang, S. Recyclable Naturally Derived Magnetic Pyrrhotite for Elemental Mercury Recovery from Flue Gas. *Environ. Sci. Technol.* **2016**, *50*, 10562–10569. [CrossRef]

21. Li, R.; Kelly, C.; Keegan, R.; Xiao, L.; Morrison, L.; Zhan, X. Phosphorus removal from wastewater using natural pyrrhotite. *Colloids Surf. A Physicochem. Eng. Asp.* **2013**, *427*, 13–18. [CrossRef]

22. Janzen, M.P.; Nicholson, R.V.; Scharer, J.M. Pyrrhotite reaction kinetics: Reaction rates for oxidation by oxygen, ferric iron, and for nonoxidative dissolution. *Geochim. Et Cosmochim. Acta* **2000**, *64*, 1511–1522. [CrossRef]

23. Hu, G.; Dam-Johansen, K.; Wedel, S.; Hansen, J.P. Decomposition and oxidation of pyrite. *Prog. Energy Combust. Sci.* **2006**, *32*, 295–314. [CrossRef]

24. Sanhueza, V.; Kelm, U.; Cid, R. Synthesis of molecular sieves from Chilean kaolinites: 1. Synthesis of NaA type zeolites. *J. Chem. Technol. Biotechnol.* **1999**, *74*, 358–363. [CrossRef]

25. Widler, A.M.; Seward, T.M. The adsorption of gold(I) hydrosulphide complexes by iron sulphide surfaces. *Geochim. Et Cosmochim. Acta* **2002**, *66*, 383–402. [CrossRef]

26. Duan, S.; Liu, X.; Wang, Y.; Shao, D.; Alharbi, N.S.; Alsaedi, A.; Li, J. Highly efficient entrapment of U(VI) by using porous magnetic $Ni_{0.6}Fe_{2.4}O_4$ micro-particles as the adsorbent. *J. Taiwan Inst. Chem. Eng.* **2016**, *65*, 367–377. [CrossRef]

27. Liu, P.; Yuan, N.; Xiong, W.; Wu, H.; Pan, D.; Wu, W. Removal of Nickel(II) from Aqueous Solutions Using Synthesized β-Zeolite and Its Ethylenediamine Derivative. *Ind. Eng. Chem. Res.* **2017**, *56*, 3067–3076. [CrossRef]

28. Sun, Y.; Lou, Z.; Yu, J.; Zhou, X.; Lv, D.; Zhou, J. Immobilization of mercury (II) from aqueous solution using Al_2O_3-supported nanoscale FeS. *Chem. Eng. J.* **2017**, *323*, 483–491. [CrossRef]

29. Jin, Z.; Wang, X.; Sun, Y.; Ai, Y.; Wang, X. Adsorption of 4-n-Nonylphenol and Bisphenol-A on Magnetic Reduced Graphene Oxides: A Combined Experimental and Theoretical Studies. *Environ. Sci. Technol.* **2015**, *49*, 9168–9175. [CrossRef]

30. Weng, C.H.; Pan, Y.-F. Adsorption characteristics of methylene blue from aqueous solution by sludge ash. *Colloids Surf. A Physicochem. Eng. Asp.* **2006**, *274*, 154–162. [CrossRef]

31. Liu, N.; Charrua, A.B.; Weng, C.H.; Yuan, X.L.; Ding, F. Characterization of biochars derived from agriculture wastes and their adsorptive removal of atrazine from aqueous solution: A comparative study. *Bioresour. Technol.* **2015**, *198*, 55. [CrossRef] [PubMed]

32. Xu, Y.; Zeng, X.; Zhang, B.; Zhu, X.; Zhou, M.; Zou, R.; Sun, P.; Luo, G.; Yao, H. Experiment and Kinetic Study of Elemental Mercury Adsorption over a Novel Chlorinated Sorbent Derived from Coal and Waste Polyvinyl Chloride. *Energy Fuels* **2016**, *30*, 10635–10642. [CrossRef]

33. Deze, E.G.; Papageorgiou, S.K.; Favvas, E.P.; Katsaros, F.K. Porous alginate aerogel beads for effective and rapid heavy metal sorption from aqueous solutions: Effect of porosity in Cu^{2+} and Cd^{2+} ion sorption. *Chem. Eng. J.* **2012**, *209*, 537–546. [CrossRef]

34. Yang, S.; Yan, N.; Guo, Y.; Wu, D.; He, H.; Qu, Z.; Li, J.; Zhou, Q.; Jia, J. Gaseous elemental mercury capture from flue gas using magnetic nanosized $(Fe_{3-x}Mn_x)1\text{-delta}O_4$. *Environ. Sci. Technol.* **2011**, *45*, 1540–1546. [CrossRef] [PubMed]

35. Andac, M.; Mirel, S.; Senel, S.; Say, R.; Ersoz, A.; Denizli, A. Ion-imprinted beads for molecular recognition based mercury removal from human serum. *Int. J. Biol. Macromol.* **2007**, *40*, 159–166. [CrossRef] [PubMed]

36. Zhu, Y.; Chen, T.; Liu, H.; Xu, B.; Xie, J. Kinetics and thermodynamics of Eu(III) and U(VI) adsorption onto palygorskite. *J. Mol. Liq.* **2016**, *219*, 272–278. [CrossRef]

37. Gong, Y.; Liu, Y.; Xiong, Z.; Zhao, D. Immobilization of mercury by carboxymethyl cellulose stabilized iron sulfide nanoparticles: Reaction mechanisms and effects of stabilizer and water chemistry. *Environ. Sci. Technol.* **2014**, *48*, 3986–3994. [CrossRef] [PubMed]

38. Wang, A.; Jolliff, B.L.; Haskin, L.A. Raman spectroscopic characterization of a Martian SNC meteorite: Zagami. *J. Geophys. Res. Planets* **1999**, *104*, 8509–8519. [CrossRef]

39. Frost, R.L.; Martens, W.N.; Kloprogge, J.T. Raman spectroscopic study of cinnabar (HgS), realgar (As_4S_4), and orpiment (As_2S_3) at 298 and 77K. *Neues Jahrb. Für Miner. Mon.* **2002**, *2002*, 469–480. [CrossRef]

40. Szuskiewicz, B.W.W. Raman spectroscopy of cubic HgS Fe. *Acta Phys. Pol. A* **1995**, *87*, 416–418.

41. Sun, Y.; Yang, S.; Chen, Y.; Ding, C.; Cheng, W.; Wang, X. Tuning the chemistry of graphene oxides by a sonochemical approach: Application of adsorption properties. *Environ. Sci. Technol.* **2015**, *49*, 4255–4262. [CrossRef] [PubMed]

42. Bone, S.E.; Bargar, J.R.; Sposito, G. Mackinawite (FeS) reduces mercury(II) under sulfidic conditions. *Environ. Sci. Technol.* **2014**, *48*, 10681–10689. [CrossRef] [PubMed]

43. Pratt, A.R.; Muir, I.J.; Nesbitt, H.W. X-ray photoelectron and Auger electron spectroscopic studies of pyrrhotite and mechanism of air oxidation. *Geochim. Et Cosmochim. Acta* **1994**, *58*, 827–841. [CrossRef]

44. Knipe, S.W.; Mycroft, J.R.; Pratt, A.R.; Nesbitt, H.W.; Bancroft, G.M. X-ray photoelectron spectroscopic study of water adsorption on iron sulphide minerals. *Geochim. Et Cosmochim. Acta* **1995**, *59*, 1079–1090. [CrossRef]

45. Philippe, B.; Pascale, B.G.; Marc, A.; Renaud, R.; Jean, J.E. XPS and XAS Study of the Sorption of Hg(II) onto Pyrite. *Langmuir* **2001**, *17*, 3970–3979.

46. Bootharaju, M.S.; Pradeep, T. Uptake of Toxic Metal Ions from Water by Naked and Monolayer Protected Silver Nanoparticles: An X-ray Photoelectron Spectroscopic Investigation. *J. Phys. Chem. C* **2010**, *114*, 8328–8336. [CrossRef]

47. Temgire, M.K.; Joshi, S.S. Optical and structural studies of silver nanoparticles. *Radiat. Phys. Chem.* **2004**, *71*, 1039–1044. [CrossRef]

Article

One-Step Synthesis of Hydroxysodalite Using Natural Bentonite at Moderate Temperatures

Bo Liu [1,*], **Hongjuan Sun** [2,*], **Tongjiang Peng** [2] **and Qian He** [2]

1   School of National Defence Science and Technology, Southwest University of Science and Technology, Mianyang 621010, Sichuan, China
2   Key Laboratory of Ministry of Education for Solid Waste Treatment and Resource Recycle, Southwest University of Science and Technology, Mianyang 621010, Sichuan, China; tjpeng@swust.edu.cn (T.P.); heqiantq@163.com (Q.H.)
*   Correspondence: liubo@swust.edu.cn (B.L.); sunhongjuan@swust.edu.cn (H.S.); Tel.: +86-08166089471 (B.L.)

Received: 19 September 2018; Accepted: 6 November 2018; Published: 9 November 2018

Abstract: Ca-bentonite was used as the feedstock material for the synthesis of hydroxysodalite due to its high Al, Si content, good chemical reactivity, and natural abundance. A one-step method is proposed here to fabricate hydroxysodalite in a water bath at moderate temperature. The effects of the Na/Si molar ratio, Si/Al molar ratio, reaction time, and reaction temperature on the synthesis of hydroxysodalite have been systematically investigated here. The crystallizing phases and morphology of the synthetic products were characterized using X-ray diffraction (XRD) and scanning electron microscopy (SEM), respectively. The results showed that the Na/Si molar ratio and reaction temperature both played important roles in controlling the degree of crystallinity of the synthetic hydroxysodalite. The Si/Al molar ratio and reaction time both affect the purity of the synthetic hydroxysodalite. Optimum conditions for synthesizing hydroxysodalite using a one-step water-bath method at moderate temperature are as follows: a Na/Si molar ratio of 12, a Si/Al molar ratio of 1.0, a reaction temperature of 90 °C, and a reaction time of 12 h.

Keywords: preparation; hydroxysodalite; bentonite; alkali-activation; water-bath

---

1. Introduction

Sodalite is a microporous tectosilicate with the chemical formula $Na_8[AlSiO_4]_6(Cl, OH, I)_2 \cdot nH_2O$ [1]. It has a framework with a cubic symmetry that is constructed from the vertex-linking of SiO_4 and AlO_4 tetrahedrons into four- and six-membered oxygen-rings that form β cages [2]. Owing to its framework flexibility, sodalite can accommodate anions with different geometries, leading to the formation of different types of sodalities, which are each suited to different applications [3,4].

An interesting member of the sodalite family is hydroxysodalite, a porous hydrophilic functional mineral with the same framework structure as sodalite. When inserting non-stoichiometric amounts of metal ions into the β cages, hydroxysodalite could be invoked as the photochromic material [5,6]. Due to its small aperture size (0.28 nm), hydroxysodalite can be used to separate small molecules such as helium, hydrogen and water [7]. In addition, due to its desirable antioxidation properties, moisture resistance, and chemical stability, hydroxysodalite can also serve as a substrate for catalyst carriers and membrane materials [8,9]. However, the storage of natural hydroxysodalite is quite low, and it is difficult to obtain sufficient amounts to meet the demands of industry. Therefore, the synthesis of hydroxysodalite is of great practical significance.

There are numerous literature reports on the synthesis of hydroxysodalite by a variety of different methods. Xu et al. [10] used aluminum foil and a silica sol as the raw materials to synthesize a high-quality, pure hydroxysodalite membrane via a microwave-assisted hydrothermal method. Nabavi et al. [11] obtained a hydroxysodalite membrane via a hydrothermal method using sodium aluminate and silica foam that was then used for the separation of H_2/CH_4. Jiang et al. [12] synthesized sub-micron crystals of hydroxysodalite from natural palygorskite clay via a two-step approach. Passos et al. [13] used kaolin as a source of aluminium and silicon to perform a hydrothermal synthesis of hydroxysodalite, and discussed the optimization of the synthesis conditions. Golbad et al. [14] employed fly ash to synthesize hydroxysodalite using a hydrothermal activation process for the removal of lead ions from water. Based on the above research, the chemical materials or organic additives that were used were not economically-viable for industrial-scale production. The use of kaolin minerals or fly ash for synthesis of hydroxysodalite requires further activation with calcination and acid activation. Therefore, it is urgent to seek a straightforward and economical method to synthesize hydroxysodalite.

Bentonite is considered to have a high potential for use as an industrial raw material due to its large-scale availability and good physicochemical characteristics [15]. Bentonite is essentially composed of Al_2O_3 and SiO_2, in addition to other oxides such as MgO, Fe_2O_3, Na_2O, and CaO. The relatively high Al_2O_3 and SiO_2 content in bentonite makes it an inexpensive raw material for the synthesis of hydroxysodalite. In this paper, a one-step water-bath method at moderate temperatures was used to obtain the hydroxysodalite, and batch experiments were conducted. This research focused on evaluating how the Na/Si molar ratio, Si/Al molar ratio, reaction time, and reaction temperature each affected the synthesis of hydroxysodalite.

2. Experiments and Methods

2.1. Materials

Raw bentonite collected from bentonite deposits in the city of Fuxin, China was used as a starting material for synthesis of hydroxysodalite. From the XRD pattern (Figure 1), the essential mineral components were found to be Ca-montmorillonite, cristobalite, and albite. The SEM images (Figure 2) revealed that the dominant morphological forms occurring in bentonite were primarily flake accumulations. The major oxides were found to be SiO_2, Al_2O_3, Fe_2O_3, MgO, CaO, Na_2O, and TiO_2 and are listed in Table 1.

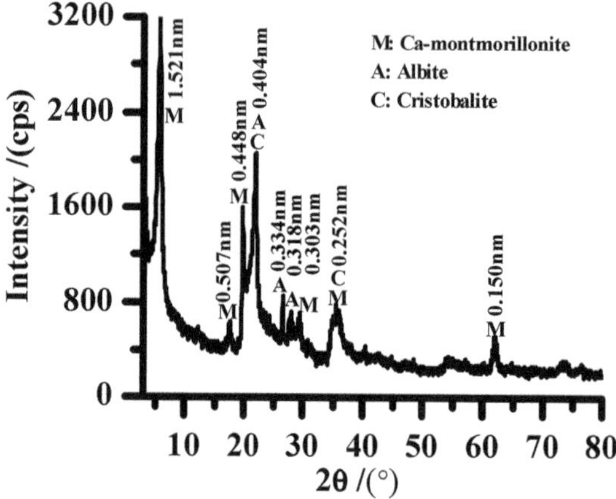

Figure 1. XRD of bentonite samples.

Table 1. X-ray fluorescence (XRF) data of bentonite samples after normalizing (wt. %).

Sample	SiO$_2$	Al$_2$O$_3$	Fe$_2$O$_3$	MgO	CaO	Na$_2$O	TiO$_2$	K$_2$O	LOI
Bentonite	67.73	11.54	1.75	1.64	1.74	0.01	0.09	0.28	15.30

Figure 2. SEM of bentonite samples.

The bentonite samples were crushed and dried at 100 °C overnight and then ground to an average particle size less than 150 microns as determined according to the standard sieving analysis. According to the data in Table 1, the molar ratio of Si/Al in the bentonite samples is about 4.98, which is higher than that of the theoretical Si/Al molar ratio of hydroxysodalite. Therefore, additional $NaAlO_2$ was added.

All reagents such as NaOH and $NaAlO_2$ used for synthesis were all analytical purity and were purchased from the Chengdu Kelong Chemical Reagent Company (Chengdu, China). They were used as received without any further purification. The water used in the experiment was ultrapure water with a resistivity of 18.25 mΩ·cm.

2.2. Synthesis

Four groups of bentonite-based hydroxysodalite samples were synthesized in this work with different Na/Si molar ratios, Si/Al molar ratios, reaction times, and reaction temperatures, respectively. The hydroxysodalite was synthesized via the route shown in Figure 3. One gram of bentonite was added to 35 mL of ultrapure water. Then, certain dosages of NaOH and $NaAlO_2$ were added to the mixture with continuous stirring for 3 h at a certain temperature, according to the synthetic scheme (Table 1). After a certain reaction time at a constant temperature, the sample was filtered and washed repeatedly with ultrapure water until the solution pH was less than 8. Finally, the washed sample was dried at 90 °C for 24 h. Each experiment was conducted in triplicate.

Figure 3. The flow diagram for the synthesis of hydroxysodalite.

2.3. Characterization

X-ray diffraction (XRD) analysis of the samples was conducted using a PANalytical X'Pert PrO multifunctional powder diffractometer with an X'Celerator scintillation detector (PANalytical B.V., Almelo, The Netherlands) operating at 30 mA and 40 kV with Cu Kα radiation (λ = 0.154060 nm), with a recording range of 3–80°. The crystallinity was estimated by comparing the sum of typical characteristic peak intensities (d_{110} at 13.89°, d_{211} at 24.30°, d_{310} at 31.74°, d_{222} at 34.83°, and d_{330} at 43.06°) of samples and the reference hydroxysodalite synthesized with analytically pure sodium silicate and sodium aluminate. The crystallinity was calculated by Equation (1):

$$\text{Crystallinity} = \frac{\sum \text{peak intensities of sample}}{\sum \text{peak intensities of reference}} \times 100\% \tag{1}$$

The chemical compositions of the sample were determined using an Axios Panalytical X-ray fluorescence (XRF) spectrometer (PANalytical B.V., Almelo, The Netherlands) fitted with an X-ray tube containing Rh anode. The maximum power of the X-ray tube was 2.4 kW, with a maximum voltage of 60 kV and a maximum current of 100 mA. The samples were prepared by mixing them with a flux material and melting into glass beads. To determine the loss on ignition (LOI), the samples were heated to 1000 °C for 3 h. The morphology of the samples was examined using a ZEISS Ultra 55 scanning electron microscope (SEM). Prior to SEM analysis, the surface of the samples was sputtered with gold. The thermal analysis (TG) of the materials was performed using a TA SDT-Q6000 thermal analyzer by flowing nitrogen and heating sample from room temperature to 1000 °C at a ramp rate of 10 °C/min. Raman spectra were obtained using a RM 2000 Renishaw InVia Raman spectrometer with a scanning range of 1000–4000 cm^{-1}. Fourier transform infrared (FTIR) spectra were obtained using a Nicolet-5700 Fourier transform infrared spectrometer. The KBr pellet compression method was used, and the scanning range was 400–4000 cm^{-1}.

3. Results and Discussion

3.1. Effect of Na/Si Molar Ratio

To determine the influence of the Na/Si molar ratio on the synthesis of hydroxysodalite, the Si/Al molar ratio was chosen to be 1.0 at a reaction temperature of 90 °C, and a reaction time of 12 h. The XRD patterns of the obtained samples and the crystallinity of hydroxysodalite are shown in Figures 4 and 5, respectively.

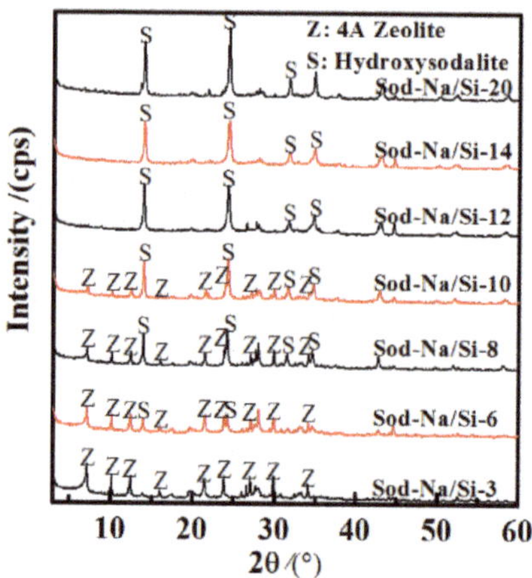

Figure 4. XRD of synthetic products obtained from different Na/Si molar ratios.

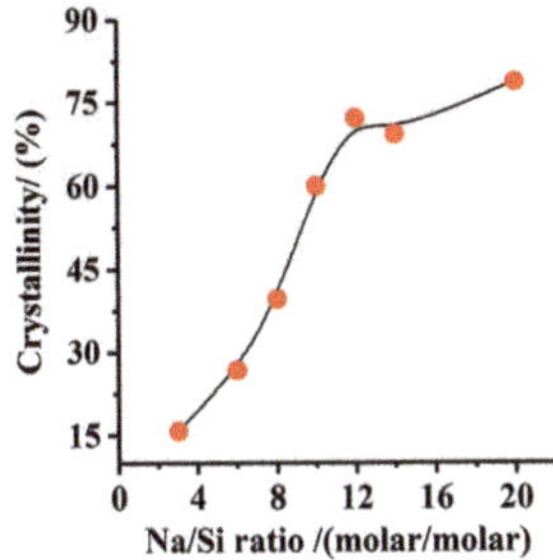

Figure 5. Crystallinity of synthetic products obtained from different Na/Si molar ratios.

Due to the fixed Si/Al molar ratio and water content, the Na/Si molar ratio was modified by adding different amounts of NaOH (Table 2). Therefore, the Na/Si molar ratio also represents the alkalinity of the reaction system, which influenced the activation of bentonite and controlled the polymerization of SiO_4 and AlO_4 monomer in the reaction system [16–18]. When the Na/Si molar ratio was 3, the diffraction peaks of Ca-montmorillonite, cristobalite, and albite in bentonite were remarkably reduced, and the main crystalline phase in the product was 4A zeolite. 4A zeolite was a type of crystalline aluminosilicate with uniform micro-channels and it had a pore size of 4 Å [19]. The pore framework of 4A zeolite was composed of sodalite cages and Na^+, which was used to compensate the negative charge of the skeleton [20,21]. When the Na/Si molar ratio was increased to 6, the diffraction peaks of hydroxysodalite appeared. Further increasing the Na/Si molar ratio to 10, hydroxysodalite became the predominant phase in the obtained sample, indicating that 4A zeolite was transformed to hydroxysodalite as the amount of NaOH increased, which can be ascribed to the high structural stability and framework density. When the Na/Si molar ratio reached 12, the diffraction peaks of 4A zeolite vanished, and only the diffraction peaks of hydroxysodalite were observed, indicating that the strong alkalinity was helpful for the formation of hydroxysodalite. As shown in Figure 5, the crystallinity of hydroxysodalite was gradually increased as the Na/Si molar ratio increased, which is similar to a result reported by Hu et al. [6].

Table 2. Synthetic scheme of hydroxysodalite.

NO.	Bentonite (g)	NaOH (g)	NaAlO$_2$ (g)	T (°C)	t (h)	H$_2$O (mL)
Sod-Na/Si-3	1	0.99	0.74	90	12	35
Sod-Na/Si-6	1	2.35	0.74	90	12	35
Sod-Na/Si-8	1	3.25	0.74	90	12	35
Sod-Na/Si-10	1	4.15	0.74	90	12	35
Sod-Na/Si-12	1	5.06	0.74	90	12	35
Sod-Na/Si-14	1	5.96	0.74	90	12	35
Sod-Na/Si-20	1	8.67	0.74	90	12	35
Sod-Si/Al-0.5	1	4.61	1.67	90	12	35
Sod-Si/Al-0.75	1	4.91	1.05	90	12	35
Sod-Si/Al-1.0	1	5.06	0.74	90	12	35
Sod-Si/Al-1.5	1	5.21	0.43	90	12	35
Sod-Si/Al-2.0	1	5.28	0.28	90	12	35
Sod-T-60	1	5.06	0.74	60	12	35
Sod-T-70	1	5.06	0.74	70	12	35
Sod-T-80	1	5.06	0.74	80	12	35
Sod-T-90	1	5.06	0.74	90	12	35
Sod-T-100	1	5.06	0.74	100	12	35
Sod-t-4	1	5.06	0.74	90	4	35
Sod-t-8	1	5.06	0.74	90	8	35
Sod-t-10	1	5.06	0.74	90	10	35
Sod-t-12	1	5.06	0.74	90	12	35
Sod-t-16	1	5.06	0.74	90	16	35
Sod-t-20	1	5.06	0.74	90	20	35
Sod-t-24	1	5.06	0.74	90	24	35

Figure 6 displays the SEM images of the products synthesized under different Na/Si molar ratios. As shown in Figure 6a, the morphology of the products synthesized under a Na/Si molar ratio of 3 was mainly composed of cubes with a particle size of 1–2 μm, which conformed to the crystal structure of 4A zeolite. The generation of 4A zeolite was attributed to the dissolution of the bentonite and the polymerization of SiO_4 and AlO_4 monomers which occurred under the weakly alkaline conditions [22]. When the Na/Si molar ratio reached 6, some spherical agglomerates of hydroxysodalite started to form on the surface of the 4A zeolite, as shown in Figure 6b, which was in agreement with XRD results. Notably, when the Na/Si molar ratio reached 10, spherical agglomerates with less imperfect cubes appeared, as shown in Figure 6c, suggesting that the main phase of the product was hydroxysodalite, with similar trends having been reported in the literature [23,24]. By further increasing the Na/Si molar ratio to 12, nearly all of the pseudo-cubic grains of 4A zeolite were destroyed, and only spherical agglomerates appeared (Figure 6d). Fan et al. [25] reported that nano-sized hydroxysodalite crystals could be synthesized at high NaOH concentrations in the absence of organic additives. Based on the aforementioned analysis, it was considered that low alkalinity was not conducive to the crystallinity of hydroxysodalite, and high alkalinity could promote the transformation of 4A zeolite (metastable phase) into hydroxysodalite (stable phase). Therefore, relatively pure hydroxysodalite could be synthesized when the Na/Si molar ratio was fixed to 12.

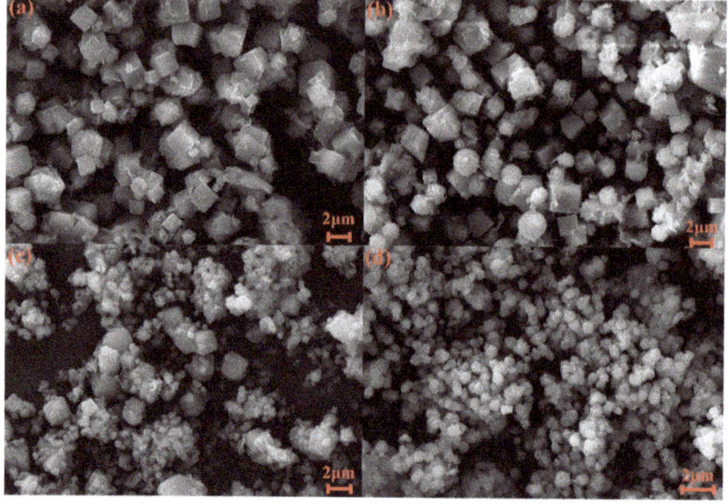

Figure 6. SEM of synthetic products obtained from different Na/Si molar ratios. (**a**) Na/Si = 3; (**b**) Na/Si = 6; (**c**) Na/Si = 10; (**d**) Na/Si = 12.

3.2. Effect of Si/Al Molar Ratio

As indicated by the XRD patterns of the synthetic products with different Si/Al molar ratios in Figure 7, hydroxysodalite could be synthesized when the Si/Al molar ratio was 0.5 or 0.75. However, an obvious 4A zeolite phase was detected, which indicated that the synthesized products from bentonite were a mixture of 4A zeolite and hydroxysodalite. When the Si/Al molar ratio reached the ratio of theoretical hydroxysodalite (Si/Al = 1), the diffraction peaks of 4A zeolite disappeared, and hydroxysodalite became the predominant phase in the synthetic products. Hydroxysodalite could be also synthesized by further increasing Si/Al molar ratio to 1.5, but a small number of diffraction peaks from 4A zeolite was detected, suggesting that high Si/Al molar ratios were unfavorable for the crystalline phase of hydroxysodalite. The crystallinity of the synthetic products in Figure 8 showed that the crystallinity of the prepared hydroxysodalite reached the optimal yield of around 72.1% with a Si/Al molar ratio of 1.0.

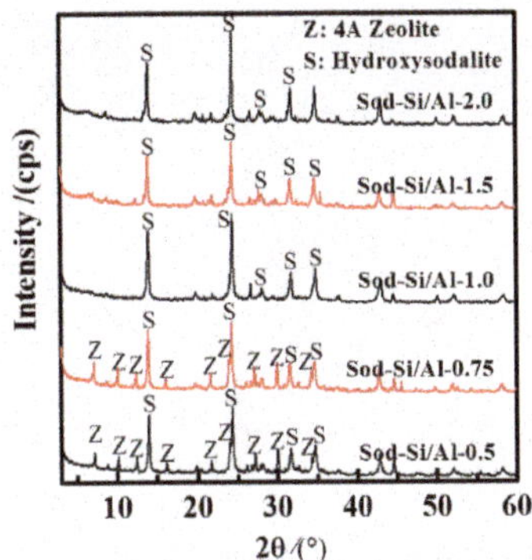

Figure 7. XRD of synthetic products obtained from different Si/Al molar ratios.

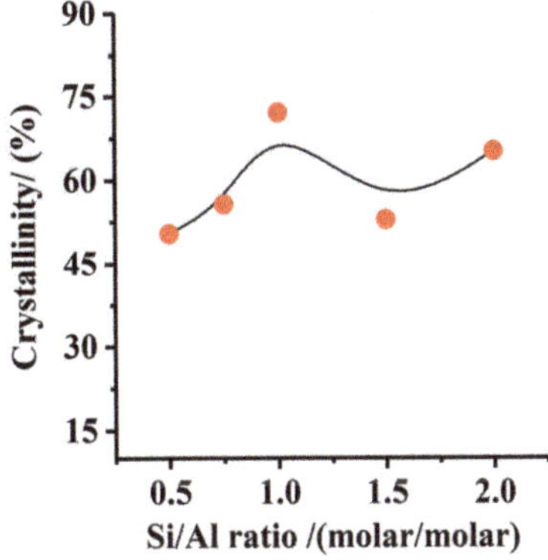

Figure 8. Crystallinity of synthetic products obtained from different Si/Al molar ratios.

Figure 9 presents the SEM images of the synthetic products obtained from different Si/Al molar ratios. When the Si/Al molar ratio was 0.5, the synthetic product exhibited both pseudo-cubic agglomerates (4A zeolite phase) and small spherical agglomerates (hydroxysodalite), shown in Figure 9a, which is in agreement with XRD results (Figure 7). When the Si/Al molar ratio reached 0.75, pseudo-cubic agglomerates were collapsed and decreased with the addition of the spherical agglomerates. By further increasing the Si/Al molar ratio to 1.0, most of the pseudo-cubic 4A zeolite was converted to hydroxysodalite with a spherical morphology (Figure 9c). However, when the Si/Al molar ratio exceeded 1, the spherical particle agglomerates of hydroxysodalite were not uniform, indicating a lower crystallinity of hydroxysodalite, which is also confirmed by the crystallinity results shown in Figure 8.

3.3. Effect of Reaction Temperature

The reaction temperature has a strong influence on the nucleation and the crystal growth process of hydroxysodalite [26]. To determine the influence of temperature, individual synthetic experiments were run at 60, 70, 80, 90, and 100 °C, respectively. The Na/Si molar ratio was chosen to be 12, at a reaction time of 12 h, and the Si/Al molar ratio was chosen to be 1.0. The XRD patterns of the synthetic products obtained from reactions run at these different temperatures are displayed in Figure 10.

Figure 9. SEM of synthetic products obtained from different Si/Al molar ratios. (**a**) Si/Al = 0.5; (**b**) Si/Al = 0.75; (**c**) Si/Al = 1.0; (**d**) Si/Al = 1.5.

As shown in Figure 10, a well-crystallized 4A zeolite was formed when the reaction temperature was 60 °C. With an increase of temperature, the intensity of the diffraction peaks of hydroxysodalite gradually increased, which agrees with a similar result reported by Li et al. [27]. Increasing the temperature to 90 °C caused the diffraction peaks of 4A zeolite to vanish, leaving only the diffraction peaks of hydroxysodalite, suggesting that relatively pure hydroxysodalite was prepared at this temperature. In addition, the crystallinity of hydroxysodalite gradually increased as the reaction temperature increased, as shown in Figure 11. The crystallinity of the prepared hydroxysodalite reached the optimal yield with reaction temperature of 90 °C.

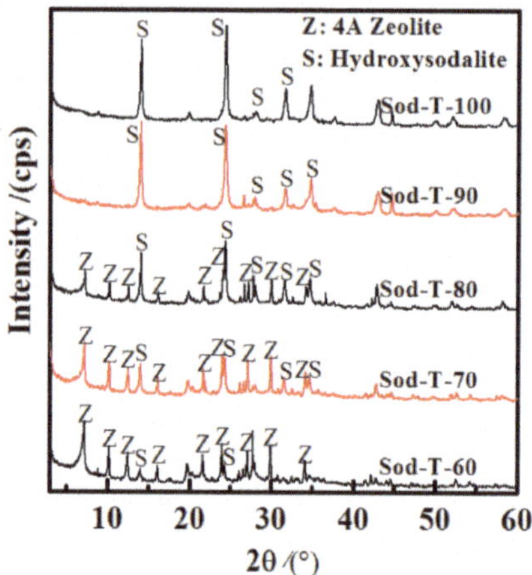

Figure 10. XRD of synthetic products obtained from different reaction temperatures.

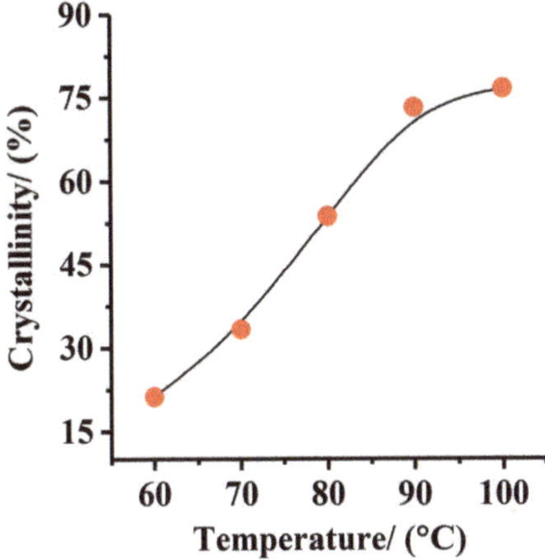

Figure 11. Crystallinity of synthetic products obtained from different reaction temperatures.

Figure 12 shows the SEM images of the synthetic products obtained at different reaction temperatures. When the reaction temperature was 60 °C, a greater amount of 4A zeolite with a pseudo-cubic morphology was identified in Figure 12a. As the temperature increased, the pseudo-cubic morphology was reduced, and more spherical aggregates began to appear, as shown in Figure 12b. This could be attributed to the 4A zeolite having a lower thermodynamic stability compared to hydroxysodalite, and increasing the reaction temperature appears to promote the transformation of 4A zeolite into the more thermodynamically stable hydroxysodalite [28]. Further increasing the reaction temperature to 90 °C caused the morphology of the synthetic product to exhibit uniform spherical aggregates with an average particle size of 2 μm, as shown in Figure 12d.

Figure 12. SEM of synthetic products obtained from different reaction temperatures. (**a**) reaction temperature = 60 °C; (**b**) reaction temperature = 70 °C; (**c**) reaction temperature = 80 °C; (**d**) reaction temperature = 90 °C.

3.4. Effect of Reaction Time

To determine the influence of reaction time on hydroxysodalite synthesized from bentonite, the samples were synthesized with reaction times of 4, 8, 10, 12, 16, 20, and 24 h. The Na/Si molar ratio was chosen to be 12, at a reaction temperature of 90 °C, and the Si/Al molar ratio was chosen to be 1.0. The XRD patterns of the synthetic products obtained from different reaction times are displayed in Figure 13.

It was found that hydroxysodalite was the predominant crystalline phase with a small amount of the 4A zeolite phase in the product synthesized at 90 °C for 4 h. Further increasing the reaction time to 8 h, the diffraction peaks of 4A zeolite disappeared, and only diffraction peaks ascribed to hydroxysodalite remained in the XRD patterns. In addition, the crystallinity of synthetic hydroxysodalite increased as the reaction time increased (Figure 14).

Figure 15 shows the SEM images of the synthetic products obtained from the different reaction times. The final products exhibited spherical aggregates and a small amount of pseudo-cubic aggregates as shown in Figure 16a when the reaction time was 4 h. When the reaction time was prolonged, the amount of pseudo-cubic crystals decreased, along with a significant increase of spherical aggregates (Figure 15b–d), which was consistent with the XRD results. Although prolonging the reaction time could potentially further improve the crystallinity of hydroxysodalite, the optimal reaction time for synthesizing hydroxysodalite was fixed to 12 h based on economic cost.

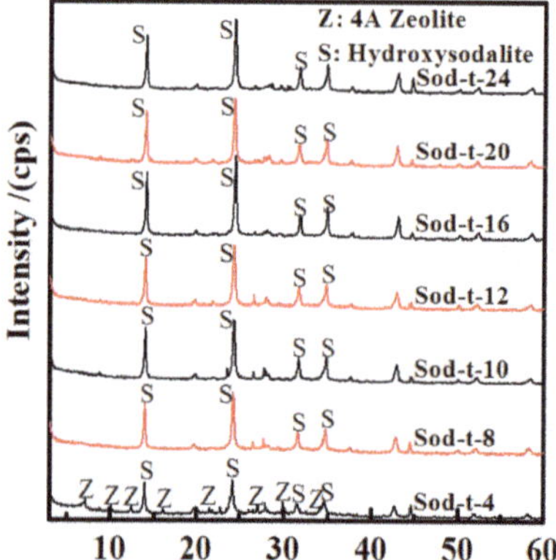

Figure 13. XRD of synthetic products obtained from different reaction times.

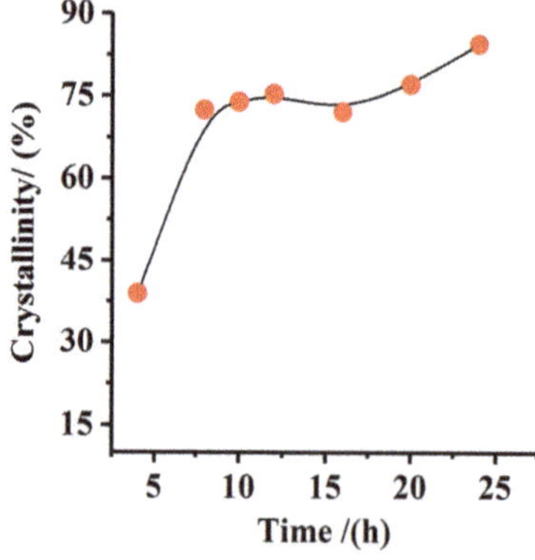

Figure 14. Crystallinity of synthetic products obtained from different reaction times.

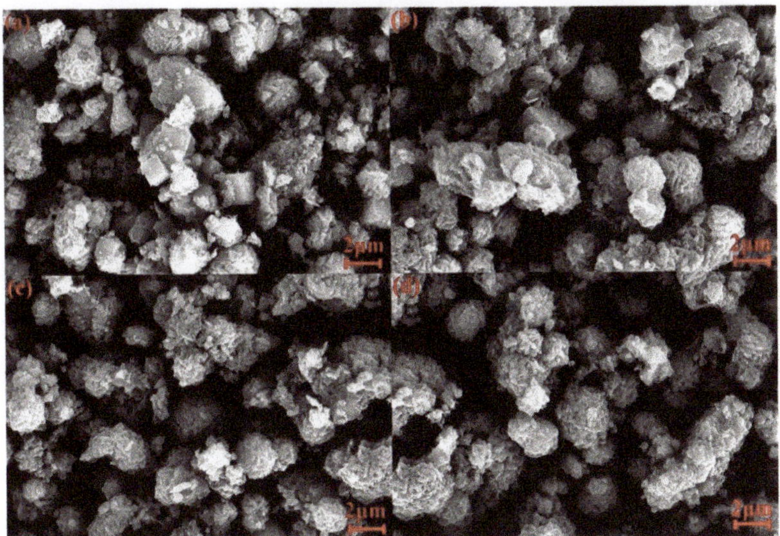

Figure 15. SEM of synthetic products obtained from different reaction times. (**a**) reaction time = 4 h; (**b**) reaction time = 8 h; (**c**) reaction time = 10 h; (**d**) reaction time = 12 h.

3.5. Characterization of Hydroxysodalite in Optimum Conditions

The FTIR spectrum of hydroxysodalite is displayed in Figure 16a. The bands at 437 cm^{-1} and 464 cm^{-1} are attributed to the bending vibration of O-T-O (T = Al, Si). The bands at 664 cm^{-1} and 732 cm^{-1} are caused by the symmetric stretching of T-O-T, while the band at 989 cm^{-1} is caused by the antisymmetric stretching of T-O-T. The bands at 3644, 3434, 1643, and 1629 cm^{-1} are assigned to the stretching vibration of OH and bending vibration of water in the synthetic hydroxysodalite. This FTIR spectrum is in good agreement with previously-reported spectra [4]. In addition, the Raman spectrum (Figure 16b) of the synthetic product is consistent with the experimental and theoretical results of hydroxysodalite [29]. The combined FTIR and Raman results confirm that the hydroxysodalite was successfully synthesized.

Thermal analysis was performed, and, in Figure 16c, a total mass loss of 11.79% was observed in three stages. The mass loss of 0.93% in the range from room temperature to 85 °C was attributed to the elimination of adsorbed water. The mass loss of 7.52% in the range of 85 °C to 445 °C corresponded to the elimination of structural water molecules of hydroxysodalite, indicating the formation of anhydrous hydroxysodalite Na$_8$[AlSiO$_4$]$_6$(OH)$_2$ [30]. Above 445 °C, the mass loss was 3.24%, which could be due to the decomposition of hydroxysodalite [31,32].

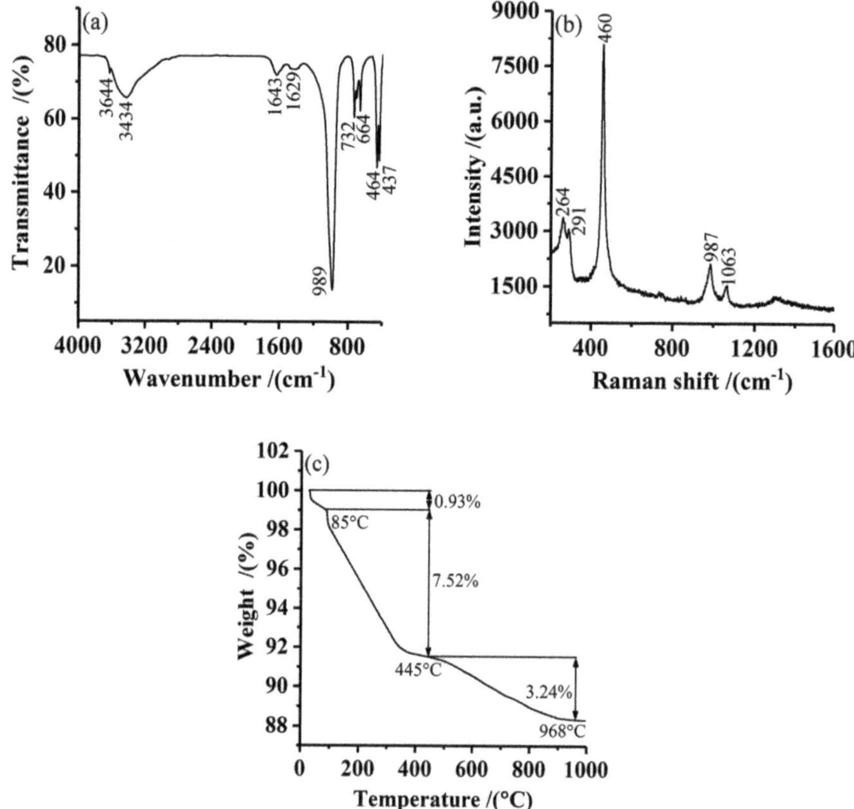

Figure 16. Characterization of synthetic hydroxysodalite. (**a**) Fourier transform infrared (FTIR) spectrum; (**b**) Raman spectrum; (**c**) Thermogravimetry (TG) curve.

4. Conclusions

Hydroxysodalite was synthesized via a one-step water-bath method at a moderate temperature by controlling the synthesis parameters such as the Na/Si molar ratio, Si/Al molar ratio, reaction time, and reaction temperature. Hydroxysodalite could be synthesized at a high Na/Si molar ratio and a high reaction temperature. On the contrary, a low Na/Si molar ratio at a temperature of 90 °C was beneficial for the synthesis of 4A zeolite. Both the Na/Si molar ratio and reaction temperature played important roles in controlling formation of the crystalline phase of the synthetic hydroxysodalite. The Si/Al molar ratio and reaction time were shown to influence the purity of the synthetic hydroxysodalite. The optimum synthesis conditions were determined to be a Na/Si ratio of 12, an Si/Al ratio of 1.0, a reaction temperature of 90 °C, and a reaction time of 12 h. The one-step water-bath method, when performed at moderate temperatures, can potentially be a promising method for the synthesis of other zeolites derived from bentonite clay.

Author Contributions: B.L. and H.S. conceived and designed the experiments; Q.H. performed the experiments; B.L. analyzed the data; T.P. contributed reagents and materials; B.L. and Q.H. wrote the paper.

Funding: This research was funded by the doctoral foundation of Southwest University of Science and Technology (Grant No. 18zx7104), the National Natural Science Foundation of China (Grant: 41372052) and the Sichuan innovation team project (Grant: 14TD0012).

Conflicts of Interest: The authors declare no conflict of interest.

References

1. Bibby, D.M.; Dale, M.P. Synthesis of silica-sodalite from non-aqueous systems. *Nature* **1985**, *317*, 157. [CrossRef]
2. Li, J.; Zeng, X.; Yang, X.; Wang, C.; Luo, X. Synthesis of pure sodalite with wool ball morphology from alkali fusion kaolin. *Mater. Lett.* **2015**, *161*, 157–159. [CrossRef]
3. Yao, J.; Zhang, L.; Wang, H. Synthesis of nanocrystalline sodalite with organic additives. *Mater. Lett.* **2008**, *62*, 4028–4030. [CrossRef]
4. Naskar, M.K.; Kundu, D.; Chatterjee, M. Effect of process parameters on surfactant-based synthesis of hydroxy sodalite particles. *Mater. Lett.* **2011**, *65*, 436–438. [CrossRef]
5. Johnson, G.M.; Mead, P.J.; Weller, M.T. Synthesis of a range of anion-containing gallium and germanium sodalites. *Microporous Mesoporous Mater.* **2000**, *38*, 445–460. [CrossRef]
6. Hu, T.; Qiu, J.; Wang, Y.; Wang, C.; Liu, R.; Meng, C. Synthesis of low Si/Al ratio hydroxysodalite from oil shale ash without pretreatment. *J. Chem. Technol. Biotechnol.* **2015**, *90*, 208–212. [CrossRef]
7. Van den Berg, A.W.C.; Bromley, S.T.; Jansen, J.C. Thermodynamic limits on hydrogen storage in sodalite framework materials: A molecular mechanics investigation. *Microporous Mesoporous Mater.* **2005**, *78*, 63–71. [CrossRef]
8. Khajavi, S.; Jansen, J.C.; Kapteijn, F. Application of hydroxy sodalite films as novel water selective membranes. *J. Membr. Sci.* **2009**, *326*, 153–160. [CrossRef]
9. Khajavi, S.; Jansen, J.C.; Kapteijn, F. Application of a sodalite membrane reactor in esterification-coupling reaction and separation. *Catal. Today* **2010**, *156*, 132–139. [CrossRef]
10. Xu, X.; Bao, Y.; Song, C.; Yang, W.; Liu, J.; Lin, L. Microwave-assisted hydrothermal synthesis of hydroxy-sodalite zeolite membrane. *Microporous Mesoporous Mater.* **2004**, *75*, 173–181. [CrossRef]
11. Nabavi, M.S.; Mohammadi, T.; Kazemimoghadam, M. Hydrothermal synthesis of hydroxy sodalite zeolite membrane: Separation of H_2/CH_4. *Ceram. Int.* **2014**, *40*, 5889–5896. [CrossRef]
12. Jiang, J.; Gu, X.; Feng, L.; Duanmu, C.; Jin, Y.; Hu, T.; Wu, J. Controllable synthesis of sodalite submicron crystals and microspheres from palygorskite clay using a two-step approach. *Powder Tech.* **2012**, *217*, 298–303. [CrossRef]
13. Passos, F.A.C.M.; Castro, D.C.; Ferreira, K.K.; Simões, K.M.A.; Bertolino, L.C.; Barbato, C.N.; Garrido, F.M.S.; Felix, A.A.S.; Silva, F.A.N.G. Synthesis and characterization of sodalite and cancrinite from Kaolin. In *Characterization of Minerals Metals and Materials*; Springer: Berlin, Germany, 2017; pp. 279–288.
14. Golbad, S.; Khoshnoud, P.; Abu-Zahra, N. Hydrothermal synthesis of hydroxy sodalite from fly ash for the removal of lead ions from water. *Int. J. Environ. Sci. Technol.* **2017**, *14*, 135–142. [CrossRef]
15. Bouberka, Z.; Kacha, S.; Kameche, M.; Elmaleh, S.; Derriche, Z. Sorption study of an acid dye from an aqueous solutions using modified clays. *J. Hazard. Mater.* **2005**, *119*, 117–124. [CrossRef] [PubMed]
16. Karnland, O.; Olsson, S.; Nilsson, U.; Sellin, P. Experimentally determined swelling pressures and geochemical interactions of compacted Wyoming bentonite with highly alkaline solutions. *Phys. Chem. Earth Parts A/B/C* **2007**, *32*, 275–286. [CrossRef]
17. Simonsen, M.E.; Sonderby, C.; Sogaard, E.G. Synthesis and characterization of silicate polymers. *J. Sol-Gel Sci. Technol.* **2009**, *50*, 372. [CrossRef]
18. Simonsen, M.E.; Sonderby, C.; Li, Z.; Søgaard, E.G. XPS and FT-IR investigation of silicate polymers. *J. Mater. Sci.* **2009**, *44*, 2079. [CrossRef]
19. Tang, W.; Han, J.; Zhang, S.; Sun, J.; Li, H.; Gu, X. Synthesis of 4A zeolite containing La from kaolinite and its effect on the flammability of polypropylene. *Polym. Compos.* **2018**, *39*, 3461–3471. [CrossRef]
20. Wu, Z.; Chen, T.; Liu, H.; Wang, C.; Cheng, P.; Chen, D.; Xie, J. An insight into the comprehensive application of opal-palygorskite clay: Synthesis of 4A zeolite and uptake of Hg^{2+}. *Appl. Clay Sci.* **2018**, *165*, 103–111. [CrossRef]
21. Meng, Q.; Chen, H.; Lin, J.; Lin, Z.; Sun, J. Zeolite A synthesized from alkaline assisted pre-activated halloysite for efficient heavy metal removal in polluted river water and industrial wastewater. *J. Environ. Sci.* **2017**, *56*, 254–262. [CrossRef] [PubMed]
22. Itani, L.; Liu, Y.; Zhang, W.; Bozhilov, K.N.; Delmotte, L.; Valtchev, V. Investigation of the physicochemical changes preceding zeolite nucleation in a sodium-rich aluminosilicate gel. *J. Am. Chem. Soc.* **2009**, *131*, 10127–10139. [CrossRef] [PubMed]

23. Barnes, M.C.; Addai-Mensah, J.; Gerson, A.R. The mechanism of the sodalite-to-cancrinite phase transformation in synthetic spent Bayer liquor. *Microporous Mesoporous Mater.* **1999**, *31*, 287–302. [CrossRef]
24. Liu, Q.; Navrotsky, A. Synthesis of nitrate sodalite: An in situ scanning calorimetric study. *Geochim. Cosmochim. Acta* **2007**, *71*, 2072–2078. [CrossRef]
25. Fan, W.; Morozumi, K.; Kimura, R.; Yokoi, T.; Okubo, T. Synthesis of nanometer-sized sodalite without adding organic additives. *Langmuir* **2008**, *24*, 6952–6958. [CrossRef] [PubMed]
26. Geon, J.K.; Wha, S.A. Synthesis and characterization of iron-modified ZSM-5. *Appl. Catal.* **1991**, *71*, 55–68. [CrossRef]
27. Li, Y.; Peng, T.; Man, W.; Ju, L.; Zheng, F.; Zhang, M.; Guo, M. Hydrothermal synthesis of mixtures of NaA zeolite and sodalite from Ti-bearing electric arc furnace slag. *RSC Adv.* **2016**, *6*, 8358–8366. [CrossRef]
28. Tounsi, H.; Mseddi, S.; Djemel, S. Preparation and characterization of Na-LTA zeolite from Tunisian sand and aluminum scrap. *Phys. Procedia* **2009**, *2*, 1065–1074. [CrossRef]
29. Mofrad, A.M.; Peixoto, C.; Blumeyer, J.; Liu, J.; Hunt, H.K.; Hammond, K.D. Vibrational spectroscopy of sodalite: Theory and experiments. *J. Phys. Chem. C* **2018**, *122*, 24765–24779. [CrossRef]
30. Khajavi, S.; Kapteijn, F.; Jansen, J.C. Synthesis of thin defect-free hydroxy sodalite membranes: New candidate for activated water permeation. *J. Membr. Sci.* **2007**, *299*, 63–72. [CrossRef]
31. Sharp, Z.D.; Helffrich, G.R.; Bohlen, S.R.; Essene, E.J. The stability of sodalite in the system $NaAlSiO_4$-NaCl. *Geochim. Cosmochim. Acta* **1989**, *53*, 1943–1954. [CrossRef]
32. Gaidoumi, A.E.; Benabdallah, A.C.; Bali, B.E.; Kherbeche, A. Synthesis and characterization of zeolite HS using natural pyrophyllite as new clay source. *Arabian J. Sci. Eng.* **2018**, *43*, 191–197. [CrossRef]

Article

Enhanced Potential Toxic Metal Removal Using a Novel Hierarchical SiO$_2$–Mg(OH)$_2$ Nanocomposite Derived from Sepiolite

Qizhi Yao [1], Shenghui Yu [2,3,*], Tianlei Zhao [2,4], Feijin Qian [2,4], Han Li [2,4], Gentao Zhou [2,4,*] and Shengquan Fu [5]

[1] School of Chemistry and Materials Science, University of Science and Technology of China, Hefei 230026, China; qzyao@ustc.edu.cn
[2] CAS Key Laboratory of Crust-Mantle Materials and Environments, School of Earth and Space Sciences, University of Science and Technology of China, Hefei 230026, China; zhaotian@mail.ustc.edu.cn (T.-L.Z.); fjqian@mail.ustc.edu.cn (F.-J.Q.); lihan211@mail.ustc.edu.cn (H.L.)
[3] School of Environmental Science and Engineering, Shaanxi University of Science and Technology, Xi'an 710021, China
[4] CAS Center for Excellence in Comparative Planetology, Hefei 230026, China
[5] Hefei National Laboratory for Physical Sciences at Microscale, University of Science and Technology of China, Hefei 230026, China; fusq@ustc.edu.cn
* Correspondence: yushenghui@sust.edu.cn (S.-H.Y.); gtzhou@ustc.edu.cn (G.-T.Z.)

Received: 9 April 2019; Accepted: 14 May 2019; Published: 15 May 2019

Abstract: Clays are widely used as sorbents for heavy metals due to their high specific surface areas, low cost, and ubiquitous occurrence in most soil and sediment environments. However, the low loading capacity for heavy metals is one of their inherent limitations. In this work, a novel SiO$_2$–Mg(OH)$_2$ nanocomposite was successfully prepared via sequential acid–base modification of raw sepiolite. The structural characteristics of the resulting modified samples were characterized by a wide range of techniques including field emission scanning electron microscopy (FESEM), transmission electron microscopy (TEM), energy dispersive X-ray spectroscopy (EDX), X-ray diffraction (XRD), and nitrogen physisorption analysis. The results show that a hierarchical nanocomposite constructed by loading the Mg(OH)$_2$ nanosheets onto amorphous SiO$_2$ nanotubes can be successfully prepared, and the nanocomposite has a high surface area (377.3 m^2/g) and pore volume (0.96 cm^3/g). Batch removal experiments indicate that the nanocomposite exhibits high removal efficiency toward Gd(III), Pb(II), and Cd(II), and their removal capacities were greatly enhanced in comparison with raw sepiolite, due to the synergistic effect of the different components in the hierarchical nanocomposite. This work can provide a novel route toward a hierarchical nanocomposite by using clay minerals as raw material. Taking into account the simplicity of the fabrication route and the high loading capacities for heavy metals, the developed nanocomposite also has great potential applications in water treatment.

Keywords: hierarchical nanocomposite; sepiolite; clay mineral; heavy metals; water treatment

---

1. Introduction

Heavy metals or potential toxic metals are significant environmental pollutants, and their toxicity is a problem of increasing significance for ecological, evolutionary, nutritional, and environmental reasons. The term "heavy metals" refers to any metallic element that has a relatively high density and is toxic or poisonous even at low concentration [1]. Heavy metals, such as Pb(II), Hg(II), Cd(II), Co(II), Ni(II), and Cr(VI), are known to be prominent pollutants and even carcinogenic agents, and may represent a serious threat to the living population because of their non-degradable, persistent, and accumulative nature [2–4]. The notorious itai-itai disease occurring in Japan during the 1960s and 1970s is one of the most famous accidents caused by chronic cadmium contaminated rice fields [5]. In recent decades, however, rare earth elements (REEs) were also widely exploited and used in industrial and high-tech fields as a result of their irreplaceable roles in designing magnetic, luminescent, catalytic, hydrogen storage, and superconductive materials [6–8]. More and more rare earth elements would inevitably enter into the environmental waters and work places, thus causing adverse health effects [9]. For instance, trivalent La(III) and Gd(III) ions can interfere with calcium channels in human and animal cells, also alter or even inhibit the action of various enzymes, and regulate synaptic transmission, as well as block some receptors (for example, glutamate receptors) when they are found in neurons [10,11]. Removal of the REEs entering the environment due to human activities has become a new concern [12–14]. Moreover, the small global reserve of REEs of no more than 99 million tons limits the wide use of REEs [15]. Hence, the recycling of REEs is also an urgent task [7].

The existing methods for the removal of heavy metals from the environment can be grouped into biotic and abiotic [16]. The biotic routes are based on the accumulation of the heavy metal by plants or microorganisms, while the abiotic ones mainly include physicochemical processes such as precipitation, co-precipitation, ion exchange, solvent extraction, electrolysis, reverse osmosis, and adsorption by a suitable sorbent. Because biotic accumulation by the present generation of plants and microorganisms is time consuming, fast removal of heavy metals from the environment and decontamination of drinking water still require abiotic or physicochemical methodologies [2,17]. Among the abiotic techniques, the adsorption process is cost effective, flexible, and easy to design and operate [4]. Clay minerals such as montmorillonite, kaolinite, vermiculite, sepiolite, etc., as dominant adsorptive materials, are widely used to capture metal ions from solutions, due to their high specific surface areas associated with their small particle size, low cost, and ubiquitous occurrence in most soil and sediment environments [4,16,18–25], and their adsorption for heavy metal ions proceeds mainly via ion exchange reactions and the formation of inner-sphere complexes through $\equiv$Si–O– and $\equiv$Al–O– groups at the clay particle edges [4,26]. However, the inherent limitations of clays as adsorbents of heavy metals are their low loading capacity, relatively small metal ion binding constants, and low selectivity to the type of metal [27].

To circumvent the inherent limitations, the modification of clay minerals with reagents containing metal chelating functional groups has become a prevalent and versatile means to enhance both their binding capacity and selectivity to heavy metals or organic pollutants [21,28]. After such modification, the adsorption of organic contaminants by clays is remarkably improved. However, their adsorption ability to heavy metals still remains relatively inefficient or limited. For example, Mercier and Pinnavaia [29] reported that when thiol groups were grafted to the interlamellar surface of magadiite and kenyaite, negligible heavy metal ion binding was observed. Celis et al. [16] found that 3-mercaptopropylsilyl-sepiolite and 2-mercaptoethylammonium-montmorillonite display higher adsorption capacity for Hg^{2+}, but significantly low adsorption capacity for Pb^{2+} and Zn^{2+} relative to their pure counterparts. Especially, the adsorption of montmorillonite functionalized by quaternary ammonium cations (QACs) toward heavy metals dramatically decreases compared with the raw montmorillonite, since the interlayer QACs are not readily exchangeable or the hydrophobic interlayer environment restrains the adsorption affinity for the hydrated metal cations [30,31]. Moreover, the modification processes are relatively complex, and organic solvents are usually needed [21,27,28,32]. Therefore, seeking facile clay modification methods to enhance the removal capacities for heavy metals is still a challenge.

Herein, a facile and environmentally friendly method was developed for modifying clay mineral sepiolite sequentially with acid and alkali to form a novel SiO_2–$Mg(OH)_2$ nanocomposite. Sepiolite was selected because it is fibrous hydrated magnesium silicate with the ideal chemical formula $Mg_4Si_6O_{15}(OH)_2 \cdot 6H_2O$ and is characterized by a high porosity and specific surface area [33]. Furthermore, mineral sepiolite has an alternation of blocks along its crystallographic [010] direction and tunnels that grow up along its crystallographic [100] direction. Each structural block is built by two tetrahedral silica sheets with a central octahedral magnesia sheet [34,35]. The acid treatment investigations for sepiolite showed that variable amounts of structural Mg^{2+} ions can be removed, depending on the intensity of the acid treatment, while the tetrahedral silica sheets can form a free insoluble amorphous silica gel [36–39]. Therefore, it can be excepted that the in situ alkalization of the acid-treated sepiolite may produce the $Mg(OH)_2$-loaded amorphous silica composite. Moreover, nanostructured $Mg(OH)_2$ has also been proven to be an ideal water treatment agent to remove dyes and soluble heavy metals [7,40,41]. Therefore, the composite may possess improved removal performance to heavy metals and can be potentially applied in water treatment.

2. Materials and Methods

2.1. Materials and Chemicals

All chemical reagents were of analytical grade and used as received without any further purification. Hydrochloric acid (HCl), sodium hydroxide (NaOH), gadolinium oxide (Gd_2O_3), lead nitrate ($Pb(NO_3)_2$), and cadmium nitrate ($Cd(NO_3)_2$) were purchased from Sinopharm Chemical Reagent Co., Ltd. Raw sepiolite (Mg: ~13 wt %) was purchased from Sigma-Aldrich Chemical Reagent Co., Ltd. Deionized water was used in all the experiments.

2.2. Preparation of the SiO_2–$Mg(OH)_2$ Nanocomposite

Typical preparation processes of the SiO_2–$Mg(OH)_2$ nanocomposite were as follows: 5.0 g of sepiolite was first added into 100 mL of 5% (*v/v*) HCl solution (ca. 0.6 mol/L, pH = 0.22) at 60 °C for acid activation at a stirring speed of 300 r/min for 12 h, then 1.0 mol/L of NaOH solution was added dropwise to the solution under magnetic stirring until the pH of the solution reached 10.5. After 2 h of stirring, the suspension was centrifuged at 10,000 rpm for 10 min, and followed by washing with deionized water and absolute ethanol several times. Finally the precipitate was collected and dried at 50 °C under vacuum.

2.3. Characterizations

Several analytical techniques were used to characterize raw sepiolite and the treated products. X-ray diffraction (XRD) patterns were collected using a Japan Map XHF X-ray diffractometer equipped with graphite-monochromatized Cu Kα irradiation (λ = 0.154056 nm). FESEM images (FESEM) were obtained on a field-emission scanning electron microscopy JEOL JSM-2010. The size and structure of the samples were determined by a JEM-2010 transmission electron microscope (TEM) operating at 200 kV. Energy dispersive X-ray spectroscopy (EDX) analyses were obtained with an EDAX detector installed on the same TEM. N_2 adsorption–desorption isotherms were performed with a Micromeritics Coulter (USA) instrument utilizing Barrett–Emmett–Teller (BET) for specific surface area calculation.

2.4. Heavy Metal Removal

The removal abilities of the SiO_2–$Mg(OH)_2$ nanocomposite for rare earth element Gd(III) and traditional heavy metals Pb(II) and Cd(II) were studied by batch experiments. Stock solution with 10 mmol/L Gd(III) was prepared by dissolving gadolinium oxide (Gd_2O_3) in diluted hydrochloric acid. The Pb(II) or Cd(II) stock solution with a concentration of 10 mmol/L was obtained by dissolving $Pb(NO_3)_2$ or $Cd(NO_3)_2$ in deionized water. A series of solutions used during the removal experiments were prepared by diluting the stocks to the desired concentrations. Their actual concentrations were also measured using inductively coupled plasma atomic emission spectroscopy (ICP-AES, optima 7300 DV). Batch removal experiments for Gd(III), including the effects of pH and contact time were conducted in a 100 mL beaker equipped with a magnetic stirrer at a speed of 300 r/min. All of the beakers were covered with Parafilm. In a typical run, 50 mg of sorbent was mixed with 50 mL of solution containing Gd(III) at room temperature (293 K). The pH of the solution was adjusted with HCl and/or NaOH solution before the sorbent addition, and measured using a pH meter (inoLab WTW series pH 740). The removal ability of the nanocomposite for Pb(II) and Cd(II) was investigated by the same procedure used for Gd(III) removal. The supernatant of the suspension was collected by centrifugation at 10,000 rpm for 10 min, and then filtered using a 0.22 μm pore size membrane filter. The concentration of Gd(II), Pb(II), or Cd(II) in the filtrate was determined by ICP-AES. The experiments were conducted in triplicate, and averaged values were reported. The content of heavy metal Gd(II), Pb(II), or Cd(II) removed at time t, q_t (mg/g), uptake percentage, U%, and the content of the heavy metal removed at equilibrium, q_e (mg/g), were calculated according to the following equations, respectively:

$$q_t = \frac{(C_o - C_t) \times V}{W}, \tag{1}$$

$$U\% = \frac{(C_o - C_t) \times 100\%}{C_o}, \tag{2}$$

$$q_e = \frac{(C_o - C_e) \times V}{W}, \tag{3}$$

where C_o (mg/L), C_t (mg/L), and C_e represent the liquid phase concentration of the heavy metal initially, at any time t, and at equilibrium, respectively. V is the volume of the solution (mL) and W is the mass of the sorbents added (mg).

3. Results and Discussion

3.1. Characterization of the SiO_2–$Mg(OH)_2$ Nanocomposite

The morphology and textures of raw sepiolite, acid-activated sepiolite, and sepiolite treated by sequential acid–base were first observed by FESEM and TEM techniques. The raw sepiolite had a nanofiber-like structure with a length of several micrometers and a width of ca. 30–40 nm (e.g., FESEM image in Figure S1a, TEM images in Figure 1a,b). The EDX spectrum showed that the rod-like sepiolite contains elements O, Si, and Mg, as well as C and Cu, here the elements C and Cu come from the carbon-coated Cu grid (inset in Figure 1b). After the acid activation of sepiolite in 5% (*v/v*) HCl solution at 60 °C for 12 h, a white colloid-like precipitate was obtained. Corresponding FESEM analyses show that the acid-activated sepiolite still exhibited the rod-like nanostructures (e.g., Figure S1b), but significant aggregation occurred compared to the raw sepiolite (e.g., Figure S1a). The TEM analyses further revealed that the morphological texture was still inherited from the raw sepiolite (e.g., Figure 1c). However, the magnified TEM image in Figure 1d unambiguously exhibits the distinct contrasts between the black lateral part and light middle part of the nanorods, unveiling that the space in the nanorods was significantly increased. Furthermore, the EDX result shows that no Mg^{2+} existed in the acid-activated sepiolite (Figure 1d inset), further confirming that the complete leaching of Mg^{2+} in octahedral sheets of sepiolite yielded the silica nanotubes (Figure 1d). For the products after the sequential acid–base treatment, one can find from the FESEM image (e.g., Figure S1c) that the surface of the product became rough with respect to raw sepiolite (Figure S1a) and acid-activated sepiolite (Figure S1b). The TEM results (Figure 1e,f) show that many neo-formed nanoneedles or nanosheets were anchored onto the silica nanotubes, distinct from the structures obtained after only the acid treatment (Figure 1c,d). The further-magnified TEM image (Figure 1f) reveals that the nanotubes were covered with many wrinkle nanosheets, and their width significantly increased. The EDX results (e.g., inset in Figure 1f) show that besides element O and Si, extra element Mg can be detected, suggesting that the neo-formed nanosheets were an Mg-bearing material.

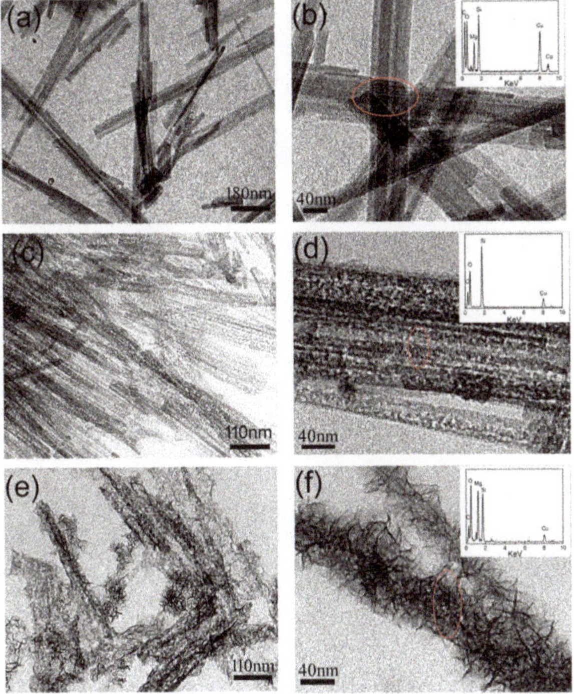

Figure 1. Typical TEM images of raw sepiolite (**a,b**), acid-activated sepiolite (**c,d**), and sepiolite modified by sequential acid–base treatment (**e,f**). Insets in panels b, d, and f are the corresponding energy dispersive X-ray spectroscopy (EDX) spectra. The red ellipses in Figure 1b,d,f indicate the areas of EDX analysis.

Figure 2 presents typical XRD patterns of raw sepiolite, acid-activated sepiolite, and sepiolite treated by sequential acid–base. The raw sepiolite exhibited the typical powder XRD pattern of pure sepiolite (JCPDF: 13-0595) (Figure 2a). For the acid-activated sepiolite, the characteristic peaks belonging to sepiolite disappeared, and a broad hump peak could be observed (Figure 2b), indicating that an amorphous phase formed. Combining with the EDX results (e.g., inset in Figure 1d), it can be concluded that the aggressive acid treatment led to the formation of amorphous silica nanotubes. However, after the sequential treatment of sepiolite with HCl and NaOH solutions, two broadened diffraction peaks in the XRD pattern could be identified, and these peaks can be indexed to the characteristic (001) and (100) diffractions of brucite ($Mg(OH)_2$, JCPDF: 07-0239) (Figure 2c), confirming that the leached Mg^{2+} from sepiolite can be precipitated as brucite after adding NaOH solution. The broadened diffraction peaks also support that the precipitated $Mg(OH)_2$ has a nanostructure (Figure 1e,f). Furthermore, the content of the $Mg(OH)_2$ in the SiO_2–$Mg(OH)_2$ nanocomposite was determined by ICP-AES after the sample was dissolved by HNO_3 solution, the result shows that up to 23.09 wt % of $Mg(OH)_2$ was loaded on the surface of the rod-like silica, indicating that the leached Mg^{2+} was precipitated as $Mg(OH)_2$ nanosheets anchored to the SiO_2 nanotubes, finally forming a hierarchical $Mg(OH)_2$–SiO_2 nanocomposite.

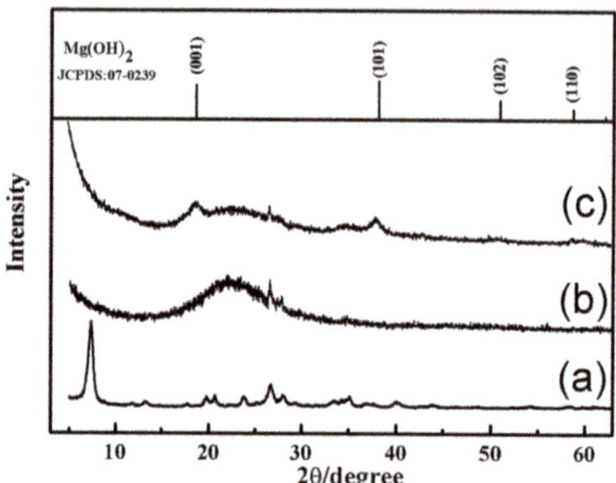

Figure 2. Representative XRD patterns of raw sepiolite (**a**), acid-activated sepiolite (**b**), and sepiolite modified by sequential acid–base treatment (**c**).

Figure 3 further shows the representative FTIR spectra of raw sepiolite, the amorphous SiO_2 nanotubes, and the $Mg(OH)_2$–SiO_2 nanocomposite. As shown in Figure 3a, the vibrational bands located in the 4000–400 cm^{-1} range can all be assigned to the characteristic vibrations of clay mineral sepiolite, i.e., the vibrations of the Mg–OH group (3690 cm^{-1}), coordinated water (3568 cm^{-1}), zeolitic water (3422 cm^{-1}, 1650 cm^{-1}); and the bonds of Si–O–Si (1016 and 460 cm^{-1}), Si–O (1215, 1076 cm^{-1} (shoulder), and 973 cm^{-1}), Si–O–Mg (437 cm^{-1}), and Mg–O (690 and 637 cm^{-1}) [37,38]. After the acid treatment, the intensity of bands in the 4000–3000 cm^{-1} range and at 1650 cm^{-1} were decreased. The wide band centered at 1016 cm^{-1} (actually consisting of four different bands at 1215, 1076, 1016, and 973 cm^{-1}) changed its form, and shifted to higher wave numbers, from 1016 to 1085 cm^{-1}, indicating the textural changes in the solids (Figure 3b) [37]. The peaks at 1081, 800, and 460 cm^{-1} agreed with the Si–O–Si bond [42,43], confirming that the natural sepiolite transformed into silica after the acid activation, consistent with the EDX results (e.g., inset in Figure 1d). As for the $Mg(OH)_2$–SiO_2 composite, one can find from Figure 3c that the peak locations have no obvious changes relative to the silica (Figure 3b), except for the increase in the intensities of the bands in the 4000–3000 cm^{-1} and 700–600 cm^{-1} ranges, which usually are indicative for Mg–OH bond vibrations [37], confirming the formation of $Mg(OH)_2$ precipitate after the base treatment.

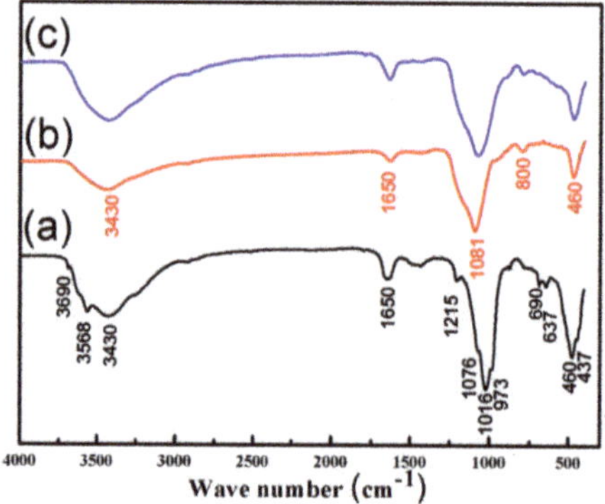

Figure 3. FTIR spectra of raw sepiolite (**a**), amorphous SiO_2 nanotubes (**b**), and $Mg(OH)_2$–SiO_2 composite (**c**).

Representative N_2 adsorption–desorption isotherms are shown in Figure 4 for the raw sepiolite, amorphous SiO_2, and SiO_2–$Mg(OH)_2$ nanocomposite, respectively. It can be seen that all the specimens showed a type IV N_2 adsorption isotherm with an evident hysteresis loop, suggesting the presence of mesopores in the materials [44]. After the acid activation of raw sepiolite, the BET surface area and total pore volume of the formed amorphous SiO_2 increased from 256.5 m^2/g and 0.47 cm^3/g to 488.5 m^2/g and 0.87 cm^3/g, this could be attributed to the acid leaching of Mg^{2+} from the sepiolite and the formation of silica with central channels (Figure 1e,f). For the hierarchical SiO_2–$Mg(OH)_2$ nanocomposite, the surface area and total pore volume were 377.3 m^2/g and 0.96 cm^3/g, respectively. The relative lower surface area of the nanocomposite with respect to the amorphous silica nanotubes should arise from the formation of nanosized $Mg(OH)_2$ on the silica nanotube surfaces (Figure 1e,f).

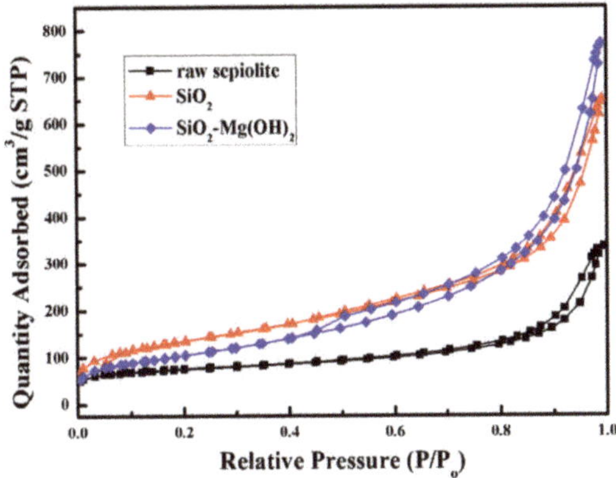

Figure 4. N_2 adsorption/desorption isotherms of raw sepiolite (black), amorphous SiO_2 nanotubes (red), and $Mg(OH)_2$–SiO_2 composite (blue).

3.2. Removal of Gd(III) by SiO_2–$Mg(OH)_2$ Nanocomposite

It is well known that $Mg(OH)_2$ would dissolve under a low-pH circumstance [45], and the precipitation of heavy metals hydroxides can also occur at higher pH values. Therefore, the parameter pH is usually believed to be one of the most important factors affecting the adsorption process. Here, in order to avoid the dissolution of $Mg(OH)_2$ and the precipitation of Gd(III) hydroxide, $Ksp_{Gd(OH)3}$ is 8.2×10^{-23} at 25 °C [46], a pH range from 3.0 to 7.5 was selected, and the concentration of Gd(III) solution 0.1 mmol/L (ca. 15.2 mg/L) was used. The desired initial pH values were adjusted before the addition of the sorbent into the Gd(III) solution. One can find from Figure 5a that the Gd(III) was almost completely removed by the nanocomposite over the wide pH range. For the raw sepiolite, however, the removal percentages of Gd(III) were lower than the nanocomposite, especially at low pHs, indicating that the nanocomposite has excellent removal efficiency over the raw sepiolite. Usually, the surface charge of the adsorbent is believed to play a crucial role in adsorption [14,35]. As is known, the point of zero charge (PZC) is an important property of the adsorbent as the surfaces acquire negative charge at pH values higher than the pH at PZC (pHpzc). The pHpzc for sepiolite is about 7.4 [37]. Therefore, the low Gd(III) uptakes of raw sepiolite at low pH values were most probably due to the protonation of the active sites in sepiolite, which inhibited their binding toward Gd(III) [35]. As pH increases, however, the surface positive charge of adsorbent sepiolite decreases, and thus the uptake of Gd(III) increases. As for the SiO_2–$Mg(OH)_2$ nanocomposite, the pHpzc value was close to 1.5 (Figure S2), measured by a Zeta potential analyzer, and the nanocomposite had higher surface area than the raw sepiolite. These dominate the almost complete removal of Gd(III) by the nanocomposite over the used pH range. Figure 5b further shows their pH evolution relationships before and after adsorption. It can be found that the final pH values rose significantly after removal treatment by the SiO_2–$Mg(OH)_2$ nanocomposite relative to the raw sepiolite. This was possible because the $Mg(OH)_2$ in the nanocomposite has a basic nature, thereby producing a level-off effect on the pH of the Gd(III) removal system. As a result, a high final pH 9.7 was always achieved over the initial pH range

3–7.5; as was the case in the $Ag_2O–Mg(OH)_2$ nanocomposite for I^- removal [41]. According to the pH-dependent experiments, a neutral pH 7.0 was selected in the following removal experiments.

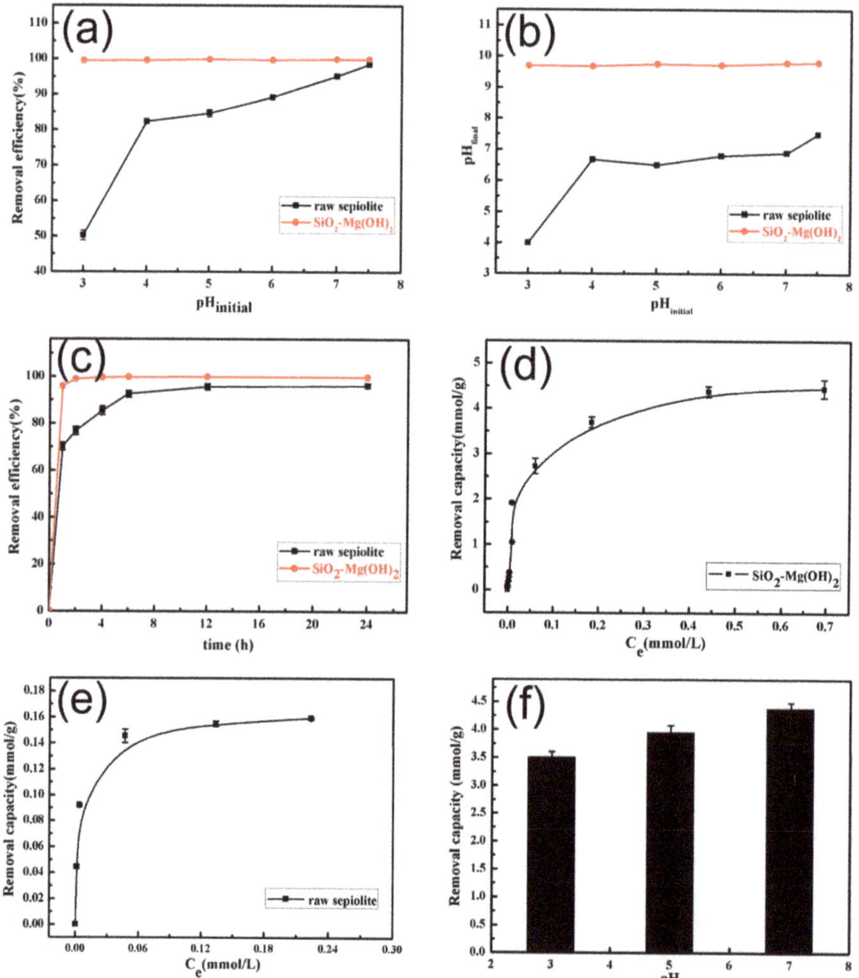

Figure 5. Effect of pH on the removal efficiency of Gd(III) by raw sepiolite and the $SiO_2–Mg(OH)_2$ nanocomposite (**a**); relationship between the initial and final pH of the adsorption system (**b**); effect of contact time on the Gd(III) removal by raw sepiolite and the nanocomposite (**c**); removal capacity of the nanocomposite (**d**) and raw sepiolite (**e**) for Gd(III); removal capacity of the nanocomposite for Gd(III) at different pHs (**f**).

Figure 5c presents the performance of Gd(III) removal by the $SiO_2–Mg(OH)_2$ nanocomposite and raw sepiolite versus contact time at the initial Gd(III) concentration of 0.1 mmol/L. It is not difficult to find that the Gd(III) removal rapidly increased in the presence of the nanocomposite, and almost all Gd(III) was removed in the first 120 min. In contrast, only 78% of Gd(III) was removed for the raw sepiolite, and the final removal efficiency for Gd(III) was also lower than that by the nanocomposite. In most wastewater, the REE concentrations are as low as one to hundreds of mg/L, despite the high total annual wastewater generation of more than 7.2 million tons [7]. Traditional adsorbents, e.g., zeolite, clay, and active carbon, donate high adsorption capacities only at high metal ion concentrations, but they are usually not suitable for dilute metal ion solutions because of their weak extraction force [47]. In this regard, the SiO_2–$Mg(OH)_2$ nanocomposite derived from sepiolite has great potential for the removal and enrichment treatment of low-concentration REEs and even other heavy metals. Moreover, the removal capacities of the nanocomposite and raw sepiolite for Gd(III) were also tested at neutral pH and room temperature. As shown in Figure 5d, the nanocomposite had an amazing removal capacity for Gd(III), ca. 4.45 mmol/g (ca. 698.65 mg/g), but only 0.159 mmol/g (ca. 24.96 mg/g) for the raw sepiolite (Figure 5e), indicating that the Gd (III) removal ability of the sepiolite treated by sequential acid—base was greatly enhanced. Especially at pH 3.0 and 5.0, the removal capacity toward Gd(III) of the nanocomposite still reached 3.52 and 4.1 mmol/g (e.g., Figure 5f), indicating its superb removal ability for Gd(III) over the wide pH range. As the REEs and some trivalent actinides have similar chemical behavior, this removal reagent should also be useful for other lanthanide and actinide elements.

3.3. Removal Abilities of SiO_2–$Mg(OH)_2$ Nanocomposite for Pb(II) and Cd(II)

Lazarević et al. [37] studied the adsorption of Pb(II) and Cd(II) on natural sepiolite and the influence of acid activation treatment on the adsorption capacity. Their results indicated that raw sepiolite is more effective for the adsorption of Pb(II) and Cd(II) than acid-activated sepiolite, i.e., the adsorption ability of natural sepiolite for Pb(II) and Cd(II) is 0.35 and 0.23 mmol/g, respectively; whereas the sepiolite activated by a 4 M of HCl solution for 10 h at room temperature only has 0.22 mmol/g for Pb(II), and 0.21 mmol/g for Cd(II). They proposed that the retention of Pb(II) and Cd(II) occurs dominantly by specific adsorption and exchange of Mg^{2+} ions from the sepiolite structure, and the lowered removal performance of the acid-activated sepiolite can be ascribed to the decrease in the number of Mg–OH groups as main centers for specific adsorption and the number of Mg^{2+} ions available for ion exchange with cations [37]. Herein, the removal ability of the SiO_2–$Mg(OH)_2$ nanocomposite for Pb(II) and Cd(II) was also tested by similar removal experiments at pH 5.6, as described by Lazarević et al. [37]. As shown in Figure 6, the removal capacity of the SiO_2–$Mg(OH)_2$ nanocomposite for Pb(II) and Cd(II) was 6.84 and 4.88 mmol/g, respectively, which are much higher than the removal capacities of the natural and acid-activated sepiolite [37]. This is possible because the $Mg(OH)_2$ nanosheets anchored onto the nanocomposite increased the Mg^{2+} availability and exchangeability, leading to the high removal capacities. The current results also indicate that after the facile modification, the removal ability of sepiolite for other common heavy metals would be greatly enhanced.

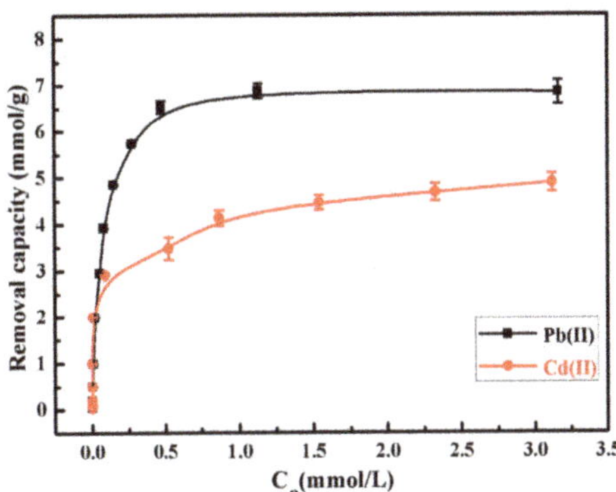

Figure 6. Removal capacities of the SiO_2–$Mg(OH)_2$ nanocomposite for Pb(II) and Cd(II) at pH 5.6.

3.4. Removal Mechanism of SiO_2–$Mg(OH)_2$ Nanocomposite for Heavy Metals

In order to understand the removal mechanism of heavy metal cations (Me^{n+}) by the SiO_2–$Mg(OH)_2$ nanocomposite, numerous techniques including FESEM, TEM, EDX, XRD, and FTIR, were used to characterize the samples after the heavy metal immobilization. As shown in Figure 7a, the Gd(III)-loaded nanocomposite still exhibited rod-like structures, and numerous small particles adhering on the nanocomposite surface could be found (as indicated by the red circle). The TEM image further revealed that the tube-like SiO_2 were covered with many isolated nanoparticles (Figure 7b), and the element Gd was also detected by the EDX analysis (e.g., inset in Figure 7b), indicating that the particles on the nanocomposite surfaces were Gd(III)-bearing precipitate. Further, the XRD analyses exhibited an obvious hump peak, which was similar to the pattern of acid-activated sepiolite (e.g., Figure 2b), and no diffraction peaks belonging to the known gadolinium compounds, especially gadolinium hydroxide were detected. However, the FTIR spectrum displayed two extra bands at 1520 and 1412 cm^{-1} (Figure 7d) in comparison with the FTIR spectrum of the SiO_2–$Mg(OH)_2$ nanocomposite (Figure 3c), which could be assigned to the bidentate asymmetric band of C–O–O and the bidentate symmetric band of C–O–O, respectively [48–50]. These results indicated that the Gd(III)-bearing compound formed in the removal processes may be Gd(III) carbonates. Therefore, the removal mechanism of Me^{n+} may be via sorption and the subsequent precipitation of Me-containing carbonates on the composite surface. To verify the deduction, the samples after the Pb(II) and Cd(II) removal by SiO_2–$Mg(OH)_2$ nanocomposite were also characterized by FESEM and XRD. Similarly, many small grains could be observed from the FESEM images after the Pb and Cd removal by the nanocomposite (Figure S3a,b), indicating the formation of Me-bearing compounds. The corresponding XRD results clearly revealed that all the peaks in XRD patterns could be indexed to hydrocerussite ($Pb_3(CO_3)_2(OH)_2$, JCPDS: 13-0131) (Figure 7e) and otavite ($CdCO_3$, JCPDS: 42-1342) (Figure 7f), respectively. The results confirmed the removal mechanism of SiO_2–$Mg(OH)_2$ nanocomposite to heavy metals via a heavy metal carbonatation on its surfaces. Therefore, the superb removal capacity of SiO_2–$Mg(OH)_2$ nanocomposite toward heavy metals could be attributed to the synergistic effect of nano-$Mg(OH)_2$ and amorphous SiO_2 nanotubes in SiO_2–$Mg(OH)_2$ composite. The SiO_2 nanotubes, obtained by complete acid-activated of sepiolite, possessed high surface area and acted as a support substrate. The $Mg(OH)_2$ nanosheets, formed by the reprecipitation of the leached Mg ions from sepiolite, acted as a reactive reagent in heavy metal removal. During the removal processes, the heavy metals were first adsorbed onto the SiO_2–$Mg(OH)_2$ composite by Mg^{2+} exchange with Me^{n+}, and then the nano-$Mg(OH)_2$ with basic nature promoted the intake of CO_2, leading to Me-containing carbonates on the nanocomposite.

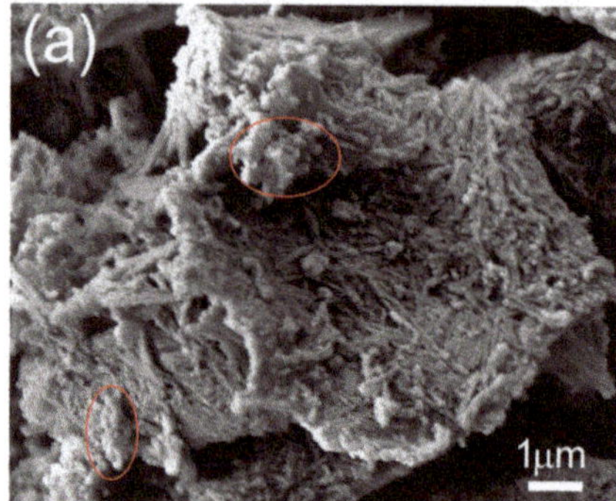
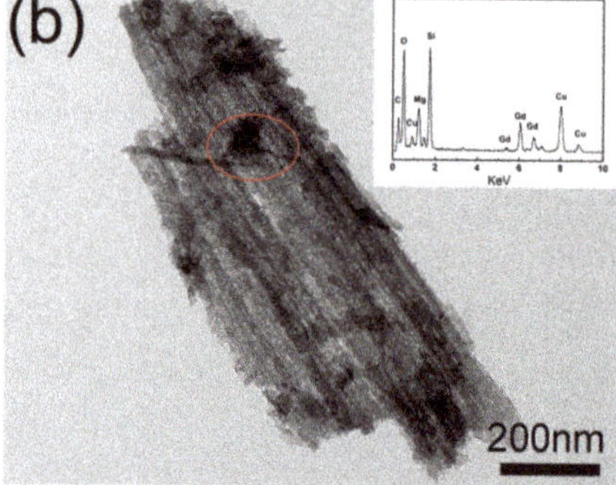

Figure 7. *Cont.*

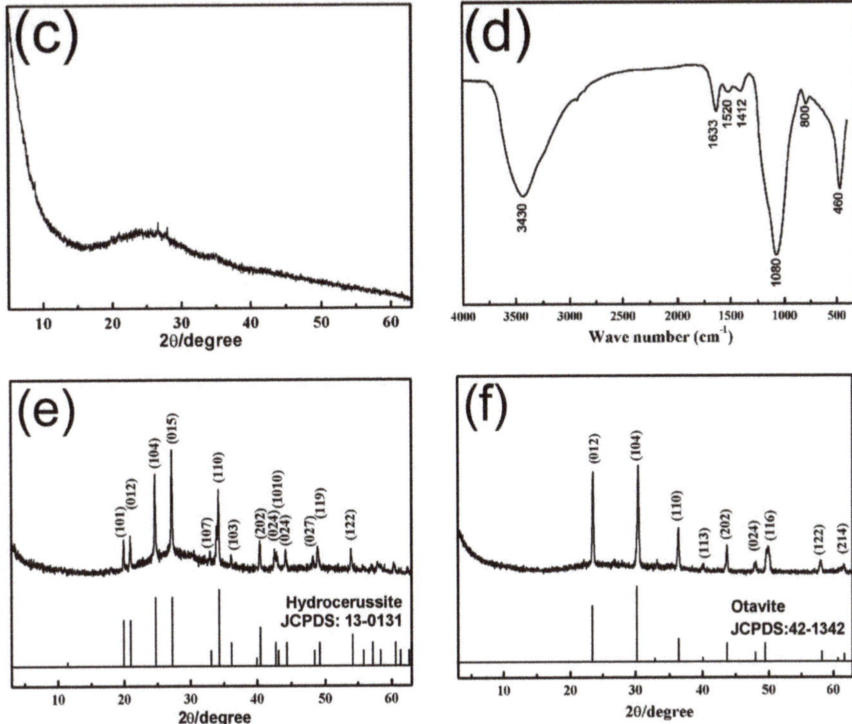

Figure 7. FESEM (**a**) and TEM (**b**) images, XRD pattern (**c**), and FTIR spectrum (**d**) of Gd(III)-loaded SiO_2–$Mg(OH)_2$ nanocomposite; XRD patterns of Pb(II)-loaded SiO_2–$Mg(OH)_2$ nanocomposites (**e**); and Cd(II)-loaded SiO_2–$Mg(OH)_2$ nanocomposites (**f**). The red ellipses in Figure 7a,b highlight the small particles adhering to the nanocomposite surface.

4. Conclusions

In summary, the hierarchical SiO_2–$Mg(OH)_2$ nanocomposite was successfully obtained by sequential acid–base treatment of natural sepiolite. First, the acid treatment led to the rod-like amorphous silica with central channels due to the Mg^{2+} leaching out of sepiolite. After adding the base agent, the leached Mg^{2+} further precipitated into sheet-like $Mg(OH)_2$ on the silica nanotube surfaces, finally producing the hierarchical nanocomposite. The SiO_2–$Mg(OH)_2$ nanocomposite with high surface area and pore volume exhibited an amazing removal capacity for Gd(III) (ca. 4.45 mmol/g), which was about 28 times of that of raw sepiolite (0.159 mmol/g). Moreover, this nanocomposite also showed exceptional removal capacity for common heavy metals (6.84 mmol/g for Pb^{2+}, 4.88 mmol/g for Cd^{2+}). The superb removal capacity of the SiO_2–$Mg(OH)_2$ nanocomposite toward heavy metals can be attributed to the synergistic effects of the two nanostructure components in the nanocomposite, i.e., the removal of the heavy metals proceeds via the initial sorption and subsequent carbonatation of heavy metal ions on the nanocomposite surface. The removal capacity of sepiolite to heavy metals was greatly enhanced after the sequential acid–base modification. As the modification method is facile and suitable for large scale preparation, the SiO_2–$Mg(OH)_2$ nanocomposite derived from sepiolite has real potential for applications in water treatment.

Supplementary Materials: The following are available online at http://www.mdpi.com/2075-163X/9/5/298/s1, Figure S1: Typical SEM images of raw sepiolite, acid-activated sepiolite, and sepiolite modified by sequential acid–base treatment; Figure S2: Zeta potential-pH profile for the SiO_2–$Mg(OH)_2$ nanocomposite in deionized water; Figure S3: SEM images of Pb(II)-loaded SiO_2–$Mg(OH)_2$ nanocomposite and Cd(II)-loaded SiO_2–$Mg(OH)_2$ nanocomposite.

Author Contributions: Conceptualization, Q.-Z.Y. and S.-H.Y.; methodology, Q.-Z.Y.; validation, S.-H.Y. and G.-T.Z.; formal analysis, T.-L.Z.; F.-J.Q., H.L., and S.-Q.F.; investigation, Q.-Z.Y and S.H.Y.; data curation, Q.-Z.Y., S.-H.Y., and G.-T.Z.; writing—original draft preparation, Q.-Z.Y.; writing—review and editing, S.-H.Y. and G.-T.Z.; supervision, G.-T.Z.; funding acquisition, Q.-Z.Y., H.L., S.-H.Y., and G.-T.Z.

Funding: This work was partially supported by the National Natural Science Foundation of China (Nos. 41672034, 41702038, and 41772030), and the Specialized Research Fund for the Doctoral Program of Higher Education (No. 20133402130007).

Conflicts of Interest: The authors declare no conflict of interest.

References

1. Nagajyoti, P.C.; Lee, K.D.; Sreekanth, T.V.M. Heavy metals, occurrence and toxicity for plants: A review. *Environ. Chem. Lett.* **2010**, *8*, 199–216. [CrossRef]

2. Kocaoba, S. Adsorption of Cd(II), Cr(III) and Mn(II) on natural sepiolite. *Desalination* **2009**, *244*, 24–30. [CrossRef]

3. Fabbricino, M.; Ferraro, A.; Luongo, V.; Pontoni, L.; Race, M. Soil washing optimization, recycling of the solution, and ecotoxicity assessment for the remediation of Pb-Contaminated sites using EDDS. *Sustainability* **2018**, *10*, 636. [CrossRef]

4. Uddin, M.K. A review on the adsorption of heavy metals by clay minerals, with special focus on the past decade. *Chem. Eng. J.* **2017**, *308*, 438–462. [CrossRef]

5. Kaji, M. Role of experts and public participation in pollution control: the case of Itai-itai disease in Japan. *Ethics Sci. Environ. Polit.* **2012**, *12*, 99–111. [CrossRef]

6. Ali, O.I.M.; Osman, H.H.; Sayed, S.A.; Shalabi, M.E.H. The removal of some rare earth elements from their aqueous solutions on by-pass cement dust (BCD). *J. Hazard. Mater.* **2011**, *195*, 62–67. [CrossRef] [PubMed]

7. Li, C.; Zhang, Z.; Huang, F.; Wu, Z.; Hong, Y.; Lin, Z. Recycling rare earth elements from industrial wastewater with flowerlike nano-$Mg(OH)_2$. *ACS Appl. Mater. Interfaces* **2013**, *5*, 9719–9725. [CrossRef] [PubMed]

8. Charalampides, G.; Vatalis, K.I.; Apostoplos, B.; Ploutarch-Nikolas, B. Rare earth elements: Industrial applications and economic dependency of Europe. *Procedia Econ. Finance* **2015**, *24*, 126–135. [CrossRef]

9. Aghayan, H.; Mahjoub, A.R.; Khanchi, A.R. Samarium and dysprosium removal using 11-molybdo-vanadophosphoric acid supported on Zr modified mesoporous silica SBA-15. *Chem. Eng. J.* **2013**, *225*, 509–519. [CrossRef]

10. Pałasz, A.; Czekaj, P. Toxicological and cytophysiological aspects of lanthanides Action. *Acta Biochim. Pol.* **2000**, *47*, 1107–1114. [PubMed]

11. Chen, Z.Y.; Zhu, X.Z. Accumulation of rare earth elements in bone and its toxicity and potential hazard to health. *J. Ecol. Rural Environ.* **2008**, *24*, 88–91.

12. Tan, X.; Fan, Q.; Wang, X.; Grambow, B. Eu(III) sorption to TiO_2 (anatase and rutile): batch, XPS, and EXAFS studies. *Environ. Sci. Technol.* **2009**, *43*, 3115–3121. [CrossRef] [PubMed]

13. Yesiller, S.U.; Eroğlu, A.E.; Shahwan, T. Removal of aqueous rare earth elements (REEs) using nano-iron based materials. *J. Ind. Eng. Chem.* **2013**, *19*, 898–907. [CrossRef]

14. Yu, S.H.; Yao, Q.Z.; Zhou, G.T.; Fu, S.Q. Preparation of hollow core/shell microspheres of hematite and its adsorption ability for samarium. *ACS Appl. Mater. Interfaces* **2014**, *6*, 10556–10565. [CrossRef] [PubMed]

15. Chen, Z.J. Global rare earth resources and scenarios of future rare earth industry. *J. Rare Earth* **2011**, *2*, 1–6. [CrossRef]

16. Celis, R.; Hermosin, M.C.; Cornejo, J. Heavy metal adsorption by functionalized clays. *Environ. Sci. Technol.* **2000**, *34*, 4593–4599. [CrossRef]

17. Lothenbach, B.; Furrer, G.; Schulin, R. Immobilization of heavy metals by polynuclear aluminium and montmorillonite compounds. *Environ. Sci. Technol.* **1997**, *31*, 1452–1462. [CrossRef]

18. Bailey, S.E.; Olin, T.J.; Bricka, R.M.; Adrian, D.D. A review of potentially low-cost sorbents for heavy metals. *Water Res.* **1999**, *33*, 2469–2479. [CrossRef]

19. Yuan, P.; Li, D.; Fan, M.; Yang, D.; Zhu, R.; Ge, F.; Zhu, J.; He, H. Removal of hexavalent chromium [Cr(VI)] from aqueous solutions by the diatomite-supported/unsupported magnetite nanoparticles. *J. Hazard. Mater.* **2010**, *173*, 614–621. [CrossRef]

20. Arancibia-Miranda, N.; Baltazar, S.E.; García, A.; Muñoz-Lira, D.; Sepúlveda, P.; Rubio, M.A.; Altbir, D. Nanoscale zero valent supported by Zeolite and Montmorillonite: Template effect of the removal of lead ion from an aqueous solution. *J. Hazard. Mater.* **2016**, *301*, 371–380. [CrossRef] [PubMed]

21. Zhu, L.; Wang, L.; Xu, Y. Chitosan and surfactant co-modified montmorillonite: A multifunctional adsorbent for contaminant removal. *Appl. Clay Sci.* **2017**, *146*, 35–42. [CrossRef]

22. Doğan, M.; Turhan, Y.; Alkan, M.; Namli, H.; Turan, P.; Demirbaş, Ö. Functionalized sepiolite for heavy metal ions adsorption. *Desalination* **2008**, *230*, 248–268. [CrossRef]

23. García, N.; Guzmán, J.; Benito, E.; Esteban-Cubillo, A.; Aguilar, E.; Santarén, J.; Tiemblo, P. Surface modification of sepiolite in aqueous gels by using methoxysilanes and its impact on the nanofiber dispersion ability. *Langmuir* **2011**, *27*, 3952–3959. [CrossRef] [PubMed]

24. Hu, B.; Hu, Q.; Chen, C.; Sun, Y.; Xu, D.; Sheng, G. New insights into Th(IV) speciation on sepiolite: Evidence for EXAFS and modeling investigation. *Chem. Eng. J.* **2017**, *322*, 66–72. [CrossRef]

25. Kara, M.; Yuzer, H.; Sabah, E.; Celik, M.S. Adsorption of cobalt from aqueous solutions onto sepiolite. *Water Res.* **2003**, *37*, 224–232. [CrossRef]

26. Sheng, G.; Xu, S.; Boyd, S.A. A dual function organoclay sorbent for lead and chlorobenzene. *Soil Sci. Soc. Am. J.* **1999**, *63*, 73–78. [CrossRef]

27. Mercier, L.; Detellier, C. Preparation, characterization, and applications as heavy metals sorbents of covalently grafted thiol functionalities on the interlamellar surface of montmorillonite. *Environ. Sci. Technol.* **1995**, *29*, 1318–1323. [CrossRef] [PubMed]

28. Yang, J.; Yu, K.; Liu, G. Chromium immobilization in soil using quaternary ammonium cations modified montmorillonite: Characterization and mechanism. *J. Hazard. Mater.* **2017**, *321*, 73–80. [CrossRef] [PubMed]

29. Mercier, L.; Pinnavaia, T.J. Heavy metal ion adsorbents formed by the grafting of a thiol functionality to mesoporous silica molecular sieves: Factors affecting Hg(II) uptake. *Environ. Sci. Technol.* **1998**, *32*, 2749–2754. [CrossRef]

30. Theng, B.K.G.; Churchman, G.J.; Gates, W.P.; Yuan, G.D. Organically Modified Clays for Pollutant Uptake and Environmental Protection. In *Soil Mineral Microbe-Organic Interactions*; Huang, Q., Huang, P., Violante, A., Eds.; Springer: Berlin/Heidelberg, Germany, 2008; pp. 145–174.

31. Ma, L.; Chen, Q.; Zhu, J.; Xi, Y.; He, H.; Zhu, R.; Tao, Q.; Ayoko, G.A. Adsorption of phenol and Cu(II) onto cationic and zwitterionic surfactant modified montmorillonite in single and binary systems. *Chem. Eng. J.* **2016**, *283*, 880–888. [CrossRef]

32. Liang, X.; Xu, Y.; Sun, G.; Wang, L.; Sun, Y.; Sun, Y.; Qin, X. Preparation and characterization of mercapto functionalized sepiolite and their application for sorption of lead and cadmium. *Chem. Eng. J.* **2011**, *174*, 436–444. [CrossRef]

33. Al-Ani, A.; Gertisser, R.; Zholobenko, V. Structural features and stability of Spanish sepiolite as a potential catalyst. *Appl. Clay Sci.* **2018**, *162*, 297–304. [CrossRef]

34. Wan, C.Y.; Chen, B.Q. Synthesis and characterization of biomimetic hydroxyapatite/sepiolite nanocomposites. *Nanoscale* **2011**, *3*, 693–700. [CrossRef] [PubMed]

35. Yu, S.H.; Li, H.; Yao, Q.Z.; Zhou, G.T.; Fu, S.Q. Microwave-assisted preparation of sepiolite-supported magnetite nanoparticles and their ability to remove low concentrations of Cr(VI). *RSC Adv.* **2015**, *5*, 84471–84482. [CrossRef]

36. Aznar, A.J.; Gutiérrez, E.; Díaz, P.; Alvarez, A.; Poncelet, G. Silica from sepiolite: Preparation, textural properties, and use as support to catalysts. *Microporous Mater.* **1996**, *6*, 105–114. [CrossRef]

37. Lazarević, S.; Janković-Častvan, I.; Jovanović, D.; Milonjić, S.; Janaćković, D.; Petrović, R. Adsorption of Pb^{2+}, Cd^{2+} and Sr^{2+} ions onto natural and acid-activated sepiolites. *Appl. Clay Sci.* **2007**, *37*, 47–57. [CrossRef]

38. Myriam, M.; Suaárez, M.; Martín-Pozas, J.M. Structural and textural modifications of palygorskite and sepiolite under acid treatment. *Clay Clay Miner.* **1998**, *46*, 225–231. [CrossRef]

39. Franco, F.; Pozo, M.; Cecilia, J.A.; Benítez-Guerrrero, M.; Pozo, E.; Martín-Rubí, J.A. Microwave assisted acid treatment of sepiolite: The role of composition and crystallinity. *Appl. Clay Sci.* **2014**, *102*, 15–27. [CrossRef]

40. Liu, W.; Huang, F.; Wang, Y.; Zou, T.; Zheng, J.; Lin, Z. Recycling $Mg(OH)_2$ nanoadsorbent during treating the low concentration of Cr^{VI}. *Environ. Sci. Technol.* **2011**, *45*, 1955–1961. [CrossRef]

41. Chen, Y.Y.; Yu, S.H.; Yao, Q.Z.; Fu, S.Q.; Zhou, G.T. One-step synthesis of $Ag_2O@Mg(OH)_2$ nanocomposite as an efficient scavenger for iodine and uranium. *J. Colloid. Interf. Sci.* **2018**, *510*, 280–291. [CrossRef] [PubMed]

42. Venkatathri, N.; Srivastava, R.; Yun, D.S.; Yoo, J.W. Synthesis of a novel class of mesoporous hollow silica from organic templates. *Micropor. Mesopor. Mat.* **2008**, *112*, 147–152. [CrossRef]

43. Shi, J.Y.; Yao, Q.Z.; Li, X.M.; Zhou, G.T.; Fu, S.Q. Formation of asymmetrical structured silica controlled by a phase separation process and implication for Biosilicification. *PLoS ONE* **2013**, *8*, e61164. [CrossRef] [PubMed]

44. Liu, H.C.; Chen, W.; Liu, C.; Liu, Y.; Dong, C.L. Magnetic mesoporous clay adsorbent: Preparation, characterization and adsorption capacity for atrazine. *Microporous Mesoporous Mater.* **2014**, *194*, 72–78. [CrossRef]

45. Pokrovsky, O.S.; Schott, J. Experimental study of brucite dissolution and precipitation in aqueous solutions: Surface speciation and chemical affinity control. *Geochim. Cosmochim. Acta* **2004**, *68*, 31–45. [CrossRef]

46. Moeller, T.; Fogel, N. Observations on the rare earths. LXI. precipitation of hydrous oxides or hydroxides from perchlorate solutions. *JACS* **1951**, *73*, 4481. [CrossRef]

47. Babel, S.; Kurniawan, T.A. Low-cost adsorbents for heavy metals uptake from contaminated water: A review. *J. Hazard. Mater.* **2003**, *97*, 219–243. [CrossRef]

48. Yanagihara, N.; Vemulapalli, K.; Fernando, Q.; Dyke, J.T. Synthesis of lanthanide carbonates. *J. Less Common Met.* **1991**, *167*, 223–232. [CrossRef]

49. Wu, Y.; Xu, X.; Tang, Q.; Li, Y. A new type of silica-coated $Gd_2(CO_3)_3$: Tb nanoparticle as a bifunctional agent for magnetic resonance imaging and fluorescent imaging. *Nanotechnology* **2012**, *23*, 205103. [CrossRef]

50. Nasution, E.Y.; Ahab, A.; Nuryadin, B.W.; Haryanto, F.; Arif, I.; Iskandar, F. Synthesis of gadolinium carbonate-conjugated-poly(ethylene)glycol ($Gd_2(CO_3)_3$@PEG) particles via a modified solvothermal method. *AIP Conf. Proc.* **2016**, *1710*, 030001–030007.

Article

The Characterization and Amoxicillin Adsorption Activity of Mesopore CaCO$_3$ Microparticles Prepared Using Rape Flower Pollen

Lvshan Zhou [1,2], Tongjiang Peng [1,*], Hongjuan Sun [1], Xiaogang Guo [3] and Dong Fu [2]

[1] Key Laboratory of Ministry of Education for Solid Waste Treatment and Resource Recycle, Institute of Mineral Materials & Application, Sichuan Engineering Lab of Nonmetallic Mineral Powder Modification & High-quality Utilization, Center of Forecasting and Analysis, Southwest University of Science and Technology, Mianyang 621010, China; zhoulvshan@126.com (L.Z.); sunhongjuan@swust.edu.cn (H.S.)

[2] School of Chemistry and Chemical Engineering, Eastern Sichuan Sub-center of National Engineering Research Center for Municipal Wastewater Treatment and Reuse, Sichuan University of Arts and Science, Dazhou 635000, China; fudongemail@163.com

[3] College of Chemistry and Chemical Engineering, Yangtze Normal University, Chongqing 408100, China; gxg_cqu@hotmail.com

* Correspondence: tjpeng@swust.edu.cn; Tel.: +86-0816-2419276

Received: 29 March 2019; Accepted: 21 April 2019; Published: 25 April 2019

Abstract: A precipitation reaction method was employed to prepare mesopore calcium carbonate (CaCO$_3$) using rape flower pollen as the template. CaCO$_3$ adsorbent was characterized using X-ray diffraction (XRD), scanning electronic microscopy (SEM), and Brunner−Emmet−Teller measurements (BET). The equilibrium adsorption data on amoxicillin were explained using Langmuir, Freundlich, and Temkin adsorption isotherm models. The pseudo-first order, second order, pseudo-second order, and intra-particle diffusion kinetic models were used to explore adsorption kinetics. Equilibrium adsorption of as-prepared CaCO$_3$ was better depicted using the Langmuir adsorption model with an R^2 of 0.9948. The separation factor (R_L) was found to be in the range of $0 < R_L < 1$, indicating the favorable adsorption of amoxicillin. The adsorption capacity of mesopore CaCO$_3$ reached 13.49 mg·g^{-1} in 0.2 g·L^{-1} amoxicillin solution. The values of adsorption thermodynamic parameters (ΔH^θ, ΔS^θ, ΔG^θ) were obtained. In addition, the adsorption process turned out to be endothermic and spontaneous for the CaCO$_3$ product at 298 K, 308 K, and 318 K.

Keywords: calcium carbonate; mesopore; amoxicillin; adsorption; kinetics; thermodynamics

---

1. Introduction

Calcium carbonate is an environmentally friendly and widely used material [1]. In nature, limestone, marble, and chalk are widely disseminated. Calcite, aragonite, and vaterite are three main kinds of anhydrous crystalline form [1]. Porous calcium carbonate is a modern multi-functional material and has developed quickly in last decade. An imposing specific surface area, good biocompatibility and non-toxic are the obvious features of porous calcium carbonate. Therefore, porous calcium carbonate is widely applied into all walks of life, such as drug carriers [2–4], bioceramics [5,6], biomicrocapsules [7,8], biosensors [9], bone repair [10], biomimetic mineralization [6,11], and so on. Generally, material structures and crystal forms, affected strongly by preparation and operation conditions, have rather significant effects on their application. Therefore, the fabrication of porous calcium carbonate has been a promising area of research. Up until now, the template-assisted method [12–14], mainly based on surfactants, polymers, and natural plant ingredients as a template agent, has become the mainstream method of porous calcium carbonate preparation. At the same time, coprecipitation [15], emulsion liquid membrane [16], solvent/hydrothermal method [17], and other methods have been improved, providing more preparation references for calcium carbonate.

Antibiotics are produced in bulk and used all over the world. A considerable amount of antibiotic wastewater is drained into rivers and lakes. Amoxicillin, a kind of beta-lactam-containing antibiotic, has been widely used in both human and veterinary medicine to treat infections caused by Gram-positive or Gram-negative bacteria. Though amoxicillin exhibits a short half-life, studies found that the amoxicillin degradation products remained toxic and may become more dangerous [18]. Thus, coagulation, adsorption, biofilm, and advanced oxidation were put forward and applied in order to effectively solve the antibiotic pollution. Of all the treatment methods, the adsorption process is highly beneficial due to the advantage of not being toxic or having the ability to degrade contaminants. Nevertheless, many adsorbents are difficult to recycle and reuse. In 2014, Shinya Yamanaka [19] and Kai Yin Chong [20] reported that porous calcium carbonate was used to treat formaldehyde vapor and disodium 3,3'-[biphenyl-4,4'-diyldi(E)diazene-2,1-diyl]bis(4-aminonaphthalene-1-sulfonate) (Congo red) solution, and the adsorption capacity of porous calcium carbonate was demonstrated to be excellent. It is reported that the structure of the rape flower pollen is a regular ellipsoidal shape with uniform particle size, a transparent porous structure on the surface, and the pores have high openness [21]. However, porous calcium carbonate used to treat antibiotic wastewater, being feasible based on these achievements, has not been reported.

In this work, rape flower pollen as a template agent was employed to synthesize mesopore calcium carbonate from eggshells at a specified temperature. The structure and morphology of calcium carbonate samples were characterized using XRD, SEM, and BET. The data for synthetic calcium carbonate adsorbing amoxicillin in an aqueous solution was studied using adsorption isotherms, thermodynamics analysis, and kinetic modeling with the aim of improving the data regarding its application.

2. Materials and Methods

2.1. Materials

All reagents except amoxicillin and eggshells were analytical reagents and supplied by Chengdu Kelon Chemical Reagent Co., Ltd., Chengdu, China. Amoxicillin (minimum 99.5%) was purchased from Shanghai Aladdin Biochemical Technology Co., Ltd., Shanghai, China. Eggshells were collected from the markets of Sichuan province in China. Distilled water was used for the preparation of all test solutions.

2.2. Characterization

The phase of the obtained calcium carbonate was determined using powder X-ray diffraction (D8 Advance, BRUKER AXS, Karlsruhe, Germany) with a range scan of $10° < 2\theta < 90°$. The estimated standard uncertainty of the 2θ measurement was $0.01°$. The morphological features were determined using SEM (MERLIN Compact, Carl Zeiss AG, Oberkochen, Germany). The specific surface area, pore size, and other parameters were obtained using BET (V-Sorb 2800TP, Gold APP Instrument Corporation, Beijing, China).

3. Experimental

The waste eggshell as a calcium resource was used to produce calcium carbonate microparticles using rape flower pollen as a template at room temperature. The template containing the calcium carbonate microparticles was roasted to obtain a porous adsorbent for adsorbing amoxicillin in an aqueous solution. The flow chart is given in Figure 1.

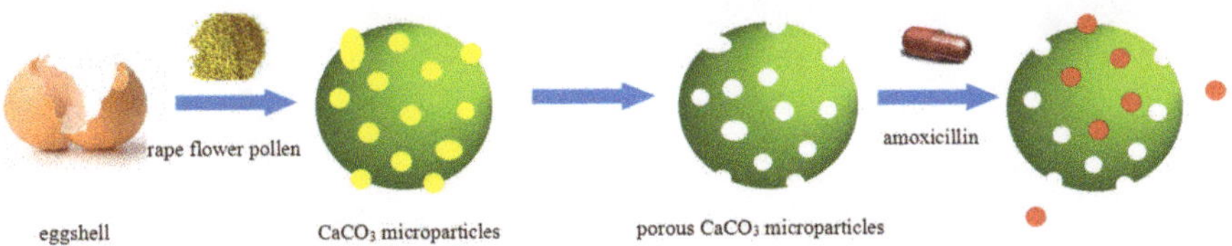

Figure 1. The flow chart of porous $CaCO_3$ preparation and adsorption.

3.1. Calcium Carbonate Fabrication

Mesopore calcium carbonate microparticles were prepared using eggshells. Before the experiment, a series of processes were carried on the eggshells as follows. First, the eggshells were cleaned using distilled water and the membranes were artificially removed. Then, the treated eggshell was dried in a drying oven at 105 °C for 120 min. Finally, an eggshell powder was obtained via grinding. The purchased rape flower pollen was dried at 50 °C in a drying oven for 24 h. The 0.1 g·L^{-1} solution was prepared by adding 0.05 g dry rape flower pollen in 500 mL distilled water. Five grams of eggshell powder was dissolved using 75 mL 10% (volume ratio) hydrochloric acid. After filtering, 4.4 g ammonium carbonate and 2 mL 0.1g·L^{-1} rape flower pollen aqueous solution were added into the filtrate to react for 45 min. At last, white precipitation was collected via filtration and dried at 105 °C for 60 min, and roasted at 500 °C for 90 min.

3.2. Amoxicillin Adsorption Experiment

A batch of amoxicillin adsorption experiments were carried out by adding 0.5 g prepared calcium carbonate with strong stirring for different contact times in a 200 mL amoxicillin aqueous solution under darkness for 30 min; the amoxicillin initial concentration varied from 0.01 to 0.2 g·L^{-1}, and the reaction temperature was 298 K, 308 K, or 318 K. The chemical oxygen demand (COD) values were determined using a potassium permanganate method for the amoxicillin solution before and after adsorption, which was filtered using a 0.45 μm filter membrane. The adsorption capacity (Γ) was calculated using the followed formula:

$$\Gamma(\mathrm{mg}\cdot\mathrm{g}^{-1}) = \frac{(c_0 - c_t) \times V}{m} \tag{1}$$

where c_0 and c_t are the concentrations of amoxicillin at initial and time t of the contact time, respectively (g·L^{-1}); V represents the volume of amoxicillin solution (mL); and m is the mass of calcium carbonate (g).

4. Results and Discussion

4.1. Characterization of Calcium Carbonate ($CaCO_3$) Microparticles

4.1.1. Composition Analysis

The crystal structures of as-prepared calcium carbonate were studied using XRD as shown in Figure 2. As compared with the standard powder diffraction pattern, the diffraction peaks of roasted and unroasted calcium carbonate belonged to vaterite (PDF#05-0586) and (PDF#33-0268). However, there is an unknown peak (29.49°) in Figure 2b, and some calcite peaks at 38.86°, 40.76°, 51.05°, 59.92°, 62.99°, and 68.71° that are extremely weak. The existence of rape flower pollen maybe affected the crystalline conversion of calcium carbonate in some aspect. When the as-synthesized product was roasted at 500 °C for 90 min, the crystal structure changed from vaterite to calcite, which showed that vaterite is unstable at high temperatures.

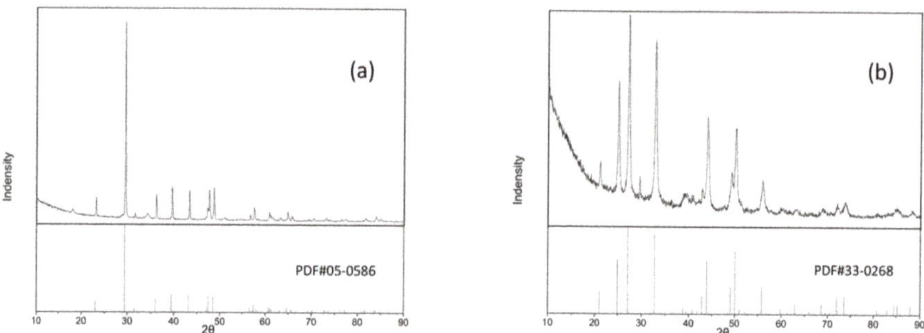

Figure 2. The XRD plots of calcium carbonate: (**a**) roasted calcium carbonate and (**b**) unroasted calcium carbonate.

4.1.2. Morphological Features

The surface morphology of products were measured using SEM images as shown in Figure 3. As seen in Figure 3a,b, products possessed a spherical morphology with diameters of 0.06–1.5 μm. After roasting, the surface of the calcium carbonate microspheres began to gelatinize and melded together. The number of internal channels became fewer, but the aperture had become larger, as shown in Figure 3c,d.

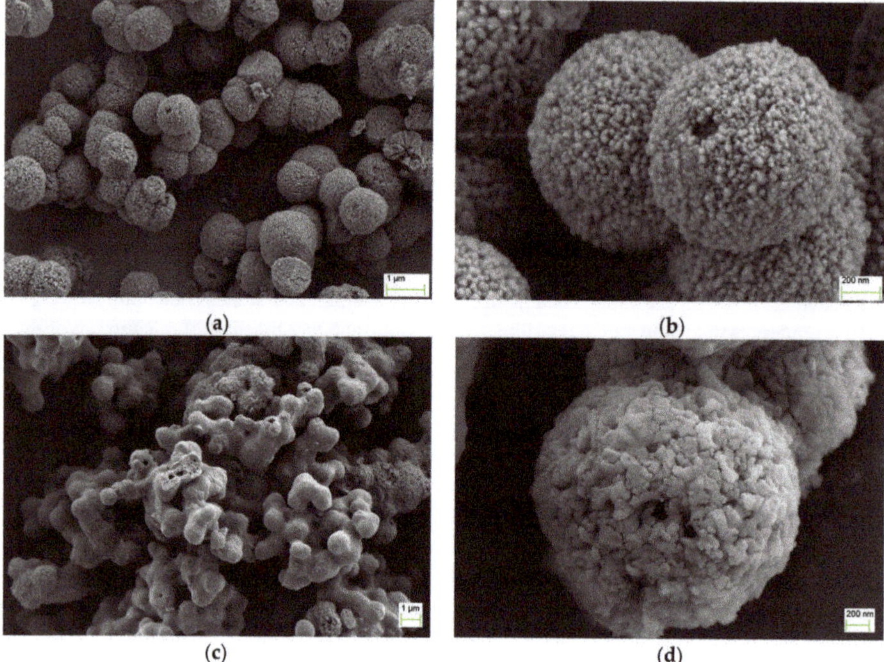

Figure 3. The SEM images of calcium carbonate: (**a,b**) unroasted calcium carbonate and (**c,d**) roasted calcium carbonate.

4.1.3. N$_2$ Adsorption/Desorption Isotherm

Pore size distribution and N$_2$ adsorption/desorption isotherms (inset) are shown in Figure 4. Based on the International Union of Pure and Applied Chemistry (IUPAC) classification, the N$_2$ adsorption/desorption isotherm of as-prepared product fits the type IV curve. The inflection point in the isotherm at low p/p$_0$ (<0.1) indicated the accomplished monolayer adsorption of N$_2$. The multilayer adsorption initiated the next stage. The adsorption hysteresis loop corresponds to the capillary condensation phenomenon of mesopores [20]. The pore size distribution showed that the diameter of the roasted product varied mainly from 3 to 15 nm (belonging to mesopores), which was consistent with the N$_2$ adsorption/desorption isotherms. However, the over 100 nm pores were rare owing to the roasting at high temperature. The BET surface area was 144.38 m$^2 \cdot$g^{-1}.

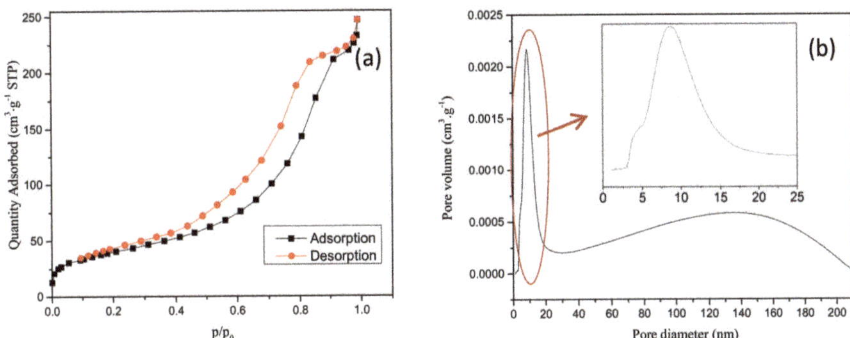

Figure 4. N$_2$ adsorption/desorption isotherms (**a**) and pore size distribution curves (**b**) of the roasted calcium carbonate.

4.2. Amoxicillin Adsorption

A series of amoxicillin standard solutions (0.01 g·L^{-1}, 0.02 g·L^{-1}, 0.05 g·L^{-1}, 0.1 g·L^{-1}, 0.15 g·L^{-1}, 0.2 g·L^{-1}) were prepared using an amoxicillin standard reagent. The corresponding COD was measured according to the potassium permanganate method. By plotting the COD versus *c* values (Figure 5), the linear relationship between amoxicillin concentration and COD was y = 668.91x + 5.1268, R^2 = 0.9992.

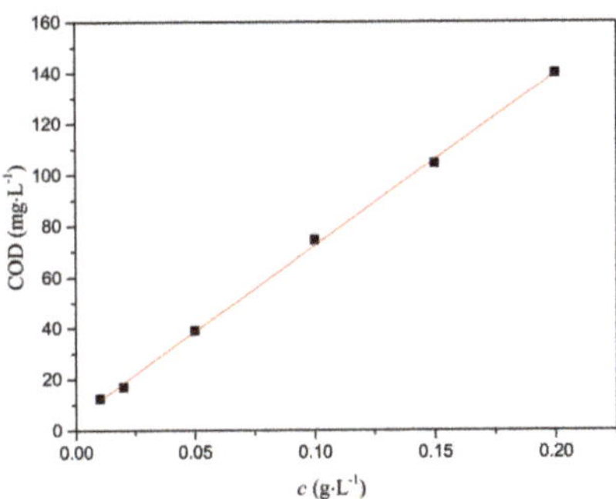

Figure 5. The plot of COD versus *c* (amoxicillin concentration) for the amoxicillin standard solution.

Homemade calcium carbonate was used to deal with the amoxicillin aqueous solution. Tables 1–3 exhibit the adsorption capacity of calcium carbonate at different contact times, amoxicillin concentrations, and temperatures, respectively. As given in Table 1, the balanced adsorption amount reached 13.49 mg·g^{-1} after 300 min of contact time, which is higher than 333.6 µg·g^{-1} reported in Reference [22]. Porous calcium carbonate was prepared using an microemulsion method with hexadecyl trimethyl ammonium bromide (CTAB) as a template as in Li's work [22]. In different concentration solutions, the balanced adsorption amount constantly increased with the growth of the concentration in Table 2. Meanwhile, when the increase of temperature, a similar trend appeared in a certain concentration in Table 3.

Table 1. The adsorption capacity of calcium carbonate at different contact times in 0.2 g·L^{-1} amoxicillin solution at 298 K.

t (min)	c_t (g·L^{-1})	q_t^1 (mg·g^{-1})	t (min)	c_t (g·L^{-1})	q_t (mg·g^{-1})
5	0.1944	2.23	10	0.1905	3.82
15	0.1870	5.20	20	0.1847	6.14
25	0.1815	7.42	30	0.1778	8.88
40	0.1727	10.92	50	0.1704	11.83
60	0.1688	12.49	90	0.1676	12.96
120	0.1673	13.07	150	0.1669	13.25
180	0.1665	13.39	240	0.1664	13.43
300	0.1663	13.49	600	0.1663	13.49

^{1} The amount of adsorbate adsorbed at time t of the reaction time.

Table 2. The adsorption capacity of calcium carbonate in different amoxicillin concentrations for 300 min at 298 K.

c_0 (g·L^{-1})	c_e (g·L^{-1})	q_e (mg·g^{-1})	c_0 (g·L^{-1})	c_e (g·L^{-1})	q_e (mg·g^{-1})
0.01	0.0027	2.91	0.05	0.0270	9.22
0.02	0.0072	5.11	0.1	0.0727	10.93
0.03	0.0128	6.89	0.2	0.1663	13.49
0.04	0.0205	7.80	-	-	-

Table 3. The adsorption capacity of calcium carbonate at 298 K, 308 K, or 318 K for 300 min in 0.2 g·L^{-1} amoxicillin solution.

T (K)	c_0 (g·L^{-1})	c_e (g·L^{-1})	q_e (mg·g^{-1})
298	0.2001	0.1664	13.48
308	0.2001	0.1615	15.68
318	0.2001	0.1569	17.28

4.3. Adsorption Isotherms

At a specific temperature, the adsorption isotherm curve refers to the relationship between the concentration of the solute molecules in the two phases when the adsorption process reaches equilibrium. The separated substance concentration relationship in liquid and solid phase can be expressed by the adsorption equation, Langmuir adsorption isotherm (Equation (2)), Freundlich adsorption isotherm (Equation (4)), and Temkin equation (Equation (5)), which are commonly used to describe it [23,24].

$$q_e = \frac{k_L c_e}{1 + \alpha_L c_e} \text{ or } \frac{c_e}{q_e} = \frac{1}{k_L} + \frac{\alpha_L}{k_L} c_e \tag{2}$$

where q_e the amount of adsorbate adsorbed (mg·g^{-1}), c_e is the equilibrium concentration (mg·L^{-1}), and α_L (L·mg^{-1}) and k_L (L·g^{-1}) are the adsorption equilibrium constants. The theoretical monolayer saturation capacity q_{max} (mg·g^{-1}) can be evaluated using the Langmuir equilibrium constants α_L (L·mg^{-1}) and k_L (L·g^{-1}) using Equation (3):

$$q_{max} = \frac{k_L}{\alpha_L} \tag{3}$$

$$q_e = k_F c_e^{\frac{1}{n}} \text{ or } \log q_e = \frac{1}{n}\log c_e + \log k_F \tag{4}$$

where k_F $((\text{mg·g}^{-1})\cdot(\text{mg·L}^{-1})^{-1/n})$ and n are the Freundlich constants that give the relative capacity and adsorption intensity, respectively.

$$q_e = \frac{RT}{b}\ln(Ac_e) \text{ or } q_e = B\ln A + B\ln c_e \tag{5}$$

where $B = RT/b$, B is the Temkin constant related to the heat of sorption (J·mol^{-1}), A is the Temkin isotherm constant (L·g^{-1}), R is the gas constant (8.314 J·mol^{-1}·K^{-1}), and T is the temperature (K).

In this work, the adsorption capacities of the as-synthesized mesopore calcium carbonate were studied using Langmuir, Freundlich, and Temkin isotherm models for the adsorption of amoxicillin. The result from plots (Figure 6) indicate R^2 = 0.9948, 0.9937, and 0.9399 for Langmuir (Figure 6a), Freundlich (Figure 6b), and Temkin (Figure 6c), respectively. In terms of R^2, Langmuir was more favorable than Temkin and Freundlich. To determine whether the adsorption of amoxicillin on calcium carbonate was favorable, the values of R_L (R_L = 1/(1 + α_L·c_0)) and n for Langmuir and Freundlich were evaluated. The results indicate that R_L = 6.94 × 10^{-4} and n = 2.8. The work of N.K. Goel et al. [24] reported that if R_L is less than unity, then adsorption is favorable. Hence, the as-synthesized mesopore calcium carbonate is a good adsorbent for amoxicillin.

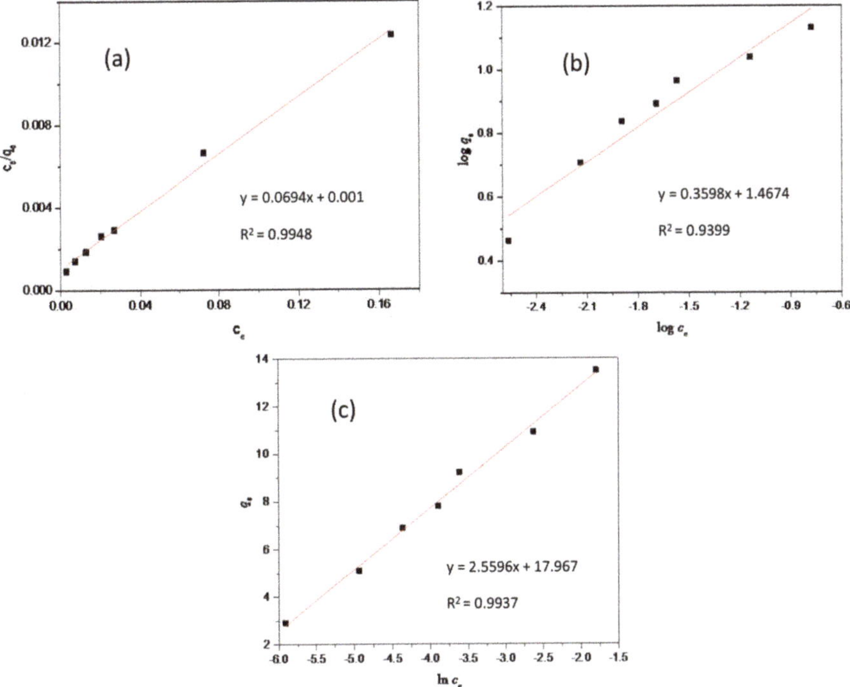

Figure 6. Adsorption isotherms of amoxicillin using calcium carbonate: (**a**) Langmuir isotherm, (**b**) Freundlich isotherm, and (**c**) Temkin isotherm.

4.4. Thermodynamic Parameters Analysis

A series of experiments on temperature dependence of as-prepared calcium carbonate were carried out to obtain thermodynamic adsorption parameters, such as standard Gibbs free energy (ΔG^θ,

J·mol^{-1}), enthalpy ($\Delta H^\ominus$, J·mol^{-1}), and entropy ($\Delta S^\ominus$, J·mol^{-1}·K^{-1}) [24], which can be determined using Equations (6) and (7):

$$\ln k_D = \frac{\Delta S^\ominus}{R} - \frac{\Delta H^\ominus}{RT} \tag{6}$$

$$\Delta G^\ominus = \Delta H^\ominus - T\Delta S^\ominus \tag{7}$$

where k_D is the distribution coefficient of the adsorbent and is equal to q_e/c_e, T is the absolute temperature, and R is the gas constant. The plot of $\ln k_D$ versus $1/T$ (figure is not shown) is linear with the slope and the intercept giving values of $\Delta H^\ominus$ and $\Delta S^\ominus$.

The values of adsorption thermodynamic parameters are listed in Table 4. The enthalpy change is positive implying that the adsorption process is endothermic. The negative value of $\Delta G^\ominus$ confirms the feasibility of the adsorption process and also indicates spontaneous adsorption of amoxicillin on the product.

Table 4. Thermodynamic parameters for the adsorption of amoxicillin on as-prepared calcium carbonate.

T (K)	k_D	$\Delta H^\ominus$ (J·mol^{-1})	$\Delta S^\ominus$ (J·mol^{-1}·K^{-1})	$\Delta G^\ominus$ (J·mol^{-1})
298	81.01	-	-	−10,897.10
308	96.02	12,108.51	77.20	−34,674.69
318	110.13	-	-	−59,224.29

4.5. Adsorption Kinetics

The kinetics of adsorption was investigated using the pseudo-first order, second order, pseudo-second order, and intra-particle diffusion model, as represented in Equations (8)–(11):

$$\text{Pseudo-first-order model } \ln(q_e - q_t) = -kt + \ln q_e \tag{8}$$

$$\text{Second-order model } \frac{1}{q_e - q_t} = kt + \frac{1}{q_e} \tag{9}$$

$$\text{Pseudo-second-order model } \frac{t}{q_t} = \frac{t}{q_e} + \frac{1}{kq_e^2} \tag{10}$$

$$\text{Intra-particle diffusion model } q_t = kt^{0.5} + C \tag{11}$$

According to the data in Table 1, linear plots were drawn based on the kinetics models mentioned above, as shown in Figure 7. It is apparent that the pseudo-second-order model was more favorable in terms of R^2, showing that the adsorption phenomenon followed the second order. The intra-particle diffusion plot (Figure 7d) indicates multi-linearity represented by two different stages, suggesting involvement of two different adsorption processes, which is similar to Reference [24]. In the first and fast-changing stage, the mass transfer through the boundary layers of liquid and adsorption on the available occurred on the external surface of the adsorbent of amoxicillin. After this stage, the amoxicillin molecules entered the interior of the adsorbent through the porous structures, and eventually were adsorbed on the internal active sites. Owing to increasing diffusion resistance of the amoxicillin molecule transportation from the external surface to the bulk of the adsorbent, the adsorption rate slowed down, which is demonstrated in the second stage. However, the fitted straight line that does not pass through the origin displays that intra-particle diffusion was not the only step controlling the adsorption process [23,24].

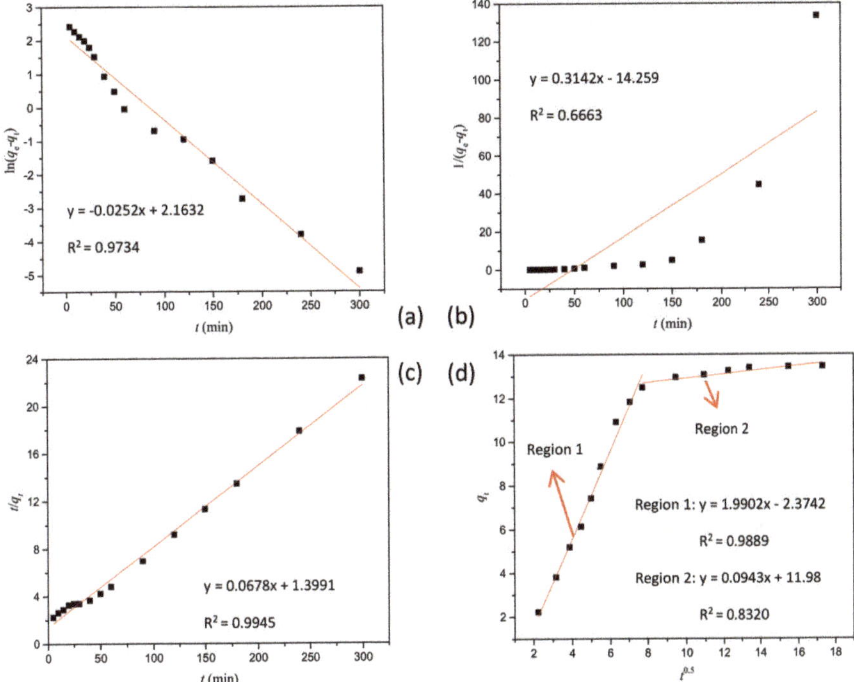

Figure 7. The plot of different kinetics models: (**a**) pseudo-first-order model, (**b**) second-order model, (**c**) pseudo-second-order model, and (**d**) intra-particle diffusion model.

5. Conclusions

Mesopore calcium carbonate with a BET surface area of 144.38 $m^2 \cdot g^{-1}$ was produced successfully using rape flower pollen as a template for removing amoxicillin in aqueous solution. The results indicated that it is a favorable adsorbent, and its adsorption capacity reached 13.49 $mg \cdot g^{-1}$. Langmuir, Freundlich, and Temkin isotherm models were used to interpret the adsorption phenomenon of the adsorbent. The Langmuir isotherm was the best fitting model for the equilibrium adsorption data gathered from the product. For low pressure ($p/p_0 < 0.1$), the monolayer adsorption was dominant, and in the following stage, multilayer adsorption dominated instead. Adsorption kinetics was demonstrated to follow a pseudo-second-order model. The negative value of ΔG^θ signified that the adsorption reaction was a spontaneous adsorption, and the positive value of ΔH^θ indicated the adsorption process was endothermic.

Author Contributions: L.Z. contributed to the literature search, study design, experimental study, data analysis, and writing—review and editing; T.P. contributed to the design of the work and data analysis; H.S., X.G., and D.F. contributed to the work of writing—review and editing.

Funding: This research was funded by the Opening Project of Key Laboratory of Green Catalysis of Sichuan Institutes of Higher Education (Grant No. LYJ1605); the program projects funded of Sichuan University of Arts and Science (Grant No.2018SCL001Y) and the Scientific and Technological Innovation Team Foundation of Southwest University of Science and Technology, China (Grant No.17LZXT11).

Conflicts of Interest: The authors declare no conflict of interest.

References

1. Zhou, L.; Lai, C.; Wang, F.; He, P. Progress in fabrication and applications of porous calcium carbonate. *Chem. Ind. Eng. Prog.* **2018**, *37*, 159–167.

2. Kurapati, R.; Raichur, A.M. Composite cyclodextrin–calcium carbonate porous microparticles and modified multilayer capsules: Novel carriers for encapsulation of hydrophobic drugs. *J. Mater. Chem. B* **2013**, *1*, 3175. [CrossRef]

3. Zhang, Y.; Ma, P.; Wang, Y.; Du, J.; Zhou, Q.; Zhu, Z.; Yang, X.; Yuan, J. Biocompatibility of Porous Spherical Calcium Carbonate Microparticles on Hela Cells. *World J. Nano Sci. Eng.* **2012**, *2*, 25–31. [CrossRef]

4. Preisig, D.; Haid, D.; Varum, F.J.O.; Bravo, R.; Alles, R.; Huwyler, J.; Puchkov, M. Drug loading into porous calcium carbonate microparticles by solvent evaporation. *Eur. J. Pharm. Biopharm.* **2014**, *87*, 548–558. [CrossRef] [PubMed]

5. Xiong, L.; Tan, J.; Fang, F.; Yang, F. Biomechanical Evaluation on Bone Defect Repairing with Porous Carbonate Calcium Ceramics. *Med. J. Wuhan Univ.* **2008**, *29*, 13–16.

6. Zhong, Q.; Li, W.; Su, X.; Li, G.; Zhou, Y.; Kundu, S.C.; Yao, J.; Cai, Y. Degradation pattern of porous $CaCO_3$ and hydroxyapatite microspheres in vitro and in vivo for potential application in bone tissue engineering. *Colloids Surfaces B Biointerfaces* **2016**, *143*, 56–63. [CrossRef]

7. He, F.; Ren, W.; Tian, X.; Liu, W.; Wu, S.; Chen, X. Comparative study on in vivo response of porous calcium carbonate composite ceramic and biphasic calcium phosphate ceramic. *Mater. Sci. Eng. C* **2016**, *64*, 117–123. [CrossRef] [PubMed]

8. Fujii, A.; Maruyama, T.; Ohmukai, Y.; Kamio, E.; Sotani, T.; Matsuyama, H. Cross-linked DNA capsules templated on porous calcium carbonate microparticles. *Colloids Surfaces A: Physicochem. Eng. Asp.* **2010**, *356*, 126–133. [CrossRef]

9. He, G. To prepared fluorescent chitosan capsules by in situ polyelectrolyte coacervation on poly(methacrylic acid)-doped porous calcium carbonate microparticles. *Polym. Bull.* **2012**, *69*, 263–271. [CrossRef]

10. Gong, J.; Liu, T.; Song, D.; Zhang, X.; Zhang, L. One-step fabrication of three-dimensional porous calcium carbonate–chitosan composite film as the immobilization matrix of acetylcholinesterase and its biosensing on pesticide. *Electrochem. Commun.* **2009**, *11*, 1873–1876. [CrossRef]

11. Dey, A.; De With, G.; Sommerdijk, N.A.J.M. In situ techniques in biomimetic mineralization studies of calcium carbonate. *Chem. Soc. Rev.* **2010**, *39*, 397–409. [CrossRef]

12. Wei, Q.; Lu, J.; Ai, H.; Jiang, B. Novel method for the fabrication of multiscale structure collagen/hydroxyapatite-microsphere composites based on $CaCO_3$ microparticle templates. *Mater. Lett.* **2012**, *80*, 91–94. [CrossRef]

13. Zhang, T.; Pang, Y.; Feng, J. Facile Synthesis and Morphology Control of Porous $CaCO_3$ Microparticles. *J. Chem. Eng. Chin. Univ.* **2015**, *29*, 377–381.

14. Yue, L.; Zheng, Y.; Jin, D. Spherical porous framework of calcium carbonate prepared in the presence of precursor PS–PAA as template. *Microporous Mesoporous Mat.* **2008**, *113*, 538–541. [CrossRef]

15. Trushina, D.B.; Sulyanov, S.N.; Bukreeva, T.V.; Kovalchuk, M.V. Size control and structure features of spherical calcium carbonate particles. *Crystallogr. Rep.* **2015**, *60*, 570–577. [CrossRef]

16. Yang, H.; Li, H. The preparation of porous superfine calcium carbonate microspheres. *J. Shaanxi Univ. Sci. Technol.* **2013**, *31*, 111–115.

17. Zhao, Z.; Zhang, L.; Dai, H.; Du, Y.; Meng, X.; Zhang, R.; Liu, Y.; Deng, J. Surfactant-assisted solvo- or hydrothermal fabrication and characterizationof high-surface-area porous calcium carbonate with multiple morphologies. *Micropor. Mesopor. Mat.* **2011**, *138*, 191–199. [CrossRef]

18. Liu, H.; Hu, Z.; Liu, H.; Xie, H.; Lu, S.; Wang, Q.; Zhang, J. Adsorption of amoxicillin by Mn-impregnated activated carbons: performance and mechanisms. *RSC Adv.* **2016**, *6*, 11454–11460. [CrossRef]

19. Yamanaka, S.; Oiso, T.; Kurahashi, Y.; Abe, H.; Hara, K.; Fujimoto, T.; Kuga, Y. Scalable and template-free production of mesoporous calcium carbonate and its potential to formaldehyde adsorbent. *J. Nanoparticle* **2014**, *16*, 2266. [CrossRef]

20. Chong, K.Y.; Chia, C.H.; Zakaria, S.; Sajab, M.S. Vaterite calcium carbonate for the adsorption of Congo red from aqueous solutions. *J. Environ. Chem. Eng.* **2014**, *2*, 2156–2161. [CrossRef]

21. Gao, X.; Liu, Y.; Wang, Y.; Liu, X.; Tang, Y.; Li, J. Preparation of zinc oxide nanomaterial by biotemplated method. *Chemistry* **2019**, *82*, 63–67. (In Chinese)

Minerals **2019**, *9*, 254

22. Li, S. Study on the preparation and adsorption on amoxicillin for CaCO$_3$ with different morphologies. Master's Thesis, Kunming University of Science and Technology, Kunming, China, 2014.

23. Homem, V.; Alves, A.; Santos, L. Amoxicillin removal from aqueous matrices by sorption with almond shell ashes. *Int. J. Environ. Anal. Chem.* **2010**, *90*, 1063–1084. [CrossRef]

24. Goel, N.K.; Kumar, V.; Pahan, S.; Bhardwaj, Y.K.; Sabharwal, S. Development of adsorbent from Teflon waste by radiation induced grafting: Equilibrium and kinetic adsorption of dyes. *J. Hazard. Mater.* **2011**, *193*, 17–26. [CrossRef] [PubMed]

MDPI
St. Alban-Anlage 66
4052 Basel
Switzerland
Tel. +41 61 683 77 34
Fax +41 61 302 89 18
www.mdpi.com

Minerals Editorial Office
E-mail: minerals@mdpi.com
www.mdpi.com/journal/minerals